03454265
CHEMISTRY

Library of Congress Cataloging-in-Publication Data

Šulcek, Zdeněk.
Methods of decomposition in inorganic analysis.

Translated from the Czech manuscript.
Bibliography: p.
Includes index.
1. Chemistry, Analytic. 2. Solution (Chemistry)
3. Decomposition (Chemistry) I. Povondra, Pavel.
II. Title.
QD75.3.S84 1989 543 88-10572
ISBN 0-8493-4963-X

This book represents information obtained from authentic and highly regarded sources. Reprinted material is quoted with permission, and sources are indicated. A wide variety of references are listed. Every reasonable effort has been made to give reliable data and information, but the author and the publisher cannot assume responsibility for the validity of all materials or for the consequences of their use.

All rights reserved. This book, or any parts thereof, may not be reproduced in any form without written consent from the publisher.

Direct all inquiries to CRC Press, Inc., 2000 Corporate Blvd., N.W., Boca Raton, Florida, 33431.

© 1989 by CRC Press, Inc.

International Standard Book Number 0-8493-4963-X

Library of Congress Card Number 88-10572
Printed In the United States

Methods of Decomposition in Inorganic Analysis

Authors

Zdeněk Šulcek, Ph.D.
Senior Research Chemist
Department of Chemical Laboratories
Central Geological Institute
Prague, Czechoslovakia

Pavel Povondra, D.Sc.
Associate Professor
Department of Mineralogy and Geochemistry
Charles University
Prague, Czechoslovakia

CRC Press, Inc.
Boca Raton, Florida

AUTHORS

Dr. Zdeněk Šulcek graduated in chemistry and geology in 1950 from the Faculty of Natural Sciences of Charles University in Prague, obtaining the degree ''Rerum Naturalium'' Doctor (RNDr). After postgraduate studies in analytical chemistry he received his Ph.D. from the same university in 1961.

Dr. Šulcek has been employed as a chemist at the department of chemical laboratories of the Geological Survey of Czechoslovakia in Prague since 1948 and was appointed Head of the laboratories in 1956. Presently, Dr. Šulcek is working in these same laboratories as senior research chemist. He had the opportunity to spend limited periods of time at the universities of Oslo, Moscow, and Ljubljana and at Geological Surveys laboratories of many European countries.

Dr. Šulcek is a member of the Czechoslovak Society for Mineralogy and Geology. In the period 1972 to 1986 he acted as a chairman of the specialized group of Industrial Analytical Chemistry organized in the framework of the Czechoslovak Scientific — Technical Society. His main field of interest is in analytical chemistry applied to geological materials and in geochemistry, and in the occurrence of certain rare trace elements in rocks, minerals, and waters. His activity concentrates on enrichment and separation procedures by utilizing various sorbents (ion-exchangers, silica gel, etc.) and on decomposition techniques of inorganic solids.

Dr. Šulcek, together with Jan Doležal and Pavel Povondra, is co-author of the book *Decomposition Technique in Inorganic Analysis,* which was published in 1966 and was translated into English and Russian in 1968. His publications include more than 100 papers published in scientific chemical and geological journals.

Dr. Pavel Povondra graduated from the Faculty of Science, Charles University, Prague, in 1950, receiving his Rer. Nat. Doctorate at the same time. He defended his CsC (Ph.D.) Thesis at the Institute of Polarography, Czechoslovak Academy of Sciences, in 1961 and his Doctorate of Science in 1981 at Charles University.

After graduation, Dr. Povondra worked as a chemist in the Geological Survey of Czechoslovakia, later in the Institutes of Polarography and Geology of the Czechoslovak Akademy of Sciences. In 1975 he defended his Associate Professorship Thesis at Charles University. In 1969 to 1970, Dr. Povondra worked as a visiting scientist at the Ruhr Universität (FGR) and spent periods of time at the University of Glasgow (1978, 1979) and Centre National de la Recherche Scientifique, Orléans, France. Presently, he is employed as Associate Professor at the Faculty of Science, Charles University, Prague, Czechoslovakia.

Dr. Povondra is a Member of the Czechoslovak Society for Mineralogy and Geology and the Czechoslovak Chemical Society. In 1964 he was awarded the Prize of the Czechoslovak Akademy of Science for his work in ion-exchange chromatography. He has published over 130 research papers on analytical chemistry, mineralogy, and geochemistry. His main current research topics are the crystallochemistry of cyclosilicate group minerals.

QD75
.3
S841
1989
CHEM

TABLE OF CONTENTS

Chapter 1

INTRODUCTION

The present trend in the development of the methods of instrumental chemical analysis has brought about a discrepancy between the time requirements of the determination itself and those of the sample pretreatment prior to analysis. Reliability of most methods of determination depends on rapid and quantitative conversion of solids into homogeneous solutions, either liquid or solid. This sample decomposition should be as simple as possible and should satisfy the following conditions: (1) all the solids should be completely dissolved and all the elements should be quantitatively retained in the homogeneous medium obtained; and (2) the solution should be pure, uncontaminated from the chemicals, laboratory equipment, the vessels, and the laboratory atmosphere.

The present situation is peculiar in that the development is again being concentrated on the classical decomposition reagents and their applications, although they seem to be obsolete. The materials and instrumentation for decomposition techniques have been improved, but little has changed in the development of the reagents. Automated fusion apparatuses are progressively more employed, especially in laboratories dealing with the X-ray fluorescence spectrometry (XRF) and inductively coupled plasma-optical emission spectrometry (ICP-OES) methods, using fusion with lithium borates or sintering with sodium peroxide. Platinum, as a classical material for fusion crucibles, has been replaced by unwettable platinum alloys or glassy carbon. The introduction of vessels made of fluorinated hydrocarbon polymers has enabled successful application of pressure decompositions. The effect of microwaves has considerably shortened the reaction time for dissolution of materials in acids. However, in spite of all these improvements, a greater part of the laboratory personnel activity must be concentrated on sample pretreatment than on the final determination, as the methods of determination are largely automated and have a very high sample throughput.

In our practice we have learned a lot about the problems of decomposition of inorganic compounds, especially about different effects of the classical reagents in dependence on the crystallochemical composition of solids and the interactions in heterogeneous mixtures of solids such as rocks, ores, and other mineral materials. We have always felt that the problems of decomposition of solids have not received systematic attention and that many valuable findings have been widely scattered in many journals and internal reports of various institutions. In 1968 we attempted to summarize these findings in a monograph that was originally published in Czech, but soon translated into English and Russian. The book has been accepted favorably by the specialized public. An analogous monograph was published by R. Bock in 1972 and in 1979; the book also treats decompositions of organic materials to a considerable extent. In view of the importance of the field, we were pleased by the invitation from CRC Press to review again the present state of the decomposition techniques for inorganic materials.

The present book retains the classical classification into the three groups of methods, namely, (1) wet decompositions, (2) fusion and sintering, and (3) decomposition by the action of gases and high temperatures. The relative sizes of the individual parts have been determined so as to reflect the present trends and an increase in the scope of application. For example, emphasis has been placed on the use of phosphoric and hydrofluoric acid in wet decomposition and experience with acid dissolution in microwave field has also been discussed. Among the fusion methods, particular attention has been paid to the use of borates as universal fusion agents in the preparation of glasses for XRF and master solutions for various spectrometric methods of determination. Fire assay has also been treated in some detail, as it retains an important role in analyses of raw materials and the products in obtaining the precious metals. Among the procedures for trace and ultratrace analysis, volatilization in gas streams is discussed to a greater extent. We have also given detailed information on the determination of various valence forms of compounds, mainly those of iron, in natural materials. We primarily deal with decomposition of inorganic materials and discuss those of natural organic substances (bitumens, coal, oil, environmental samples) only in connection with their inorganic carriers; this discussion deals with degradation and removal of organic substances, rather than with their analyses.

We present this monograph to the professional audience and hope that it will help analysts in their work and will assist in bridging the gap existing in the literature on the decomposition techniques. We have prepared the manuscript with great care, but, undoubtedly, the text is not perfect; we will be grateful to all who will bring any errors and omissions to our attention. With the given size of the book it has been impossible to treat the topic exhaustively: the book does not deal with the techniques used in phase analysis, electrochemical dissolution, and some other special procedures.

We would like to thank all the people who have helped us in the preparation of the monograph. First of all, we thank CRC Press and, further, the publishers of *Analytical Chemistry, Journal of the American Ceramic Society,* the *Analyst, Analytica Chimica Acta, Talanta, Zhurnal Aualiticheskoi Khimii, Zavodshaya haboratoriya, and Freseniuses' Zeitschrift für Analytische Chemie* who kindly permitted us to use figures and tables published in these journals. We are grateful to the reviewer of the manuscript for all the critical comments and to Dr. Karel Štulík of the Department of Analytical Chemistry, Charles University, Prague, for the translation of the Czech manuscript into English. We are equally thankful to our families for their great patience and understanding of our work; we dedicate this book to them.

Zdeněk Šulcek
Pavel Povondra

Chapter 2

DECOMPOSITION — THE MAIN SOURCE OF ERROR IN INORGANIC ANALYSIS

I. INTRODUCTION

Most modern analytical methods, e.g., emission and absorption flame spectrometry, plasma spectrometry, flow injection analysis (FIA), and high-performance liquid chromatography (HPLC), require a solution of the test substance. The main advantage of soutions is ideal homogeneity, even in separate microliter portions. Solids, that are never homogeneous on microscale, are transferred into a homogeneous liquid phase by using various decomposition techniques.

II. CAUSES OF ERROR IN DISSOLUTION OF SOLIDS

The process of dissolution of solids always brings about the danger of losses in some components and contamination of the reaction mixture by impurities from the reagents or leached from the walls of the decomposition vessels. The character of the test sample and the decomposition technique decide which of the processes affects the reliability of the data obtained. The composition of the solution reflects all the errors committed in the solution preparation connected with heterogeneity of the distribution of the test element in the solid analyzed, chemical changes and contamination of the sample during crushing of the material to the required grain size, and the decomposition process itself. The principal causes of error in dissolution of solids are summarized below.

A. Incomplete Dissolution of the Test Element or Element Association

The greatest error arises in decompositions of polyphase materials of an unknown composition, where the individual solids resist to a varying degree the effect of the decomposition agents. The degree of resistivity of the solid depends on the kind of solvent used. Some substances cannot be quantitatively dissolved even by prolonged action of acids in an autoclave, under elevated pressure and temperature (cubic boron nitride, silicon carbide, topaz, etc.). The resistivity of substances toward solvents is also affected by thermal pretreatment (Al_2O_3, PuO_2, BN, etc.). A decrease or increase in the solubility is usually connected with a change in the crystal structure of the test substance, its surface area, and porosity.

B. Losses in the Test Components in the Form of Volatile Compounds

Elements may volatilize in the elemental form as oxides, hydrides, and most often as halides. Volatile compounds may be contained in the sample (e.g., organometallic compounds) or are formed by interaction with decomposition agents. They are produced by the action of acids or melts, but first of all, they appear during thermal pretreatment of the sample. With polyvalent elements, the losses depend on the oxidation state of the compound formed (compounds of sulfur, phosphorus, arsenic fluorides, chromium chlorides, germanium oxides, etc.). Losses through volatilization during sample ignition are the main source of error in elemental analysis of coal. The main problem is the preparation of the ash that should contain all the admixtures, including the compounds of the halogens, S, Se, As, Ag, Cd, Ge, Pb, and other elements.

C. Formation of Sparingly Soluble Compounds

The formation of these compounds complicates dissolution of solids. The contents of various elements vary in the precipitates formed, depending on changes in the composition of the test solid and in the decomposition conditions. The surface of the sparingly soluble compounds often contains active sites capable of further reactions with the solvent and the solutes. A concentration gradient is formed between the solution and the precipitate. Equilibration usually takes several days; thus, the composition of the solution sampled in the vicinity of the precipitate changes with time.

The main components of precipitates involve $BaSO_4$, $PbSO_4$, AgCl, complex aluminum fluorides, titanium and zirconium phosphates, and silicic acid gel. The formation of surface active baryum sulfate causes complications not only in the determination of baryum, but also in isolation of lead in the form of lead(II) sulfate. The presence of chlorides in nitric acid complicates the decomposition of sulfidic ores for determination of silver. Production of complex aluminum fluorides causes difficulties in decomposition of silicate rocks by hydrofluoric acid. Precipitation of titanium phosphates is often the main source of negative error in determination of thorium in apatite concentrates. The gel of silicic acid, separating in decomposition of silicates, sorbs compounds of Sn, Ti, Zr, Nb, and Ta, even in strongly acidic solutions.

D. Formation of Soluble Complex Compounds

This phenomenon has primarily been observed in decompositions of materials with hydrofluoric and phosphoric acids. The fluorides remaining in the solution after the decomposition mask some metal ions and interfere in chelometric or spectrophotometric determinations of Zr, Ti, Al, and other elements. Phosphates and polyphosphates interfere analogously.

E. Solvent Reaction with Vessel Walls

The extent of the "permissible" contamination of the solution from solvent reactions with the vessel walls depends on the analysis level. It may be substantially higher for routine analyses of major components than in determination of admixtures in substances of a high purity.

The greatest problems are associated with fusion. Highly reactive fusion agents (e.g., Na_2O_2) corrode the vessels, which limits the applicability of the agents capable of decomposing resistant substances, such as silicon carbide, boron nitride, chromite, etc. Higher nickel oxides produced during fusion with sodium peroxide in nickel crucibles are surface active and bind a substantial part of stannates, tungstates, and molybdates that would otherwise be leached with water from the melt. In fusion of silicate rocks with alkali carbonates, compounds of iron, lead, tin, etc. are reduced, and the metals are alloyed with the vessel walls.

For preparation of solutions for trace analysis, quartz glassy carbon and Teflon® vessels are suitable. Fusion is usually impossible to use. The test material is dissolved in a small volume of very pure acids, best in a closed system. In strongly acidic solutions, the reactions connected with leaching of elements from the vessel walls predominate over the losses caused by sorption processes. The amount of the leached impurities can be substantially decreased when decomposing substances by acid vapors.

F. A Change in the Oxidation State of the Test Component During Dissolution

During dissolution, the ions of polyvalent elements undergo valence changes through interactions among the solutes or by oxidation by atmospheric oxygen. These problems are most often met in organic analysis in determination of the ratios Fe(III)/Fe(II), Mn(III)/Mn(II), U(VI)/U(IV), Ti(IV)/Ti(III), Ce(IV)/Ce(III), etc. On dissolution in complexing acids, variously stable complexes are formed, and the reactivity of the redox pair components changes. The redox effects of acids may differ in open and closed systems. By a suitable choice of the decomposition conditions, the interactions among the ions liberated can be suppressed or completely eliminated (e.g., those between Fe^{2+} and Mn^{3+} or Fe^{3+} and V^{3+}).

G. Magnitude and Fluctuations of the Blank

The total blank value involves the effects of the impurities introduced with the decomposition agents, by contamination from the laboratory atmosphere, and by leaching from the vessel walls. The content of impurities in liquid decomposition agents (acids) is found relatively easily, provided that a sufficiently sensitive analytical method is available. The blank value is substantially decreased in decompositions in closed systems, as the amount of the acids is decreased, and the contact of the reaction mixture with the laboratory atmosphere is eliminated. However, for analyses of high-purity substances, even this technique produces blank values that are too large and fluctuate too much. Minimal blank values are attained in decompositions by vapors, e.g., in decomposition of pure silicon dioxide by vapors of hydrofluoric acids.

The contamination of the sample in fusion decompositions is most difficult to determine, as the amounts and composition of impurities liberated from the walls of the fusion vessels vary considerably in fusion of the sample and of the fusion agent alone. Some components present in the test material may highly increase the corrosion of the crucible and the exchange of trace elements between the vessel wall and the melt.

In analytical practice, test materials often occur that require a combination of several analytical methods and variously difficult decomposition techniques. The results obtained are subject to the error of the final determination of the test component, but also to the errors stemming from the heterogeneity of the sample and the dissolution procedure. In a direct determination of the element (without separation), the total variance of the results, s_t^2, is given by

$$s_t^2 = s_m^2 + s_h^2 + s_s^2 \tag{1}$$

where the partial contributions to the total variance, s_m^2, s_h^2, and s_s^2, are the variances corresponding to the analytical method, sample heterogeneity, and the sample dissolution, respectively. The variance of the results in decomposition of a solid is significant only when the material is difficult to dissolve or the solution is contaminated to a variable extent, so that s_s^2 is comparable with s_m^2. To determine the s_s^2 values, samples with negligible heterogeneity are used, mainly, some SRM. Equation 1 is then simplified to

$$s_s^2 = s_t^2 - s_m^2 \tag{2}$$

from which the s_s^2 value can be obtained.

Chapter 3

CHEMICAL AGENTS AND VESSELS USED FOR DECOMPOSITION

I. INTRODUCTION

When dissolving solids, analysts usually prefer dissolution in acids to fusion decompositions. To accelerate the reaction, excess acids are used and the mixture is heated. However, most of the acid volatilizes in open systems and does not take part in the dissolution process. The amount of impurities in the acid volume added is often many times higher than the contents of the required components in the test material.[1] This problem has been solved by gradually decreasing the impurity contents in the solvents used. The decomposition process has been transferred to closed vessels and carried out in laboratories where the atmosphere purity can be controlled.

Distillation performed at the boiling point is now insufficient for purification of strong inorganic acids, as microdroplets and aerosols are formed during boiling and pass into the condensate. A substantial improvement in the acid quality has been achieved using isothermal (isopiestic) distillation. The acid vapors volatilizing from acid concentrated solutions at 25°C are absorbed at the same temperature in pure water in a separate vessel. The experimental equipment is very simple and usually consists of a glass vessel or a closed polypropylene or Teflon® box in which the vessels with the acid and water are placed.[2-4] The purification procedure takes several days. The concentration of the product can be controlled by varying the concentration of the feed acid and the volume of the water phase. Highly volatile acids, such as HCl and HF, or acetic acid among organic acids, are purified in this way. Using isothermal distillation, highly pure 10*N* HCl and 40% HF can be prepared,[4] with lead contents of less than 0.03 and 0.05 ng/mℓ, respectively. Acids with higher boiling points cannot be purified by this method.

An important improvement in the methods of purifying acids is subboiling distillation,[5] whose efficiency has further been enhanced by application of more sophisticated quartz, polypropylene, or Teflon® distillation apparatuses permitting a controlled increase in the temperature of the feed acid. The apparatus contains a tightly closed heating element in a quartz casing, which is protected by a Teflon® mantle for distillation of hydrofluoric acid.[4,6-12] The acid vapors condense on a water-cooled finger and are collected in carefully cleaned polypropylene or Teflon® vessels. To avoid accidental contamination and further improve the purity of the acids obtained, two distillation apparatuses are connected in series.

The simplest design of a distillation apparatus involves two interconnected polyethylene bottles with a water cooler.[5] A more efficient variant consists of two polypropylene or Teflon® bottles, connected by a Teflon® block with screw-on closings.[13-15] The bottle containing the feed acid is heated by an external infrared radiator, and the acid vapors condense in the other bottle that is cooled by water with ice. The device has yielded good results in the preparation of hydrochloric and nitric, as well as hydrofluoric, acids with low lead contents[13] (0.002 ng Pb per gram). As the feed material, both concentrated hydrofluoric acid and pressurized hydrogen fluoride can be used.[16-19] The gas, purified by passage through filters, condenses in a plastic vessel cooled with liquid nitrogen and is distilled into another container with water. Repeated distillation yields hydrofluoric acid with a concentration exceeding 50%.[17]

Distillation apparatuses are placed in clean laboratories in a fume cupboard with a laminar stream of purified air[6,20] and usually operate continuously. The amount of the pure acid prepared should not be greater than immediately required, as the highly pure acid cannot be stored for a long time. The quality of the feed acid does not affect that of the product, but the cleanliness of the vessels and the laboratory environment is essential. Highly pure 32% HCl, 70% HNO_3, 48% HF, 70% $HClO_4$, and 96% H_2SO_4 are obtained without difficulties. The main impurities usually involve common admixtures, e.g., Na, K, Si, Ca, and Fe. The content of lead in hydrofluoric acid is especially carefully monitored, mainly because of high demands on the precision of analyses of geological and cosmic objects for dating purposes.[4,16] The content of chromium in perchloric acid cannot be decreased by distillation, as it apparently volatilizes as chromyl chloride. Hydrofluoric acid with decreased contents of chlorides and bromides can be obtained by distillation in the presence of a small amount of silver nitrate;[21,22] volatilization of boron trifluoride is prevented by adding mannitol.[23,24] The above procedures have so far been insufficiently efficient for removal of volatile compounds of arsenic from hydrofluoric and hydrochloric acids and those of selenium from sulfuric acid. Phosphoric acid cannot be purified by distillation, and the pure acid must be prepared from phosphorus pentoxide purified by triple sublimation.[1]

It is much more difficult to purify fusion agents than acids. If the agents are not decomposed on dissolution in water (as, e.g., the alkali peroxides), many impurities can be removed by repeated crystallization. The contents of certain ions in fusion agents can be substantially decreased by sorption from concentrated solutions on chelating ion exchangers with iminodiacetate functional groups (Chelex 100 type). Sodium carbonate can also be prepared from sodium nitrate, after removing the impurities by a combination of extraction and ion exchange.[25] The final crystallization employs the less soluble sodium hydrogencarbonate. The alkali hydroxides are difficult to purify, and pure solutions of them can be prepared from the alkali metals. The amounts of sodium and potassium in lithium hydroxide can be decreased by more than 50 times when the liquid phase obtained after dissolving the compound in a mixture of ethanol, isopropanol, and water is separated.[26] The alkali peroxides are usually considerably contaminated. Small amount of pure preparations can be synthesized by burning the alkali metal vapors in a mixed oxygen-helium atmosphere.[27] The best procedure for the obtaining of fusion agents with minimal impurity contents is zone melting.[1,28] This procedure is advantageous for preparation of borax with less than 0.5 ppm of tin.

II. SELECTION OF VESSELS FOR DECOMPOSITION

One of the main sources of contamination during decompositions is the laboratory ware. Dissolution of substances in concentrated acids is accompanied by leaching of some components from glass. Interaction of melts with the crucible material causes corrosion of the latter leading to a high and fluctuating blank value.[1,29]

The selection of the material of the decomposition vessels depends on the composition of the test substances, the components to be determined, and the requirements placed on the analysis.[10] For routine analyses vessels of resistant boron-silicate glass or porcelain provided with a resistant glaze often suffice. For determinations of the alkali and the alkaline earth metals, especially at low concentrations, platinum crucibles and dishes, as well as vessels of plastics, quartz, or glassy carbon, have been found most convenient. In ultratrace analysis, the selection of the vessels is limited to the polymers of fluorinated hydrocarbons, quartz glass, and glassy carbon. Closed vessels with a small surface area are used to suppress the amount of impurities liberated during the decomposition. The time of the contact of the solvent with the vessels should be minimized. The surface of the vessels must be carefully cleaned prior to analysis, e.g., using acid vapors.[7,8,10,12,30-32]

A. Vessels of Platinum and Its Alloys

Vessels made of platinum and its alloys are still indispensable equipment of all analytical laboratories. They resist inorganic acids, provided that the reaction mixture contains no free chlorine or bromine. Free halogens may be contained in the acids added or be formed by the oxidation of hydrochloric or hydrobromic acid by the substance to be dissolved (e.g., MnO_2, CeO_2, or V_2O_5). On evaporation of hydrofluoric, nitric, or sulfuric acid in platinum dishes, the amount of the platinum dissolved does not exceed 10 μg. Higher corrosion (30 to 80 μg), observed during heating of hydrochloric acid, is probably caused by the presence of small amounts of chlorine.[29,33] Evaporation of hydrobromic acid solutions exhibits highly corrosive effects, as bromine is produced by the photochemical oxidation of bromides by atmospheric oxygen. Prolonged exposure of vessels to phosphoric acid at temperatures above 250°C also leads to increased corrosion. The microstructure of platinum crucibles and dishes affects their behavior during decomposition processes. An addition of 0.08% of ZrO_2 stabilizes these structures and prevents softening of the vessels at high temperatures. Alloys of platinum with 0.3 to 1.0% of iridium have better mechanical properties than pure platinum. Platinum vessels also contain admixtures of the other platinum metals (especially palladium and rhodium), gold, silver, and varying amounts of other components at trace concentrations.[28] Some impurities, such as iron, calcium, and magnesium, are leached from the vessel walls during decompositions with acids. After prolonged use, the crucibles and dishes often recrystallize and exhibit a greater porosity. Components of the solutions penetrate into the pores and deeper subsurface layers. Vessels in which silicates were repeatedly dissolved in hydrofluoric acid cannot be used for decompositions of materials for a trace determination of fluoride (using alkaline fusion or sintering with a mixture of sodium carbonate and zinc oxide), even after careful cleaning of the vessel.

Platinum resists sodium peroxide up to a temperature of 500°C, but exceeding this limit leads to serious damage to the vessel.[29,34,35] The vessel corrosion is minimal at 480 to 495°C. Sintering with this reagent within the above temperature interval is often successfully used in analytical practice for decomposition of resistant phases.[36] Fusion with the alkali hydroxides must be carried out very cautiously and at a low temperature; otherwise, the crucible is strongly corroded.[29,33] A melt of sodium or potassium carbonate corrodes vessels with losses in the vessel material of a few milligrams at temperatures above 1000°C. It is assumed that the corrosion is caused by thermal dissociation of the reagent

$$Me^{I}_2CO_3 \rightarrow CO_2 + Me_2O \qquad (1)$$

with formation of the highly reactive alkali metal oxide. The corrosion is especially pronounced in fusion with lithium carbonate (m.p. 723°C). Although cesium carbonate begins to decompose already at 610°C, the amount of platinum dissolved after fusion or sintering of silicates (1 g sample and 1.8 to 2.2 g of the fusion agent, 20 min fusion at 850°C) has not exceeded 0.3 mg. The cesium oxide formed is consumed in interactions with the silicates present; compared with lithium carbonate, its molar excess over SiO_2 used is substantially lower.[37]

A melt of disulfate dissolves some oxides (rutile, corundum) that resist acids and many fusion agents. From 0.5 to 5.0 mg of platinum is dissolved, and these losses fluctuate in dependence on the sample composition, procedure, and the time of fusion.[29,38] Melts of potassium hydrogenfluoride behave analogously.

A serious complication arises from alloying of platinum with some metals from the melt, such as iron, lead, tin, antimony, and bismuth. The main part of alloyed iron is removed from the platinum by oxidative ignition in the air and dissolution of the oxides formed in hydrochloric acid. This operation is repeated several times and combined with disulfate fusion, but still the removal of the iron from the crucible walls is incomplete.[38] Therefore, in analyses of materials with low iron contents (glass sands and aluminum oxides) there is always the danger of contamination of the reaction mixture when platinum crucibles and dishes are used that have been in use for a long time, even if they are carefully cleaned prior to analysis. Similar difficulties arise in determinations of submicrogram amounts of tin in rocks. Platinum crucibles, used for routine analyses of silicates, liberate the tin alloyed in the walls during fusion of rocks with a mixture of potassium carbonate and borax;[39] lead behaves analogously. Therefore, acid decompositions are preferred in analyses of uranium ores, dusts, and wastes (determination of ^{210}Pb).[40] Platinum vessels are primarily alloyed with metals from alkaline fusion agents. Acidic agents, on the other hand, gradually liberate the alloyed metals.[38] Interaction of some metals with platinum in alkaline melts causes a ''memory effect'' of the vessels used. The composition of the contaminants and their amounts liberated are determined by the history of the vessel. From this point of view, sintering in platinum vessels is less risky than fusion. However, platinum vessels can only rarely be used for determinations of elements in submicrogram amounts. Substances containing larger amounts of sulfur and arsenic cannot be ignited in platinum vessels.

Alloys of platinum with gold (95 + 5) have been successfully used for the preparation of borax glasses for XRF determinations of major and minor components of silicates.[41] The melt of lithium borates does not wet the crucible walls and can be completely transferred to a forming platinum mold or dissolved in dilute nitric or hydrochloric acid. The whole procedure is usually automated, and commercial devices are used. The unwettability of the crucible walls by the melt is enhanced by adding cesium iodide, best immediately before completing the fusion. The corrosion of Pt-Au vessels in borate fusion is similar to that of platinum crucibles.[38] In fusion with lithium borates, it is recommended to maintain a mildly oxidizing character of the melt.

B. Zirconium Vessels

On introduction of zirconium in the manufacture of fusion crucibles, application of alkaline-oxidizing fusion agents to decomposition of solids has been facilitated.[42] Zirconium metal forms a hexagonal lattice that is converted into a cubic one at 862°C. It is resistant toward hot, concentrated acids, e.g., hydrochloric, nitric, perchloric, and sulfuric. It is readily dissolved only in hydrofluoric acid containing oxidants. Pure zirconium crucibles are marketed with satisfactory shapes and quality. This metal is at present (together with glassy carbon) the most suitable material for fusion with sodium peroxide at temperatures from 500 to 600°C.[29,34,35,42,43] An average loss in the crucible material is *circa* 5 mg Zr during a single fusion operation. In contrast to platinum crucibles, the fusion must be carried out in a reducing gas flame or in an electric furnace in an argon atmosphere. Fusion in a furnace in contact with the air leads to rapid oxidation of the crucible and the conversion of zirconium metal into the dioxide. Zirconium vessels are thus unsuitable for determinations of ashes of carbonaceous materials. The use of zirconium vessels has facilitated decompositions of resistant inorganic compounds, such as silicon carbide, boron nitride, refractory ceramics, and chromium ores. Small amounts of the corrosion products do not complicate the subsequent analysis. Zirconium crucibles and dishes are convenient for fusion and sintering decompositions with the alkali hydroxides, even if they contain oxidizing admixtures (KNO_3, $KClO_3$), borates, and the alkali fluorides. Bisulfates corrode the vessels on prolonged reactions at elevated temperatures.[29]

C. Pure Nickel Vessels

Pure nickel vessels have been successfully used mainly in metallurgical laboratories. The material is usually contaminated by cobalt and iron. Nickel crucibles are strongly corroded in fusion with sodium peroxide, with formation of dark, surface-active higher nickel oxides that liberate chlorine during dissolution in hydrochloric acid. In strongly alkaline solutions, stannates and, to a lesser extent, molybdates, tungstates, and probably other oxygen-containing anions are sorbed on these oxides. The sorption properties of the oxides make it impossible to separate stannates from many accompanying elements in an aqueous extract of a melt of an alkali peroxide. This seriously complicates analyses of poor cassiterite ores.

On the other hand, the interaction of an alkali hydroxide (NaOH) melt with the walls of a nickel crucible is minimal, and the amount of nickel transferred into the melt mostly does not exceed 1 mg.[29] Nickel vessels are equally suitable for sintering decompositions with the alkali carbonates. Thick-walled nickel dishes resist phosphoric acid and its condensation products; this property has not yet been fully appreciated in analytical laboratories.

D. Silver Vessels

Silver vessels behave analogously as nickel ones. The loss in weight in a single fusion with sodium hydroxide in the air amounts to *circa* 10 mg. In the hydroxide melt, compounds of iron do not interact with silver.

E. Iron, Molybdenum, and Tantalum Vessels

Iron crucibles were mainly used for fusion with sodium peroxide in decompositions of tin ores and in determination of sulfur in pyrite ores. A high extent of corrosion and an insufficient purity of the material, however, limit the applicability of iron vessels to decompositions of solids.

Molybdenum and tantalum vessels are used in inorganic analysis for thermal treatment of samples at high temperatures, as the two metals exhibit exceptionally high melting points (2610°C for molybdenum and 2996°C for tantalum). Molybdenum vessels have found use, e.g., in fusion of samples of granitic rocks without fusion agents. The reaction must be performed in an inert atmosphere or in a vacuum, as molybdenum is rapidly oxidized to the volatile trioxide at temperatures exceeding 400°C in the presence of oxygen.

F. Glassy Carbon Vessels

Glassy carbon vessels are also suitable for fusion with corrosive agents. Materials of satisfactory quality are produced by controlled, high-pressure pyrolysis of organic substances (e.g., some cross-linked resins) and a high-temperature treatment of the product.[29,44-46] Glassy carbon is a compact material, relatively hard, and readily polishable, with a very low porosity. The material permeability for helium is 10^{-10} to 10^{-12} cm^2sec^{-1}, which is similar to the values for glass. On heating in the air, the surface is oxidized with formation of acidic carboxylic or phenolic groups that are thermolabile and begin to decompose at temperatures above 400°C, producing carbon monoxide and dioxide. Glassy carbon crucibles and dishes are unsuitable for determination of ashes or of the loss on ignition, as their weight is unstable because of the formation of gaseous oxidation products. On the other hand, they are advantageous for treatment of substances in an inert atmosphere at temperatures higher than 1200°C (up to 3000°C, depending on the quality of the product).[29]

Glassy carbon vessels are used in analytical laboratories for decompositions of samples by oxidizing mixtures of acids (e.g., nitric acid, aqua regia, and hydrochloric acid with bromine or chlorate) in the presence of hydrofluoric acid. Because of minimal porosity and stability at elevated temperatures, glassy carbon has been proposed as a material for internal vessels in autoclaves, instead of Teflon®.[47] Microphotographs have shown that the vessel surface is minimally corroded after a 10-hr action of concentrated nitric acid in a closed system at 160°C. The amount of impurities leached from the crucible can be substantially decreased by repeated high-pressure leaching. Evaporation to dryness with perchloric acid causes strong corrosion of the vessel walls.[29,48]

Glassy carbon is primarily used in decompositions of resistant materials, mainly with alkaline fusion agents with low melting points, e.g., mixtures of the alkali hydroxides with carbonates, sodium peroxide, or the alkali nitrates. The losses in the vessel weight in fusion with various agents[49] are surveyed in Table 1. The loss is mainly dependent on the reaction mixture temperature. If the external surface of the crucible is protected by an asbestos mantle against atmospheric oxygen, the crucible lifetime is substantially prolonged.[50] In the Plas-

Table 1
THE MASS LOSS OF GLASSY CARBON CRUCIBLE BY FUSION[49]

Fusion agent	Fusion temperature (°C)	Mass loss (%)
KHF_2	500	0.01
$K_2S_2O_7$	500	0.02
NaOH	500	0.04
	600	0.23
	700	0.90
Na_2O_2	500	0.80
Mixture Na_2O_2 and Na_2CO_3 (3 + 1)	500	0.80
Na_2CO_3 or K_2CO_3	700	4.40
$Na_2B_4O_7$	900	3.20
Mixture $Na_2B_4O_7$ and Na_2CO_3 (2 + 1)	900	3.90
Na_2CO_3	900	6.60
K_2CO_3	900	8.60

Note: Fusion time is 10 min in furnace, with 2 g of melting agent.

masol automatic fusion apparatus, a glassy carbon vessel is inserted into a platinum crucible, which suppresses the surface oxidation of carbon and permits an increase of the temperature of a borate melt up to 1100°C.[51] However, in the nonoxidizing medium of this melt, losses in cobalt and iron have occurred due to reducing effect of the carbon crucible.[52]

Thick-walled, glassy carbon crucibles have mostly been used for decompositions of raw materials and technological intermediates in iron metallurgy, using fusion with sodium peroxide. However, the oxidation reaction was often explosive, and the crucible was strongly corroded by the active oxygen liberated from the melt.[53,54] The corrosive effect of the melt was decreased by adding alkali carbonates. The procedures developed have been primarily useful in determination of phosphorus in iron ores.[55] A fusion agent of a mixture of sodium peroxide and carbonate (4 + 1 to 1 + 1) decomposes many technologically important products, e.g., slags, nitrides, carbides, and some ferroalloys, at 550 to 650°C within 3 to 15 min.[48,56,57] Sodium peroxide can be replaced by an alkali nitrate; the nitrogen oxides liberated during fusion attack the crucible walls less than the oxygen from the sodium peroxide melt, and thus the vessels resist 60 to 70 fusion operations.[50,58]

Glassy carbon crucibles are very important in analyses of baryte concentrates containing heavy metals. In determination of sulfates and most trace components, the test material must be completely dissolved using fusion with alkali carbonates and their mixtures with alkali hydroxides and oxidants. Platinum, nickel, and iron crucibles are unsuitable, as they are attacked to various extents. The corrosion products prevent reliable blank determination and complicate the separation and determination of the test components. In contrast, the corrosion products of glassy carbon crucibles are gases and do not interfere in the analytical procedure.

G. Pyrolytic Graphite Vessels

Pyrolytic graphite vessels have been rarely used in decomposition techniques. In contrast to glassy carbon, pyrolytic graphite is anisotropic, due to the preparation method — carbonization of hydrocarbons at a decreased pressure followed by a thermal treatment.[1,45] Oriented crystallites are formed in parallel on a support, mostly polycrystalline graphite. The basal planes with the graphite structure are highly chemically resistant. However, aqueous solutions or gases must not penetrate among the parallel structural planes. Pyrolytic graphite is more readily oxidized than glassy carbon by atmospheric oxygen in a dry state. It is usually highly pure. A crucible of this material was used, e.g., in decompositions of rich ores of the rare earths by fusion with sodium peroxide.[59]

Sintered corundum crucibles resist acids, but are rapidly corroded in fusion with bisulfates and are damaged by Na_2O_2 melts. The corrosion can be somewhat suppressed by adding an alkali hydroxide to the fusion agent.

H. Quartz Glass Vessels

The quality of quartz glass depends on that of the initial materials and on the technology used.[1,60,61] Quartz glass containing less than 50 ppm of metals can be made from natural, purified quartz. A perfectly transparent glass with an impurity content of less than 1 ppm is obtained from purified silicon dioxide, prepared by hydrolysis of silicon tetrachloride. The glass consists of amorphous, solidified silicon dioxide, and its structural units are not silicon dioxide molecules, but silicon atoms interconnected by oxygen atoms in stable, irregular six-membered rings.

The properties of quartz glass (e.g., transparence or density) depend on the material purity. Fused quartz exhibits the lowest thermal expansion of all the glass materials. The glass surface is covered with silanol and siloxane groups at various densities. The silanol groups have the properties of a weakly acidic cation exchanger. The quartz vessel surface is minimally corroded by acid solutions, except for hydrofluoric acid and hot, concentrated phosphoric acid, and the amount of silicon dioxide dissolved does not exceed some tenths of a milligram. However, the vessel walls are rapidly attacked by hot, concentrated solutions of the alkali hydroxides.

Transparent quartz glass crucibles are suitable for fusion of samples with alkali bisulfates, and the whole process can be readily observed. Quartz glass tubes are advantageous for decompositions of materials in gas streams. However, glass recrystallizes on repeated heating with formation of polymeric molecules $(SiO_2)_n$. Its volume increases, and it is mechanically damaged. Quartz glass is an ideal material for pressure decompositions in sealed ampules. The amount of silicon dioxide leached from the ampule wall does not exceed 3.5 mg SiO_2 per 100 cm^2, even after a 46-hr treatment with hydrochloric acid at 350°C (see Chapter 5, Table 1).

I. Boron-Silicate Glass and Porcelain Vessels

Chemical boron-silicate glass (Pyrex® type) contains the oxides of silicon, boron, and sodium as the major components and varying amounts of potassium and aluminum oxides. Boron-silicate glass vessels are used for dissolution of substances in solutions of hydrochloric, nitric, sulfuric, and perchloric acids. Hydrofluoric acid readily dissolves glass and hot, concentrated, 85% phosphoric acid corrodes the vessel walls.[1,29] Acids leach from glass—mainly silicon dioxide, sodium ions, small amounts of boric acid, and many trace elements. In hot, alkaline solutions, the amount of silicon dioxide dissolved increases up to tens of milligrams. Hence, glass vessels are unsuitable for leaching of melts of alkali hydroxides, peroxides, or carbonates. Thick-walled test tubes of hard boron-silicate glass can also be used for fusion with potassium bisulfate under field conditions. This simple decomposition has given good results in analyses of soil and lithogeochemical samples in geochemical prospecting.

The resistance of glass increases with increasing content of silicon dioxide. The Vycor-type silicate glasses contain *circa* 96% of SiO_2, small amounts of boron and aluminum oxides, and admixtures of alkali metals and exhibit an increased resistance, especially toward acids. Bottles of this material are used for commercial distribution of acids (e.g., perchloric acid).[1]

Porcelain vessels have not found wide use in decomposition techniques, although their resistance toward most acids is greater than that of glass. Aluminum is the predominant component leached by acids from glazed vessels. Porcelain dishes, crucibles, and tubes have been used in some thermal decompositions and in fusion of ores with a carbonate and sulfur. Special fireclay crucibles find use in reducing fusion and belong to basic equipment of laboratories dealing with determinations of precious metals.

J. Vessels of Organic Polymers

Over the last three decades, many organic polymers have been successfully applied to decomposition techniques. Plastic vessels have made it possible to use even highly corrosive mixtures of acids for decomposition without a substantial damage to the vessel walls and, thus, without contamination of the reaction mixture. Polymeric material dishes and crucibles are used more and more for dissolution of substances in acids in open or closed systems. The most common materials involve polyethylene, polypropylene, polytetrafluoroethylene, polycarbonate, and recently, polysulfone.

Polyethylene belongs among the materials that have been used for the longest time. Its physical properties and chemical resistance depend on the structure of its macromolecules.[1,29] The macromolecule contains linear $-CH_2-$ chains, and the formation of side chains decreases the mechanical strength and chemical resistance of the polymer. A polymer with a minimal number of side chains (linear polyethylene) has an increased mechanical strength, higher density, better thermal stability, and chemical resistance. In decomposition techniques, dishes, beakers, and bottles with screw caps are made of this material. As the material has a poor thermal conductivity, the reaction mixtures are usually heated in a water bath and less often on a temperature-controlled hot plate. Polyethylene is sufficiently resistant toward the alkali hydroxides and hydrochloric and hydrofluoric acids, even in hot, concentrated solutions. It is not attacked by dilute sulfuric acid, but is readily oxidized by dilute nitric acid and aqua regia. The maximal permissible temperatures for brief exposures of the linear polymer are 105 to 120°C, while for conventional polyethylene, they do not exceed 80°C.[1,62] Polyethylene vessels should not be used for prolonged storage of solutions, as the liquids penetrate into the vessel walls, and undesirable memory effects may occur. The material is also permeable for gases, e.g., nitrogen, bromine, carbon dioxide, and ammonia; the permeability decreases with increasing density of the polymer.

The behavior of polypropylene is similar to that of polyethylene. The material is translucent and sufficiently rigid and thus, is suitable for the manufacture of beakers and volumetric vessels. The thermal stability of the polymer permits an increase in the decomposition temperature up to 125 to 135°C.[1,28,62] The principal unit of the polymer is the

$$\begin{array}{c} -CH_2-CH- \\ \quad\quad | \\ \quad\quad CH_3 \end{array}$$

group. Because of this branched structure, it is less resistant toward oxidizing acids than linear polyethylene. The material becomes yellow or even brown by the action of dilute nitric acid or aqua regia and by prolonged action of concentrated hydrochloric acid, but in spite of this, polypropylene containers have been successfully used for storage of highly pure hydrochloric acid. It is resistant toward dilute sulfuric and phosphoric acid, even at elevated temperatures. Hydrofluoric acid does not corrode the vessel walls, even in concentrated solutions. Closed, wide-necked polypropylene bottles performed well, e.g., in decompositions of silicates for determination of ferrous oxide[63,64] in resistant minerals and of tin in soils, sediments, and rocks.[65] Organic compounds cannot be oxidized by hydrogen peroxide in mixtures of hydrofluoric and sulfuric acids, as the walls of polypropylene vessels are rapidly damaged.

Polymethylpentene is almost equally chemically resistant as polypropylene and its structure is also similar, but contains the isobutyl group in the side chain. Transparent vessels made of polymethylpentene permit visual observation of the dissolution process and resist temperatures up to 175°C. However, they have, so far, found little use in decompositions of solids.

Polycarbonates consist of macromolecules of the polyester type with the basic structural unit formed by phenolic nuclei interconnected by carbonate groups –O–CO–O–. The material is used to make beakers, bottles, or conical flasks with polypropylene or Teflon® screw-on caps. The vessels are transparent, and thus, the decomposition can readily be observed. The plastic has a high tensile strength, but the polymeric molecule is easily damaged by concentrated inorganic acids at the location of the carbonate group.[62] The polycarbonate is also attacked by 48% hydrofluoric acid, and thus, decompositions must be rapid, using the acid of a lower concentration (glass sands, readily decomposable silicates).[71] The maximal temperature should not exceed 135°C. A closed polycarbonate bottle placed in a boiling water bath functions as a pressure vessel, in which, e.g., slags, mattes, claystones, and phosphate rocks can be decomposed by a mixture of concentrated hydrochloric and hydrofluoric acids, with an addition of concentrated nitric acid. The lifetime of the vessel under these conditions is at least eight decompositions.

Polysulfone has more favorable physical properties than polycarbonate. The main structural units of its macromolecule are phenolic nuclei connected by hydrocarbon chains with an etheric bond and a sulfone group. The material is transparent and equally rigid as polycarbonate, but resists temperatures of up to 170°C and is more resistant toward oxidizing acids.[62]

By mastering polymerization reactions and introducing new groups into macromolecules, the properties of plastics are continuously being improved. Some plastics that have been used in military and space research are gradually made available to civil use.

K. Vessels of Fluorinated Polymers

Decisive progress in decomposition techniques was brought about by the introduction of vessels made of fluorinated polymers of aliphatic hydrocarbons. Due to the particular structure of the macromolecules, these substances are highly resistant toward hydrofluoric acid, as well as toward oxidizing mixtures of acids, such as aqua regia, hydrochloric acid with potassium chlorate, and nitric with perchloric acid. The resistance of the material is explained by the high energy of the C–F bonds in the polymer and by a protective effect of the fluorine atoms on the basic hydrocarbon skeleton.[1] Among these materials, polytetrafluoroethylene (PTFE) with the basic group $-CF_2-CF_2-$ in the linear chain of the macromolecule has been most useful for the manufacture of decomposition vessels. In the analytical literature it is mostly denoted as PTFE or Teflon®, which is the trademark of the Du Pont Company. Other structurally analogous polymers have also found use, e.g., the tetrafluoroethylene and tetrafluoropropylene copolymer (Teflon® FEP from Du Pont) or the ethylene and tetrafluoroethylene copolymer containing more than 75% of the latter component (e.g., Tefzel® from Du Pont). Other polymeric derivatives with different structures (e.g., Teflon® PFA) are becoming progressively more common in the production of vessels and apparatuses for analytical chemistry and chemical industry.

Teflon® vessels are almost universally applicable to dissolution of solids in acids. Except for glassy carbon, there is no material available to analysts that would be so resistant to permit work in highly corrosive mixtures, such as hydrofluoric acid with aqua regia or nitric acid with perchloric acid. Teflon is strongly corroded by molten alkali metals and some fluorine compounds at elevated temperatures.

However, Teflon® cannot be considered as a material perfectly resistant toward acids. Prolonged evaporation of reaction mixtures with sulfuric acid (e.g., in dissolution of monazites) causes corrosion of Teflon® dishes,[66] as does pressure decomposition of substances with nitric acid.[67] Organic compound leached from the vessels interfere in subsequent determinations of trace elements by differential pulse anodic stripping voltammetry.[68]

A PTFE vessel made by a perfect technology permits a short-time heating of a reaction mixture in a metallic autoclave mantle up to 285°C. However, common working temperatures lie below 250°C. The maximal working temperatures[62] for the Tefzel® and Teflon® FEP polymers amount to only 180 and 250°C, respectively. Deformation of the vessels during pressure decompositions are discussed in Chapter 5, Section III.

The PTFE-based polymeric materials have a low thermal conductivity. Therefore, various devices have been constructed to accelerate the reaction mixture heating and the evaporation of excess acids.[69] A heated metallic block with holes fitting the teflon vessels is best suited for the purpose.

Teflon® FEP is advantageous as the material for bottles for storage of high-purity acids.[6,8] The presence of fluorides from the degraded polymer, observed by Zief[1] in concentrated solutions of hydrochloric and perchloric acids stored in containers made of this plastic, has not later been confirmed.

Plastics based on PTFE are permeable for certain gases, such as oxygen, carbon dioxide, nitrogen oxides, acid vapors, and vapors of some organic solvents and of water. The permeability is caused by mobility of the polymeric chains (especially at elevated temperatures) and by the material porosity. The size and density of the pores can be affected by the technological treatment and the degree of crystallinity of the plastic. The gases absorbed in the walls of vessels made of polymeric materials may complicate the subsequent analytical procedure: In a determination of vanadium(III) oxide in rozcoelite, the empty Teflon® vessel is first heated in an evacuated oven to remove oxygen. Only then is the mineral decomposed by a hydrofluoric acid solution under an increased pressure and temperature.[70]

III. ANALYZING IMPURITIES FROM VESSEL MATERIALS

Using sensitive analytical methods, total amounts of impurities have been determined in the vessel materials, as well as the fraction that can be leached from the walls by hydrochloric, nitric, and hydrofluoric acids. Trace impurities have been determined in polyethylene, polypropylene, PTFE, polycarbonate, and polysulfone, as well as in platinum, glassy carbon, quartz, and chemical glass.[7,10,12,28,32,47,72-78] Insufficient attention has so far been paid to the organic compounds that pass into solution during decompositions of substances in plastic vessels. The presence of these compounds in solutions has been demonstrated by differential pulse polarography.[68]

REFERENCES

1. **Zief, M. and Mitchell, J. W.,** *Contamination Control in Trace Element Analysis,* John Wiley & Sons, New York, 1976, 28.
2. **Irving, H. and Cox, J. J.,** *Analyst,* 83, 526, 1958.
3. **Kwestroo, W. and Visser, J.,** *Analyst,* 90, 297, 1965.
4. **Arden, J. W. and Gale, N. H.,** *Anal. Chem.,* 46, 2, 1974.
5. **Coppola, P. P. and Hughes, R. C.,** *Anal. Chem.,* 24, 768, 1952.
6. **Kuehner, E. C., Alvarez, R., Paulsen, P. J., and Murthy, T. J.,** *Anal. Chem.,* 44, 2050, 1972.
7. **Dabeka, R. W., Mykytink, A., Berman, S. S., and Russell, D. S.,** *Anal. Chem.,* 48, 1203, 1976.
8. **Moody, J. R. and Beary, E. S.,** *Talanta,* 29, 1003, 1982.
9. **Mitchell, J. W.,** *Talanta,* 29, 993, 1982.
10. **Tschöpel, P., Kotz, L., Schulz, W., Veber, M., and Tölg, G.,** *Fresenius Z. Anal. Chem.,* 302, 1, 1980.
11. **Gaivoronskii, P. E. and Pimenov, V. G.,** *Zavod. Lab.,* 50(6), 20, 1984.

12. **Gretzinger, K., Kotz, L., Tschöpel, P., and Tölg, G.,** *Talanta,* 29, 1011, 1982.
13. **Mattinson, J. M.,** *Anal. Chem.,* 44, 1715, 1972.
14. **Little, K. and Brooks, J. D.,** *Anal. Chem.,* 46, 1343, 1974.
15. **Lange, J.,** *Silikattechnik,* 31, 44, 1980.
16. **Tatsumoto, M.,** *Anal. Chem.,* 41, 2088, 1969.
17. **Lancet, M. S. and Huey, J. M.,** *Anal. Chem.,* 46, 1360, 1974.
18. **Stegmann, H.,** *Fresenius Z. Anal. Chem.,* 154, 267, 1957.
19. **Coleman, M. L.,** *Anal. Chim. Acta,* 60, 426, 1972.
20. **Moody, J. R.,** *Anal. Chem.,* 54(13), 1358A, 1982.
21. **Heumann, K. G., Beer, F., and Weiss, H.,** *Microchim. Acta,* 95, 1982.
22. **Heumann, K. G., Schrödl, W., and Weiss, H.,** *Fresenius Z. Anal. Chem.,* 315, 213, 1983.
23. **Vasilevskaya, L. S., Kondrashina, A. I., and Shifrina, G. G.,** *Zavod. Lab.,* 28, 674, 1962.
24. **Vláčil, F. and Drabal, K.,** *Chem. Listy,* 62, 1371, 1968.
25. **Mitchell, J. W.,** *Int. Lab.,* January/February, 12, 1982.
26. **Goguel, R.,** *Anal. Chim. Acta,* 169, 179, 1985.
27. **Rigin, V. I.,** *Zh. Anal. Khim.,* 40, 253, 1985.
28. **Mizuike, A.,** *Enrichment Technique for Inorganic Trace Analysis,* Springer-Verlag, Berlin, 1983, 14.
29. **Trofimov, I. V. and Busev, A. I.,** *Zavod. Lab.,* 49(3), 5, 1983.
30. **Tschöpel, P.,** *Pure Appl. Chem.,* 54, 913, 1982.
31. **Tölg, G.,** *Pure Appl. Chem.,* 55, 1989, 1983.
32. **Kosta, L.,** *Talanta,* 29, 985, 1982.
33. **Oelschläger, W.,** *Fresenius Z. Anal. Chem.,* 246, 376, 1969.
34. **Blake, H. E. and Holbrook, W. F.,** *Chemist Analyst,* 46, 42, 1957.
35. **Belcher, C. B.,** *Talanta,* 10, 75, 1963.
36. **Rafter, T. A.,** *Analyst,* 75, 485, 1950.
37. **Šulcek, Z. and Huka, M.,** *Sklar. Keram.,* 28, 204, 1978.
38. **Russell, B. G., Spangenberg, J. D., and Steele, T. W.,** *Talanta,* 16, 487, 1969.
39. **Weiss, D.,** unpublished results, 1977.
40. **Sill, C. W.,** *Health Phys.,* 33, 397, 1977.
41. **Exnar, P.,** *Chem. Listy,* 78, 920, 1984.
42. **Petretic, G. J.,** *Anal. Chem.,* 23, 1183, 1951.
43. **Dodson, E. M.,** *Anal. Chem.,* 34, 966, 1962.
44. **Yamada, S. and Sato, H.,** *Nature,* 193, 261, 1962.
45. **Štulíková, M. and Štulík, K.,** *Chem. Listy,* 68, 800, 1974.
46. **Huettner, W. and Busche, C.,** *Fresenius Z. Anal. Chem.,* 323, 674, 1986.
47. **Kotz, L., Henze, G., Kaiser, G., Pahlke, S., Veber, M., and Tölg, G.,** *Talanta,* 26, 681, 1979.
48. **Dymova, M. S., Kozina, G. V., and Titova, T. V.,** *Zavod. Lab.,* 50(10), 18, 1984.
49. **Mashkovich, L. A., Kuteinikov, A. F., Pekaln, L. A., Kirevina, T. P., Tashchilova, L. P., and Litvinov, V. F.,** *Zh. Anal. Khim.,* 37, 1528, 1982.
50. **Vokhrysheva, L. E. and Gladysheva, K. F.,** *Zavod. Lab.,* 50(10), 16, 1984.
51. **Wittmann, A. A. and Willay, G. M.,** Paper presented at the 1984 Winter Conf. on Plasma Spectrometry, San Diego, Calif., June 1 and 2, 1984.
52. **Bennett, H. and Oliver, G. J.,** *Analyst,* 96, 427, 1971.
53. **Bhargava, O. P.,** *Analyst,* 101, 125, 1976.
54. **Bhargava, O. P. and Hines, W. G.,** *Anal. Chem.,* 48, 1701, 1976.
55. **Bhargava, O. P., Gmitro, M., and Hines, W. G.,** *Talanta,* 27, 263, 1980.
56. **Kuteinikov, A. F., Kirevina, T. P., Mashkovich, L. A., and Stepanova, A. N.,** *Zavod. Lab.,* 50(6), 16, 1984.
57. **Kuteinikov, A. F., Mashkovich, L. A., Kirevina, T. P., Pekaln, L. A., and Gryukan, V. S.,** *Zavod. Lab.,* 44, 666, 1977.
58. **Kustova, L. V., Larkina, A. N., and Smirnova, N. V.,** *Zavod. Lab.,* 52(4), 15, 1986.
59. **Brenner, I. B., Steele, T. W., Watson, A. E., and Jones, E. A.,** *Spectrochim. Acta,* 36B, 785, 1981.
60. **Horká, M., Janák, K., and Tesřík, K.,** *Chem. Listy,* 79, 840, 1985.
61. **Fanderlík, I., Ed.,** *Křemenné sklo,* State Publishers of Technical Literature, Praha, C.S.S.R., 1986, 101.
62. **Farrell, R. F., Matthes, S. A., and Mackie, A. J.,** *U.S. Bur. Mines Rep. Invest.,* p. 8480, 1980.
63. **French, W. J. and Adams, S. J.,** *Anal. Chim. Acta,* 62, 324, 1973.
64. **French, W. J. and Adams, S. J.,** *Analyst,* 97, 828, 1972.
65. **Smith, J. D.,** *Anal. Chim. Acta,* 57, 371, 1971.
66. **Schärer, W. and Allegre, C. J.,** *Earth Planet Sci. Lett.,* 63, 423, 1983.
67. **Tölg, G.,** *Fresenius Z. Anal. Chem.,* 283, 257, 1977.
68. **Oehme, M.,** *Talanta,* 26, 913, 1979.

69. **Kaigorov, V. A. and Churlina, E. F.,** *Zavod. Lab.*, 39, 159, 1973.
70. **Wanty, R. B. and Goldhaber, M. B.,** *Talanta,* 32, 295, 1985.
71. **Langmyhr, F. J. and Paus, P. E.,** *Anal. Chim. Acta,* 43, 397, 1968.
72. **Tamenori, H. and Inoue, J.,** *Bunseki Kagaku,* 32, 337, 1983.
73. **Vasilevskaya, L. S., Muravenko, V. P., and Kondrashina, A. I.,** *Zh. Anal. Khim.*, 20, 540, 1965.
74. **Moody, J. R. and Lindström, R. M.,** *Anal. Chem.*, 49, 2264, 1977.
75. **Tölg, G.,** *Fresenius Z. Anal. Chem.*, 294, 1, 1979.
76. **Tölg, G.,** *Pure Appl. Chem.*, 50, 1075, 1978.
77. **Karin, L. W., Buono, J. A., and Fasching, J. L.,** *Anal. Chem.*, 47, 2296, 1975.
78. **Heydorn, K. and Damsgaard, E.,** *Talanta,* 29, 1019, 1982.

Chapter 4

DECOMPOSITION IN OPEN SYSTEMS

I. INTRODUCTION

Dissolution in acids is the most common method of decomposition of solids. Strong inorganic acids, such as hydrochloric, nitric, perchloric, and sulfuric, are most often employed for the purpose. Medium strong (H_3PO_4) and weak (HF) acids exhibit complexing properties that enhance the decomposing effect. Substances with polyphase composition are rarely dissolved in a single acid, and various acids are combined in order to increase the dissolving power. Mutual interactions in acid mixtures produce unstable, intermediate compounds that substantially accelerate the dissolution process. A certain acid is indispensable for decomposition of a particular material and must be present in the mixture. This predominating acid not only determines the rate and completeness of the decomposition, but often also affects the method of separation of the analyte and the selection of the final analytical method. Such an acid is, e.g., hydrofluoric acid, in analyses of silicates and silicon dioxide or condensed phosphoric acids in dissolution of resistant oxides.

However, it is often impossible to decide which of the acids in a mixture plays the role of the dominant solvent. For this reason, the structure of the text of this chapter should be considered as a subjective opinion of the authors of this book. Acid mixtures with hydrofluoric and phosphoric acid are treated in independent, extensive chapters. Some mixtures have been used for a very long time and, thus, have been given special names, e.g., aqua regia, Leffort (Lunge) mixture, or the Dixon mixture in soil analyses. Other favored acid combinations involve, e.g., mixtures of nitric and perchloric acids, aqua regia with perchloric acid, nitric and sulfuric acids, and hydrofluoric acid mixed with all the above acids.

II. DECOMPOSITION BY HYDROCHLORIC AND NITRIC ACIDS

A. Decomposition by Hydrochloric Acid

Hydrochloric acid, alone or in mixtures with other strong acids, belongs to solvents for solids that are used most often. It is commercially available as a 36 to 38% solution (11.6 to 12.4 *M*). It forms an azeotropic mixture with water, containing 20.4% HCl and boiling at 109.7°C. It reacts with most transition metal cations with formation of chloride complexes that are suitable for mutual separations by extraction or anion exchange. The chlorides are soluble in water, except for silver and thallous chloride. Hydrochloric acid exhibits weak reducing properties during dissolution of solids; e.g., selenates and tellurates are reduced to tetravalent salts on heating. Analogously, vanadates are reduced to tetravalent salts on heating. Analogously, vanadates are reduced to compounds of tetravalent vanadium and in closed systems even to V^{3+} ions. Higher oxides of manganese are reduced to stable manganese(II) salts. Hydrochloric acid is suitable for dissolution of melts of samples with alkali hydroxides, carbonates, and peroxides. Some metals (e.g., Zn, Cd, Fe, and Sn) can be dissolved in dilute hydrochloric acid, but the decomposition is usually accelerated by addition of another acid or an oxidant. In decomposition of steels, heterogeneous admixtures are important; graphitic carbon, resistant carbides, and nitrides usually remain undecomposed. Steel doped with yttrium is completely dissolved in 6 *M* HCl. The same kind of steel with a scandium admixture is dissolved with a residue containing up to 80% of the admixture.[1]

Hydrated oxides are easier to dissolve than anhydrous, high-temperature ignited oxides. With oxidic compounds of iron, the solubility increases with increasing Fe(II) content. Hematite is dissolved in 6 to 11 *M* HCl more slowly than limonite, goethite, or magnetite.

Boiling of hematite single crystals for a long time with dilute hydrochloric acid leads to perceptible reduction of ferric ions.[2,3] Corundum, the oxides of tetravalent actinoids, cassiterite, and rutile are minimally attacked by the acid. On the other hand, fine anatase impregnations can be leached from claystones almost quantitatively, even with dilute hydrochloric acid. The rare earth oxides are readily dissolved, but cerium dioxide, ignited at a high temperature, resists hydrochloric acid in an open system. The SiO_2 gel is readily soluble in the acid, but the solution is unstable. Higher oxides of nickel and manganese oxidize hydrochloric acid to elemental chlorine. Spinels are highly resistant toward the acid, except for magnetite. The rates of dissolution of ilmenites from various locations are different.

Among natural carbonates, calcite and aragonite react most rapidly. The difference in the reaction rates of calcite and dolomite can be utilized for orientative determination of the two phases.[4] In 1 + 1 dilute hydrochloric acid, carbon dioxide begins to be evolved from dolomite only after 5 min, whereas calcite is dissolved immediately, even in cold solution. Magnesite and siderite are dissolved slowly. Heavy metal carbonates are dissolved without difficulties, but fluorocarbonates, such as bastnesite and parisite, react slowly even in the hot, concentrated acid, leaving a small undecomposed residue containing rare earths. Carbon dioxide cannot be quantitatively liberated from silicocarbonates, e.g., scapolite, without destruction of the silicate structure. After dissolution of carbonates in dilute hydrochloric acid, the undecomposed fraction of the rock contains resistant mineral phases, as well as graphite, amorphous carbon, and variously metamorphosed organic substances, e.g., kerogene. The noncarbonate carbon can at present be readily determined on an automated analyzer, after isolation of the insoluble residue. This separation need not always be quantitative for all rock types. Some organic compounds are dissolved in the form of protonated compounds (amino acids) or colloidal solutions.

Hydrochloric acid displaces hydrogen sulfide from its salts. Natural simple sulfides, e.g., sphalerite, galena, pyrrhotite, and alabandine, are easily decomposed by boiling with 6 to 12 *M* HCl. Antimonite and most of its sulfosalts react slowly and often incompletely with dilute hydrochloric acid. Chalcopyrite, molybdenite, and cinnabar are not attacked, as are pyrite and marcasite. A removal of a major part of hydrogen sulfide from the reaction mixture simplifies the subsequent stage of decomposition, in which resistant sulfides are dissolved on addition of oxidants. In a preliminary decomposition with hydrochloric acid, the amount of sulfates formed is decreased, including sparingly soluble lead(II) sulfate that adsorbs many metal ions, especially Bi^{3+}, Sb^{3+}, etc. After dissolution of galena, crystals of lead(II) chloride precipitate from the cold solution and also contain part of the silver present. Lead(II) chloride is dissolved on addition of sodium or ammonium chloride, or on an increase in the acid concentration to 8 *M*, when the soluble chlorocomplex is formed. The solubility of silver chloride also increases with increasing chloride concentration, due to the formation of the complex anions $AgCl_n^{1-n}$. According to the values of the stability constants for the complexes $AgCl_2^-$, $AgCl_3^{2-}$, and $AgCl_4^{3-}$, the concentration of the dissolved silver ions should increase[5] to 8.1 and 130 mg/ℓ in 1 and 3 *M* HCl, respectively. However, it has been found that solutions with these concentrations of silver ions are not stable. Silver chloride precipitates[6,7] even at an initial concentration of 10^{-4} *M* Ag^+. On decreasing the concentration to 10^{-5} *M*, a suitably stable solution is obtained, for determination of the metal in poor ores and rocks.[8-10] The stability of silver salts in hydrochloric acid media is increased by diethylenetriamine,[7,11-13] which reacts with silver and lead(II) ions to form complex compounds and prevents precipitation of insoluble chlorides. In solutions of 10% hydrochloric acid and 1% reagent, the silver ions (concentrated $Ag^+ \leqslant 1$ mg ℓ^{-1}) are stable at least 2 weeks,[12] and this time is increased to several months in 6.2% acid with an addition of a polyamine.[7]

When ores containing galena and barite are unsuitably decomposed (e.g., with concentrated H_2SO_4), the lead(II) ions liberated are built, in the form of a binary sulfate, into the crystal structure of baryum sulfate. The losses in lead are minimal with the $PbCl_n^{2-n}$ complex anions.[14] A 0.3-g sample of a Pb ore is decomposed by heating with 20 mℓ HCl and 2 g NaCl. The hot solution is filtered off from the remaining solid. The lead(II) ions are determined chelatometrically, gravimetrically, or by AAS.

Iron disulfides, pyrite, and marcasite, can be converted into the readily soluble monosulfide in a strongly reducing medium, or can be decomposed by hydrochloric acid containing an oxidant (HNO_3, Br_2, an alkali chlorate, or hydrogen peroxide). The decomposition in a reducing medium has been utilized for differentiation among various forms of sulfur in oil shales. After dissolution of pyrrhotite and wurtzite by boiling with dilute HCl, the remaining solid is treated with the concentrated acid containing chromous chloride. The hydrogen sulfide liberated corresponds to the contents of pyrite and marcasite.[15]

A mixture of hydrochloric acid with bromine has been employed to dissolve resistant sulfides and sulfosalts, by mixing 10 to 20 mℓ of 6 to 11 *M* HCl with 0.5 to 3.0 mℓ Br_2. Most sulfides react already in cold solution or at mildly increased temperatures. The decomposition is completed by heating, and the excess bromine is thus removed. Pentavalent compounds are formed from sulfides and sulfosalts of As(III) and Sb(III). In contrast to the chloride, arsenic pentabromide (and the corresponding antimony compound) is volatile. Molybdenite and arsenopyrite are also dissolved.[16] The mixture also oxidizes natural or synthetic arsenides, in cool solution or at a mildly increased temperature.[17,18] A 0.1-g sample of gallium(III) arsenide with a clean surface is dissolved in 2 mℓ HCl and 1 mℓ Br_2 at 45°C. After 10 min, the reaction mixture is heated to 60°C, and the bromine volatilizes. Then 1 mℓ HCl and 1 mℓ 10% hydroxylammonium chloride are added, and the solution is maintained at the boiling point. Arsenic bromides volatilize on evaporation to dryness. A mixture of 6 *M* HCl with bromine has given good results in determination of impurities in tin metal. The major component volatilizes as $SnBr_4$ during the dissolution, and thus the subsequent analysis is simplified.[19]

In addition to bromine, nascent chlorine and its oxides are readily accessible oxidants. They are obtained, e.g., by the interaction of hydrochloric acid with alkali chlorates.

$$ClO_3^- + 6H^+ + 5Cl^- \rightarrow 3Cl_2 + 3H_2O \quad (1)$$

This procedure has found use in geochemical prospecting, in determination of metal sulfides in the bedrock and weathered rock mantle, especially in prospecting for deposits of porphyry copper ores.[20] Combined with AAS, it can also be carried out under field conditions and is suitable for analyses of soils, stream sediments, and rocks. The elements bound in the silicate matrix are liberated only to a small extent: to a 1-g sample, 1 g $KClO_3$ and, in several portions, 4 mℓ of concentrated HCl are added. The contents of the vessel are heated for 30 min on a hotplate until the yellow chlorine vapors escape. The decomposition can be combined[21,22] with an extraction preconcentration, e.g., using Aliquat 336 for Ag, Bi, Cd, Cu, Pb, and Zn. The behavior of pyrite, chalcopyrite, galena, auripigment, antimonite, tetrahedrite, cinnabar, and molybdenite in this mixture has been studied in detail, by determining the residual sulfur in the undecomposed fraction. The isolated sulfide was first mixed with clay. All the above minerals were completely decomposed, except for pyrite, chalcopyrite, and molybdenite. The latter sulfides were completely decomposed only when the reaction mixture was finally heated[23] with 4 *M* HNO_3.

Losses in arsenic, antimony, and molybdenum have been observed during boiling of reaction mixtures containing hydrochloric acid and potassium chlorate, which partially offset the importance of the method for prospecting.[24,25] The losses of these elements in decompositions of soils, hot spring sediments, and rocks (GXR 1 to 6 standards) cannot be satisfactorily explained by volatilization or a poor decomposability of the mineral (especially with As). Isolated minerals, such as arsenopyrite, löllingite, sulfoarsenides of nickel, and cobalt from the safflorite-skutterudite group, are readily dissolved in a mixture of hydrochloric acid with chlorate. The dissolution of the isolated minerals and that of these phases in complex systems, e.g., soils or rocks, has apparently somewhat different mechanisms. By using the same mixture cold, small amounts of sulfides scattered in rocks can be quantitatively converted into sulfates: a 0.2-g sample is mixed with 0.2 g $NaClO_3$, 5 mℓ HCl are added, and the mixture is allowed to stand overnight. The excess acid and the oxidant is removed by evaporation on a water bath.[26]

Strongly reactive chlorine is also formed by interaction of hydrochloric acid with hydrogen peroxide. The oxidizing power of the solution increases with its increasing acidity, and thus a mixture of concentrated hydrochloric acid with 30% H_2O_2 is used:

$$2\ HCl + H_2O_2 \rightarrow Cl_2 + 2\ H_2O \tag{2}$$

Sulfides are readily dissolved in this mixture. As the hydrogen peroxide is reduced to water by hydrochloric acid, the reaction mixture does not contain undecomposed residues of oxidants, except for the chlorine. The procedure has given good results in determinations of metals (including As, Sb, and Mo) in soils and rocks;[25] a 1-g sample is decomposed in a test tube, adding 5 mℓ HCl and 1 mℓ H_2O_2 in several portions. The solution is set aside for 30 min and then heated for 20 min on a boiling water bath. Prior to AAS determinations of the metals, the solution is suitably diluted.

Dilute hydrochloric acid, without addition of oxidants or hydrofluoric acid, has been successfully used for leaching of some elements from complex polyphase mixtures. A high leaching yield has been demonstrated by comparing the results with those of parallel decompositions involving total destruction of the test material.[27-31] With soil and rock samples, the efficiency of 6 to 9 *M* HCl for arsenic and lead is comparable with that of aqua regia or a mixture of nitric, perchloric, and hydrofluoric acids. Arsenic compounds have been dissolved in 6 *M* HCl for 2 to 3 hr at 90°C and sandy soils[29] in 9.6 *M* HCl for 12 to 24 hr at laboratory temperature. However, it has not yet been decided whether this experience also holds for other rock and soil types, with different genetical origin and the element and phase composition.

Magnetite iron ores are readily dissolved in hydrochloric acid.[32] Various natural modifications of ferric oxide (itabirites, specularites, and fibrous hematites) are slowly dissolved, and the reaction is accelerated by adding stannous chloride that reduces the ferric ions to ferrous ones during the dissolution.[33] A 0.5- to 1-g ore sample is dissolved in 6 *M* HCl with several drops of $SnCl_2$ (5% $SnCl_2{\cdot}2H_2O$ in concentrated HCl), sometimes adding a small amount of HF, depending on the ore character. After the decomposition, the solution should always contain a small amount of chloroferrate(III). Excess stannous chloride must not be added. Thioglycolic acid exerts the same effect during dissolution of hematite, and its excess is eliminated by oxidation with nitric acid.[34] To a 0.5-g sample, 20 mℓ HCl and 0.5 mℓ thioglycolic acid are added; the solution formed is diluted with water and oxidized with 1 mℓ of concentrated HNO_3.

The decomposability of bauxites varies greatly in dependence on their origin. Hydrochloric acid is not particularly suitable for decomposition of compounds with high contents of aluminum oxide, e.g., gibbsite and diaspore, apparently because of a poor solubility of aluminum chloride that is formed at the interface.

An addition of a reductant during decomposition with hydrochloric acid serves another purpose with precious metal ores. The stannous chloride added prevents dissolution of the precious metals, whereas the compounds of common metals, especially oxides of iron, are dissolved.[35] A 2- to 5-g sample of the roasted ore is heated to the boiling point with 50 mℓ of a 15% $SnCl_2$ solution in concentrated HCl. Gold is dissolved from the isolated insoluble residue using either aqua regia, or a mixture of hydrochloric acid with bromine. The solution of the auric ions in hydrochloric acid must not be evaporated to dryness, to prevent the reduction to the aurous salt or the gold metal.[36] For this reason, the solution is stabilized by adding sodium chloride prior to evaporation. Ores of the platinum metals, their technological products, and alloys are freed of accompanying metals analogously.[37-39] A 1- to 25-g sample of the material is heated for 0.5 to 2 hr with 100 mℓ of concentrated HCl to which 20 g of NH_4Cl may be added. The platinum metals remain in the undecomposed fraction. The material is sometimes preroasted and leached with hydrochloric acid to which manganese(II), tin(II), and iron(II) salts are added.[40]

The rate of dissolution of scheelites ($CaWO_4$) varies greatly. Some scheelites undergo the reaction within a mere 5 min; others resist hydrochloric acid for several hours.[41] The rate of dissolution may also be affected by rare earths that are typically contained in this mineral. Wolframite can be decomposed by hydrochloric acid, provided that it is very finely pulverized (below 300 mesh). The reaction at laboratory temperature leads to the precipitation of the white dihydrate of α-tungstic acid that is converted into the β-form ($WO_3 \cdot H_2O$) on heating. Tungstic acid is dissolved in the concentrated acid with the formation of oxychloride complexes, but reprecipitates on prolonged boiling of the solution. If a temperature of 70°C is not exceeded during dissolution of the ore, the solution obtained is sufficiently stable for rapid, orientative AAS determination of tungsten.[36] Leaching of tungsten ores with hydrochloric acid has also been utilized technologically, to treat tin-tungsten concentrates.[42]

The rate of decomposition of oxidic uranium ores depends on the U(VI)/U(IV) ratio. The so-called uranium blacks with minimal contents of U(IV) are even dissolved in dilute hydrochloric acid.[43] Uraninite requires an oxidizing mixture of hydrochloric acid with hydrogen peroxide, chlorate, or nitric acid for dissolution. Uranium and thorium compounds cannot be completely leached from granites by hydrochloric acid.[44] Natural and synthetic thorium dioxides are highly resistant toward hydrochloric acid and must be decomposed in a pressure vessel. Binary phosphates of uranyl and divalent cations, e.g., autunite and torbernite, are dissolved without difficulties. On the other hand, phosphates of thorium, tetravalent uranium, and the rare earths (monazite and xenotime) are only negligibly attacked, even with the concentrated acid. Apatites of magmatic and sedimentary origins — the main raw materials for phosphate fertilizers — are dissolved in hydrochloric acid; however, an undecomposed fraction, mostly containing Ti or Zr minerals, remains from apatite concentrates ("Kola phosphates") and must be treated by fusion. On evaporation of chloride solutions of this mineral to dryness, sparingly soluble compounds are formed, containing phosphates of titanium and zirconium, together with small amounts of thorium, uranium, and rare earths. If silicon dioxide is to be determined in phosphates, the fluorides present are bound by adding aluminum chloride, and the decomposition is carried out at a mildly elevated temperature. Decomposition of fluorite concentrates is accelerated analogously.[45] A 0.1-g sample of the raw material containing 90 to 98% CaF_2 is decomposed by 10 mℓ of an 8% $AlCl_3$ solution in HCl (2 + 1). The reaction mixture is evaporated to dryness and dissolved in hot water containing some hydrochloric acid. The aluminum chloride can be replaced by a boric acid solution,[46] e.g., in dissolution of the alkaline earth fluorides doped with rare earths.[47] A 30-mg sample is dissolved by heating with 1 mℓ of 6 *M* HCl saturated with boric acid on a water bath.

The principal ores of boron — colemanites — are dissolved by refluxing with dilute hydrochloric acid, best in a quartz glass apparatus. Datolite concentrates are dissolved similarly. Among sulfates, gypsum and its dehydration product, plaster, are dissolved without difficulties. Anhydrite requires a longer reaction time; refluxing with the acid yields the best results. Synthetic lead(II) sulfate is readily converted into a soluble chlorocomplex by the concentrated acid, and anglesite should, according to phase diagrams, even be dissolved in a 25% solution of sodium chloride.[48] However, X-ray diffraction measurements indicated that anglesite is contained in the insoluble residue isolated from the ore, even when the material was pretreated by boiling with concentrated hydrochloric acid. Barite is highly resistant toward hydrochloric acid, but barytocelestite and celestite are perceptibly attacked (up to 10%).

Destruction of silicate structures in hydrochloric acid solutions leads to precipitation of a hydrated gel of silicon dioxide. Minerals from the zeolite group, some Ni-Mg hydrosilicates (pimelite), copper, zinc, manganese, and magnesium silicates (chrysocolla, dioptase, willemite, hemimorphite, rhodonite, and olivine), and silicates containing rare earths and actinoids (gadolinite, orthite, cerite, and thorite) are dissolved.[49] Repeated evaporation with hydrochloric acid suffices for the purpose. Most cements are degraded by the acid even at laboratory or a mildly increased temperature. Portland-type cements are dissolved with formation of a small residue (usually less than 1%). Aluminate cements are difficult to degrade, and the undecomposed fraction must be fused, e.g., with disulfates or borates. Calcium silicates with the wollastonite structure are dissolved in dilute hydrochloric acid, but most synthetic and natural silicates resist the acid.[44,50]

The solubility of synthetic oxidic compounds with the garnet structure depends on their composition: a 0.1-g sample of a Ga-Gd garnet is dissolved by boiling with concentrated hydrochloric acid for several hours.[51] Garnets doped with chromium are even dissolved in the dilute acid, but the same compounds containing cobalt or nickel must be decomposed at elevated temperature and pressure.[52] Barium titanate is degraded[53,54] even in 6 *M* HCl. Uranium carbides are dissolved in the concentrated acid; irradiated samples require a prolonged reaction time with 3 *M* HCl.[55] Plutonium nitride, PuN, is dissolved with formation of ammonium salts, while uranium nitride, UN, is dissolved slowly, with an addition of a small amount of hydrofluoric acid.[56]

Some chlorides, especially those of As(III), Sb(III), Sn(IV), Se(IV), Ge(IV), and Hg(II), are volatile and can be separated from the reaction mixture by distillation. The volatility of halides can also be utilized in determination of impurities in tin[19] and gallium(III) arsenide.[16] Germanium(II) present in sphalerites is oxidized to the volatile germanium tetrachloride during the dissolution. Materials can be decomposed in a mixture of hydrochloric acid and permanganate and the germanium tetrachloride distilled off using a single vessel.[57] On evaporation of solutions with dilute or concentrated HCl on a water bath or a hotplate, 82 to 100% of Se(IV) compounds added are lost,[58,59] while there are no losses in Te(IV) under the same conditions.[60] However, on evaporation of a solution of Te alloys in a distillation apparatus, tellurium has been found in the condensate.[61] The losses increase on evaporation of the solution to dryness. Losses in indium also occurred under these conditions. The losses are suppressed by adding potassium chloride or zinc chloride (from the dissolved sphalerite).[62] However, it is not clear whether evaporation on a sand bath did not lead to the formation of sparingly soluble compounds. An addition of an alkali chloride considerably decreases volatility of tin tetrachloride. The reactivity of tin(IV) ions and the volatility of tin tetrachloride depend on the hydrochloric acid concentration. At low acidities (less than 0.7 *M* HCl), poorly reactive, low-molecular polymers are formed.[63]

B. Decomposition by Mixtures of Hydrochloric and Nitric Acids

A mixture of hydrochloric and nitric acid has been used as a solvent for the precious metals for more than 1000 years. The ratio of the two acids is adjusted in dependence on the test material composition. The 3 + 1 mixture of hydrochloric and nitric acid is called aqua regia, whereas the mixture with the reversed ratio of the components is called the Leffort or Lunge mixture. The interaction of the two acids is substantially more complex than indicated by the equation,

$$3\,HCl + HNO_3 \rightarrow NOCl + Cl_2 + 2\,H_2O \qquad (3)$$

Both elemental chlorine and the trivalent nitrogen in nitrosyl chloride exhibit oxidizing effects, as do other unstable products.

Gold is dissolved in aqua regia to yield auric salts. Platinum and palladium react very rapidly, especially when finely dispersed,[64] with formation of Pt(IV) salts and unstable chloropalladates that are converted into Pd(II) salts. Compact iridium and rhodium cannot be dissolved in aqua regia, but in Pt-Ir and Pd-Ir alloys, part of the Ir is dissolved in the form of the poorly reactive $Ir\ NOCl_5^-$ ion.[65] Lead buttons obtained by reducing fusion contain the precious metals in the form of intermetallic compounds or very fine dispersions. They can be dissolved in a mixture of hydrochloric and nitric acid,[66] and the solutions are suitable for AAS determinations of Pt, Pd, Rh, Au, and Ag. The mixture of the two acids is a common solvent for various alloys and special steels[67,68] and is suitable for AAS determinations of gold in ores, concentrates, and their treatment products. Minerals containing gold (mostly pure Au or complex tellurides) are readily soluble in aqua regia. The extent of leaching of gold from geological materials primarily depends on the sample weight and the particle size, the contents of reductants, the solvent volume, and the dissolution procedure. The mode of occurrence of the gold and the degree of homogeneity of its distribution within the test material influence the sample weight and the leaching rate. The sample amount is from 1 to 500 g, and large samples are difficult to treat with most ore types. Very fine dispersions of gold in sulfidic minerals are easier to leach, even from coarser fractions than gold trapped in quartz and even when the material is finely pulverized. The dissolution is preceded by oxidizing roasting to oxidize sulfides and carbonaceous substances that exhibit reducing properties,[69-82] at 500 to 650°C for 1 to 3 hr, with a sufficient supply of the air. The sample (with a grain size of less than 200 mesh) is spread in a thin layer over a porcelain dish and is stirred during roasting. If arsenic is present, the sample is first roasted at 480°C and then at 550 to 600°C, to prevent volatilization of the gold, probably in the form of low-melting Au-As alloys.[70,74] The efficiency of the oxidizing roasting can be improved by intimate mixing of the samples with ammonium nitrate (10 g per 25-g sample).[83] Incomplete removal of reductants decreases the recovery in the leaching of gold.[84] To improve the efficiency of the solvent, up to 100 mℓ of aqua regia should be used, or repeated leaching carried out with several portions of the acid mixture.[74,75,78,85] Samples rich in iron oxides are easier to dissolve in hydrochloric acid than in aqua regia; the leaching is completed by adding nitric acid to the reaction mixture only after decomposition of the iron oxides. The insoluble residue is best separated by vacuum filtration[75] or centrifugation,[85] and hot 2 *M* HCl is best suited for the washing.[72] Some procedures for dissolution of rocks and ores of gold are listed in Table 1. The efficiency of the individual operations in decomposition with aqua regia has been repeatedly checked[75,79,81] using ^{195}Au. The results have been compared with those of fire assay or those after total decomposition of the sample. The amount of gold leached usually exceeds 90% of the total content.[70,72,75,78,81,82]

Silver is often determined together with gold. The main problem of the decomposition is completeness of leaching of the silver from the ignited material and the stability of silver salts in aqua regia. It has been shown experimentally that a 10^{-4} *M* Ag^+ solution is sufficiently stable in 2% aqua regia[6] and silver bromide does not precipitate even in the presence of bromine and bromides[5] (for the conditions see Table 1). Conversion of silver chloride into the soluble ammincomplex is not particularly suitable in sample analysis. To prevent precipitation of the hydroxides, a further complexing agent must be added, which may often exhibit undesirable reducing effects.[86,87]

In sample roasting, the silver interacts with the rock components, forming sparingly soluble compounds that are insoluble in aqua regia.[9,80,88,89] Silver can be liberated from most samples only after destruction of the silicate matrix by hydrofluoric acid or by dissolving the material without thermal pretreatment. However, as gold cannot be completely leached from the sample under such conditions, the oxidizing effect of aqua regia is enhanced by adding bromine.[5] Some selected procedures are given in Table 1.

Table 1
DECOMPOSITION OF GEOLOGICAL MATERIALS FOR DETERMINATION OF THE PRECIOUS METALS BY A MIXTURE OF HCl AND HNO_3

Test element and material	Procedure	Ref.
Au; Cu ores	10-g sample roasted at 480 and 600°C for 2 hr, dissolution in 80 mℓ 5 *M* HCl at 90°C for 2 hr; 20 mℓ HNO_3 added and evaporated to dryness	74
Au; rocks and minerals	10-g sample roasted 2 hr at 600°C; gradual leaching with 30 and 20 mℓ of aqua regia; check with ^{195}Au	78
Au; geochemical samples	10-g sample roasted at 600°C; gradual decomposition with 10 mℓ HCl and 5 mℓ HNO_3; heating to NO_x vapors; dissolution of salts in 1 *M* HCl	77, 89
Au; Ag; geochemical samples	10-g sample evaporated with 20 mℓ HF to damp salts; 15 mℓ aqua regia added and evaporation to 7 mℓ	
Au; granites	50-g sample roasted for 2 hr at 600°C; dissolved in 100 mℓ aqua regia; vacuum filtration; check with ^{195}Au	81
Au; rocks	30-g sample dissolved in 70 mℓ aqua regia without roasting; evaporation to dryness and again leached with 40 mℓ aqua regia; the residue centrifuged off; the insoluble residue leached again with 40 mℓ aqua regia	85
Au; antimonites, sulfidic concentrates	2-g sample roasted at 550°C; dissolution in 50 mℓ Lefort mixture; evaporation to 10 mℓ	73
Au; Ag; sulfidic ores and concentrates, rocks	1-g sample ($\leq$0.45 mg Ag) dissolved in 10 mℓ aqua regia with a few drops of Br_2; evaporation to 5 mℓ	5
Au and Pd ores	25-g sample mixed with 10 g NH_4NO_3; gradual increase in temperature and roasting for 30 min at 650°C; gradual dissolution in 60 mℓ HCl and 20 mℓ HNO_3; heating to boiling point and evaporation to damp salts	83
Au; Pt, Pd, Rh; ores, concentrates	25-g sample roasted for 1 hr at 600°C; leaching for 4—12 hr with 100 mℓ aqua regia; evaporated five times with 50 mℓ HCl; residue dissolved in 1*M* HCl	95

Note: If not stated otherwise, concentrated acids are used.

In determinations of the platinum metals, the decomposition technique is determined by the phase composition of the test material. The platinum metals are present in mineral raw materials as alloys, sulfides, tellurides, and arsenides, mostly together with ores of copper, nickel, cobalt, iron, and arsenic. Because the platinum metal minerals rarely occur, reliable data on their decomposability are scarce. Sperrylite, $PtAs_2$, cooperite, PtS, braggite, (Pt,Pd,Ni)S, and many other solids are not dissolved in aqua regia. Roasting in the air should convert most of these minerals into forms soluble in aqua regia.[90] The degree of leaching of the platinum metals from rocks and poor ores varies, and, thus, the insoluble residue after decomposition with aqua regia is fused, e.g., with sodium peroxide.[91-93] Platinum and palladium are most easily dissolved, while the leaching of rhodium is substantially lower. The other platinum metals are minimally dissolved. Samples of rocks and poor ores are first treated with hydrofluoric acid, to degrade the basic matrix.[37,92,94] Analyses of the isolated residues insoluble in aqua regia have shown that with some ore types the dissolution in a mixture of hydrochloric and nitric acid is more efficient than the classical fire assay with lead.[95] The amount of platinum in the aqua regia insoluble fraction varied from 0.8 to 2.0%, that of palladium from 2.1 to 5.2% of the total content; the losses in rhodium amounted to 1.3 to 11%. The smallest fraction is leached from chromites, in which platinum, palladium, and, especially, rhodium are built into the lattice of this resistant mineral. Gold is usually also determined, together with platinum and palladium, in aqua regia extracts.[95-97]

Virtually all sulfides can be decomposed by aqua regia. Pyrite, pyrrhotite, sphalerite, realgar, antimonite, chalcopyrite, and other copper sulfides react rapidly, as do natural and synthetic sulfosalts of heavy metals. Sulfides can be quantitatively converted into the sulfates by oxidation at laboratory temperature (or, better, with cooling of the reaction mixture), which has been utilized for determination of sulfur in pyrite concentrates. The Leffort mixture has given good results in this decomposition, and its oxidizing effects are enhanced by adding 0.1 to 0.3 g $KClO_3$ or, better, 0.5 to 1.0 mℓ Br_2 that facilitate the conversion of any unstable sulfur compounds that might be formed (including elemental sulfur) into sulfates. To 0.25 g of pyrite, 15 mℓ of cooled Leffort mixture are added; the mixture is set aside for 12 hr and then heated for 30 min on a water bath. The solution is evaporated to the formation of damp salts.[98] The same reaction mixture has also been successfully used in determination of selenium in zinc ores.[99] A 0.4-g sample is treated with 32 mℓ of the Leffort mixture, accelerating the decomposition by heating on a water bath for 1 hr. Then 10 mℓ of 11.5 *M* $HClO_4$ is added, and the reaction mixture is evaporated at 150°C until the boiling ceases. The solution is heated for 2 hr with 25 mℓ 12 *M* HCl.

An analogous dissolution procedure has been employed for determination of this element in copper concentrates and tellurium metal.[100] No loss in the ^{75}Se added has been observed. The results obtained are at variance with the data of Bock et al.,[59] who have found a loss of 70 to 100% of $^{75}Se(IV)$ during evaporation of aqua regia solutions on a water bath.

Direct decomposition of galena with aqua regia is unsuitable, as sparingly soluble lead(II) sulfate is precipitated. Cinnabar is dissolved in aqua regia, and the decomposition is carried out at a temperature below 80°C in a beaker covered with a lid on a water bath, to prevent volatilization of mercury compounds. The reaction mixture must not be evaporated to dryness in decomposition of tetrahedrite, cinnabar, and sphalerite; otherwise, part of the mercury volatilizes.

Molybdenite is only slowly dissolved in aqua regia and must be very finely pulverized prior to analysis. One of the causes is the sheet structure of the mineral, exposing mainly the basal plane of the crystal to the acids. The dissolution is fastest using the Leffort mixture containing bromine. This mixture is universally applicable to sulfidic minerals.

Compounds of As(III), arsenides, and sulfoarsenides are rapidly oxidized to arsenates by aqua regia. Here primarily belong the minerals of the safflorite-skutterudite group, niccolite, rammelsbergite, löllingite, arsenopyrite, gersdorphite, and cobaltite.[101] Gallium arsenide is also readily attacked by mixtures of hydrochloric and nitric acids.[102,103]

The main minerals of tungsten ores — wolframite and scheelite — are dissolved in aqua regia. The mineral is usually first dissolved in hydrochloric acid, with precipitation of tungstic acid, and complete precipitation of this acid is attained on an addition of nitric acid, thus decomposing soluble chlorocomplexes of tungsten. Pyrite and other sulfides, if present, are dissolved simultaneously. Oxidic uranium ores with uraninite and synthetic mixed oxides (U_3O_8) are dissolved in aqua regia, with the oxidization of the uranium(IV) to UO_2^{2+} ions. However, this decomposition procedure is insufficient for poor ores; the resistant, insoluble fraction is further attacked, e.g., by fusion with sodium peroxide or borates, or dissolution in a mixture of hydrofluoric, nitric, and perchloric acids. Chemically prepared barium sulfate and barite are only little attacked by aqua regia, and the latter is, rather, used for their isolation from accompanying phases. Ammonium sulfate is added to the solution to suppress the solubility of barite.[104] Celestite is substantially more soluble under these conditions, and the dissolved fraction may attain 20% of the total amount, especially with synthetic preparations.

With polyphase materials of varying composition (soils, river and sea sediments, and waste sludges), the contents of elements are determined in acid extracts primarily for the purposes of environmental protection and geochemical prospecting. The selection of the element association depends on the purpose of the analysis. The acid mixture is selected so that the extraction is rapid and efficient. To check the decomposition efficiency, the samples are dissolved in various ways in parallel, including total destruction of the test material.[105]

A mixture of hydrochloric and nitric acids is a proven solvent. Leaching is carried out both at laboratory and an elevated temperature, best by refluxing.[106-112] A 2-g soil sample or a 3-g sludge sample is mixed with 2 mℓ of water and dissolved in 7.5 mℓ of 6 *M* HCl and 2.5 mℓ of concentrated HNO_3. The reaction mixture is set aside for 16 hr and then is refluxed for 2 hr.[111] The degree of leaching of Cu, Cd, Cr, Mn, Pb, Zn, etc. is checked by reference analyses after total destruction of the material (fusion with $LiBO_2$ or decomposition by hydrofluoric acid with other acids) and by parallel analyses of standard (SRM) samples. As further solvents, hydrochloric acid, sulfuric acid with nitric acid, or nitric acid alone are mostly used. The test materials always contain a large fraction of organic matter that is only partially oxidized during dissolution. The extracted fraction of the metals varies from 50 to 90%, and the poorest efficiency is exhibited by aqua regia for chromium and nickel and sometimes also for lead. The acid extract recovery is higher for waste sludges than for soils, where it depends on the genetic type of the soil and the quality of the rock bed.[111] Extractability of some rocks and rock-forming minerals has been studied analogously and amounts to up to 70% of the total content for biotite, amphibole, pyroxene, and Cu, Pb, Zn, and Ni. The procedures used for soils and geological materials, important for geochemical prospecting, have been evaluated by Fletcher in his book.[112]

The metal contents in waste sludges are a criterion for a further use of the sludge in agrochemistry. Analogous extracts of soils, sediments, and rocks have been obtained for determination of arsenic, antimony, and mercury. In view of the volatility of arsenic and mercury compounds, the test material is dissolved either at laboratory temperature or at a slightly elevated temperature (water bath), usually under reflux.[27,28,113-117] The degree of extraction of these components approaches that obtained after total decomposition. The data obtained are evaluated both from the point of view of geochemistry and from that of environmental protection. For the same purposes, lead is determined in soils close to highways.[31] The material is leached by cold aqua regia to prevent the formation of a gel of silicon dioxide which blocks dissolution of lead. A 0.1-g sample is dissolved in 2 mℓ of aqua regia. After 16 hr the solution is heated and evaporated to a wet residue. Then 2 mℓ of 0.1 *M* EDTA is added, and the solution is diluted to 10 mℓ after 2 hr. The lead content is determined by anodic stripping voltammetry. Molybdenum compounds cannot be quantitatively leached from soils by aqua regia, but the efficiency is sufficient for differentiation of the regions with anomalously increased contents of the element.[118] An analogous procedure has been used for determination of thalium.[119]

C. Decomposition by Nitric Acid

Nitric acid is similar to aqua regia in its oxidizing properties. The main reduction products are nitrogen oxide and nitrous acid (sometimes also hydrogen or ammonia). Nitrogen oxide is rapidly oxidized to nitric dioxide in the air, and the latter dimerizes. Commercial solutions of nitric acid have compositions close to that of the azeotropic mixture (69.2% HNO_3, i.e., 15.5 *M*, bp 121.8°C). The acid with a concentration higher than 85% is sometimes used for dissolution and is termed *fuming nitric acid*. It is unstable, photosensitive, and highly reactive. The acid may be manipulated only by using protective gloves and goggles, in a properly operating fume cupboard. The complexing properties of nitric acid are substantially less pronounced than those of hydrochloric acid and are primarily utilized in ion exchange or extraction separations of the actinoids and the lanthanoids from accompanying metals.

Solids are usually decomposed in vessels made of glass, porcelain, quartz, platinum, Teflon®, and glassy carbon. Both the acid alone and its mixtures are used, usually with HCl, $HClO_4$, H_2SO_4, and HF. The oxidizing properties of nitric acid are enhanced by adding chlorate, permanganate, hydrogen peroxide, and bromine. Nitric acid is a good solvent for many metals (Cu, Pb, Zn, Cd, Mo, etc.) and their alloys. Volatile hydrides may be formed from arsenides and phosphides present in the test metals.[120] Therefore, decomposition is carried out in the presence of potassium permanganate, gradually increasing the reaction mixture temperature. A 0.4-g sample of a Cu-Ni alloy is dissolved in 10 mℓ of HNO_3 with 1 mℓ of 0.1 *M* $KMnO_4$. After 30 min, the solution is heated and evaporated down to 1 mℓ on a water bath. The reaction mixture is heated with 2 mℓ of water to remove nitrogen oxides. Nitric acid alone is not particularly suitable for iron metal and most ferroalloys.[121] Ferromolybdenum is dissolved in nitric acid,[122] and the reaction mixture is evaporated with sulfuric acid to the appearance of SO_3 vapors.

Among the precious metals, only silver and palladium are dissolved in nitric acid. Silver can be leached from an alloy with gold only when it is present in an at least fourfold excess. This dissolution procedure (parting) is used for precious metal beads obtained in fire assay, with nitric acid diluted at ratios from 1 + 4 to 2 + 1. Isolated undissolved gold often contains silver as an impurity. Alloys of Ag with Pd, with an excess of silver, are most easy to dissolve, whereas the presence of platinum, iridium, and ruthenium makes the dissolution of palladium more difficult.[123,124] Nickel metal containing 10^{-4} to 10^{-1}% Ir, Ru, and Os is attacked by nitric acid, while an analogous Al alloy is decomposed with formation of a black residue containing the platinum metals.[125]

Some industrially important metal compounds — silicides,[126] borides,[127] and nitrides[128] — can be dissolved in nitric acid, e.g., the compounds Mo_5Si_3 and VN. The vanadium is partially oxidized to the pentavalency during the dissolution. Boron phosphides are attacked by a mixture of fuming nitric and sulfuric acids, and elemental α-boron can also be dissolved in this way. Boron carbides and cemented carbides are attacked minimally.[127,129] Dissolution of the carbides UC and PuC in HNO_3 is connected with liberation of graphitic carbon and the formation of many organic compounds; hot 8 *M* HNO_3 should then be used, as graphite and the solid organic compounds are not formed.[55,56]

Decomposition of thorium and uranium dioxides is catalytically accelerated[130,131] by adding 0.05 to 0.1 *M* HF. A 20-g sample of burnt ThO_2-UO_2 fuel is refluxed with 300 g of 13 *M* HNO_3 and 0.05 *M* HF for 12 to 40 hr. A solid solution of the mixed oxides $(Pu,U)O_2$ behaves analogously.[55,56,132] Plutonium dioxide, ignited at temperatures below 800°C, is also decomposed under these conditions. An addition of sulfuric acid to the reaction mixture accelerates the decomposition of a preparation ignited at 1500°C.[55,133] Up to 90% of the test oxide is dissolved in a mixture of 12 *M* HNO_3, 0.1 *M* HF, and 0.05 *M* H_2SO_4. Fluorides and ceric salts have similar effects as sulfuric acid. Pure ceric oxide is dissolved in nitric acid only with continuous addition of hydrogen peroxide. Nitric acid with a few drops of hydrofluoric acid is more advantageous, as a ceric salt solution is formed.[134] To 1 g of CeO_2, 7 mℓ of concentrated HNO_3, and 0.1 to 0.2 mℓ of HF are added. The reaction mixture is heated until a clear solution is obtained.

Nitric acid alone is not suitable for dissolution in total analyses of hematite, pyrolusite, wad, and manganite, but the 0.3 *M* acid has been used as a selective solvent in phase analysis for determination of iron and manganese bound in carbonates. These components can also be determined in the presence of the above oxidic minerals.[135] Dilute 5 to 7 *M* nitric acid containing hydrogen peroxide dissolves higher Mn oxides, and this has been utilized in analyses of Mn nodules and synthetic Mn(III) and Mn(IV) oxides.

Nitric acid has found principal use in analyses of sulfidic minerals. Sulfates are formed during dissolution, together with varying amounts of elemental sulfur and further compounds that are unstable under the given conditions. The amount of sulfur liberated varies in dependence on the kind of the test material and on the dissolution procedure. The reaction is substantially faster at higher temperatures, but the amount of the sulfur liberated increases. Rhombic marcasite reacts with precipitation of sulfur, whereas pyrite with the same chemical composition is dissolved almost without a residue. This decomposition procedure is unsuitable for determination of sulfur. The same hold for galena, with which sparingly soluble lead(II) sulfate is formed. Pyrrhotite, sphalerite, chalcosite, chalcopyrite, and bornite are dissolved without difficulties. Natural and synthetic selenides and tellurides are oxidized by nitric acid to the selenates and tellurates. Se(VI) and Te(VI) compounds are not formed, even when nitric acid is mixed with bromine or HCl, as demonstrated in detailed studies[136,137] employing ^{75}Se. On prolonged evaporation of the mixture with perchloric acid, selenate should partially be formed;[138,139] however, the valence states in the Se compounds have not yet been unambiguously clarified (participation of catalytic effects?). The first stage of the decomposition proceeds at laboratory or a mildly elevated temperature, to prevent separation of elemental sulfur and consequent losses in the selenium.[140-142] The oxidation of sulfides is also catalyzed[143] by adding 0.1 g of KI. The following procedure has been recommended for determination of selenium in Cu-Zn ores. A 0.5- to 1-g ore sample is dissolved in 15 mℓ of a cold, 7 + 3 mixture of nitric and perchloric acids. At the end of the reaction, the temperature is gradually increased, and the reaction mixture is evaporated several times to the appearance of perchloric acid vapors.[142] A solution obtained by dissolution of semiconductor crystals of compounds of the type $Me^{II}Cr_2S_4$ or lead(II) telluride has been used for determination of admixtures in these physically important materials.[144,145] The degree of extraction of tellurium compounds from silicate rocks depends on the phase composition of the rocks[146] (only the rare ferrotellurite, $FeTeO_4$, is insoluble in nitric acid) and on the degree of weathering of the rocks.[147] For some rocks, leaching with nitric acid and evaporation with perchloric acid suffice, but with granodiorite GSP 1 required preliminary destruction of the silicate by hydrofluoric acid.

The presence of the residues of nitric acid and its decomposition products interferes in most analytical procedures (e.g., in hydride generation) and, thus, they must be removed, usually by repeated evaporation with sulfuric or perchloric acid. More simply, these oxidizing compounds are removed by evaporation with formic acid.[146] These products must be removed not only when decomposing by nitric acid, but also when aqua regia is used. In generation of the selenium and arsenic hydrides, residues of nitrous acid especially interfere[148] and are converted into inert nitrogen by adding sulfamic acid,

$$NO_2^- + HSO_3NH_2 \rightarrow N_2 + HSO_4^- + H_2O \quad (4)$$

The remaining nitrates react more slowly and nitrous oxide is formed

$$NO_3^- + HSO_3NH_2 \rightarrow N_2O + HSO_4^- + H_2O \quad (5)$$

Nitrous acid also reacts with sulfanilamide with formation of a diazonium salt that does not interfere in the hydride generation.[149] Decomposition with a mixture of nitric and sulfuric acid is complicated by the formation of nitrosylsulfuric acid. The nitronium cation NO_2^+ is formed, first in a mixture of the concentrated acids and then as a decomposition product, nitrosyl NO^+, which also retains oxidizing properties. Nitrosylsulfuric acid is decomposed by dilution and repeated evaporation. Analogous phenomena occur in the presence of perchloric acid. Oxidizing products can also be eliminated[150] by adding 0.5 g of urea and 10 mℓ of HCl.

A mixture of nitric and sulfuric acid has long been used for laboratory dissolution in determination of arsenic. The oxidation is enhanced by adding a bromine solution in carbon tetrachloride.[151] To a 1-g ore sample, 10 mℓ of 20% Br_2 in CCl_4 and 15 mℓ of HNO_3 are added. The solution is set aside and then slowly heated until the bromine volatilizes. Then 10 mℓ of HCl and 25 mℓ of H_2SO_4 (1 + 1) are added, and the cold solution is heated to disappearance of nitrogen oxides. This procedure has been successful in determination of arsenic in Pb, Cu, Zn, and Mo ores, concentrates, and matte. The arsenic is oxidized to the pentavalency, but no loss in arsenic has been observed, even in the presence of bromine. The same oxidation procedure has given good results in determination of Ag, Sb, Bi, Cu,

Cd, and In in the above raw materials. However, the final stage of the decomposition procedure differs in dependence on the kind of the test material and the elements to be determined.[12,152] Decomposition with a mixture of nitric and sulfuric acid has also been found highly effective in leaching of arsenic from soils.[27,153,154] It is more reliable than a pressure decomposition. According to Reference 27, it is universally applicable to various soil types. A 0.5-g sample is heated in a conical flask with 20 mℓ of HNO_3 and 10 mℓ of H_2SO_4 (1 + 1). The reaction mixture is evaporated to the appearance of SO_3 vapors, several drops of nitric acid are added, and the whole procedure is repeated until a clear solution is obtained. The solution is then diluted to 100 mℓ, and arsenic is determined by AAS.

Nitric acid dissolution presents some problems in determination of antimony and tin, as sparingly soluble, surface-active products are formed, called "metaantimonic" and "metastannic" acids. Precipitation of these compounds is prevented by adding tartaric or citric acid. Even then, an insoluble phase rich in antimony separated in decomposition of 0.3 g of antimonite by nitric acid with 2 g of tartaric acid.[155] The precipitate has not appeared with lower Sb contents (tetrahedrite, tenantite).[156] A mixture of nitric and sulfuric acids does not oxidize antimony completely to the pentavalency, in contrast to arsenic. A varying amount of a nonreactive product with the composition Sb_4O_8 is formed and contains Sb(III) and Sb(V) at a constant ratio.[157,158] The formation of this compound causes a negative error wherever a uniform valence of antimony is required. The compound cannot be oxidized by chlorate, permanganate, and perchloric acids, or reduced by stannous chloride, sulfite, and hydrazine, but it can be converted into antimony(V) compounds by aqua regia[158] and then reduced to Sb(III), if required.

High-percentage molybdenite ores and concentrates can be dissolved in a mixture of nitric and sulfuric acids,[122,159,160] but, at present, dissolution in nitric and hydrobromic acids is preferred, with the oxidizing effect of the nascent bromine. A 0.3-g sample of molybdenite is oxidized with 15 mℓ of HNO_3 and 5 mℓ HBr. The reaction mixture is allowed to stand for several hours, is heated, 5 mℓ of H_2SO_4 (1 + 1) are added, and the mixture is evaporated to the appearance of SO_3 vapors. A similar effect is exerted by nitric acid in the presence of a 20% bromine solution in carbon tetrachloride.[161] Dissolution in a mixture of nitric acid and potassium chlorate has also given good results.[158] In the interaction, explosive chlorine dioxide is formed in addition to chlorine.

Nitric acid media are not particularly suitable for determination of silver. Chloride impurities, introduced into the reaction mixture with the reagents added and the vessels used, cause uncontrollable precipitation of silver chloride adhering to the vessel walls and the insoluble residue. If mercuric nitrate (*circa* 30 mg Hg^{2+}) is added to the acid, the silver chloride precipitation is prevented.[155] Cinnabar is not attacked by nitric acid solutions. If this decomposition procedure is selected for environmental materials (sediments, waste sludges, and fly ashes), cinnabar must not be present in the test sample. Mercuric sulfide is not dissolved, even in the presence of strong oxidants ($KMnO_4$, $K_2S_2O_8$), but the oxidants facilitate destruction of organometallic compounds. However, cinnabar is dissolved on addition of a minimal amount of hydrochloric acid.[162,163] In view of the volatility of mercury compounds, the temperature is controlled during the decomposition. With claystones, no difference in the results has been observed for decompositions with a mixture of nitric and sulfuric acids and with the same mixture containing also hydrochloric acid. However, a mixture of nitric and sulfuric acids did not liberate all the mercury from pyrite concentrates and thus the material was decomposed with an addition of hydrochloric acid. A 0.2-g sample of a sediment or concentrate is dissolved with cooling in 15 mℓ of a mixture of HNO_3 and H_2SO_4 (2 + 1). After the reaction, 2 mℓ of HCl is added and the reaction mixture is heated for 2 hr at 50 to 60°C with constant shaking. The solution is cooled and oxidizes with 10 mℓ of 6% $KMnO_4$ overnight. The excess oxidant is reduced by 10 mℓ of 6% hydroxylammonium chloride in 6% NaCl. The undecomposed residue is centrifuged off.[162]

Decomposition of soils with nitric and sulfuric acids, with an addition of permanganate, has been useful in determination of mercury in soils.[164] The results obtained did not differ from those obtained after dissolution in aqua regia. A mixture of nitric acid with vanadium pentoxide has been used for the same purpose.[165]

Dilute 7 *M* nitric acid dissolves natural phosphates analogously as hydrochloric acid of the same concentration. The rare earth elements are partly dissolved. Natural and synthetic aluminum phosphates can also be dissolved without problems. Xenotime and monazite resist nitric acid, but rhabdophan (a hydrated rare earth phosphate) is readily soluble in dilute nitric acid. Oxidic uranium and tungsten ores behave in nitric acid solutions in the same way as in aqua regia. Fluorite is dissolved slowly, and synthetic aluminum fluoride is resistant toward nitric acid. Freshly precipitated rare earth fluorides can be dissolved in a mixture of nitric and perchloric acids, but their solubility decreases on recrystallization.

The main problems connected with leaching of metals from polyphase materials are described above.[44,106-112,166-168] These procedures have mostly been employed as screening methods for rapid evaluation of impurity contents and of decomposability of samples with acids.

III. DECOMPOSITION BY HYDROBROMIC OR HYDROIODIC ACID

The principal use of hydrobromic acid as a solvent for solids is based on the properties of the acid. Pure hydrobromic acid, without degradation products, exhibits reducing properties during dissolution of substances. On the other hand, in the presence of the product of its oxidation (elemental bromine), some of its effects are similar to those of aqua regia. The acid is commercially available in the form of an aqueous solution whose composition corresponds to the azeotropic mixture with water (47.6% HBr, bp 124.3°C). The acid has found use in separation techniques, in ion exchange chromatography, and especially in the extraction of bromide complexes of Au(III) into oxonium solvents; this procedure is substantially more selective than the extraction from a hydrochloric acid medium.

Hydrobromic acid has been used for decomposition of iron oxides, especially magnetite.[169] In our experience, it is the best solvent for cerium dioxide preparations that are freed of water and carbon dioxide by ignition at 950 to 1000°C. A defined stock solution of a cerous salt can be prepared in this way:[170,171] a 1-g sample of CeO_2, wetted with water, is dissolved in 10 mℓ of HBr. The solution is evaporated to a volume of a few milliliters. Then 5 mℓ HCl and 5 mℓ of 30% H_2O_2 are added, and the bromine formed is expelled from the reaction mixture by heating. Ceric ions are reduced to cerous ones by hydrobromic acid:

$$2\,CeO_2 + 8\,HBr \rightarrow 2\,CeBr_3 + 4\,H_2O + Br_2 \qquad (6)$$

Hydrobromic acid also reduces perrhenate formed by fusion of geological samples with a mixture of sodium peroxide and hydroxide; heating of the solution with 8 *M* HBr under a reflux leads to quantitative reduction to Re(IV) bromide.[172]

Hydrobromic acid has been especially suitable as a solvent for materials with high contents of arsenic, antimony, and tin, as well as those with mercury, germanium, and selenium. The bromides of these metals (including those of pentavalent arsenic and antimony) volatilize during decomposition, and, thus, their possible interference in determinations of trace elements is prevented. The ore or alloy is dissolved in hydrochloric acid with an addition of nitric acid, or in aqua regia. Then 10 mℓ of HBr is added to the solution, together with 10 mℓ of $HClO_4$, and the reaction mixture is evaporated to obtain a thick paste. If the above elements are present in excess, the evaporation is repeated after another addition of hydrochloric, hydrobromic, and perchloric acids. The procedure is suitable, e.g., for a determination of lead, cadmium, and iron in ores containing arsenopyrite and stannite or antimony sulfides. It has also been used to eliminate the effect of arsenic in a spectrophotometric determination of phosphorus in iron ores.[173] If the temperature is not controlled during the evaporation, bismuth(III) bromide may also volatilize.[174] In evaporation of solutions containing hydrobromic and sulfuric acids with sodium sulfate, the loss in ^{210}Bi is very small, only 2.5%, but it can increase up to 35% when the amount of sulfuric acid is insufficient. Tin admixture is removed in this way during preparation of radiochemically pure ^{210}Bi.[175]

Solutions of hydrobromic acid with bromine are outstanding solvents for sulfides, selenides, and tellurides, with formation of the sulfates, selenites, and tellurites.[176-178] For determination of tellurium, a 2.5-g sample of a rock, soil, or a stream sediment is mixed with 15 mℓ of the mixture (20 mℓ Br_2 in 1000 mℓ HBr). Samples with macroscopically perceptible sulfides are oxidized by adding another 2 mℓ of Br_2. After the reaction the mixture is heated for 30 min at 140°C, the bromide complex of Te(IV) is extracted into methylisobutyl ketone and determined by AAS.

Similar to aqua regia, a mixture of hydrobromic acid with bromine also dissolves minerals of gold.[179-182] A 10-g rock sample is ignited for 2 hr at 600°C and leached by 20 mℓ of the mixture (5 mℓ Br_2 in 1000 mℓ HBr) for 15 to 30 min in vigorously stirred cold solution. The reaction mixture is diluted with water, and tetrabromoaurate is extracted into methylisobutyl ketone, even in the presence of an undecomposed fraction. The procedure[180] represents a modification of the method of Ward et al.[179] The decomposition efficiency (more than 90% of the Au is leached) has been verified by comparing the results with those of fire assay. An analogous leaching procedure has also been employed for determination of platinum and palladium in rocks.[183] However, platinum and some of its alloys with metals resist hydrobromic acid mixed with bromine. On the other hand, powdered rhodium, freed of the surface oxide layer by reduction in a hydrogen atmosphere, is dissolved.[184]

Hydroiodic acid has only rarely been used for dissolution of solids. It acts as a strong reductant and reduces ferric ions to ferrous ones in acidic solutions. Solutions of the acid are readily oxidized by atmospheric oxygen, and the iodine liberated reacts with iodide to form the I_3^- ion. The acid is commercially available usually as a 55 to 57% aqueous solution, corresponding to a constant-boiling mixture (bp 127°C). It is mostly stabilized by an addition of hypophosphorous acid.

The most important application of this acid in decomposition techniques is the determination of sulfur and of its isotopic composition in barites, some oxides, and metals. Barite reacts in a medium of hydroiodic, hydrochloric, and hypophosphorous acids; the latter reduces the iodine produced by the oxidation to iodide. Baryum sulfide, which is easily decomposed by acids, is the reaction product.

$$BaSO_4 + 8\,I^- + 10\,H^+ \rightarrow H_2S + Ba^{2+} + 4\,I_2 + 4\,H_2O \qquad (7)$$

In view of a possibility of sulfur isotope fractionation during the decomposition, more than 95% of the mineral must be decomposed.[185] A 0.2-g baryte sample is decomposed in a distillation flask with a reflux by 50 mℓ of the mixture (500 mℓ HI, d = 1.7; 816 mℓ HCl, and 245 mℓ 50% H_3PO_4) in a nitrogen stream. The hydrogen sulfide liberated is collected in a solution of cadmium acetate.[185,186] Hydrogen sulfide can be quantitatively displaced by 50% HI from natural and synthetic sulfides of Pb, Zn, Co, Ni, and Ag, at laboratory or a mildly elevated temperature.[187] Under these conditions, pyrite and chalcopyrite are only slowly dissolved. Only 14% of the pyrite sulfur can be converted into H_2S. The reaction is accelerated by adding a droplet of mercury to the reaction mixture.[188] A droplet of mercury and 10 mℓ of 50% HI are added to a 0.1-g mineral sample. The decomposition is carried out with mild heating in a distillation flask in a hydrogen stream that transports the hydrogen sulfide to absorption vessels containing an aqueous suspension of cadmium hydroxide. In addition to pyrite and chalcopyrite, realgar and auripigment are also decomposed. The procedure has given good results in determination of sulfur in soils with pyrite.[189]

The reducing medium of hydroiodic, hypophosphorous, and hydrochloric acids has also been found useful in determination of traces of sulfur in pure arsenic and tellurium. The test material is first dissolved in hydrochloric acid with bromine. The bromine is then removed and only then the reducing acid mixture is added. The reduction to elemental arsenic and tellurium does not prevent quantitative liberation of the hydrogen sulfide.[190]

Under certain conditions, even cassiterite surface is corroded by hydroiodic acid, but the mineral is inert to short-time exposure. Tin bound in silicate minerals can be converted into the reactive oxide by the action of hydroiodic and nitric acids, followed by evaporation with sulfuric acid. Tin tetraiodide, extractable into benzene, is already formed after a 5-min interaction with hydroiodic acid. This procedure has been utilized to differentiate between the tin bound to silicate minerals (mostly micas) and cassiterite.[191]

IV. DECOMPOSITION BY HYDROFLUORIC ACID

Hydrofluoric acid belongs among the solvents that are most extensively used in inorganic analysis. Although it is one of the hydrogen halide acids, its effects during dissolution of solids are quite different and cannot be compared with the effect of other inorganic acids. From the point of view of dissociation of hydronium ions,[192,193] it is a weak acid ($pK_a \doteq 3.1$ at 25°C). Its aqueous solutions contain F^- ions that react with undissociated HF molecules in concentrated solutions, with formation of HF_2^- anions. In laboratories, the acid is commonly used in the form of 38 to 48% aqueous solutions (21.5 to 29 *M* HF) that are stored in polyethylene, polypropylene, or PTFE bottles provided with tight screw-on caps.

Gaseous hydrogen fluoride is readily soluble in water; the hydrofluoric acid formed gives rise to a constant-boiling azeotropic mixture with water (38.3% HF, bp 112°C at 0.10 MPa). The purity of the acid depends on the preparation method and on the storage conditions. In view of the low boiling point of the azeotropic mixture, the acid can be simply purified by distillation in a PTFE apparatus (Chapter 3), thus obtaining solutions with minimum amounts of impurities (especially Pb^{2+}) that can be used for highly demanding analyses of, e.g., extraterrestrial materials.

Hydrofluoric acid and fluorides exhibit pronounced complexing properties that can be used in all fields of analytical chemistry. Hydrofluoric acid prevents the formation of sparingly soluble hydrolytic products in solution, especially of compounds of the elements from the IVth to VIth groups of the periodic system. Soluble hydrolytic products that are often polymeric depolymerize in the presence of fluoride, with formation of reactive monomeric species suitable for further analytical operations. In this way, it is possible to avoid the formation of colloidal solutions and increase the stability of solutions even with compounds of the elements that are readily hydrolyzed in aqueous solution (e.g., Si, Sn, Ti, Zr, Hf, Nb, Ta, and Pa).

Mixed with an oxidizing acid (most often HNO_3), hydrofluoric acid hastens dissolution of tungsten alloys and prevents the formation of sparingly soluble tungstic acids. Metallic Zr, Hf, Ti, Nb, and Ta, as well as their alloys, are dissolved analogously. The solutions obtained contain the above metals in the form of negatively charged fluoride complexes that are suitable for their separation in a column with a strongly basic anion exchanger. As the eluting agent, mixtures of hydrofluoric and hydrochloric acids are mostly used, with a variable concentration of ammonium fluoride or chloride.

Hydrofluoric acid does not exhibit oxidizing effects, but affects the formal oxidation-reduction potential through the formation of complexes with different stabilities with the components of a redox couple. These changes are pronounced, e.g., in the Fe^{3+}/Fe^{2+} system, where the oxidation-reduction potential shifts[193] from 0.77 V (NHE) to 0.10 V (0.5 *M* NH_4F

and 0.5 *M* HF). Similar changes have been found in the Mn^{3+}/Mn^{2+}, Sn^{4+}/Sn^{2+}, and UO_2^{2+}/U^{4+} systems. This decrease in the oxidizing strength of ferric salts has been utilized in numerous analytical applications, e.g., in iodometric titrations. In polarographic determinations of many metal ions, the interference of ferric ions is eliminated by an addition of fluoride to the test solution, causing a shift of the half-wave potential of the ferrifluoride complex wave toward negative values.

In strongly acidic solutions of hydrofluoric acid, an addition of HF prevents the formation of colloidal solutions of niobium salts during reduction in metal reductors, thus permitting quantitative reduction of Nb(V) to Nb(III) compounds.[193] Analogously, tungsten and molybdenum compounds can quantitatively be reduced to the corresponding tungsten(III) and molybdenum(III) salts and determined by titration.

Fluorides have also found use in extraction spectrophotometric determinations of tantalum and boron, based on the formation of intensely colored ion associates of TaF_6^- and BF_4^- with the cations of basic dyes.

The dissolution of solids in HF solutions leads to the formation of soluble fluorides, accompanied by the formation of sparingly soluble compounds (the fluorides of the alkaline earths, rare earths, and actinoids). Although the formation of sparingly soluble fluorides causes many complications in analyses of various materials (e.g., the determination of silicon dioxide in isolated oxides of the lanthanoids or in the determination of aluminum), the procedure is still used to separate the rare earth elements (actinoids) from the major components that interfere in the final determination. A serious problem in analyses of various materials is the undesirable formation of complex, sparingly soluble aluminum fluorides.

It should be pointed out that uncontrolled presence of fluorides in the test solution may often complicate the whole subsequent procedure. During the storage of the solutions in glass vessels, the vessel walls are corroded, and the liquid phase is contaminated with the glass components, especially silicon dioxide, the alkali metals, boric acid, etc. A number of reactions (e.g., the formation of colored compounds of the Zr^{4+}, Ti^{4+}, Al^{3+}, Be^{2+}, and other cations with some reagents, some chelatometric titrations, and some separations) are hindered by fluorides to such an extent that they cannot be used for analytical purposes.

The basic property in which hydrofluoric acid differs from the other mineral acids is the interaction with silicon dioxide, connected with the formation of fluorosilicic acid or silicon fluoride. The reaction is usually described by the simplified equation

$$SiO_2 + 6\,HF \rightarrow H_2SiF_6 + 2\,H_2O \qquad (8)$$

The fluorosilicic acid formed dissociates into gaseous silicon fluoride and hydrogen fluoride on heating:

$$H_2SiF_6 \rightarrow SiF_4 + 2\,HF \qquad (9)$$

This principle is used not only for the determination and the preparation of standard solutions of silicon dioxide, but, also, in the presence of oxidants, for dissolution of elemental silicon and its alloys. On treatment with hydrofluoric acid, isomorphously bonded admixtures and components contained in liquid or solid inclusions are also liberated from quartz and silicates. The reaction between the acid and silicon dioxide always yields the same products, but the rate of dissolution depends on the structure of the test oxide. The most common, stable, low-temperature modification of silicon dioxide, α-quartz with the ordered crystal structure, reacts more slowly than a quartz glass formed by glass-like, amorphously solidified silicon dioxide. The highest resistance toward the effects of hydrofluoric acid is exhibited by high-pressure, polymorphous modifications of SiO_2 — the monoclinic coesite and the tetragonal stishovite that are formed in rocks during a shock metamorphism under high pressures, e.g., in meteoric craters. Because of their low reactivity with dilute hydrofluoric acid, they can be separated from quartz and silicates that accompany them in rock materials.[194,195]

A number of works[196-199] have dealt with the kinetics of dissolution of silicon dioxide. The dissolution rate was studied using a quartz glass plate suspended in hydrofluoric acid solutions.[5] At low HF concentrations, the dissolution rate is proportional to the acid concentration. An increase in the rate at higher concentrations has been ascribed to the effect of the HF_2^- ions that predominate in the solution under the given conditions. The temperature dependences of the dissolution rate have been found to be the same for all the concentrations studied. A two-step reaction mechanism is assumed,[196] the first step being hydration of the glass surface:

$$SiO_2 + 2\ H_2O \rightarrow Si(OH)_4 \tag{10}$$

with the formation of a compound that is soluble in HF.

$$Si(OH)_4 + 4\ HF \rightarrow SiF_4 + 2\ H_2O \tag{11}$$

Low values of the activation energy and the experimental temperature dependences indicate that the reaction rate is controlled by the transport of the products from the glass surface into the solution.

The dissolution procedure, serving for dissolution of silicon dioxide or silicates, depends on the purpose of the analysis. If silicon dioxide is to be determined together with other components, volatilization of silicon tetrafluoride during the decomposition must be prevented. The HF-H_2SiF_6-H_2O system formed on dissolution of silicon dioxide in hydrofluoric acid (with a substantial simplification of the reaction mechanism)[200] forms a constant-boiling ternary mixture containing 10% HF, 36% H_2SiF_6, and 54% H_2O (bp 112°C, 0.10 MPa). The equilibrium diagram of the liquid — vapor system indicates that this ternary mixture, whose composition, close to the region of constantly boiling HF, is in equilibrium with a gaseous phase in which the H_2SiF_6 concentration is substantially lower than in the liquid.[201] The diagram demonstrates the theoretical conditions for decomposition of a material without silicon tetrafluoride volatilization. Solutions of fluorosilicic acid can also be heated in open systems, with HF and H_2O volatilizing until the above composition of the ternary mixture is attained, with 36% H_2SiF_6. Losses in silicon dioxide only occur in this stage.

Parallel experiments with decompositions in open and closed vessels have confirmed that this procedure is suitable for determination of silicon dioxide in, e.g., cement, slags,[202,203] and other products.[204-207] The volatility of fluorosilicic acid and silicon tetrafluoride has been utilized for separation of silicon from complex mixtures. The composition of the individual components and their mutual ratios in the gaseous phase are much more complicated[208] than indicated by the above study.[201] Distillation of silicon tetrafluoride is usually carried out in a mixture of hydrofluoric acid and perchloric, phosphoric, or sulfuric acid.[209-211] Under suitable reaction conditions, the volatilizing silicon tetrafluoride can be selectively trapped in the form of potassium fluorosilicate in a mixture of solid potassium nitrate and perchlorate (1 + 1) placed in a special Teflon® vessel. Vapors of hydrofluoric acid are, under these conditions (heating in an aluminum block at 200°C), also trapped in the crystalline layer, but are then displaced by nitric acid vapors.[212] The residue in the vessel can be dissolved in hydrochloric acid and used, e.g., for determination of main rock-forming elements.

Quartz used for the manufacture of quartz optics is analyzed after dissolution in hydrofluoric acid, without pulverizing small lumps of the material, to avoid contamination of the sample during the preparation procedure. Prior to the decomposition, the surface of the lumps is briefly rinsed with dilute hydrofluoric acid (to which hydrochloric acid may be added) and thoroughly washed with distilled water. Because of the large samples used (up to 5 g), the first step of dissolution is carried out, after wetting the sample with water, in the cold acid, in a Teflon® vessel placed in a fume cupboard in which the purity of the atmosphere is controlled. The decomposition is completed by heating and evaporating the reaction mixture.

The determination of the major contaminant of glass sands, ferric oxide, involves special problems in the decomposition step. The sample must be completely dissolved, as glass sands from certain localities contain compounds of iron (and sometimes other color-forming oxides) that are bound to resistant minerals which are not decomposed by hydrofluoric acid, such as tourmalines, chromite, staurolite, etc. In most analytical procedures, it is thus recommended to start the decomposition of quartz in hydrofluoric acid in the presence of sulfuric or perchloric acid and to complete the decomposition by fusing the residue after evaporation with a mixture of an alkali carbonate and borate, disulfate, etc. in the same platinum dish. The principal source of error of these methods is contamination of the reaction mixture from the laboratory atmosphere and the fusion process. The melt reacts with the walls of the platinum vessel, from which fluctuating amounts of iron are transferred into the melt. This decomposition procedure cannot be recommended for demanding analyses; dissolution in a closed system should be preferred.

Dissolution of substances in hydrofluoric acid is most often applied to analyses of natural and industrial silicates. The acid destroys the silicate bonds, and the individual components are transferred to the solution in the form that is suitable for the subsequent analytical treatment. Solutions free of silicon dioxide are obtained. Numerous interferences, caused by uncontrolled precipitation of gel-like species or by the presence of silicon atoms and radicals in plasma or flame spectrometry, are thus avoided.

The decomposition procedure underwent several stages in development. In the first stage, modifications were brought about by the introduction of the methods of flame spectrometry. To suppress certain interferences, the dry residue after evaporation of hydrofluoric and sulfuric acids is further thermally treated. Heating to 550 to 700°C leads to the conversion of the sulfates of iron, aluminum, titanium, and other elements into surface-active oxides.[213-216] The alkaline earth and the alkali sulfates (except for magnesium sulfate that is partially thermally destroyed) are stable under these conditions. Perchlorates are thermally destroyed in an analogous way,[217,218] obtaining the corresponding alkaline earth and alkali chlorides. The sulfates and chlorides of the alkali metals are leached with water or highly dilute hydrochloric acid from the residue after ignition. This operation substantially simplifies the solution composition and, thus, also the conditions for the final determination. On the other hand, these procedures are subject to a negative error, due to incomplete leaching of the alkali metal salts, especially of lithium salts, from the reaction mixture.[214-217] With the development of flame spectrometry, the thermal treatment of the residue is gradually being omitted from the procedure. On the contrary, stock solutions suitable for a simultaneous determination of a maximal number of components are now usually prepared.

The decomposition with hydrofluoric acid is not a universal technique yielding always clear solutions, without insoluble residues. A number of minerals are difficult to decompose, especially zircon, topaz, all the polymorphous modifications of Al_2SiO_5, some kinds of tourmaline, and staurolite.[219-222] Topaz even resists hydrofluoric acid in a closed system[222] at a temperature of 250°C. The solubility of garnets and alkali amphiboles of the glaucophane group depends on their chemical composition.[219] Almandine and pyrope dissolve sufficiently rapidly. Among micas, biotite reacts faster than muscovite and phlogopite. Mullites, formed, e.g., by a thermal conversion of kaolinite, are highly resistant. The resistivity of zircon toward hydrofluoric acid can be substantially decreased by a metamict transformation of the mineral or by the presence of heterogeneous phases (baddeleyite). Hydrozircon, $Zr[Si_{1-x}O_{4-4x}(OH)_{4x}] \cdot nH_2O$, formed in sedimentary rocks on recrystallization of gels with varying composition, is gradually degraded in hydrofluoric acid solutions.[223] Certain complications have been encountered when decomposing spodumene from some localities.[224,225] In our experience, the decomposition of this mineral, contained, e.g., in South Bohemian pegmatites, presented no difficulty; petalite and pollucite were simultaneously dissolved.[226]

After a 2- to 3-hr heating of a cyanite suspension with a mixture of hydrofluoric and perchloric acids, only 1 to 2% of the mineral amount is dissolved.[227] In the absence of other resistant phases, this procedure is applicable to isolation and determination of cyanite in metamorphed rocks and migmatites. Only 0.97% of topaz was dissolved under the same conditions. Possible reasons for this high resistivity of the two minerals toward the above acids have been given by Tewari.[227] Both the minerals have compact crystal structures in which the aluminum and silicon atoms are accessible with difficulty for hydrofluoric acid solutions. The structural position of the aluminum atoms in cyanite and topaz differs from that in common aluminosilicates. All the aluminum atoms occur in the five- and sixfold coordination and not in tetrahedra.

However, the decomposability of solid silicate phases cannot directly be related to their structure. Minerals with analogous structural arrangements (e.g., with independent SiO_4^{4-} tetrahedra) exhibit great differences during dissolution, as can be seen in different resistivities of zircon and olivine. According to an empirical rule, silicates containing large amounts of bulky univalent and divalent cations or of bulky cations of heavy metals are decomposed more easily than structures rich in aluminum.

Silicon dioxide and silicates are decomposed either by hydrofluoric acid alone or by its mixtures with other mineral acids, most often sulfuric, perchloric, and phosphoric, but sometimes also nitric acid and aqua regia. The effect of the acid added on the dissolution of solids has not yet been unambiguously explained and apparently depends both on the kind of the acid and on the composition of the substance to be dissolved. No synergistic effect has been observed on addition of a further acid. On the contrary, the dissolution of quartz is slowed down in the presence of perchloric, sulfuric, or hydrochloric acid. The results obtained with epidote and staurolite cannot at present be unequivocally interpreted.[222] The reaction time required to dissolve pieces of diabase is prolonged many times on addition of nitric acid.[228] The kind of acid added is determined by the requirements on the oxidation properties of the mixture. Nitric or perchloric acid and aqua regia permit oxidative destruction of sulfides and organic compounds trapped in quartz or silicate grains. The composition of the reaction mixtures changes with time, in dependence on the temperature program of the decomposition. The acids interact among one another, and thus their redox properties change; mixed complexes may also be formed.[229]

The rate of dissolution of quartz and silicates in hydrofluoric acid solutions depends on the grain size. A common size is 0.07 mm or less. The preparation of finely pulverized material for analysis often leads to sample contamination from the crusher, to selective segregation of light fractions, and to irreversible chemical changes in the sample composition. For these reasons, rare materials (meteorites, rare minerals, lunar rocks, and samples from very deep bores) are dissolved in the form of small lumps, without further pulverization. However, the reaction time is then longer, even more than 24 hr, and the isolated, undecomposed fraction must be redissolved in fresh portions of acids.[228]

The phase composition of the insoluble fraction after decomposition of fresh or lateritically weathered rocks by HF was examined microscopically and using X-ray diffraction methods. The results obtained are important even for petrographic characterization of the studied rocks, as the presence of resistant accessoric minerals is easily overlooked in the mineralogical study of the material. Corroded grains of resistant minerals can be microscopically differentiated from newly formed precipitates of complex fluorides.[230] Comparison analyses carried out with the original rocks using XRF, or with the solutions obtained using AAS, indicated the greatest differences in the contents of chromium and vanadium. The two methods yielded

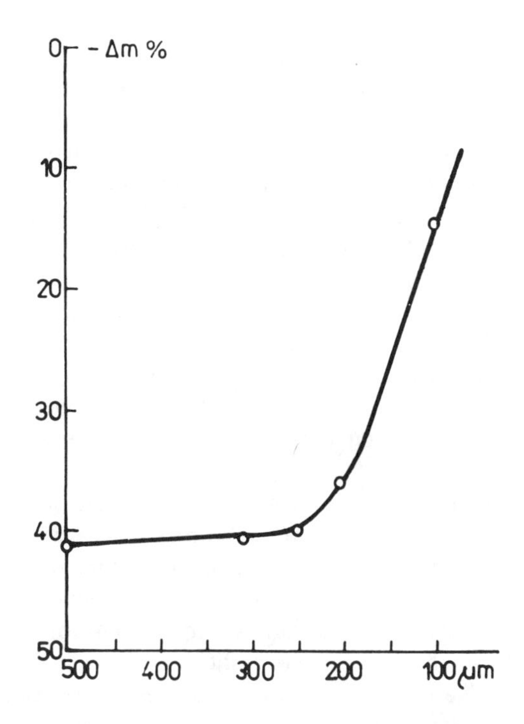

FIGURE 1. Dissolution of sodium-calcium glass in hydrofluoric acid, depending on the particle size. 100 mg of the glass was heated in a platinum dish with 5 mℓ 40% HF at a hotplate temperature of 110°C for 45 min. The insoluble residue was filtered off, washed, dried, and weighed. The y-axis shows the decrease in the glass weight, in % wt.

identical results for chromium only when the chromium was contained in chromium(III) muscovite-fuchsite.[231]

The errors caused by the presence of solid phases that are resistant toward hydrofluoric acid are usually negligible in determinations of the alkali metals. These resistant minerals and their synthetic analogues mostly contain only trace amounts of the alkali metals. Moreover, their contents in the rocks analyzed are usually very low. In some metamorphosed rocks, the content of alkaline amphiboles of the glaucophane group may exceptionally be higher, causing a negative error of determination of Na_2O. Similar errors may occur in analyses of rocks rich in tourmaline. Tourmalines contain up to 2.5% Na_2O, 1% K_2O, and especially elbaites may contain up to 2% Li_2O. Tourmalines of the dravite-schorl series contain[232] less than 0.05% Li_2O.

An accurate determination of potassium for geochronological purposes involving the $^{40}K/^{40}Ar$ method requires careful control of the whole decomposition process. It has been concluded from comparison studies that the results obtained after dissolution of the rock material in hydrofluoric acid are almost identical with those obtained with another decomposition technique, e.g., fusion with lithium metaborate.[233,234]

Industrial glasses are dissolved in hydrofluoric acid solutions without great difficulties because of their unordered structures. Only aluminum-phosphate silicate glasses resist this acid and must be decomposed using a combination of wet decomposition and fusion methods. The rate of dissolution of a sodium-calcium glass was followed in dependence on the sample surface area (i.e., on the size of the sample particles) that was in direct contact with the decomposing agents.[235] The rate significantly increases with decreasing grains size, down to 200 μm; a further decrease in the grain size has no pronounced effect (Figure 1). A procedure recommended for rapid decomposition of industrial glass has been verified by analyses of standard glasses (SRM).[235] To suppress contamination, the decomposition is performed by heating the reaction mixture in a vessel placed on a metal plate covered with a thin layer of PTFE. The plate is divided into two parts, the temperature of each of which can be independently regulated with a precision of ±3°C. The following dissolution procedure has yielded good results. A 0.1-g glass sample is wetted with 2 mℓ of water in a platinum dish and heated with 2 mℓ 70% $HClO_4$ and 5 mℓ 40% HF at 95°C until a wet residue is obtained. To the residue, 2 mℓ 70% $HClO_4$ is added, and the walls are rinsed with water. The temperature of the heating plate is gradually increased to 130°C, and the liquid is allowed to evaporate. In this stage, insoluble fluorides are decomposed, and losses of the material may occur due to solution splashing; therefore, the reaction mixture must be stirred occasionally. The addition of perchloric acid and the whole subsequent operation are repeated. Another 0.5 mℓ $HClO_4$ and 20 to 30 mℓ of water are added to the residue, and the dish with the sample is heated for 30 min at 95°C. The content is transferred to a 100-mℓ volumetric flask, and the individual glass components are determined by AAS or plasma spectrometry.

Precise determinations of low contents of the alkali metals place high demands on the decomposition procedure, especially in ultrabasic rocks, such as periodite PCC-1, dunite DTS-1, or in some minerals and products (topar, mullite). First of all, contamination of the sample during decomposition must be prevented. Small amounts of the alkali metals may be contained in the test materials in the form of liquid inclusions or may compensate the charge in heterovalent, isomorphous cation substitution. The alkali metals are often carried by pyroxenes, with the alkali metal content characterized by the exchange reaction,

$$2\,Me^{2+} \rightleftharpoons Al^{3+} + Na^{+} \tag{12}$$

The isolated undecomposed fraction must then be further treated. The procedure for this treatment is selected according to the phase composition of the residue and the kind of the test component, e.g.,

1. Dissolution in polyphosphoric acids
2. Fusion with cesium salts, especially cesium carbonate, disulfate, hydrogenfluoride, or borates
3. Fusion with lithium metaborate, especially for the determination of Na, K, and Rb
4. Separation of the residue by centrifugation followed by decomposition with acids in an autoclave, at elevated pressure and temperature

Decomposition with the above reagents is described in detail in the appropriate chapters. However, it is usually shorter to directly decompose the material by these procedures, without preliminary dissolution. A disadvantage of a direct decomposition is, however, a large amount of the reagents used, higher amounts of impurities introduced, and the presence of silicon in the solution to be analyzed. The same conditions as for minerals hold for their artificial, resistant analogues. If even a small amount of the alkali metals strongly affects the physical properties (e.g., the corundum ceramics), then the test material must usually be completely dissolved.

In view of the difficulties connected with automation of the dissolution of solids in acids, there has been a trend to use the solution obtained for measurement of a maximal amount of analytical data. Together with the major and minor components, some trace elements are also determined, which affects the selection of the sample weight, the decomposition technique, and the instrumental analytical method. Parallel experiments are used to employ various decomposition techniques and find out whether an oxidizing mixture of acids suffices for the liberation of the trace elements from the test material, or whether treatment with hydrofluoric acid or fusion is needed to destroy the silicate bonds. The solution of the problem primarily depends on the analysis level. It is necessary to decide whether a complex, precise analysis is required (rare extraterrestrial materials, standard reference materials used for control of analytical methods, etc.), or routine silicate analysis or orientative analysis for geochemical prospection and control of technological processes are involved. The decomposition of rocks with hydrofluoric acid mixed with another acid always involves a risk in determining total contents of chromium, tin, zirconium, and titanium.[236] If these elements are carried by chromite, cassiterite, zircon, and rutile, the results obtained are subject to a considerable negative error. Although this procedure yields useful results with some kinds of silicate rocks (e.g., for Zr carried by eudyalite), it cannot be generally recommended for rock analysis. With materials that underwent a thermal conversion (slag, fly ashes, some kinds of atmospheric fallout), heavy metals cannot be leached with mixtures of nitric and perchloric acids or aqua regia. To liberate the metal components, the silicate bonds must be destroyed, best by an oxidizing mixture containing hydrofluoric acid. For fly ashes and atmospheric fallout with varying amounts of surface-active carbonaceous substances, a mixture of hydrofluoric acid with hydrogen peroxide, containing hydrochloric or nitric acid, has given good results.[237,238] The isotopic composition of some elements was determined mass spectrometrically, after dissolving carbon ashes in hydrofluoric acid in a closed polypropylene vessel[239] at 25°C. The elemental composition of the insoluble fraction was also determined and even chloride, bromide, and iodide were found. An analogous variant of decomposition (a mixture of HF and HCl) of small, 50-mg rock samples can be used to determine major, minor, and trace components of rocks.[240] According to Uchido,[240] small samples are completely dissolved as early as after 4 hr of treatment in a closed Teflon® vessel, even at 25°C. Larger samples (500 mg) require a reaction time of 16 hr under the same conditions. The problems of decomposition and the determination of major and trace components in natural and synthetic, industrially important silicates (cements, slags, and building materials) have recently been dealt with.[241,242]

The difficulties connected with dissolution of the test material increase disproportionately with increasing sample weight. With 40-g samples the addition of acids (HF, HCl, aqua regia) must be repeated several times. Even if the data[243] indicate that clear solutions are mostly obtained in this procedure, it cannot be recommended for determination of trace elements. The decomposition is very sluggish and almost always incomplete. The undecomposed fraction must be transferred to solution by fusion. Complete liberation[244,245] of small amounts of lead from rocks is difficult,[246] especially if the rocks are impregnated with sulfides and baryte. The great differences in the results of routine analyses can be explained by the formation of insoluble binary Ba-Pb sulfates or by trapping of Pb^{2+} in the fluoride precipitate. However, the explanation cannot so far be found in the behavior of lead compounds after decomposition with a mixture of hydrofluoric and nitric acids. In the presence of barium (*circa* 0.4% or higher), this procedure always yielded lower results than those obtained after decomposition by aqua regia, in which the silicate matrix of the sample was not attacked.[246] On the other hand, HF and HNO_3 were much more efficient than aqua regia for leaching lead from ignited soil samples.[247] To liberate nickel from lateritic, weathered serpentines, and the soils connected with them, the nickel-magnesium hydrosilicates present must be degraded by treatment with HF, as some members of this group cannot be decomposed by strong mineral acids.

Nitrogen compounds contained in rocks cannot be determined after sample fusion, as they are thermally degraded during the fusion. After dissolution in a mixture of hydrochloric and hydrofluoric acids, they can be reduced with Devarda alloy to produce ammonium ions that are separated by distillation.[248] An interesting application from the point of view of technical decompositions is the determination of chlorides in rocks. A 0.2-g sample is dissolved in a closed PTFE vessel in a mixture of 3 $m\ell$ 48% HF and 5% $KMnO_4$ in 15% H_2SO_4 (4 + 1). The chlorine produced by the oxidation is trapped in 16% KOH containing 0.8% Na_2SO_3

and determined by an ion-selective electrode.[249] Organic compounds from claystones are more readily oxidized with hydrogen peroxide after distilling off silicon tetrafluoride.[250] Analogously, the carbonaceous pigment of lydites cannot be removed by ignition or by wet oxidation, if silicon dioxide is not removed from the sample. Carbon dioxide is not completely liberated from some scapolites without destruction of silicate by the action of hydrofluoric acid. Carbon dioxide is displaced from this mineral only by boiling phosphoric acid.[251] The liberation of the gaseous component is again connected with a deep structural degradation of the solid phase (see Chapter 4, Section VII). This decomposition procedure in the absence of hydrofluoric acid considerably simplifies the decomposition apparatus.

During evaporation of fluoride solutions containing degraded silicate rocks to a small volume, the component concentrations are extremely increased, and sparingly soluble compounds rich in aluminum, fluoride, and varying amounts of other ions, especially Ca^{2+}, Mg^{2+}, Fe^{2+}, Fe^{3+}, Pb^{2+}, Na^+, and K^+, precipitate from the solution.[240,241,252-259] The X-ray diffraction data of the isolated precipitates demonstrate the presence of many substances from simple fluorides to complex aluminum fluorides.[254] The precipitates are not only formed during evaporation of hydrofluoric acid solutions, but also in the presence of strong mineral acids and are not completely dissolved even by evaporation with concentrated nitric or perchloric acid. Similar compounds have also been prepared by action of hydrofluoric acid on the pure oxides and carbonates.[254] It has been found experimentally that the formation of the precipitates is mainly caused by an increased concentration of aluminum salts. Dissolution of granite rocks, diabase W-1, and pure oxides and carbonates yielded very similar compounds. In addition to calcium and magnesium fluorides, the precipitate contained $NaAlF_4$, $MgAlF_5$, and $Fe(II)(Al,Fe[III])F_5$ with various degrees of hydration, the product, $MgAlF_5$, being least soluble of the above compounds in 38% HF.[256] With shale TB (SRM-ZGI Berlin), containing 20% Al_2O_3, chemically and structurally different compounds, approximately corresponding to the formula, $Na(Mg,Al)_6(F,OH)_{18}\cdot 3H_2O$, have been identified by X-ray diffraction after decomposition by a mixture of hydrofluoric and sulfuric acids. It has been found that the composition of the precipitated phases depends on the composition of the test rock and on the temperature program of the decomposition procedure.[258] Together with the newly formed phases, the undecomposed fraction after decomposition with hydrofluoric acid contains residues of corroded grains of resistant minerals. In the above shale, the fraction of the resistant minerals was less than 0.4% (mostly rutile and tourmaline); however, the fraction of the precipitated complex fluorides (decomposition with HF and $HClO_4$) fluctuated uncontrollably[258] from 0.5 to 7%.

No reliable procedure has so far been found for simple conversion of complex fluorides into a solution. Mineral acids are not especially well suited for the purpose. Treatment with concentrated solutions of aluminum and beryllium salts leads to the conversion of complex fluorides into simpler, soluble forms.[253] The practical applicability of this method is limited by the high resultant concentration of aluminum salts, causing complications in the subsequent spectrometric determination of the individual components (mostly AAS or ICP-AES). Boric acid is less efficient than Al^{3+} and Be^{2+}, but is better suited for analytical practice. The dissolution is attained with excess saturated solution of boric acid and is hastened by heating the reaction mixture and sonication.

The dissolution is, however, sometimes very slow and may take several days. In determinations of major components, the purity of boric acid is usually sufficient, and the blank values are not increased, but in some trace analyses the purity is insufficient, in view of the large amount of the acid used. The impurity contents can be decreased by sorption on a cation exchanger.[239] To 454 g H_3BO_3 dissolved in 3 ℓ of H_2O at 80 to 90°C, 100 mℓ of the AG-50W-X8 cation exchanger (50 to 100 mesh) are added, and the reaction mixture is stirred for 15 min at the above temperature.

The formation of complex fluorides and their insufficient dissolution are, at present, among the most important sources of errors in silicate analysis. Because of varying composition of the precipitated products, various determinations are subject to this error,[239,240,253,257-260] especially those of Al^{3+}, Fe^{3+}, Fe^{2+}, Pb^{2+}, Ca^{2+}, Mg^{2+}, Na^{+}, K^{+}, and determinations of some trace elements. In gradual dissolution of the precipitated solid phase, a concentration gradient is formed in unstirred solutions, especially in analyses of microvolumes of solutions sampled from various positions in a volumetric flask.

Incomplete dissolution of fluorides causes negative errors in analyses of minerals with high contents of the alkali metals (e.g., muscovite and biotite).[258,261,262] In dissolution of these minerals in hydrofluoric and nitric acid solutions, a solid phase was formed that was richer in potassium than the initial test substance.[258] Analogous errors have been encountered in the determination of sodium in alkaline pyroxenes, separated from eclogites, and, to a lesser degree, during dissolution in a mixture of hydrofluoric and perchloric acids.[262] If a rock does not contain higher amounts of the alkaline earths, then complex fluorides are destroyed by sulfuric acid much more efficiently than by other mineral acids. Nitric acid exhibits the poorest efficiency. Very fine suspensions of precipitated fluorides may often be overlooked by the analyst or considered to be residues of resistant solid phases. Fluoride precipitates also cause difficulties in isolation of kerogen from rocks.[263] A rare case of the use of the formation of complex fluorides is a determination of arsenic in coal by AAS with electrothermal atomization.[264] The ashes obtained from the coal by combustion in the presence of magnesium and nickel(II) nitrates are decomposed by a solution of hydrofluoric and nitric acids, with formation of the complex fluoride, $MgAlF_5{\cdot}xH_2O$, by means of which the content of aluminum in solution decreases to a less than 5% of the total content. As aluminum interferes in the determination of traces of arsenic, this procedure substantially decreases the systematic error of the analytical method.

The above decomposition of sparingly soluble fluorides is primarily aimed at a transfer of the cationic components to a solution in which they can be suitably determined. In most procedures it is useful to destroy the insoluble fluorides and simultaneously remove the liberated hydrofluoric acid from the solution in order to simplify the subsequent analytical procedures. Ion-selective electrodes and spectrophotometric methods make it possible to monitor precisely the volatilization of hydrofluoric acid (and possibly also volatile fluorides) during evaporation of solutions with mineral acids.

The difficulties encountered in decomposition of sparingly soluble fluorides are apparently the main cause of slow and often incomplete removal of fluorides from the sample evaporation residues. In older procedures, fluorides were removed using oxalic acid.[261,265] By gradually increasing the reaction mixture temperature, a melt was obtained that decomposed to gaseous components. The carbonates formed in the solution were readily dissolved in a dilute mineral acid. The efficiency of this procedure has not, however, recently been verified by direct determination of the remaining fluoride.

Repeated evaporation with sulfuric acid at temperatures of 240 to 280°C has, so far, been the most efficient technique for displacement of fluorides from solutions of decomposed rocks, glasses, feldspars, and ceramic raw materials.[253,266-270] The removal of fluorides from complex mixtures containing increased concentrations of Ce^{3+}, La^{3+}, Al^{3+}, Fe^{3+}, Zr^{4+}, Ti^{4+}, Ca^{2+}, Sr^{2+}, and Mg^{2+} is substantially slower and less efficient than in hydrofluoric acid solutions in the absence of these ions. Especially difficult is the removal of fluorides from the reaction mixture formed by dissolving magnesite and refractory raw materials[268] containing primarily MgF_2. In the presence of boric and sulfuric acids, fluorides form volatile boric fluoride and, thus, their contents in the evaporation residue are considerably decreased.[271] The positive effect of boric acid in removal of fluorides can be illustrated by the following experiment. Fluorides were displaced from a 0.2-g sample of basalt BM (SRM-ZGI Berlin), dissolved in 2 to 5 mℓ hydrofluoric acid, and then evaporated to dryness by heating with 1 mℓ H_2SO_4 (d= 1.84) at 220 to 240°C. After complete evaporation of the sulfuric acid, 2 to 10 mg F^- still remained bound in the residue. If the residue was heated in the same way with 0.4 mℓ H_2SO_4 and 2.5 mℓ of a saturated H_3BO_3 solution, the fluoride content decreased below 0.1 mg. On replacing sulfuric acid by perchloric acid, the fluoride content in the residue increased (≥10 mg) and could not be decreased below 0.35 mg, even by a subsequent evaporation of the reaction mixture with the above amount of boric acid. Prolonged, intense evaporation of fluorides with perchloric acid may even lead to the formation of dehydrated, sparingly soluble titanium dioxide.[272,273] The formation of this compound can be prevented by adding the equimolar amount of zirconyl chloride.[272]

Compared with perchloric acid, nitric and hydrochloric acids exhibit a poor efficiency, and the residue contains substantially larger amounts of fluorides. Application of 1,3-disulfonic acid for this purpose is promising,[274,275] as it has no complexing properties and forms soluble salts even with the ions of the alkaline earths. The derivative used has a higher boiling point than sulfuric acid and has been successfully used, e.g., in the determination of sodium and potassium in feldspars.[275]

As the presence of sulfuric acid is undesirable in some spectrophotometric methods, the displacement of fluorides by perchloric acid was studied in detail in dissolution of various kinds of glass,[235] involving sodium-calcium glasses with 0.5 to 12% Al_2O_3, borosilicate chemical glasses with 2 to 3% Al_2O_3, and glasses for chemical fibers with 14 to 15% Al_2O_3. Samples of 0.1 g were wetted with 1 to 2 mℓ H_2O and decomposed with 7 mℓ 38% HF and 2 mℓ 60% $HClO_4$ on a hotplate with programmed surface temperature. The damp residue was evaporated 3 times with 2 mℓ $HClO_4$. The residue was then dissolved in 0.5 mℓ $HClO_4$ and 20 mℓ H_2O and the solution transferred to a 100-mℓ volumetric flask. The fluoride concentration in the solution was measured by an ion-selective electrode. The temperature program of the decomposition was selected so that the fluorides were displaced within 2 to 2.5 hr. The systematic experiments have yielded several interesting conclusions:

1. The concentration of the remaining fluoride varies from 0.1 to 2 μg F^- per milliliter.
2. The decomposition must be performed with a single addition of hydrofluoric acid. Repeated addition of HF leads to an increase in the fluoride content.
3. The amount of the remaining fluoride increases with increasing evaporation time.
4. In evaporation in Teflon® beakers, the concentration of the fluoride remaining in the solution increases by *circa* one order of magnitude compared with evaporation in platinum dishes.

The general validity of these surprising results may be distorted due to the existence of finely dispersed solid phase with a high fluoride content in the analyte. The reliability of the values obtained depends on the quality of the electrode used, as the fluoride concentrations to be determined are at the limit of applicability of the method (especially at extreme F^-/Al^{3+} ratios).

A. Decomposition for Determination of Iron(II) Oxide

The determination of elements in the same valence state as that in the crystal structure of solids places substantially higher demands on the dissolution procedure than does the common determination of the total contents of the individual components. Polyvalent ions may form stable structures in the solid phase that cannot coexist in solution. Rapid interactions among the polyvalent ions then occur, depending on the appropriate formal redox potentials. Both natural and artificial compounds are involved,[276] containing, e.g., Fe^{2+}, Fe^{3+}, Mn^{2+}, Mn^{3+}, V^{2+}, V^{3+}, VO^{2+}, VO_2^+, U^{4+}, UO_2^{2+}, Ti^{3+}, Ti^{4+}, Sn^{2+}, Sn^{4+}, etc.

In analyses of solids, the determination of ferrous oxide remains a principal problem that has not been satisfactorily solved the world over. The application of Mössbauer spectroscopy has brought about a certain progress, but the interpretation of the experimental spectra is exceptionally difficult in analyses of polyphase materials.[277,278] The situation is characterized by summary reports on the ferrous oxide contents in standard (SRM) rock samples, where the recommended value for FeO is often missing, or is replaced by an "accepted value" or even a numerical value with a question mark.[221]

All the operations, starting with the preparation of a silicate material sample, up to a final determination of ferrous oxide, require the increased attention of the analyst. The main sources of analytical errors primarily involve:

1. Oxidation of Fe(II) compounds in the solid phase, during the adjustment of the grain size in the sample
2. Incomplete dissolution of the test substance in the reaction mixture
3. Oxidation of Fe^{2+} by atmospheric oxygen during dissolution
4. Interaction of Fe^{2+} or Fe^{3+} with the components liberated during the dissolution, or with the solvent
5. Trapping of Fe^{2+} or Fe^{3+} in the solid reaction products, mostly in sparingly soluble fluorides
6. Interaction of the accompanying dissolved components with the excess oxidant added during dissolution or a reaction with the titrant in a titration determination
7. Oxidation of Fe^{2+} by atmospheric oxygen after the decomposition, prior to the final determination
8. Extremely prolonged reaction times causing uncontrollable side reactions, especially at elevated temperatures
9. Catalytic effects that are difficult to predict

The sample pretreatment prior to analysis has a decisive effect on the accuracy of the results. Pulverization of samples in the presence of the air leads to rapid oxidation of ferrous compounds in some solids. Intense mechanical milling causes a local increase in the temperature and thus an acceleration of the oxidation process. Some kinds of garnets, micas, amphiboles, and staurolite are especially susceptible to the oxidation. The selection of the grain size of the test sample depends on the resistance of the solid phase toward acids and the tendency of the substance toward oxidation during milling (usual values are from 100 to 250 mesh). Separated minerals and synthetic products are pulverized under an acetone layer; ethanol is unsuitable for the purpose, as its residues may reduce the oxidant added in an excess to the mixture of acids for dissolution.[278]

The oxidation of ferrous compounds by atmospheric oxygen during dissolution belongs among the most frequent errors involved in the determination of this component.[219-221,279] In the presence of fluoride, the formal redox potential of the Fe^{3+}/Fe^{2+} system decreases[193] from 0.77 to *circa* 0.1 V, and thus fluoride facilitates the oxidation of Fe^{2+} by atmospheric oxygen. The reliability of the determination of ferrous oxide is affected by many factors, such as the sample composition, the concentrations of the acids used, suitability of the apparatus and of the vessel material, and the time and temperature of the decomposition. For this reason, the efficiencies of the individual procedures are difficult to compare. The experimental data and the recommended procedures are thus often contradictory, especially as far as the effects of interferents and the extent of Fe^{2+} oxidation during dissolution are concerned.[280,281] Oxidation of ferrous ions during decomposition has been unequivocally demonstrated, even if it may be slow under certain conditions (e.g., in dissolution of basalt) and even when the reaction mixture is aerated. This effect has been explained[280] by the formation of hypothetical, unspecified fluoride complexes that are stable in the air. The errors caused by oxidation are usually eliminated as follows:

1. By maintaining an inert atmospnere during the dissolution

2. By immediate oxidation of ferrous ions by a reagent contained in the acid mixture used
3. By conversion of ferrous ions into stable chelates during the decomposition

An inert atmosphere is maintained by a stream of purified gases (N_2, CO_2, and Ar) introduced into a suitable, best Teflon®, decomposition vessel.[280,282] All the solutions to be added should be freed of dissolved oxygen; however, this requirement is not always satisfied.[281] In the Pratt method that is still used, an inert atmosphere is provided by a cloud of vapors escaping from the vessel during heating of the reaction mixture. In dissolution of a solid in a wide-neck polypropylene flask, using a hot mixture of hydrofluoric and perchloric acids, the escaping water vapors, the acids, and silicon tetrafluoride form an effective screen preventing access of the air.[283] However, the initial reaction is very fast with some materials, and losses of the test solution may occur. If the material to be dissolved contains only ferrous compounds, with traces of ferri components, an addition of a small amount of a ferric salt improves the stability of the solution formed toward atmospheric oxidation.

Mixtures of hydrofluoric and sulfuric acids at various ratios and concentrations represent a universal agent for decomposition of solids. Only rarely have dilute perchloric and hydrochloric acids been added to hydrofluoric acid. Ferrous oxide is usually directly determined by redox titration with standard solutions of potassium bichromate and sometimes also ceric sulfate and potassium permanganate. If excess oxidant is present during the decomposition, then the unconsumed oxidant is mostly back-titrated with a ferrous salt. For details of the procedure see Table 2. On dissolution of the material in the presence of a known excess of an oxidant, the ferrous compounds are converted into ferric ions immediately after liberation from the solid phase. The danger of the atmospheric oxidation of Fe^{2+} is thus largely suppressed, and an inert atmosphere is then unnecessary in routine application of the procedures. However, high demands are placed on the oxidant used that should exhibit the following properties:

1. A maximal stability, even on prolonged reaction times and at an elevated temperature
2. Reaction with the liberated Fe^{2+} rapidly and stoichiometrically
3. Minimal undesirable side reactions and interactions with interferents, even with long reaction times

Among common oxidants, vanadate has found the broadest use because of its stability. In a simple arrangement, this procedure has become a standard method for the determination of ferrous oxide in volcanic rocks, in addition to the Pratt method. The reaction can be described by the equation

$$VO_2^+ + 2H^+ + Fe^{2+} \rightarrow VO^{2+} + Fe^{3+} + H_2O \qquad (13)$$

To improve the reliability of the analytical data, Whipple[278] studied the method in detail and found the main sources of errors decisive for the precision and accuracy of the determination:

1. The vanadate solutions must be prepared in 5 M H_2SO_4, as at lower acidities the V(IV) formed is reoxidized by atmospheric oxygen.
2. During rapid dissolution of a solid, a side reaction occurs in the close vicinity of the solid substance, connected with the oxidation of V(IV) or ferrous ions by atmospheric oxygen. The reaction is caused kinetically — by depletion of excess oxidant close to the solid phase.
3. With long reaction times (1 to 6 days), negative errors are caused by reoxidation of vanadyl (IV) by atmospheric oxygen.

Table 2
DETERMINATION OF FeO IN VARIOUS MATERIALS AFTER DECOMPOSITION WITH HYDROFLUORIC ACID IN THE PRESENCE OF A KNOWN EXCESS OF AN OXIDANT

Oxidant	Test material	Decomposition conditions	Method of determination	Ref.
NH_4VO_3	Silicate rocks	HF-H_2SO_4, 65—85°C, in polypropylene vessels, 12 hr	Titration with an Fe^{2+} salt	284
	Rocks	HF-H_2SO_4, 20—25°C, polyethylene or Pt vessels, 12 hr	Titration with an Fe^{2+} salt	285
	Minerals, rocks	HF-H_2SO_4, 20—25°C, polystyrene vessels, 4—24 hr or more	Microtitration with an Fe^{2+} salt; spectrofotometry with bipyridyl after adjustment of pH 5	259
$NaVO_3$	Minerals, rocks	HF-H_2SO_4, 20—25°C, polypropylene vessels, plastic box, 1—6 days	Excess Fe^{2+}, back-titration with $K_2Cr_2O_7$	278
	Glass	HF-H_2SO_4 in hot solution, evaporation to SO_3 fumes in a Pt dish	Extraction of $FeCl_3$ into MIBK; titration with an Fe^{2+} salt in the presence of sulfosalicylic acid	286
NH_4VO_3	Rocks	HF-H_2SO_4, 20—25°C, 12 hr	Titration with an Fe^{2+} salt	287
	Rocks	HF-H_2SO_4, heating, Pt dish	Titration with an Fe^{2+} salt	288
	Rocks, minerals	HF-H_2SO_4, reaction mixture evaporated to one half in a Pt dish	Titration with an Fe^{2+} salt	289
	Rocks, minerals	HF-H_2SO_4, 20—25°C in a polystyrene vessels, 12 hr resistant minerals at 65°C in a pressure vessel, 18 hr	Automated spectrophotometry with αα′-bipyridyl after pH adjustment to 4.6—5	290
	Allanite, biotite	HF-H_2SO_4, heating for 16—20 hr in vacuo	Titration with an Fe^{2+} salt; spectrophotometry with 1,10-phenanthroline	291
	Rocks, minerals	HF-H_2SO_4, 20—25°C in a polyethylene vessel, 12 hr; Fe(III) salt added with low FeO contents	Titration with an Fe^{2+} salt	292
	Rocks, minerals	HF-H_2SO_4, 20—25°C, in a plastic, stoppered vessel, 12 hr	Titration with an Fe^{2+} salt	221
	Ashes, slags	HF-H_2SO_4, laboratory temperature	Titration with an Fe^{2+} salt	293
	Amphiboles	HF-H_2SO_4, 20—25°C, Teflon® vessel for 12—72 hr	Titration with an Fe^{2+} salt	294
$K_2Cr_2O_7$	Silicates	HF-$HClO_4$, 20—25°C, polyethylene vessel, 24 hr	Titration with an Fe^{2+} salt	295
	Minerals, rocks, including garnets	HF-H_2SO_4 or HF-HCl in a Pt dish at 65—70°C; Fe(III) salt added to stabilize $K_2Cr_2O_7$	Titration with an Fe^{2+} salt	296
	Minerals, rocks	HF-H_2SO_4 in a Pt crucible with a stirrer, 80°C, excess $K_2Cr_2O_7$ added immediately prior to the titration	Titration with an Fe^{2+} salt	297, 298

Table 2 (continued)
DETERMINATION OF FeO IN VARIOUS MATERIALS AFTER DECOMPOSITION WITH HYDROFLUORIC ACID IN THE PRESENCE OF A KNOWN EXCESS OF AN OXIDANT

Oxidant	Test material	Decomposition conditions	Method of determination	Ref.
$Ce(SO_4)_2$	Glass	HF-H_2SO_4, decomposition with stirring, 20—25°C, Teflon® vessel or boiling water bath; up to 24 hr	Titration with an Fe^{2+} salt; U(IV) determined simultaneously	299
$KMnO_4$	Minerals, rocks	HF-H_2SO_4, 80—90°C, 12 hr, corrected for $KMnO_4$ reduction	Titration with an Fe^{2+} salt	300
Ag^+	Sedimentary rocks and minerals with carbonaceous substances	HF-$HClO_4$, 20—25°C, 12 hr, then 1 hr at 90—95°C in a plastic vessel	KBr titration	301
	Sedimentary rocks	HF-$HClO_4$, 20—25°C, 12 hr, or 1 hr at 90—95°C	KBr titration	292
	Rocks	HF-$HClO_4$, 20—25°C, then 1 hr at 90—95°C	KI titration	302
KIO_3	Rocks	HF-H_2SO_4, 12—25 min boiling in Pyrex glass	$Na_2S_2O_3$ titration of excess KIO_3, after neutralization and addition of KI	303
I^+(ICl)	Minerals, rocks	HF-HCl, 20—25°C, 0.5—10 min in a Pt dish	KIO_3 in the presence of CCl_4	304
I^+	Sedimentary rocks with carbonaceous substances	HF-H_2SO_4, 85—90°C, up to decomposition; ICl added after decomposition	KIO_3 titration in the presence of CCl_4	220, 305
	Rocks, minerals	HF-HCl, 20—25°C, in centrifugation vessel of the Tefzel® plastic, 12 hr in the presence of CCl_4	KIO_3 titration	306
	Glasses, silicates, rocks	HF-HCl in a Pt vessel, reagents and vessels cooled with ice	I_2 liberated titrated with $Na_2S_2O_3$	307

4. An unspecified catalytic effect of the apparatus material (the organic polymer of which the box for the storage of the reaction mixture was made) has been observed, leading to rapid, spontaneous oxidation of V(IV) compounds.
5. Prior to the titration, the precipitated fluorides must completely be dissolved in boric acid added, to avoid a considerable negative error of determination of ferrous oxide.
6. To improve the precision, the solutions of vanadate, the sample, and hydrofluoric acid are freed of dissolved oxygen in an atmosphere of nitrogen, purified by passage through a solution of Cr(II) salt in dilute sulfuric acid.

To obtain reliable results, the following procedure has been recommended.[278] To a sample containing at most 7 mg FeO and placed in a polypropylene or Teflon® beaker, 4 mℓ 0.05 *M* sodium vanadate in 5 *M* H_2SO_4 is added, measuring 8 mℓ 48% HF into a separate beaker. The HF solution is placed for several hours in a box purged with purified nitrogen and then is added to the sample, keeping the reaction mixture in a box (without a nitrogen atmosphere) for 12 to 60 hr at 20 to 25°C, depending on the resistivity of the test substances. The separated fluorides are dissolved by adding 105 mℓ of a saturated solution of H_3BO_3 and vigorously stirring by a magnetic stirrer. With magnesium-rich samples, the reaction mixture is allowed to stand overnight. A volume of 5 mℓ of 0.047 *M* Fe^{2+} in 1 *M* H_2SO_4 is added to the cold solution. The titration with 0.04 *M* $K_2Cr_2O_7$ is performed slowly using a sodium diphenylamine-*p*-sulfonate indicator. Immediately before the end of the titration, 5 mℓ of a H_2SO_4 and H_3PO_4 solution (400 mℓ concentrated H_2SO_4, 200 mℓ 85% H_3PO_4, and 400 mℓ H_2O) is added to the reaction mixture. In parallel with a series of samples, two blank determinations are carried out.

An increase in the reaction mixture temperature (Table 2) hastens the decomposition, but decreases the vanadate stability. Simultaneously, undesirable side reactions with interferents are accelerated. On concentration of the acid mixture when using ammonium vanadate, the ammonium ions may be oxidized through the reaction

$$6VO_2^+ + 2NH_4^+ + 4H^+ \rightarrow 6VO^{2+} + 6H_2O + N_2 \tag{14}$$

For this reason, Whipple[278] prefers sodium vanadate over the ammonium salt.

Other oxidants, e.g., potassium permanganate and ceric sulfate, are much less stable than vanadates and are suitable only for analyses of simple systems under rigorously defined conditions. Bichromate is also insufficiently stable in the presence of hydrofluoric acid, and a so far unclarified interaction occurs, leading to partial reduction of the oxidant.[296] This effect is especially pronounced with materials characterized by high Fe^{2+}/Fe^{3+} ratios. The bichromate reduction can be slowed down by adding a ferric salt, but still the calculated contents of ferrous oxide must be corrected.

Ferrous ions can also be oxidized by iodine chloride, immediately after liberation from the bonding structures.[220,304-307] The iodine, formed as a reaction product,

$$2ICl + 2Fe^{2+} \rightarrow I_2 + 2Cl^- + 2Fe^{3+} \tag{15}$$

is titrated by an iodate solution in the presence of carbon tetrachloride. The method is useful for rapid determination of ferrous oxide in minerals readily soluble in hydrofluoric acid, such as in olivines and plagioclases. When this procedure[305] is modified,[220] it is applicable to determination of ferrous oxide in slates with organogenic admixtures.

A 0.5-g sample is dissolved in a platinum dish in 10 mℓ 10 *M* H_2SO_4 and 5 mℓ concentrated HF. The rock suspension, covered with a lid, is heated for *circa* 5 min to dissolution. The cooled liquid is quantitatively transferred to a 100-mℓ beaker containing 2 g H_3BO_3 and diluted with water to 50 mℓ. In a separate flask, a solution containing 75 mℓ concentrated HCl and 6 mℓ ICl is prepared, and the two solutions are mixed together. A volume of 10 mℓ CCl_4 is added, and the iodine formed is extracted into the organic phase. It is then titrated with 0.1 *M* KIO_3 until the carbon tetrachloride layer is discolored. The ICl solution is prepared by mixing 10 g KI and 6.44 g KIO_3 dissolved in 150 mℓ 6 *N* HCl.

Rocks are also decomposed in the presence of excess iodate by boiling them for 12 to 15 min with a mixture of 15 mℓ H_2SO_4 (1 + 1) and 15 mℓ 49% HF. After neutralization, the excess oxidant is titrated with a thiosulfate solution. The iodine produced by the reaction is removed by boiling. This procedure[303] has been verified on analyses of diabase W1 and granite G1.

An addition of an oxidant to the solution of a decomposed sample suppresses atmospheric oxidation of ferrous ions during the sample pretreatment before titration and during the titration, thus, often considerably simplifying the titration.[220,297,298,305] For these purposes, even less stable oxidants can be used (e.g., permanganate, see Table 2). However, interaction of Fe^{2+} with atmospheric oxygen during the dissolution is not prevented in this way.

The large difference in the redox potentials of the Ag^+/Ag^o and FeF_6^{3-}/FeF_6^{4-} systems has permitted the use of silver ions as the excess oxidant. Silver perchlorate is used, and the procedure has found use in analyses of sedimentary rocks with bitumen and other carbonaceous substances.

In spectrophotometric determination of FeO, derivatives of 1,10-phenanthroline (Phen) with the functional group

=N–C–C–N=

have been successfully used. The reagents serve not only for initiation of a color reaction, but also for modification of the decomposition step, especially for prevention of the Fe^{2+} interaction with atmospheric oxygen. The $Fe(Phen)_3^{2+}$ chelate is highly stable and is only oxidized by strong oxidants. The redox potential of the $Fe(Phen)_3^{3+}/Fe(Phen)_3^{2+}$ system as a redox indicator amounts to 1.14 V. The chromogenic reagents are either added directly to the reaction mixture or to the dissolved sample after neutralizing the solution. An undisputed advantage of spectrophotometric methods is their sensitivity that makes it possible to determine even low contents of ferrous oxide using milligram samples of rocks or separated minerals.[259,292,308-314]

The formation of the chelate *in situ*, during dissolution of the material, has been utilized in determination of FeO in rocks containing sulfides and organic substances. To a 10-mg sample in a polyethylene flask, 20 mg Phen, 3 mℓ 4 *N* H_2SO_4, and 0.5 mℓ HF are added, and the flask is kept on a steam bath for 30 min. The spectrophotometric determination is carried out after adding H_3BO_3 and sodium citrate. The calibration solution is treated in the same way.[220,305]

The principal drawback of this procedure is a slow formation of the colored chelate and a poor stability of the chelate at pH less than 2. The solution after the decomposition is usually slightly turbid and, thus, the absorbance is measured at two different wavelengths. As the chromogen added is also partially degraded at an elevated temperature, new reagents have been synthesized that are stable toward acids during the decomposition and form chelates even in strongly acidic solutions. Some sulfonated triazine derivatives have yielded best results:[313,314]

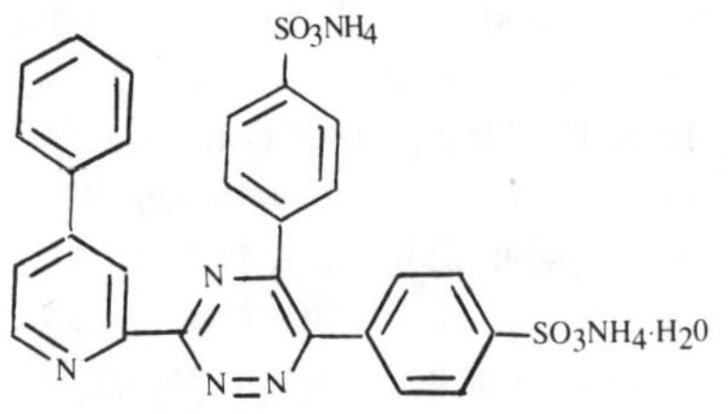

3-(4-phenyl-2-pyridyl)-5,6-(diphenyl-4,4′-disulfonic acid)-1,2,4-triazine, diammonium salt (PPDT-DAS)

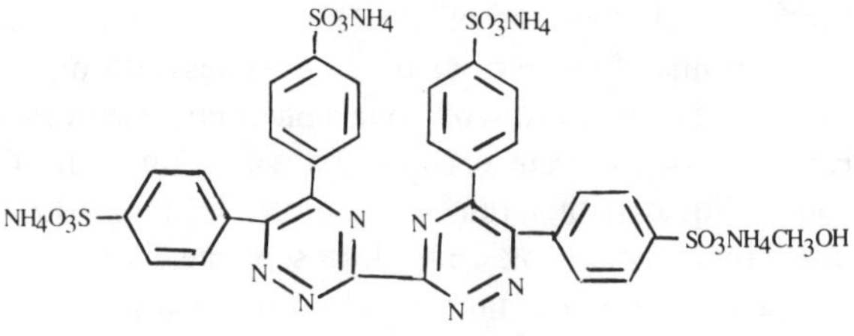

5,6,5′,6′-tetra(4-sulfophenyl)-3,3′-bis-(1,2,4-triazine), tetraammonium salt (TSBT-TAS)

Both the reagents are highly stable toward oxidants. A protective chelation effect against atmospheric oxidation of ferrous ions is exhibited by a 2% solution of PPDT-DAS in a medium of 10% HF and 5% H_2SO_4, where the colored Fe(II) chelate is formed most rapidly.[313] Acidic rocks are readily decomposed in this medium. The procedure is also applicable to dissolution of micas, feldspars, and most amphiboles and is carried out in a mixture of acids, in the dark and at laboratory temperature. The dissolution takes 12 to 24 hr. Basic (e.g., basalt BCR-1) and intermediary rocks are difficult to decompose and, thus, are dissolved by heating with 20% HF and 40% H_2SO_4 in the absence of the chromogen.

If the decomposition mixture contains the TSBT-TAS derivative, the protective chelation occurs immediately over a wide range of hydrofluoric acid concentrations, as well as those of sulfuric acid,[314] thus permitting a more suitable selection of the decomposition mixture and a faster dissolution of, e.g., basalts and some kinds of garnets. A 1- to 5-mg sample is suspended in 0.2 mℓ 20% acetone and 1 mℓ of 0.3 *M* TSBT-TAS is added and dissolved in 25% HF and 35% H_2SO_4. A holder with ten samples is immersed in an ultrasound bath for 30 min, changing the temperature from an initial 20°C gradually up to 52°C. The reaction mixture is allowed to stand for 1 hr, and then 5 mℓ of 40% $AlCl_3 \cdot 6H_2O$ and 10 mℓ of an acetate buffer are added. With garnets and spinels, the reaction time is extended by another 10 min at a temperature of 80°C. A further increase in the reaction time has an adverse effect on the stability of the chelate formed.[314]

If the color reaction is induced only after sample dissolution, the protective effect of chelation of ferrous ions is limited to the short period of the sample treatment prior to the final spectrophotometric determination.[283,290,309-312] In addition to phen, 4,7-diphenyl-1,10-phenanthroline and α,α′-bipyridyl are also used as the reagents. An interesting, very effective decomposition method is dissolution of the test substance in excess vanadate, in combination with a spectrophotometric determination. On neutralization of the solution to pH 5, the V(IV) compounds are reoxidized to vanadate, and the original ferrous ions are regenerated.[259,290,291] The atmospheric oxidation of Fe^{2+} is thus suppressed during the whole analysis. The whole procedure has been automated using a Technicon analyzer.[290] The reaction scheme is described by the equation[259]

$$VO^{2+} + Fe^{3+} + H_2O \rightleftharpoons VO_2^+ + Fe^{2+} + 2H^+ \qquad (16)$$

In all the methods based on color reactions of ferrous ions with 1,10-phenanthroline derivatives, the specified, permissible concentration of hydrofluoric acid must not be exceeded. With excess of the acid, the formation of the colored chelate is hindered, and negative error appears. Fluorides are bound using solutions of boric acid, aluminum, or, exceptionally, beryllium salts.[259,290,291,314]

To calculate the Fe(III)/Fe(II) ratio in the test material, FeO and total iron are usually determined. Total iron is usually determined in another solution, simultaneously with other components. The indirect determination of ferric oxide from the difference of the above two values is subject to the errors of both the procedures. At extreme excesses of ferrous oxide, the results obtained are only orientative. To improve the reliability of the analytical data, ferric oxide must be determined directly in addition to ferrous oxide. Surprisingly, few procedures have been proposed for this purpose. Titrations are used, with EDTA,[315] Hg(I) salts,[316] ferrocene,[317] and coulometrically generated titanium(III) ions.[318] The main source of error, similar to FeO, is the dissolution process leading to changes in the Fe(III)/Fe(II) ratio.

The solid phase is dissolved in solutions of hydrofluoric and sulfuric acids, in many variants differing in the composition of the reaction mixture, the temperature during the decomposition and the time of reaction. Solids are dissolved at various speeds and to various extents, depending on the conditions. To determine ferrous and ferric compounds, the selection of reagents for the preparation of the stock solution is very limited. In general, the medium is much less aggressive than in the determination of total iron. Solutions of hydrofluoric and sulfuric acid only lightly attack tourmaline, chromite and staurolite, and other rock-forming minerals with high Al_2O_3 contents. Contradictory data have been given on the behavior of axinite, magnetite, ilmenite, and especially garnets. The chemical composition of garnets, determined primarily by miscibility of the end members in the group, decisively affects the behavior of these minerals during decomposition.[219,279,296,319] The credibility of the published data is further complicated by easy oxidizability of FeO during pulverization of the sample. The mineral is dissolved in mixtures of hydrofluoric and sulfuric acids, over a wide range of the acid ratios, concentrations, and the times of exposure of the solid to the reaction mixture. The recommended decomposition temperature varies from 20 to 100°C. Exceptionally unequivocal are the data on the behavior of Ca garnets. The mineral with the composition $Ca_2Fe^{II}Al_2(SiO_4)_3$ is highly resistant. On isomorphous substitution of Fe^{2+} by Mn^{2+}, the reactivity of the mineral sharply increases and the garnet is even dissolved in a cold mixture.[219,278] Grossular and melanite are not completely dissolved even after a 6-day treatment with the acid at laboratory temperature, but almandine is completely dissolved under the same conditions. In the Pratt modification of the procedure, even pyrope and almandine are dissolved without substantial difficulties, as shown by our experience. Some papers do not sufficiently exactly specify the kind of garnet analyzed, and thus the data cannot be generalized.[283,297,314] A decrease in the grain size of the mineral shortens the reaction time, but the sample may be atmospherically oxidized during pulverization.

Among micas, Fe(II)-rich biotites react fastest. Lepidolite and muscovite are also dissolved. However, samples with a sufficiently small grain size are difficult to prepare. Monoclinic pyroxenes (augite, diopside) quantitatively react even in a cold solution, after standing for a few days. However, dissolution of rhombic pyroxenes is difficult, and the dissolution rate decreases[278] with increasing Mg^{2+} content. Minerals from the amphibole group behave in various ways.[294] The evaluation of their resistivity is often complicated by insufficient mineralogical characterization of the mineralized sample. Basalt amphibole reacts sufficiently rapidly with a mixture of hydrofluoric and sulfuric acids at a temperature immediately below the boiling point. Hastingsite and crossite are dissolved in a cold mixture after 12 to 20 hr of standing.[219,278]

The rate of dissolution of rocks depends on the reactivity of the individual rock-forming minerals containing iron. In a medium of 10% HF, 5% H_2SO_4, and 2% PPDT-DAS, acidic rocks (granites, granodiorites, tonalites, and andesites) react at ambient temperature.[313] Volcanic glasses are also highly reactive and undergo rapid destruction. Basic rocks, such as basalt and intermediary rocks, react slowly and incompletely at 20°C. Eclogites resist decomposition even at an elevated temperature. The problems in their dissolution follow from their mineral composition involving a garnet and an alkaline pyroxene (''omphacite''). Similar is the behavior of ultrabasic rocks containing resistant mineral phases, such as spinels and pyroxenes, in addition to olivine. Although useful results have sometimes been obtained,[313] the determination of ferrous oxide in these rocks is unreliable.

Extreme extending of the reaction time does not lead to an improvement in the reliability of the analytical data. Undesirable side reactions occur, and the stability of the oxidant added decreases, even in decompositions at normal temperature. It is thus better to separate the undecomposed fraction by filtration, dissolve complex fluorides, isolate the residue, weigh it, and take an appropriate correction.[278] The solid phase remaining in the solution is not only the original material, but also sparingly soluble fluorides retaining ferrous and ferric ions. The formation of these products is discussed above. Here, only the effect of these products on the determination of ferrous and ferric oxides will be discussed. The behavior of these precipitated fluorides, obtained after decomposition of solids of various compositions in solutions of hydrofluoric and sulfuric acid with excess vanadate, has been studied in detail. The greatest difficulties are encountered in dissolution of magnesium silicates.[278] On decomposition of rhombic magnesium pyroxene (enstatite), a gelatinous, insoluble compound is formed whose main component is MgF_2. The precipitate is dissolved with difficulty, best by stirring the mixture for 1 hr with a magnetic stirrer and then allowing it to stand for 24 hr. The product separated from phlogopite is more granular, and the dissolution in boric acid solutions is faster. The precipitates formed during analysis of magnesium garnets, amphiboles, and some ultrabasic rocks behave analogously. Mixed calcium-magnesium fluorides, obtained by dissolving augite or diopside, are easier to dissolve than magnesium fluoride alone. However, the composition of these precipitates was not studied in detail in the cited paper. These are complex fluorides containing ferrous and ferric ions, rather than simple fluorides with sorption-active surfaces. Insoluble fluorides are also formed in decomposition of ultrabasic rocks with minimal contents of aluminum oxide.[278] Trapping of Fe^{3+} in the precipitate formed has also been demonstrated in analyses of these rocks.[317]

It is somewhat surprising that substantially less attention has been devoted in dissolution of inorganic materials to substances reducing Fe^{3+} than to interferents with oxidizing effects. In commonly analyzed materials these involve interactions of ferric ions with sulfides (or H_2S), organic compounds, and their degraded residues. The extent of interference from these components depends on the acid mixture used, its temperature, and the overall experimental arrangement of the decomposition process (e.g., on the flow rate of an inert gas).

Among sulfides, the behavior of pyrites has been studied in greatest detail, as pyrites are common accessoric components of most volcanic and sedimentary rocks. In view of the decrease in the redox potential of the Fe^{3+}/Fe^{2+} system in fluoride solutions, the oxidation of pyrite is minimal in mixtures of dilute hydrofluoric and sulfuric acids both at normal and an elevated temperature and does not substantially affect the usefulness of the results obtained.[280] However, the behavior of pyrite in a reaction mixture with excess oxidant has not yet been satisfactorily elucidated. Whipple's review,[278] containing valuable information on the possible reaction mechanism and possible sources of error, deals only marginally with the extent of the reduction of vanadate by pyrite. In our experience, pyrite dissolves in an acidified vanadate solution (especially at an elevated temperature or during prolonged reaction time) substantially more than in the Pratt method. Marcasite has a greater tendency toward oxidation than pyrite. The common mineral association, pyrite-pyrrhotine, causes great complications in dissolution of some ore-mineralized basic rocks (especially amphibolites), even if the reduction of Fe^{3+} by hydrogen sulfide,

$$2Fe^{3+} + H_2S \rightarrow 2Fe^{2+} + S + 2H^+ \qquad (17)$$

is hindered by the presence of fluoride in the solution.[280] However, the H_2S liberated quantitatively reacts with excess oxidant. All readily decomposable sulfides behave analogously, e.g., sphalerite. Molybdenite and cinnabar are virtually not attacked and remain in the undecomposed fraction.[309] With materials containing small amounts of sulfides, a cupric salt is added to the reaction mixture.[292] In decompositions with excess silver ions, sulfides must not dissolve, and silver sulfide must not precipitate, which is attained by judiciously selecting the decomposition mixture. It must, however, be pointed out that the analytical data on the contents of ferrous oxide, obtained after dissolution of silicate rocks with sulfidic ore mineralization, are mostly useless.

The contradictions in the data on the reducing properties of organic compounds from soils and carbonaceous substances (more precisely, of organic substances transformed to various degrees) from rocks towards ferric ions are mainly caused by variability in the chemical composition and the structure of the interferents.

Reduction of ferric ions during decomposition causes positive errors in the determination of ferrous oxide. Similar errors may also be caused by interactions of organic compounds and the oxidant, both during the decomposition of the material and in direct titration of the ferrous ions liberated. Reduction of ferric ions by some organic compounds has been experimentally confirmed.[305,311,312] Nicholls[305] added a natural bituminous substance (elaterite) to diabase and found no reducing effects. His method, proposed for determination of ferrous compounds in sedimentary rocks (see Table 2), only eliminates the errors stemming from the interaction of organic compounds with the oxidimetric titrant. According to Pruden,[311] Nicholls' experiments are not quite convincing, because elaterite belongs among organic compounds with a low reactivity. The author[311] repeated the experiments with diabase W1 and added peat, humic soil, and dried lucerne as organic substances. He has found that ferrous oxide content increased substantially. By adding ferric compounds to humic soil or lias slate (prior to dissolution of the material in a mixture of hydrofluoric and sulfuric acid)

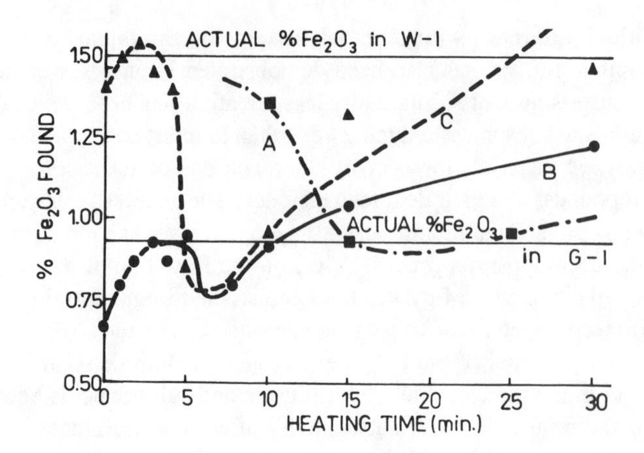

FIGURE 2. Variation of tervalent iron content with heating time. Curve (A) 0.2000-g samples of W-1 dissolved in inert atmosphere; (B) 0.2000-g samples of G-1 dissolved in uncovered crucible (exposed to air); and (C) 0.2000-g samples of W-1, the same conditions as in (B).

and determining the FeO content spectrophotometrically with bathophenanthroline, he estimated the reducing capacity of soils and the above sedimentary rocks. A decrease in the decomposition temperature causes a decrease in the extent of mutual interaction.[312] After dissolution of the material at 60 to 65°C, ferrous oxide can be determined in soils and claystones. In contrast to the data of Pruden,[311] only small effects have been found for humic acid preparations added to standard rocks G-2, BCR-1, and AGV-1. If the content of organic substances in the soil does not exceed 10%, no pronounced error is caused in the determination of FeO.[310] However, the validity of this statement is restricted to certain soil types.

The reduction of ferric ions by organic substances in the presence of reagents forming stable chelates with ferrous ions (*in situ* chelation during the decomposition) has also not been systematically studied. The formation of stable chelates with Fe^{2+} leads to an increase in the formal redox potential of the Fe^{3+}/Fe^{2+} system, which, theoretically, should facilitate reduction by organic compounds. However, there exist other factors, e.g., the existence of stable fluoride complexes and the complexing properties of the present organic substances themselves. In such complicated systems, it is difficult to predict the behavior of the compounds of di- and trivalent iron in analyses of materials with increased contents of carbonaceous substances.

The contradictions in the results of experiments illustrate the complexity of this problem. For its partial solution, it is first necessary to reliably identify and determine the organic compounds present in the test soils and rocks and to classify these substances from the point of view of their redox effects. This analysis can be carried out using the contemporary instrumentation. In any case, however, the reliability of the determination of ferrous oxide in soils and sedimentary rocks with unknown contents of organic substances is dubious. Reduction of ferric ions has also been observed in dissolution of volcanic rocks (G-1, W-1), whose elemental and phase composition has been well known.[318] The time dependence of the reduction is plotted in Figure 2. The causes of this reduction have not yet been satisfactorily explained.

Further reducing substances include iron metal that is introduced into samples of silicate materials during crushing due to abrasion of the apparatus.[320] Interaction of iron with ferric ions leads to large negative errors, similar to the effects of Ti^{3+}, V^{3+}, and Cr^{2+} ions that occur in some industrial products (slags), minerals, and extraterrestrial rocks.[320,321]

B. Decomposition of Selected Natural and Artificial Compounds

Hydrofluoric acid has been successfully used for the preparation of solutions in determination of actinoids, and the dissolution procedures have been experimentally verified not only for the determination of uranium and thorium in ores with high silicate contents, but also for other actinoids, including transuranium elements and the products of the decay series. Especially high demands follow from the necessity of systematic monitoring of radioactive pollution of the environment. The actinoids and fission products are determined in industrial dust and in soils around nuclear power plants. The global contamination level originating from test nuclear blasts is monitored through analyses of atmospheric fallout and precipitation, river and sea sediments, and other materials.

No great problem is encountered in decomposition of rich uranium ores with uranite as the main ore mineral. Secondary uranium minerals, such as uranium micas and uranium blacks, are also easy to decompose. An oxidizing acid mixture suffices for the decomposition, e.g., aqua regia or nitric acid mixed with perchloric, hydrochloric acid, or hydrogen peroxide. The leaching of uranium from poor ores and rocks is often incomplete. Therefore, silicate minerals are destroyed by hydrofluoric acid mixed with the above acids.[322-325] However, microgram amounts of uranium and thorium cannot always be completely dissolved even under these conditions. The two elements may be concentrated in accessoric, highly resistant minerals, such as zircon, monazite, euxenite, polycrase, thorite, thorianite, some titanates, and other tantaloniobates.[326] The undecomposed fraction is then isolated and decomposed by fusion with alkali borates and carbonates, sodium peroxide, or disulfate, or dissolved by acids in a closed system under increased pressure and temperature. The kind of fusion agent is selected according to the assumed composition of the insoluble residue and the purpose of the analysis. The amount of uranium in the undecomposed residue after leaching of rocks with hydrofluoric, hydrochloric, nitric, and perchloric acids varies, depending on the content of resistant minerals.[326] Among 17 test rocks, containing $n \cdot 10^{-2}\%$ of uranium, only two samples exhibited an increased content of uranium in the insoluble residue.[323] Dissolution of uranium compounds mostly requires the presence of an oxidant that converts uranium(IV) ions into stable uranyl ions. Complete dissolution of thorium compounds and solid phases containing thorium is substantially more difficult than decomposition of uranium raw materials.[322] Complications stem not only from resistivity of the solid phases present, but also from sorption and coprecipitation of the liberated Th^{4+} ions on the surface-active, poorly soluble reaction products (fluorides, phosphates, barium and calcium sulfate, silicic acid gel, etc.). Leaching of rocks with hydrofluoric acid mixed with other acids need not lead to complete liberation of thorium, although this procedure has yielded satisfactory results with many substances. Reliable results can be attained, similar to the determination of uranium, by using a suitable combination of various decomposition procedures.[220,327]

To determine the ^{234}U, ^{235}U, ^{238}U, ^{234}Th, ^{232}Th, ^{230}Th, ^{228}Th, ^{231}Pa, ^{226}Ra, ^{210}Po, and ^{210}Pb in radioactive raw materials, industrial dusts, wastes, and soils, the test material is dissolved similarly as resistant silicates.[328] The decomposing mixture contains hydrofluoric acid, together with nitric, hydrochloric or perchloric, and sulfuric acid.[329-333] The evaporation residue is then fused with alkali disulfates.[329-331] This operation is usually preceded by fusion with potassium fluoride, thus ensuring complete dissolution of the material.[328-330] The fusion with a disulfate removes fluorides from the reaction mixture. Many procedures used for the purpose have failed, just because of high resistivity of thorium and radium compounds and the products of their decay.

A combination of various decomposition procedures ensures that the elements are converted into the active, ionic state, which is a condition for subsequent radiochemical separation. Similar conditions also hold for the separation of the isotopes of some transuranium elements and their decay products. From the point of view of the environmental contamination, the extent of leaching ^{239}Pu from soils has been monitored especially carefully around nuclear power plants. The leaching solution usually contains nitric acid or aqua regia and hydrofluoric acid.[334-336] Perchloric acid is unsuitable for the purpose, because part of the plutonium may volatilize in the form of the hexafluoride during its evaporation. The leaching efficiency is evaluated after adding ^{236}Pu. To obtain an objective correction factor for the decomposition recovery, perfect isotopic homogenization of $^{239}Pu/^{236}Pu$ must occur in both the solid and the liquid phase. The ^{239}Pu present must be converted into the active, ionic form, capable of an isotopic exchange with the tracer added. This assumption is not met in most procedures. The isolated, insoluble fraction, leached with acids, often contains more ^{239}Pu than the original sample. The degree of leaching of plutonium depends on the thermal pretreatment of the test soil.[334-336] Heating up to 2400°C leads to a pronounced decrease in the solubility of plutonium dioxide. At a higher temperature, plutonium dioxide loses oxygen and approaches the composition of the substantially more soluble plutonium(III) oxide. Therefore, microparticles of plutonium oxides in global fallout produced by nuclear blasts (after heating to extremely high temperatures) are much easier to dissolve (Chapter 10) than plutonium dioxide produced in nuclear power plants. A difficult problem in decomposition is created by complex fluorides retaining plutonium, as they cannot be decomposed by evaporation with nitric acid. Complete isotopic equilibration of the ^{236}Pu spike with the ^{239}Pu present has been attained using the following procedure[334] in analysis of soil samples. A ^{236}Pu spike was added to 10 g of soil ignited at 1000°C for 4 hr, and the sample was decomposed by heating with 50 mℓ 48% hydrofluoric acid. The reaction mixture was evaporated with 50 mℓ 8 *M* HNO_3. The salts formed, including complex fluorides, were dissolved by heating the solution with 10 mℓ 8 *M* HNO_3, 25 mℓ H_2O, and 2 g H_3BO_3. An analogous procedure was selected by Hiatt et al.[337] in the determination of americium, curium, and plutonium in soils and by Knab[338] in analyses of geological samples and environmental materials.

For a simultaneous determination of all the α-emitting radionuclides, from radium to californium, 50 to 100 g of soil must be treated.[339,340] After dissolving the major sample components, the residue is further decomposed by combined fusion with potassium fluoride and sodium disulfate in a large platinum dish.[339] Decomposition of large soil samples is connected with many technical difficulties.[340] The solution foams considerably. Good results have been obtained after preliminary, repeated leaching of the sample, spiked with ^{236}Pu, by a mixture of nitric and hydrochloric acids. Only the isolated, undecomposed fraction is then further treated with a mixture of hydrofluoric and nitric acids. Compared with a fusion decomposition, the plutonium fraction extracted with acids amounted to 53 to 66%.

In the oxides of tetravalent actinoids, not only Th(IV) and U(IV), but also other transuranium elements may undergo mutual isomorphous substitutions.[341] Thorium dioxide ignited at temperatures above 1000°C is highly resistant toward acids.[342,343] A small amount of hydrofluoric acid (0.03 to 0.1 *M*) catalytically hastens dissolution of uranium and thorium dioxides, as well as of plutonium and cerium dioxides, in a medium of 7 to 14 *M* nitric acid[342,343] (Chapter 4, Section II). These procedures are primarily used in evaluation of nuclear fuels.[344] The good solubility of plutonium dioxide ignited at temperatures below 850°C in this mixture is explained by a disordered structure of the compound. At higher temperatures, the oxide becomes resistant toward these acids. The completeness of dissolution of mixed U(IV) and Pu(IV) oxides depends on the composition of the solid solution of the two compounds.

The formation of poorly soluble fluorides of thorium, tetravalent uranium, and some tetravalent transuranium elements is utilized for their isolation in analyses of complex mixtures, especially those from soils and rocks.[345-347] The sample is decomposed by excess hydrofluoric acid, and the reaction mixture is evaporated down to a few milliliters. After dilution, the fluorides formed are filtered or centrifuged off. They are dissolved in strong acids, best by heating with perchloric acid. The calcium ions present in the test substance then form a precipitate of calcium fluoride serving as an efficient collector. In soils, sediments, and other environmental samples, the fluorides of plutonium, curium, and americium separate, together with these of thorium and uranium. Ceric ions are used as the coprecipitation agent.[347]

Hydrofluoric acid is an indispensable solvent in analyses of niobium and tantalum metals, their alloys, oxides, and minerals. It is simultaneously used as a reagent for the separation of these elements from actinoids and lanthanoids.[193,226,348-355] Lanthanoids and actinoids are isolated as poorly soluble fluorides.[351,356] The solution then contains fluoride complexes of niobium and tantalum in the active, monomeric, ionic form. The filtrate, obtained after an isolation of insoluble fluorides, is suitable for subsequent treatment, e.g., for mutual separation of Nb and Ta by extraction or ion-exchange chromatography.

The precipitation of insoluble uranium(IV) and thorium(IV) fluorides has been utilized analytically to separate the elements and Pb^{2+} from natural tantaloniobates.[348,356] To determine the absolute age using the U/Pb methods, up to 50 g of a mineral must be treated, which requires maintenance of a highest possible cleanliness in dissolving the material in a mixture of hydrofluoric and hydrochloric acids. The effect of hydrochloric acid first leads to dissolution of the brown, surface oxide film that prevents penetration of hydrofluoric acid to the mineral grains.[356] Common fusion agents cannot be used for the purpose, because of high lead contents in the decomposition agents. Only 145 μg Pb^{2+} were isolated from 50 g of a mineral for the purpose of geochronological isotopic measurement, after an ion-exchange separation.

Experiments with natural tantaloniobates have demonstrated that uranyl ions are also trapped in the precipitate under certain conditions.[357] Niobium and tantalum are often combined with uranium, thorium, and the lanthanoids in nature, e.g., in the minerals betafite, euxenite, fergussonite, loparite, pyrochlor, samarskite, etc. These are mostly complex oxides with extensive isomorphous substitution among the components. Because of a great variability in the chemical composition of these minerals, the published data on decomposability of the individual phases by hydrofluoric acid are not identical. In general, minerals with tantalum predominating are more difficult to dissolve than substances with niobium predominating. Success of decomposition depends on the grain size; very small values must be attained, below 250 mesh. As the solvent, hydrofluoric acid alone is used, or in mixtures with hydrochloric, nitric, sulfuric, or even phosphoric acid. The composition of the decomposition mixture depends on the kind of compounds dissolved and the composition of the accompanying phases. Decomposition of the material with hydrofluoric acid enables group separation of the actinoids and lanthanoids and also a separation of niobium and tantalum minerals from many mineral phases, such as pyrite, molybdenite, cassiterite, and zircon.[350,357] In some procedures the solution of niobium and tantalum complex fluorides is evaporated with sulfuric acid, and the compounds of the above elements are converted into soluble peroxocomplex by adding hydrogen peroxide.[354] Leaching of tantalite concentrates with hydrofluoric acid is not only used for analytical purposes, but also in technological treatment of ores;[358,359] 20 *M* HF is used at 70°C. With concentrates containing 64% Ta_2O_5 and 8% Nb_2O_5, more than 99% of the oxides is dissolved.[358] A mixture of hydrofluoric and hydrochloric acids has also yielded good results in decompositions of poor raw materials containing pyrochlor and pandaite;[349,350] a 1- to 2-g sample is wetted with 5 to 10 mℓ of water and is dissolved in 10 mℓ each of concentrated HCl and concentrated HF by heating for 2 hr on a hotplate. The reaction mixture must not be evaporated to dryness. Then a "spectral buffer" is added, and niobium is determined by AAS.

A mixture of hydrofluoric and nitric acids also effectively attacks high-percentage niobium-tantalum ores. However, to dissolve tantalite and euxenite, the addition of the acids must be repeated several times.[226] Simpsonite is not completely dissolved under these conditions.[352,353]

Decomposition of a material by hydrofluoric acid is often combined with fusion with alkali disulfates. The dry residues after evaporation of the acids is fused, and the melt is usually dissolved in complexing solutions of tartaric or oxalic acid.[360,361] Alternative procedures start with the original ore.[355] A mixture of the raw oxides is isolated from the melt solution, and the oxides, dissolved in hydrofluoric acid, are further treated. However, the combined dissolution has no significant advantage.[349,357,362] According to the results of comparison analyses involving fusion (with alkali disulfates, borates, or carbonates), treatment with hydrofluoric acid completely dissolves the niobium and tantalum compounds.[351] On the other hand, fusion introduces additional components to the solution that may seriously complicate the subsequent analytical operations.

Complete dissolution of Nb-Ta minerals in acids is dubious in analyses of poor ores and in determinations of trace concentrations of these components. It follows from a comparison of the data obtained with various sample treatment procedures[363] that 80 to 90% of the niobium present is dissolved in solutions of hydrofluoric and nitric acids and in aqua regia when decomposing standard (SRM) rocks. The decomposition efficiency has been verified[363-365] by adding carrierless isotope ^{95}Nb. The accuracy of the results depends on isotopic equilibration between the isotope added and the solids analyzed. The correction factor obtained was independent of the amount of the test substance, which is an indirect proof of isotopic homogeneity and complete dissolution of niobium compounds in the sample.[363,364] In spectrophotometric determination of niobium using thiocyanate, platinum vessels must not be used, as traces of platinum interfere.[366,367] This interference can be circumvented by using the selective reaction with sulfochlorophenol S.[362] A 0.25-g sample is dissolved in a platinum dish with 5 mℓ concentrated nitric acid and 10 mℓ HF. The solution is evaporated to dryness on a hotplate at 125°C, and the residue is fused with 2 g $K_2S_2O_7$. The melt is dissolved in 6 *M* H_2SO_4 and 2 *M* HF. The fluoride complex of niobium is separated by extraction, and the spectrophotometric determination is carried out.

Niobium and tantalum metals are best dissolved in hydrofluoric acid to which nitric acid is added dropwise.[368,369] For 0.6 to 0.7 g of a metal, 3 mℓ 48% HF and 3 mℓ concentrated HNO_3 are sufficient.

Preconcentration of trace amounts of the rare earths is attained by precipitating insoluble fluorides in the presence of efficient collectors.[292,360,370,371] In this way, the rare earth fluorides were isolated from their minerals and some volcanic rocks. Scandium(III) ions cannot be completely separated by precipitation of the fluoride.[292,372] In the presence of excess hydrofluoric acid, soluble complex fluorides are formed, with the ScF_5^{2-} ion predominating. The fluoride precipitate contains many accompanying ions, in addition to the lanthanoids and actinoids:[292] the alkaline earths, Pb^{2+}, Al^{3+}, and Fe^{3+}. At present, this technique of group isolation ceases to be used, and ion-exchange chromatographic separation of the rare earths is preferred.

Complications are encountered in the removal of silicic acid trapped in the oxides precipitated. The silicon dioxide volatilizes during evaporation with hydrofluoric acid, but the rare earth fluorides cannot be simply converted into the initial oxide form.[373]

Modern procedures mostly determine the rare earth elements by atomic emission spectrometry with inductively coupled plasma (AES-ICP), possibly combined with HPLC. A fusion treatment of large samples is not particularly well suited for plasma spectrometry, and decomposition with acids yields better results; mixtures of hydrofluoric acid with nitric and perchloric acids are used most often.[374-377] The main drawback of this procedure is incomplete dissolution of resistant mineral phases in which the rare earths are often concentrated (zircon, monazite, some titanates, etc.). The isolated, undecomposed residue must then be fused.[377]

The natural calcium titanate, perovskite, is destroyed by hydrofluoric acid at an increased temperature, with formation of poorly soluble calcium fluoride. The precipitate also contains the rare earths (usually 3 to 5% of the mineral), and titanium is dissolved in the form of stable fluoride complexes.[378] The decomposition process is accelerated by heating in a mixture of hydrofluoric and sulfuric acids. There are great differences in the decomposability of this mineral.[220,378] Only those data that are based on analyses of mineralogically pure, perfectly defined fractions are reliable. The differences in the solubility of the minerals can be ascribed to variations in the isomorphous substitution among the components and to differences in the thermal history of the phase. Finely pulverized samples of baryum titanate single crystals that are used in microelectronics as a ferroelectric behave analogously during dissolution. The test material is, however, slowly dissolved, and baryum fluoride separates incompletely. The fraction of the barium dissolved can be increased by adding ammonium chloride.[379] Strontium titanate behaves analogously.

Titanium ores behave in various ways during decomposition, depending on the kind of ore minerals present. Ilmenites are dissolved in hydrofluoric and sulfuric acids, and hydrochloric acid is added when magnetite is present.[379-382] Again, ore minerals from various localities exhibit various solubilities. Therefore, the undissolved residue is usually fused with alkali disulfates or borates.[380,381] Ilmenite, together with titanomagnetite, magnetite, and sometimes also rutile, forms the major component of beach black sands that can sometimes be mined as titanium or iron ores. They cannot be decomposed by hydrofluoric acid, even in mixtures with other acids, and often contain further resistant minerals (such as zircon, monazite, cassiterite, and some spinels) that must be decomposed by fusion. The tetragonal form of titanium dioxide, rutile, is resistant[383] toward acids even in closed systems, at increased temperature and pressure. Another polymorphous modification of TiO_2 (anatase) is dissolved by acids, and thus the two mineral forms can be distinguished.[384] Pure titanium dioxide preparations, suitable for the preparation of stock solutions, usually contain both the above crystal modifications, depending on the thermal pretreatment. These preparations are not completely dissolved in HF; the effect of the thermal conversions on the solubility of the oxides has not yet been studied.

Small amounts of titanite ($CaTiSiO_5$) can be dissolved using a mixture of hydrofluoric and perchloric acids.[385] An amount of 40 to 50 mg of the mineral is decomposed in a platinum dish by 4 $m\ell$ 6 M $HClO_4$, adding dropwise 4 $m\ell$ of concentrated HF. If the sample contains organic compounds, 1 $m\ell$ of 8 M HNO_3 is added. The mixture is heated to the appearance of $HClO_4$ vapors. The walls are then rinsed with water, and the evaporation is repeated at a temperature of *circa* 250°C. The salts formed are dissolved in water and transferred to a 100-$m\ell$ volumetric flask containing 12 $m\ell$ 6 M $HClO_4$.

In determinations of trace amounts of the platinum metals in titanomagnetites, the main problem is the large samples that must be treated. Samples weighing up to 30 g are mostly subjected to a combined decomposition procedure; in the first stage, hydrofluoric acid alone or in a mixture with sulfuric or perchloric acid is employed in a Teflon® or glassy carbon dish. The undecomposed residue is usually fused with sodium peroxide in a corundum crucible, or with ammonium hydrogenfluoride in a glassy carbon vessel.[386,387] As titanomagnetites represent a series of solid solutions of the end members of the ulvite-magnetite series, the differences in their behavior can partially be ascribed to the fluctuating chemistry of these compounds. Some kinds of titanomagnetites are dissolved in hydrofluoric acid alone better than in the above acid mixtures.[386] A mixture of hydrofluoric and hydrochloric acid is the principal solvent for kimberlites with spinels as Pt carriers (sample weights up to 50 g). The insoluble residue is fused with sodium peroxide. All the platinum and palladium could be extracted from some basic rocks using a mixture of hydrofluoric acid and aqua regia, without the necessity of dissolving the undecomposed fraction.[388,389] For complete separation of the precious metals, reducing substances must be removed from the sample. Hydrofluoric acid is indispensable in removing the reducing carbonaceous pigment from silicites. The oxidation of the carbonaceous substances can only be carried out after destroying the SiO_2 bulk. Preignition of the samples without a chemical treatment is insufficient and causes incomplete separation of palladium and gold in the subsequent treatment with aqua regia.[390]

Hydrofluoric acid combined with nitric acid or aqua regia has been recommended[391] for leaching of silver from copper and iron ores. A decreased solubility of silver after preignition of the ore is not caused by volatilization, but by an interaction with the matrix, with formation of a species that is insoluble in aqua regia. The silver is completely dissolved in treatment with oxidizing mixtures of hydrofluoric acid with aqua regia or perchloric acid, even when treating preignited ores.[392] An analogous procedure has been recommended[393] for the determination of this metal in rocks. The residue after evaporation of perchloric acid is dissolved in 1.5 *M* sodium thiocyanate. Hydrofluoric acid acts as the reagent destroying the silicate bonds. Aqua regia, hydrobromic acid, and bromine that are also present cause simultaneous dissolution of the present gold, as well as tellurium, indium, and thallium. By this procedure, all of these metals are liberated from rocks and soils.[394]

Oxyfluoride complexes of molybdenum and tungsten play a role in dissolution of minerals and alloys of these metals.[192,193] A mixture of hydrofluoric acid with hydrochloric acid and hydrogen peroxide is suitable for decomposition of ferrotungsten; the presence of fluoride prevents precipitation of sparingly soluble tungstic acid.[193,395] Tungstite is also dissolved in this mixture; scheelite forms insoluble calcium fluoride on reaction with hydrofluoric acid, which is utilized as the collector for the rare earths present.[395] Scheelite samples from certain localities dissolve slowly. Ferberite and hübnerite are decomposed more easily than the solid solution of the two components, wolframite. However, contradictory data have been published on the solubility of the latter in hydrofluoric acid.[226,396] To decompose tungsten ores containing quartz and silicates, a mixture of hydrofluoric and nitric acids is mostly used. The soluble tungstate is extracted from the evaporation residue by an alkali hydroxide solution.[396] Trace amounts of tungsten are only liberated from silicate rocks after total destruction of the matrix by a mixture of hydrofluoric acid with nitric, perchloric, or sulfuric acid.[397,398] This procedure has recently been preferred to alkaline fusion, as the aqueous extract of the melt contains large amounts of alkali salts, including alkali silicates.[220,399,400] Moreover, it is difficult to quantitatively extract trace amounts of tungsten from the alkaline melt into water.

The most common ore mineral of molybdenum-molybdenite (MoS_2) resists even strong oxidants, unless the material is very finely pulverized. Finely dispersed molybdenite particles may be hermetically closed in the grains of quartz and silicates. For these reasons, the decomposition of molybdenum ores is carried out in a strongly oxidizing medium ($HClO_4$, HNO_3, aqua regia, Br_2, etc.) in the presence of hydrofluoric acid.[220,400,401] Trace amounts of molybdenum are dissolved analogously, as those of tungsten.[402-404] Sedimentary rocks and soils are decomposed by mixtures of hydrofluoric acid with perchloric and nitric acids. In the presence of high contents of carbonaceous admixtures, the sample is ignited with caution at temperatures of less than 500°C, to prevent volatilization of the molybdenum oxide.[402]

An addition of hydrofluoric acid accelerates the decomposition of carbides of niobium, tantalum, titanium, and zirconium by nitric acid solutions.[193,405] Decomposition of nitrides of the transition metals yields ammonium ions and also elemental nitrogen,[406] whose formation is the main cause of erroneous results in the determination of nitrogen by the Kjeldahl method. In a medium of hydrofluoric acid and hydrogen peroxide,[379] even the resistant cubic modification of boron nitride and the silicon nitride, Si_3N_4, is attacked. The oxidizing mixtures of hydrofluoric acid with nitric and hydrochloric acids have also found use for dissolution of borides from the IV to VI groups of the periodical system.[193,405,407] With the acid ratio, 5 + 3 + 3, e.g., CrB_2, WB, MoB, MoB_2, ZrB_2, and TiB_2 are destroyed.[407] Dissolution of metals, their alloys, and chemical compounds in hydrofluoric acid solutions has been surveyed by Malyutina et al.[408] Titanium metal is dissolved in this acid on gradual additions of nitric acid. Fluorides are bound in the reaction mixture by boric acid, and glycerol added prevents hydrolysis on dilution of the solution.[409] Zirconium and tungsten metals behave analogously.[408,410] Some titanium alloys react in mixtures of hydrofluoric and hydrochloric acids substantially faster than in solutions of pure hydrochloric or sulfuric acid.[408]

A mixture of fluoroboric acids and nitric acid has found use in decomposition of refractory bronzes.[411] Fluoroboric acid is used as a selective solvent for silicates in the presence of quartz. Fluorosilicic acid has similar effects. On leaching a polished section surface with 30% H_2SiF_6, layered silicates and feldspars are dissolved within 3 days at 20°C. According to microscopic observations, quartz is not attacked.[412] Similar effects are exhibited by the two acids in decomposition of simple mixtures of some silicate minerals.[413]

C. Volatility of Elements in Hydrofluoric Acid Solution

The presence of hydrofluoric acid accelerates destruction of solids and, thus, facilitates transfer of the components from the solid phase to the solution. The obtaining of solutions for determinations of some elements is, however, complicated in open systems by the formation of volatile fluorides during the decomposition process. To reliably predict the formation and the volatility of such compounds, systematic studies under the conditions that are used in decomposition of various crystal structures are needed, but are not available at present.

The published data on the volatility of fluorides from hydrofluoric acid solutions are often contradictory, even if the application of radioisotopes simplified the techniques for the control of the whole dissolution process. The tabulated data on the physical properties of the fluorides of various elements (boiling points, sublimation temperatures) are sometimes unsuitable for evaluation of the volatility, as they are related to anhydrous systems.[414] However, some of these compounds are not formed under real conditions or undergo rapid side reactions connected with the formation of nonvolatile hydrolytic products (e.g., AsF_5, SbF_5, WF_6, MoF_6, etc.). The use of radiotracers is also problematic in this respect. Experiments with pure, carrierless solutions do not correspond to the decomposition reactions in which the test component is continuously released into the reaction mixture during gradual destruction of the bonding structures. The composition of the mixture, moreover, varies in time. The valence form of the radiotracer is also important, especially with elements that form fluorides with different boiling points. Spiking of a sample of dissolved solid with a radioisotope may involve an interaction in which the valence of the radiotracer changes. More realistic simulation is the dissolution of a solid activated in a reactor, provided that the radioisotopes formed remain in the original structural positions in the irradiated material.

Most contradictory data on the losses of elements in the form of volatile fluorides are connected with determinations of arsenic and selenium. The losses, found experimentally by many authors[414] in analyses of various materials, cannot be explained on the basis of the tabulated boiling points of anhydrous arsenic fluorides. Losses occur during heating of As(III) compounds in solutions of hydrofluoric acid; on the other hand, the losses are minimal in the presence of As(V) compounds. Apparently, arsenic pentafluoride is not formed under the conditions of the decomposition process, or is immediately decomposed with formation of nonvolatile substances. Volatilization of arsenic(III) from nonoxidizing media of hydrofluoric and sulfuric acids has been unambiguously confirmed.[415-418] The losses amount to 45 to 100%, depending on the reaction conditions. It is substantially more difficult to explain losses in arsenic found during evaporation of solutions of hydrofluoric acid and oxidizing acids (nitric and perchloric, or aqua regia).[419-422] In evaporation of a mixture of hydrofluoric and nitric acids, the recovery of the ^{74}As isotope added to a sample of deep sea red clay amounted to only 70%.[416] However, the same procedure has often been recommended for analyses of ores, and no losses have been found.[423]

Some reasons for losses in arsenic during decomposition have been clarified, especially in References 422 and 424 to 429. Bajo monitored volatilization using the ^{76}As radioisotope added, in the form of an arsenous or arsenic salt, to a solution containing hydrofluoric, nitric, and perchloric acids.[424] The experiments were carried out in pure solutions and in the presence of granite with a negligible arsenic content. The losses in the As(III) added were sometimes almost 100%, although the final solution (after evaporation with nitric and perchloric acids at 150 to 220°C) only contained As(V). The losses have been explained by Bajo by a low rate of the As(III) → As(V) oxidation and by volatilization of the unoxidized compounds, most probably in the form of AsF_3. The losses were smaller in the presence of granite, which the author[424] ascribed to the oxidizing effects of some unidentified component in the rock added. The losses are also minimal during gradual, two-stage decomposition of rocks, first by nitric and perchloric acids (dissolution of a major fraction of As), followed by treatment with hydrofluoric and perchloric acids (destruction of silicates).[422,424] An analogous procedure has been applied in analyses of soils and coal ashes,[430,431] but its success depends on the amount and chemical composition of the organic substances present. In analyses of sediments and some soils, a mixture of perchloric and nitric acids is insufficient for liberation of arsenic from the organic substances present, and the oxidizing effect must be increased by adding potassium persulfate or permanganate. Excess permanganate serves as an indicator of the completeness of the oxidation of organic compounds.[422,425]

To prevent losses in arsenic, Aslin[429] has selected the opposite order of adding the reagents. The silicate minerals are destroyed first with hydrofluoric acid. In the subsequent stage, sulfides and other minerals carrying arsenic are dissolved by adding nitric and perchloric acids.

If solutions of nitric and perchloric (or sulfuric) acids are mixed with hydrofluoric acid for decomposition, a stable, oxidizing medium is maintained by adding permanganate.[427,428] No loss in arsenic has been found[432,433] during simultaneous decomposition of sulfides and silicates by mixtures of hydrofluoric acid with hydrochloric and nitric (or perchloric) acids.

Losses in arsenic can also be prevented by controlling the temperature program during dissolution.[434] In decomposition of rocks or minerals by mixtures of hydrofluoric, nitric, and hydrochloric acids, the solutions formed are evaporated at temperatures that do not exceed 80°C. To differentiate between the As(III) and As(V) contents in glass, the samples are dissolved in hydrofluoric acid at laboratory temperature.[435]

The determination of trace contents of arsenic, antimony, selenium, and other elements has recently been considerably improved by the introduction of new analytical techniques, especially the hydride generation in combination with AAS and ICP-AES. Because of the toxicity of arsenic, not only ores, rocks, glasses, and alloys are analyzed, but also environmental materials with varying composition and fluctuating contents of organic substances, e.g., lake, river, estuary and sea sediments, fly ashes, soils, and coal. As a correct decomposition of these materials is a basic condition for obtaining reliable results, the preparation of solutions from solids must satisfy the following rules:

1. The method of decomposition must be selected according to the composition of the test substance, especially in dependence on its reducing properties determined by the contents of sulfides, organic compounds, and other substances.
2. In analyses of materials rich in organic compounds, a two-step decomposition should be selected; the first stage involves treatment with nitric and perchloric acids (or with aqua regia), silicates are then destroyed by hydrofluoric and perchloric acids, followed by destruction of the residues of resistive organic compounds by potassium persulfate or permanganate.[422,424,425]
3. The two-step decomposition can also be performed with a changed order of reagent additions. Silicates and quartzes are dissolved first using hydrofluoric acid. The acid

is evaporated, and sulfides and arsenides that carry most of the arsenic are dissolved by oxidation with a mixture of nitric and perchloric acids.[429] This procedure can also be recommended for analyses of eruptive rocks. With sedimentary rocks containing large proportions of noncarbonate carbon, the liberation of arsenic need not be complete and losses may also occur.[430]

4. Eruptive rocks with small amounts of reducing admixtures can be dissolved in a single step without losses in the arsenic, using hydrofluoric acid in the presence of nitric and perchloric acids and potassium permanganate.[427,428]
5. In all the variants of decomposition of solids with hydrofluoric acid, a maximum acceleration of the oxidation of As(III) to As(V) is most important, as well as the maintenance of strongly oxidizing properties of the solution throughout the process, even in contact with the remaining undissolved solid phase.

Dissolution of solids by mixtures of acids containing hydrofluoric acids in the determination of arsenic requires substantially greater care in maintaining the prescribed reaction conditions than in simple dissolution in strong inorganic acids. Simpler procedures[436,437] are, therefore, used for routine analyses of ores (e.g., for the control of their mining and treatment) and soils. Dissolution in mixtures of nitric acid with sulfuric or perchloric acid or in aqua regia is most frequent. These oxidizing mixtures minimally attack the silicate matrix. The fraction of arsenic extracted from common silicate rocks usually amounts to 70 to 80%, which is acceptable for the purposes of geochemical prospecting.[422,427] The data on the degree of extraction of arsenic from the test soils and rocks are often contradictory. With some materials, arsenic can be completely extracted using only nitric and perchloric acid. Because of the danger of losses, some authors[402,416,436,437] have stopped using hydrofluoric acid and prefer other decomposition methods, e.g., fusion with sodium hydroxide or dissolution in nitric and sulfuric acids.[437]

Although the values given[414] for the boiling points of Sb(III) and Sb(V) fluorides suggest possible losses in the form of volatile compounds (376 and 150°C, respectively), these compounds are not formed during dissolution of the test material.[424] On the other hand, the presence of hydrofluoric acid has a beneficial effect on the whole decomposition process. Oxide and sulfidic antimony ores are more readily dissolved in a mixture of hydrofluoric and nitric acids than in nitric acid alone. The effect of hydrofluoric acid on the completeness of dissolution is especially pronounced with poor ores with a silicate matrix.[438] Similar to decomposition of arsenic materials, the oxidizing character of the solvent is increased by adding permanganate to the reaction mixture.[439] The losses in antimony, indicated by the ^{124}Sb isotope, were less than 0.2% in decomposition of deep-sea sediments by a mixture of hydrofluoric and nitric acids.[440] On dissolution of the residue in hydrochloric acid, all the antimony was transferred from the sample to the solution. Decomposition with hydrofluoric and sulfuric acids is insufficient with sedimentary rocks rich in carbonaceous substances.[441,442] The losses are again not caused by volatilization of fluorides, but rather by a limited oxidizing ability of the solvent. It is thus advisable to enhance the oxidizing effect of the mixture by adding nitric acid.[443] Organic compounds, e.g., cellulose, do not cause losses in antimony. Atmospheric fallout is trapped on a paper filter and dissolved without loss of antimony in a mixture of hydrofluoric and sulfuric acids and hydrogen peroxide.[444] No loss in the antimony content due to formation of volatile fluorides has been found in evaporation of solutions of hydrofluoric acid containing dissolved antimony oxides (high-percentage ores or slags) with sulfuric acid to the formation of dense fumes of sulfur trioxide.[445]

The presence of hydrofluoric acid in the reaction mixture does not cause losses of antimony, but actually facilitates the whole dissolution process, apparently through liberation of the antimony compounds from the particles trapped in the silicate matrix and prevention of the formation of inactive hydrolytic compounds. Losses of antimony observed during evaporation of solutions of hydrofluoric and perchloric acids[419] are more probably caused by the formation of an unreactive, poorly soluble compound than by volatilization of the fluorides. In the paper cited,[424] these losses were not found.

Ambiguous data on losses of selenium during dissolution of a solid are probably caused by various oxidation states of its compounds.[446] According to the tabulated physical constants, all selenium compounds may form volatile products under the decomposition conditions. The losses are most often explained by the formation of hydrogen selenide, the halogenides of Se(IV), and by poor thermal stability of the oxides.

In dissolution of solids in hydrofluoric acid, the losses cannot unambiguously be ascribed to the formation of volatile fluorides, even if the boiling points given and the results of some experiments suggest it.[414] The surprisingly high losses occurring even on mild heating (45°C) of a 20% HF solution to which ^{75}Se(IV) has been added can hardly be interpreted otherwise than as a result of the formation of volatile fluorides.[446] On evaporation to dryness, these losses increase from 21 to 65%.

In pure solutions, the Se(IV) and (VI) compounds retain their oxidation states, even on evaporation with hydrofluoric and perchloric acids[424] (monitored using ^{75}Se). However, in contact with granite, 10 to 80% of the Se(VI) is irreproducibly reduced to Se(IV) during evaporation of the reaction mixture. A uniform oxidation state of selenium compounds cannot be ensured, even by decomposition with a mixture of hydrofluoric, nitric, and perchloric acids. This acid mixture[447,448] should lead to the formation of Se(VI) at the temperature of evaporation of perchloric acid, which is at variance with the results of the previous study.[424] On treatment with hydrofluoric and nitric acids, copper selenides are oxidized up to selenic acid. Experiments carried out with a deep-sea clay impregnated with the ^{75}Se isotope have shown that the processes of dissolution and oxidation lead to no loss of selenium.[449,450]

Because of the bonding of selenium compounds to sulfidic minerals (or to intruded selenides and tellurides), a total decomposition of silicate rocks is not always necessary. However, the carriers of selenium may be trapped in the silicate phases, and, thus, dissolution in nitric and perchloric acids may not lead to liberation of all the selenium.[451]

To destroy silicate minerals and other components, hydrofluoric acid is used in mixtures with nitric and perchloric acids, with possible addition of potassium persulfate or permanganate, or hydrogen peroxide, especially when decomposing soils and sediments.[424,447-456] The order in which the reagents are added depends on the type of the test material. The reaction mixture should not be evaporated to dryness, as part of the selenium may volatilize.[451] However, this fact has not been unambiguously demonstrated. With a suitable decomposition process, a solution may be obtained[424] in which not only selenium, but also arsenic and antimony can be determined. A 1-g rock sample is dissolved in 10 mℓ 65% HNO_3 and 10 mℓ 60% $HClO_4$ placed in a Teflon® beaker which is gradually heated on a hotplate from 150 to 250°C within 60 to 90 min. The liquid volume is evaporated down to 2 to 3 mℓ. Then 5 mℓ 60% $HClO_4$ and 10 mℓ 40% HF are added, and the mixture is evaporated down to a volume of 2 to 3 mℓ. The salts are then dissolved in 20 mℓ of boiling water and the solution is cooled and diluted to a suitable volume.

Selenium is usually bound in glasses in various oxidation forms, and these forms can be differentiated if the decomposition conditions are suitably adjusted.[418]

In contrast to selenium, no perceptible losses of tellurium occur during evaporation of solutions, even in the presence of hydrofluoric and sulfuric acids.[457] In dissolution of glass in the above acids followed by evaporation of the solution on a platinum dish to the appearance of sulfur trioxide fumes, the loss indicated by radioisotopes amounted to only 4% of the 140 μg of Te(IV) added. All the tellurium present in silicate rocks are most readily transferred to solution by treatment with mixtures of hydrofluoric acid with nitric and perchloric acids, nitric and hydrobromic acids, or aqua regia.[458-462] The presence of hydrofluoric acid is not necessary in decomposition of weathered rocks, and the above oxidizing acids suffice for dissolution of tellurium compounds.[462] However, tellurium cannot be quantitatively extracted from unmetamorphosed rocks with these acids, and the silicate minerals must be destroyed[462] (e.g., granodiorite, GSP-1).

The bonding forms of Ge(II) and Ge(IV) compounds in the test material determine the decomposition method. Except for rare germanium minerals, the germanium content is scattered over various "nonspecific" carriers, such as quartzes, micas, feldspars, and other rock-forming minerals. Sedimentary rocks (e.g., cenomanian claystones with fossilized plant detritus) and fly coal ashes may contain part of the germanium bound to residues of degraded organic compounds.

In view of the chemical similarity between the Ge(IV) and Si(IV) compounds, hydrofluoric acid is an indispensable agent for liberation of germanium from the crystal structures of silicon, quartz and silicates. The trigonal modification of germanium dioxide is isostructural with α-quartz. The removal of silicon or silicon dioxide from the test material in the form of the volatile fluoride simplifies the following analytical operations, e.g., isolation of germanium by extraction or by distillation of germanium tetrachloride. The materials are decomposed under oxidizing conditions, to oxidize the present reducing compounds simultaneously with the decomposition. For this purpose, hydrofluoric and nitric acids are used, with an addition of phosphoric or sulfuric acid. However, losses in the germanium content have been encountered during decomposition of the test substances and have been attributed to the formation of volatile tetrafluoride.[419,421,463] The losses were especially pronounced in dissolution of Ge-Si alloys[464] and silicates in a mixture of hydrofluoric and perchloric acids.[463,465] The oxidizing action of perchloric acid on the test material also produces a small amount of hydrogen chloride as a reduction product. The decomposition of perchloric acid with formation of hydrochloric acid is catalyzed by some metals, especially manganese and, probably, also cobalt.[465] The dissolved germanium then volatilizes in the form of the tetrachloride. The losses encountered in analyses of coal ashes sometimes exceed 50% of the germanium content.

Losses of germanium have also been observed during the removal of fluorides by prolonged evaporation with a mixture of hydrofluoric, nitric, and sulfuric acids to the appearance of dense fumes of sulfur trioxide.[466] These losses cannot at present be unequivocally explained. Germanium dioxide reacts with hydrofluoric acid, forming nonvolatile complex fluorides and hydroxyfluorides with varying degree of hydration.[467] Germanium tetrafluoride also forms many nonvolatile hydrated compounds. In prolonged heating of germanium fluoride solutions with sulfuric acid, the degree of hydration decreases, which may also cause the formation of volatile germanium tetrafluoride. Complex fluorides dissociate with formation of crystalline germanium dioxide that is poorly soluble in water and in cold 9 *M* hydrochloric acid. A similar phenomenon has been observed during evaporation of solutions containing germanium(IV) salts and hydrofluoric and sulfuric acids.[468] The losses found in the extraction isolation of germanium tetrachloride may be erroneously attributed to the formation of volatile tetrafluoride. The losses of germanium are minimized if a short evaporation of fluorides with sulfuric acid is carried out under controlled temperature conditions. An addition of potassium or sodium sulfate, or of phosphoric acid and potassium dihydrogenphosphate, also leads to suppression of the losses. However, parallel experiments with decomposition of ashes using various methods have not objectively confirmed a pronounced effect of these compounds.[465] Decompositions with hydrofluoric acid mixed with phosphoric, sulfuric, and nitric acids yielded identical results, without losses of germanium. Germanium-silicon alloys can be dissolved in hydrofluoric and nitric acids, in the presence of citric acid. The latter acid must be present in an excess over the amount of hydrofluoric acid and the germanium content in the sample, otherwise germanium tetrafluoride volatilizes during evaporation.[464]

Although the published data often differ,[419,421,423,465,467,469] it is possible to recommend for determination of germanium in silicate rocks, glasses, and sulfidic ores containing quartz, a mixture of hydrofluoric, nitric, and phosphoric acids as an optimal agent for decomposition.[470-476] A possible procedure follows.[471] A 1-g sample of a sulfidic ore is cautiously dissolved by adding small portions of concentrated HNO_3. The reaction mixture is evaporated on a water bath, and the whole procedure is repeated. Then 5 mℓ 40% HF and 5 mℓ H_3PO_4 (d = 1.59) are added, the dish content is evaporated to obtain a syrupy liquid, with elemental sulfur, if formed, oxidized by more portions of HNO_3 and H_2O_2, and the residue is dissolved. Germanium is then determined, e.g., by AAS after isolation of the hydrides.

When dissolving sulfidic minerals in a mixture of hydrofluoric, nitric, and phosphoric acids, it is necessary to rapidly and completely oxidize the Ge(II) compounds; otherwise, they volatilize. An analogous problem has also been encountered with ashes, silicate rocks, natural silicates of copper, and low-melting glasses.[474,477,478] Phosphoric acid is sometimes replaced by sulfuric acid in the decomposition mixture.

Hydrofluoric acid reacts with boric acid with formation of volatile borotrifluoride and fluoroboric acids. The reaction proceeds in two steps,

$$H_3BO_3 + 3HF \rightleftharpoons HBF_3(OH) + 2H_2O \tag{18}$$

$$HBF_3(OH) + HF \rightleftharpoons HBF_4 + H_2O \tag{19}$$

The slow formation of fluoroboric acid determines the overall reaction rate.[479-482] The equilibration takes from 15 min to 18 hr, depending on the reaction conditions. The highest reaction yield is attained when the solution is cooled in an ice bath.[481] The reaction is catalyzed by hydronium,[480] ferric,[482,483] and cupric[484] ions, which is utilized in practice in the preparation of the initial solution for the determination of boron in steel. The ion associates of fluoroborate with the cations of basic dyes are used for extraction separation of boron from silicon,[480] quartz,[482] beryllium oxide,[485] and silicate rocks containing colemanite.[486] Because of interference from some ions, the extraction is preceded by distillation of boron in the form of volatile trifluoride.[482,487] The distillation is usually carried out in a hydrofluoric acid medium in the presence of sulfuric acid, at 190°C, and in a nitrogen stream, using a Teflon® apparatus. Prior to the determination of impurities in metallic boron and its compounds by the ICP-AES method, it is advantageous to remove the matrix in the form of boron trifluoride.[488] The test sample is dissolved by refluxing with nitric acid, and the solution of the boric acid formed is evaporated to dryness several times with a sufficient amount of hydrofluoric acid. Analogously, boron can be removed that interferes in the spectrophotometric determination of tantalum, using the associate of TaF_6^- with the cation of a basic dye.[489]

In the determination of boron, the formation and volatilization of boron trifluoride during the decomposition must be prevented, and thus metals,[482] alloys,[490] steels,[483] quartz,[482] natural volcanic and synthetic glasses,[491-494] and silicate rocks[486] are decomposed at laboratory temperature or at a mildly elevated temperature in closed polyethylene or polypropylene vessels. The oxidizing properties are maintained by adding hydrogen peroxide in sulfuric or nitric acid. The dissolution of silicon in a mixture of hydrofluoric acid, ammonium fluoride, and hydrogen peroxide is catalytically accelerated by cupric ions.[480] The same catalyst has found use in determination of boron in quartz.[495] An addition of mannitol to the reaction mixture prevents the formation of volatile boron trifluoride.[496,497] The solution of fluorosilicic and fluoroboric acids, obtained by selective dissolution of thin layers of glasses deposited on silicon plates in 5 *M* hydrofluoric acid, can be evaporated to dryness in the presence of mannitol, and the residue can be heated to 180°C without losses of boron.[498] Soil samples are oxidized by gradually adding nitric and perchloric acid. Mannitol and hydrofluoric acid are added after evaporation of excess nitric and perchloric acid, and the reaction mixture is slowly evaporated to dryness. However, parallel experiments did not confirm the protective effect of mannitol in this case.[499] Boron trifluoride is not volatilized together with dissolved quartz during evaporation with hydrofluoric acid, if the solution contains a small amount of phosphoric acid.[500]

The reasons for volatilization of thallium in dissolution of silicate materials have not been satisfactorily elucidated. Volatilization of thallium has been clearly demonstrated during sample fusion with alkali carbonates.[501] On the other hand, the data on losses of thallium during dissolution of silicate rocks are often contradictory. The losses are apparently not caused by the formation of thallium fluorides, but by an interaction with the acid, added to displace the fluorides from the reaction mixture. Losses only occur at elevated temperatures, during prolonged evaporation of excess sulfuric or perchloric acid[502-505] (as verified by an addition of ^{204}Tl). Using the same isotope, it has been confirmed that there is no danger of losses in thallium during decomposition of silicate rocks by solutions of hydrofluoric and nitric acids.[505]

To analyze silicate materials, mixtures of hydrofluoric acid with nitric and perchloric acid or with aqua regia are most often used in practice.[502,504-515] Thallium(III) salts are major reaction products. No loss of thallium has been observed when the reaction mixture was slowly evaporated to dryness at controlled temperature,[509,511,513] even if the mixture contained sulfuric, perchloric, or phosphoric acid.[515] This decomposition procedure has been recommended for the determination of traces of thallium in leucite,[504] zinwaldite,[510] fluorite, sulfides, silicate and carbonate rocks,[502,505,511-515] stone meteorites,[512] kaolins, slags, and environmental materials.[513] In decomposition of coal, the oxidizing effects are enhanced by adding fuming nitric acid.[513] Oxide manganese ores are readily dissolved by a mixture of hydrofluoric and hydrobromic acids.[516] By the action of the bromine liberated (and probably also through interaction with compounds of tri- and tetravalent manganese), the $TlBr_4^-$ bromide complex is formed that is suitable not only for a spectrophotometric determination, but also for an extraction or ion exchange separation.

Similar to thallium, losses of Re(VII) are encountered during evaporation of solutions containing hydrofluoric, perchloric, and sulfuric acids.[517] Isolation of boron from metallic molybdenum by distillation in the form of the trifluoride produces distillates containing small amounts of molybdenum that apparently volatilizes in the form of an oxyfluoride.[482] Unusual losses of beryllium have been observed during dissolution of stone meteorites[518] and dust particles trapped on glass fiber filters.[328] The losses encountered during dissolution of these substances in solutions of hydrofluoric, perchloric, or nitric acid have been explained by the existence of volatile beryllium compounds with an unknown composition.[518]

In monitoring contamination of soils around nuclear power plants, the reaction mixture containing hydrofluoric, nitric, and perchloric acids must not be evaporated for a long time, otherwise Pu(IV) may be oxidized to Pu(VI), forming the volatile hexafluoride, PuF_6, with a boiling point of 62°C.[334-336]

During decomposition of rock samples in quartz test tubes, interaction of phosphates with 48% HF produces volatile phosphorus compounds in addition to SiF_4, and these can be trapped in 0.005 *M* phthalic acid and determined by ion chromatography. No data have been given on the character of these compounds and the condition for their formation.[520]

V. DECOMPOSITION BY PERCHLORIC ACID

Because of its properties, perchloric acid occupies a special position among strong inorganic acids. Perchlorates are readily soluble in water, except for the ammonium, potassium, rubidium, and cesium salts. Compared with other acids, perchloric acid exhibits a minimal tendency to complexation reactions. Hot, concentrated perchloric acid belongs among the strongest oxidants, but its oxidizing properties depend on the acid concentration and the temperature of the reaction mixture.[521] Even boiling 50% acid oxidizes only slowly ferrous compounds. Boiling 60% acid readily oxidizes vanadyl ions to vanadate. A 72% solution of the acid oxidizes chromium(III) salts to compounds of chromium(VI).

The oxidation effect in hot concentrated perchloric acid is caused by liberation of active oxygen through the reaction,

$$4\ HClO_4 \rightarrow 2\ Cl_2 + 7\ O_2 + 2\ H_2O \qquad (20)$$

The actual reaction is, however, much more complicated. The results of gas chromatographic analysis indicate that the decomposition products involve O_2, Cl_2, HCl, Cl_2O, and ClO_2. The composition of the reaction products is also affected by some metals present in the substances analyzed, especially manganese and cobalt, that increase the amount of hydrogen chloride formed.[522] Compounds of vanadium, chromium, and molybdenum have also catalytic effects. Perchloric acid oxidizes organic compounds to carbon dioxide, and this reaction is explosive in concentrated, hot solutions. Therefore, the oxidation is carried out at a gradually increased temperature in the presence of catalysts or auxiliary oxidants (H_5IO_6, HNO_3) that degrade organic substances to smaller fragments, readily oxidizable by perchloric acid. On evaporation, the acid concentration attains the value determined by the composition of the azeotropic mixture with water (72.4% $HClO_4$, bp 204°C). Heating with fuming sulfuric acid causes dehydration of perchloric acid, so that its concentration may even exceed 85%. An uncontrolled increase in the perchloric acid concentration during decompositions, which is the main cause of explosive reactions, is prevented by using a reflux condenser maintaining a constant acid concentration (and, thus, also the redox potential) in the reaction mixture.

Diamond is not attacked by perchloric acid, but graphite can be oxidized up to carbon dioxide gas. This procedure has been successfully utilized in determination of admixtures (S, As, Sb, and Se) in reactor graphite, pyrolytic carbon, active charcoal, and coal.[523-526] Perchloric acid predominates in the decomposition mixture with periodic acid or sodium dichromate added. The reaction products of periodic acid are iodic acid, iodine, and iodine perchlorate. The presence of these substances in the reaction mixture suppresses the danger of volatilization of arsenic compounds. The composition of the reaction mixture and its temperature must be carefully controlled during the decomposition process. The perchloric acid concentration is regulated by various condensers (e.g., the Bethge condenser) inserted in the mouth of the flask. A scrubber, filled with 3% hydrogen peroxide solution, is used for trapping volatile sulfur compounds. Even small changes in the concentration of the perchloric acid added (the starting acid) affect the dissolution rate. For 1-g coal samples, 68% acid is recommended as safe. With strongly foaming samples, even 71.3% acid causes explosion. Therefore, coal samples with an unknown composition are first orientatively dissolved using small, 10-mg portions. For safety reasons, a small amount of sulfuric acid is added to the sample during decomposition.[526] Preliminary mineralization of the material by a mixture of nitric and sulfuric acids has also given good results.[525] The following procedure has been proposed[526] for determination of ^{90}Sr in reactor graphite. To a 1-g sample in a test tube, 3 mℓ 70% $HClO_4$ is added, with 5 mg $Na_2Cr_2O_7$ and 630 mg of $Sr(NO_3)_2$ as a carrier. The test tube is inserted into the hole in an aluminum block with regulated temperature. Heating is carried out for 1 hr at 75°C, and then the temperature is gradually increased up to 240°C. The volume of the acid is continuously replenished, and at the end the liquid is evaporated to 0.5 mℓ. Explosion may occur on complete evaporation of the acid. An advantage of this procedure is a small amount of the acid required for dissolution of 1 g of the substance. Therefore, it is possible to perform the decomposition in quartz test tubes without a condenser. Larger samples are oxidized by 5 mℓ of fuming nitric acid and 50 mℓ of 70% perchloric acid. After 1 hr of heating at 140°C, the temperature is increased up to 180 to 190°C. The condensate from the air condenser is continuously returned to the reaction mixture.[524] Coal is treated similarly — 1.5 g of H_5IO_6 and 50 mℓ of 68% $HClO_4$ suffice for the oxidation of a 1-g sample.[523]

Preliminary ignition of rock samples in an oxidizing atmosphere and following leaching with aqua regia cannot always secure complete oxidation of the carbonaceous compounds present. Sorption activity of organic matter residues causes complications in decomposition of gold ores, as dissolved gold compounds are transferred back from aqua regia to the undecomposed fraction. Leaching of samples with aqua regia has thus been replaced in determinations of gold by decomposition with a mixture of nitric, perchloric, and sulfuric acids.[527] Most of the carbonaceous compounds are oxidized in the stage of liberation of nitrogen oxides and perchloric acid vapors; a smaller part is oxidized during appearance of sulfur trioxide vapors. To prevent reducing effects of the decomposition products of sulfuric acid, a mixture of nitric and perchloric acids is added to the solution. A 10-g sample is dissolved in 60 mℓ of a 1 + 1 + 1 mixture of nitric, perchloric, and sulfuric acids. The solution is heated to the appearance of sulfur trioxide vapors. Before the elapsing of 30 min, 70 mℓ of a 1 + 1 mixture of nitric and perchloric acids is added, and the sample is heated for 20 min. The wet oxidation permits mineralization of up to 10 g of sample containing up to 30% of carbon. Sulfidic materials must be roasted before dissolution.

Difficulties in prediction of the behavior of perchloric acid, i.e., its tendency to explosions, have necessitated formulation of safety regulations for manipulation with it.[528-535] These rules mostly concern the problems of safe mineralization of organic substances, especially biological objects. Chapters on the behavior and use of perchloric acid have become an inseparable part of all handbooks on safety in chemical laboratories. In analyses of inorganic substances, certain risk is involved, especially in decomposition of samples with carbonaceous admixtures, such as various sediments, sedimentary rocks, and environmental materials. It is then recommended to dissolve minimal amounts of material in perchloric acid containing a sufficient amount of nitric acid. Explosions have been reported[533,534] in dissolution of bismuth metal and in crystallization of $Co(ClO_4)_2{\cdot}6H_2O$. However, the preparation of bismuthyl perchlorate is, in our experience, without problems, even when the amount of the initial substance (basic Bi carbonate) is as much as 10 g.

Perchloric acid acts during decomposition not only as a strong oxidant, but also as an acid with a high boiling point and dehydrating effects. Evaporation with perchloric acid converts salts of volatile acids (chlorides, nitrates, and fluorides) into perchlorates. The removal of fluorides from reaction mixtures with high concentrations of aluminum or zirconium(IV) ions is usually incomplete, and the whole procedure must be repeated several times, possibly with boric acid added (Chapter 4, Section IV).

Decompositions are mostly carried out in beakers or dishes made of chemical or quartz glass, Teflon®, and platinum. Glassy carbon vessels are unsuitable for the purpose, as their surface is damaged during evaporation.[536,537] Fume cupboards for work with perchloric acid must be made of a material resistant toward the acid (stainless steel, special ceramics, or high-silica glass, possibly protected by a Teflon® lining). The fume cupboard walls and ventilation pathways are periodically freed of perchlorate deposits by rinsing with water.

Perchloric acid is unsuitable for decomposition of materials with high contents of rubidium and cesium, as sparingly soluble perchlorates precipitate during dissolution or evaporation of reaction mixtures. However, in decomposition of potassium feldspars (using a mixture of HF and $HClO_4$), no loss of potassium through precipitation of potassium perchlorate has been observed. The sample weight must be decreased below 0.5 g and 1 to 2 mℓ $HClO_4$ used for displacement of the fluoride. Negative errors observed in determination of potassium have always been connected with formation of sparingly soluble complex fluorides (see Section IV) and not with the formation of potassium perchlorate.

The dehydrating effects of perchloric acid have been useful in isolation of small amounts of silicic acid or of the insoluble residue from pure limestones, dolomites, magnesites, apatites, Sr and Ba compounds and natural and synthetic, calcium-containing silicates (wollastonite, cement, etc.). Hydrochloric acid is not sufficient for dehydration of silicate gel, and application of sulfuric acid causes precipitation of sparingly soluble sulfates of the alkaline earth metals. Perchloric acid is well suited for decomposition of alunite ores that are difficult to dissolve in mineral acids because of a high aluminum content. An excess of the acid must be added to prevent dehydration of aluminum sulfate which is poorly soluble.[538] Lead(II) perchlorate is thermally unstable, and decomposes to hardly decomposable compounds at temperatures above 130°C. These decomposition products may adsorb up to 60% of the bismuth present. Galena is decomposed for determination of bismuth as follows:[539] a 0.1- to 0.6-g mineral sample is dissolved by boiling with 15 mℓ HCl. Hydrogen sulfide is expelled, and the solution is evaporated to one half. A volume of 5 mℓ of 70% $HClO_4$ is added, and the reaction mixture is evaporated to the formation of white fumes. It is removed from a hotplate while still liquid so that the contents of the beaker solidifies only after cooling. A volume of 10 mℓ of tartaric or citric acid is added, and the mixture is alkalized with a 10% solution of NaOH, until the lead(II) hydroxide precipitate dissolves. The solution is heated to the boiling point and acidified with 7 to 10 mℓ of 5 *M* $HClO_4$. An analogous procedure is suitable for the determination of silver in the same mineral.[540] Decomposition of galena with perchloric acid should be substantially faster than with hydrochloric or nitric acid. Some authors[541,542] have not observed an increased thermal lability of lead(II) perchlorate.

Dissolution of solid phases with high antimony and tin contents in $HClO_4$ produces poorly soluble hydrated oxides. Only exceptionally has this phenomenon been used, e.g., for separation of aluminum from tin in analyses of polycomponent Cu alloys.[543] An alloy sample dissolved in a mixture of hydrochloric acid with hydrogen peroxide is evaporated to the appearance of fumes, adding 5 mℓ $HClO_4$. The $Sn(OH)_2O$ precipitate does not contain adsorbed aluminum. Evaporation of the reaction mixture with nitric and perchloric acid, controlled by additions of ^{122}Sb and ^{124}Sb, at temperatures higher than 190°C produces a dehydrated species (probably Sb_2O_5) that adheres on the vessel walls[544-546] and is difficult to reduce to an Sb(III) compound. The formation of antimony pentoxide after this decomposition procedure is probably the main cause of negative errors in the determination of this element in soils and sea sediments. The insoluble reaction product is apparently lost during filtration.[546] Sulfuric acid has been added to prevent the formation of these compounds,[545,547] but Donaldson[548] has been unable to confirm this effect during his detailed study of determination of antimony in sulfidic ores. In decomposition of the material with a mixture of nitric and perchloric acids, the antimony was not quantitatively converted to the pentavalency, even in the presence of sulfuric acid. The poorly reactive compound formed was identified as Sb_4O_8, i.e., an equimolar mixture of the Sb(III) and Sb(V) oxides.

In contrast to antimony, a mixture of nitric and perchloric acid has been suitable for dissolution of solids for determination of arsenic. No loss in arsenic occurs in decomposition of sulfidic minerals and leaching of soils and ores.[549-551] The critical moment is the final stage of the decomposition, when a small amount of perchloric acid must remain after evaporation of the reaction mixture.[551] A 0.25-g rock sample is heated with 1 mℓ HNO_3 and 2 mℓ 70% $HClO_4$ at 150 to 200°C. The reaction mixture volume must not decrease below 0.5 mℓ. After cooling, 10 mℓ of 5 *M* HCl is added, and the undecomposed fraction is centrifuged off. The solution is used for a determination of arsenic and selenium.

The problems of losses in selenium and its valence changes during decomposition are discussed in Section II. Evaporation with perchloric acid leads to the formation of Se(VI) compounds,[552,553] but an unambiguous prediction of the selenium valence state after the decomposition is still impossible. The observed valence changes in the selenium compounds can be explained by poorly defined catalytic effects of the test substance in solutions containing hot perchloric acid.[550] No loss in selenium occurs with controlled temperature regime,[554] but the evaporation time should be as short as possible.[555]

On evaporation of mixtures of perchloric and hydrobromic acids, the bromides of arsenic, antimony, and tin volatilize, arsenic in the form of $AsBr_5$, while the species for antimony has not been specified ($SbBr_5$?). This decomposition procedure simplifies analyses of some natural and synthetic arsenides, sulfides, and sulfosalts of the above elements, including gallium arsenide, and is primarily used for trace analyses. The test substance is usually dissolved in aqua regia, hydrobromic and perchloric acids are added, and the reaction mixture is evaporated to white fumes. The procedure is given in Section III.

Perchloric acid is the only acid, in addition to sulfuric acid, capable of dissolving even highly resistant substances, e.g., monazite, chromite, and baryum sulfate. Samples of finely pulverized monazite (best under 325 mesh) or a monazite concentrate are usually less than 1 g and are decomposed either by perchloric acid alone, or by its mixtures with nitric or hydrofluoric acid or hydrogen peroxide.[556-559] The decomposition takes 3 to 8 hr, even with small samples, and the reaction mixture must be heated to 150 to 190°C, with replenishment of perchloric acid. Cerous salts are oxidized to ceric ones during dissolution. On dilution with water, a yellow precipitate separates (probably basic ceric phosphate) and is reduced to a cerous salt, e.g., by adding hydrazinium chloride. It is better to dissolve the residue in concentrated hydrochloric acid, in which ceric compounds are partially reduced on heating, and thus precipitation of poorly soluble products is avoided.[559] A 0.2-g sample is dissolved in 30 mℓ of a 4 + 1 mixture of nitric and perchloric acids, by heating for 6 hr at 120°C. The temperature is increased to 150°C, and the contents of the beaker are evaporated to dryness. To the residue, 25 mℓ concentrated HCl is added, the mixture is diluted with water, heated, and the insoluble residue is filtered off; the latter contains negligible amounts of the rare earths. On the other hand, xenotime is not quantitatively dissolved in this way, and the undecomposed residue contains 60 to 80% of the yttrium and heavy rare earths present, whereas lanthanum, cerium, and praseodymium phosphates are almost completely dissolved. The dissolution is not improved by adding hydrochloric acid. Sulfuric acid is a better solvent for xenotime, and the mineral can also be decomposed by fusion, best with a mixture of sodium carbonate and peroxide.

Spinels are not generally soluble in perchloric acid solutions, but chromite is attacked under the same conditions,[560] with formation of Cr(VI) compounds. To increase the oxidizing effects of the solvent, sulfuric acid or an alkali sulfate is added to the reaction mixture. The reaction time required varies from 3 to 10 hr, according to the mineral composition. The procedure is primarily used for determinations of major components — not chromium, because a small part of the chromium volatilizes in the form of chromyl chloride formed by the reaction with hydrogen chloride (a product of decomposition of perchloric acid). The chromic acid formed may be partially reduced on dilution of the solution, by hydrogen peroxide that is formed by the reaction,

$$2\ HClO_4 \rightarrow Cl_2 + 3\ O_2 + H_2O_2 \qquad (21)$$

This effect is eliminated by rapid cooling of the dilute reaction mixture.[521] To completely volatilize chromyl chloride, a stream of dry hydrogen chloride is introduced into a hot solution of the concentrated acid, or sodium chloride is added in small portions until brown solution coloration turns weakly yellow. Thus, the analytical procedure is simplified, e.g., in determination of Al, Ca, Mg, and the Pt metals in rocks and ores containing 20 to 98% of chromite.[561,562]

Hot, concentrated perchloric acid dissolves barium sulfate used as the collector of actinoids. The barium sulfate precipitate also adsorbs the lanthanoids under the same conditions. In analytical practice, this preconcentration procedure is mainly used in environmental control, especially for determination of the contents and isotopic composition of actinoids (all α-emitters, from radium to californium, except for Po) in dusts, soils, ores, and biological objects.[563-567] Barium sulfate, precipitated from an acid solution of an alkali sulfate (*circa* 2 to 20 mg $BaSO_4$), is heated in a test tube with 5 mℓ of 72% $HClO_4$ to dissolution. This stage of analysis is immediately followed (after medium adjustment) by extraction separation of the actinoids.[563] Dissolution of rich bastnesite-parisite ores containing barite in hot, concentrated perchloric acid leads to partial degradation of barite, followed by reprecipitation of barium sulfate on dilution of the reaction mixture, with part of the rare earth ions (especially Ce^{3+}, Ce^{4+}, and La^{3+}) trapped in the lattice of the latter. Complete dissolution of barite cannot be achieved in this way. Trapping of the lanthanoids in the barium sulfate lattice is the main cause of negative errors in their determinations in ores containing barite.

Perchloric acid, in combination with other acids (mostly nitric), has become a reliable solvent for liberation of trace elements from rocks, soils, sediments, and many environmental materials. A mixture of $HClO_4$ and HNO_3, possibly containing HF, efficiently attacks compounds of cadmium and lead in river sediments.[568] However, lower results are usually obtained for chromium than indicated by the recommended values of the standard reference materials. These errors may be caused by volatilization of chromyl chloride or by resistivity of some phases toward these acids.[569] Losses in ^{51}Cr have been measured during evaporation with $HClO_4$ and its mixture with H_2SO_4, at 110, 150, and 200°C. The greatest losses (more than 90% Cr) occurred during heating with $HClO_4$ alone (200°C, 3 hr) or with its mixture with nitric acid under the same conditions. These losses in ^{51}Cr did not occur in the presence of sulfuric acid, even at this high temperature.[570] During evaporation of perchloric acid solutions, osmium and ruthenium tetraoxides volatilize, and this medium is used for their distillation separation from the other platinum metals. Compounds of Re(VII) and substantial amounts of boric acid also volatilize.[571]

An interesting application is dissolution of lead buttons obtained in reductive fusion.[572,573] A 30- to 60-g button is dissolved by treatment with 250 mℓ 70% $HClO_4$ and 25 mℓ of anhydrous acetic acid. Platinum and palladium are separated from the solution by reduction with formic acid.

In multielemental determinations of metal admixtures in atmospheric fallout, chloric acid has also been used instead of perchloric acid.[574] Parallel analyses of ashes and a river sediment have shown that the procedure with chloric acid is more effective than that with perchloric acid. The active component of the reaction mixture is chlorine dioxide that is formed as a decomposition product, together with hydrochloric and perchloric acids and gaseous chlorine. To 25 mg of dust, 2 mℓ HF, 4 mℓ HNO_3, and 6 mℓ 20% $HClO_3$ are added. The mixture is heated in Teflon® vessels placed in an aluminum block, at temperatures of 80, 130, and 150°C, always for 1 hr. The residue in the vessel, with a volume of 1 to 2 mℓ, consists of a mixture of nitric and perchloric acids. At temperatures above 170°C, losses in chromium have been observed, apparently in the form of chromyl chloride. Commercial chloric acid is mostly a 20% solution. It is unstable at concentrations greater than 30% and decomposes explosively. The main admixture is perchloric acid.

VI. DECOMPOSITION BY SULFURIC ACID

Sulfuric acid has been extensively used in decomposition techniques, primarily because of its high boiling point; its constant-boiling mixture with water (98.7% H_2SO_4) distills at 338°C. It is commercially available as a 96 to 98% solution (18 *M*, d = 1.84). Salts of volatile acids (nitrates, chlorides, bromides, and fluorides) are readily converted into sulfates by evaporation with sulfuric acid. Hot sulfuric acid exhibits weakly oxidizing effects and oxidizes, e.g., ferrous salts to ferric ones. Sulfur dioxide is usually its reduction product, but elemental sulfur or hydrogen sulfide may also be formed, depending on the character of the substance dissolved. Hot, dilute sulfuric acid has no oxidizing properties, and thus is used, e.g., in the mixture with hydrofluoric acid, as the solvent in determination of the ferrous oxide content in silicates.

Natural and synthetic sulfates of strontium and barium are very difficult to decompose. They are dissolved in hot, concentrated sulfuric acid, probably with the formation of hydrogensulfate; the decisive role in dissolution is played by the water content in the reaction mixture.[575] Calcium sulfate is dissolved even in dilute hydrochloric acid. The formation of sparingly soluble, sorption-active lead(II) sulfate is the source of the most important error in dissolution of solids (ores, Pb-Sb alloys), not only in determination of lead, but also in that of other components, such as silver, bismuth, and antimony, that are trapped in the precipitate. In the presence of barium ions, sparingly soluble, binary Pb-Ba sulfates precipitate. In evaporation of sulfuric acid solutions, the sulfates formed are dehydrated [e.g., those of Fe(III), Cr(III), Ni(II), and Th(IV)] and the anhydrous salts are difficult to dissolve.

The boiling point of sulfuric acid and also its efficiency as a solvent are increased by addition of an alkali sulfate. After evaporation of a solution to dryness, the decomposition of resistant substances can be completed by fusion of the reaction mixture with the hydrogensulfate formed (Ti compounds, actinoids). Decompositions with sulfuric acid are carried out in glass, quartz, or platinum vessels. Platinum is highly resistant toward sulfuric acid of any concentration, at temperatures below the boiling point. The amount of SiO_2, dissolved from quartz vessels, does not exceed 1.5 mg/100 cm^2 of the vessel surface.[576] Prolonged evaporation in Teflon® dishes causes corrosion of their surface.

In dissolution of solids, mixtures of sulfuric acid with other acids or oxidants and reductants are used more often than the acid alone. An efficient solvent is, e.g., a mixture of sulfuric and nitric acids (determination of As, Sb, Se, leaching of elements from polyphase materials, etc.), or combinations of sulfuric acid with hydrofluoric, nitric, and perchloric acid in total decomposition of silicates. Mixed with perchloric acid, sulfuric acid acts as a dehydrating agent, which substantially increases the oxidizing effects of the reaction mixture (dissolution of chromites). Boiling with concentrated sulfuric acid in the presence of phosphoric acid permits dissolution of not only resistant oxides, but also resistant silicates, such as tourmaline, various garnets, and micas. In the presence of potassium permanganate, cinnabar is attacked and on addition of manganese dioxide even high-content chromium ores are dissolved. In solutions of sulfuric acid with hydrogen peroxide, many soluble peroxocomplexes are formed, and some of them have found use in decomposition techniques (especially those of W, Nb, and Ta). Certain derivatives of sulfuric acid have rarely been used, especially Caro's acid (H_2SO_5), peroxydisulfuric acid, and sulfamic acid.

Antimony and tin metals are slowly dissolved in dilute sulfuric acid, but are rapidly attacked by hot, concentrated acid and converted into antimony(III) and tin(IV) salts. As hydrochloric, nitric, and perchloric acids cannot be used for the preparation of stock solutions from these pure metals, the above procedure has often been employed in analytical practice for the preparation of stock solutions of antimony(III) and tin(IV) salts. A detailed study[577] has, however, shown that a small amount of Sb(V) salts is formed together with salts of Sb(III). The natural alloy of arsenic and antimony, allemontite, is dissolved within 5 min (0.2 g, 4 mℓ H_2SO_4), and the arsenic is dissolved to obtain tervalency. Samples of tungsten bronzes are attacked analogously.[578] The alloy is dissolved in sulfuric acid containing ammonium sulfate. After cooling, hydrogen peroxide is added to prevent precipitation of tungstic acid. An addition of hydrogen peroxide has also been found useful in the preparation of solutions of steels for determination of boron.[579] The decomposition is carried out at 60 to 70°C in 50% sulfuric acid, in the presence of potassium hydrogenfluoride in a flask with an air condenser. To determine the same element in graphite, the sample must be oxidized in a quartz crucible by a 9 + 1 mixture of sulfuric and Caro's acid at 150°C for 3 hr.[580] Elemental selenium is dissolved in the concentrated acid with formation of the polyatomic, green cation Se_8^{2+} that is converted[581] into the yellowish species Se_4^{2+}.

Sulfuric acid is suitable for dissolution of oxidic compounds. Most natural and synthetic Ti oxides are dissolved, e.g., anatase, but also brookite.[582] In our experience, natural rutile is incompletely dissolved, even in the hot, concentrated acid, with an addition of an alkali sulfate. However, the data published[583,584] on the decomposability of this material differ greatly. The existence of various crystal modifications of titanium dioxide is usually the main cause of difficulties in the preparation of a stock solution of a titanium(IV) salt from the high-purity dioxide. Most preparations contain, depending on the thermal pretreatment, variable amounts of the rutile and anatase forms, affecting the behavior of the preparation during dissolution in sulfuric acid. For destruction of titanium minerals, e.g., ilmenite, titanomagnetite, and perovskite, the reaction mixture temperature is increased by adding sodium or ammonium sulfate.[585] However, the decomposition of these phases is sometimes unsuccessful and is often supplemented with fusion with alkali sulfates. Depending on the reaction conditions, titanium(IV) or titanyl sulfate solutions are obtained in dissolution of Ti compounds. The behavior of the two ions in solution greatly differs. Whereas titanium(IV) ions readily hydrolyze even in dilute sulfuric acid, titanyl ions are stable even in 1% sulfuric acid. The formation of the individual ionic species depends on the sample pretreatment. If dissolution is preceded by sufficiently long fusion of the material with potassium disulfate (at least 40 min, after the formation of a clear melt), then the solution obtained by dissolving the melt in dilute sulfuric acid contains titanyl sulfate.[586] A combination of the decomposition with final fusion of the residue with an alkali disulfate has also been recommended for loparites, titanoniobates, and titanotantalates. The solution is stabilized by an addition of hydrogen peroxide, tartaric, or oxalic acid. In this way solutions are prepared, e.g., for determination of impurities in niobium(V) and tantalum(V) oxides.[587] Quartz is minimally attacked. Natural and synthetic zirconium dioxide is also resistant toward sulfuric acid.

The resistance of cassiterite toward sulfuric acid and its mixtures with hydrofluoric acid is the principle of phase analyses of rocks with tin ore mineralization. The tin bound to cassiterite is usually differentiated from that built into the structure of silicate minerals (micas). The decomposability of cassiterite, genetically bound to the cassiterite-quartz or cassiterite-sulfidic formation, has been studied.[588] On heating of the mineral, mixed with pure quartz at a 1 + 9 ratio with sulfuric acid at a temperature below the boiling point, only 3.7% Sn was dissolved (quartz-cassiterite formation, particle size below 4 μm). With the mineral separated from the cassiterite-sulfidic formation, the tin content in the solution amounted to mere 1.9% (particle size 4 to 70 μm). Crushing of the mineral did not lead to an increase in the reactivity of its surface and a change in the solubility.

Nuclear fuels based on plutonium dioxide and mixed Pu(IV)-U(IV)-Th(IV) oxides are dissolved by heating with sulfuric acid with ammonium sulfate added and in the presence of nitric acid.[589] The fuel behavior depends on the relative contents of the individual oxides. The dissolution takes 4 hr in a pulverized mixture of Pu and Th oxides. High-temperature ignited thorium dioxide is difficult to dissolve. The solution is diluted with water and allowed to stand for 12 hr for anhydrous thorium(IV) sulfate formed to dissolve. If the U(VI)/U(IV) ratio is to be determined in mixed uranium oxides, the fuel is dissolved in 1.5 *M* sulfuric acid containing a known excess of a ceric salt.[590]

Synthetic corundum cannot be dissolved even in hot concentrated sulfuric acid. As total dissolution of most of the crystal modifications of Al_2O_3 takes a long time, decomposition is often replaced by leaching of the substance to be analyzed, e.g., by dilute sulfuric acid (1 + 9) with an addition of HF for determination of sodium. This analytical procedure is only suitable for nonrefractory oxides,[591] but even then erroneous results may be obtained, as demonstrated by analyses of impure aluminum oxides containing 0.5% Na.

Various antimony oxides of variable composition and degree of hydration are dissolved slowly and incompletely in sulfuric acid. These substances are generally formed in technological treatment of antimony ores or during oxidation of sulfidic minerals. They are most readily dissolved using alkali fusion. Spinel structures resist sulfuric acid, depending on the contents of individual oxides and their ability to participate in redox reactions. Spinel is not attacked by the acid, even in the presence of strong oxidants. Chromite also does not dissolve in H_2SO_4, but its crystal structure is degraded in a mixture of H_2SO_4 and $HClO_4$. Owing to the oxidizing effects of perchloric acid, magnified by the presence of sulfuric acid, chromic acid and ferric ions are formed. Artificial chromites containing lanthanum, employed as refractory materials, can be dissolved in acids, e.g., in a mixture of nitric and sulfuric acids, with an addition of perchloric acid before completion of the decomposition.[592]

Dissolution of sulfidic ores in concentrated sulfuric acid leads to complex products, formed by interaction of the test material with the solvent. In addition to the components dissolved from the test substance, these products involve elemental sulfur, sulfur dioxide, and possibly other sulfur compounds which exert reducing effects on compounds of gold and the platinum metals. The effect of the reaction mixture is enhanced by adding other reductants. This phenomenon has been employed for separation of noble metals from an excess of common metals. In this way a complex ore or rock matrix can be substantially simplified for further analytical operations. Depending on the reaction conditions, the platinum metal minerals remain unattacked in the undecomposed residue, or, if dissolved, are reprecipitated by addition of reductants. In the sulfatization decomposition,[593] the sample is mixed with concentrated sulfuric acid, a reductant (starch, thiourea, etc.), and is heated to 300 to 375°C. After cooling, the accompanying metals are leached with water from the mixture. The following procedure has been employed[594] for dissolution of the test material in determination of rhodium and iridium in copper-nickel concentrates. A 5-g concentrate sample is heated at 160°C with a twofold excess of ammonium hydrogenfluoride, 10 mℓ of sulfuric acid (1 + 1) are added, and the mixture is evaporated to a wet residue. The residue with a thick paste consistency is transferred to a large beaker, 20 mℓ of concentrated sulfuric acid is added, then 2 g of thiourea, and the mixture is heated to 230°C. The solution is then diluted with 600 mℓ of water and heated until the common metal salts are dissolved. The solid fraction containing the platinum metals is filtered off, fused with sodium peroxide, and further treated. A similar procedure has given good results for rock materials with disseminated Cu-Ni ores.[586] Repeated leaching of the material with a mixture of sulfuric, hydrochloric, and perchloric acids solves the difficult problem of treatment of large samples of chromium ores (up to 30 g) for determination of the platinum metals. The silicate structures are destroyed first, and then chromite is dissolved by the action of perchloric and sulfuric acids. The hexavalent chromium compounds are then converted into volatile chromyl chloride by adding sodium chloride to a boiling mixture of perchloric and sulfuric acid. The perchloric acid is displaced, and the residue is further treated. In a suspension of manganese dioxide in H_2SO_4, unstable Mn(IV) sulfate is formed on heating, which oxidizes chromite up to chromic acid.[595,596] The decomposition is useful both for poor and rich chromium ores. A 0.1-g ore sample is mixed with 0.2 g MnO_2, 25 mℓ H_2SO_4 (1 + 1) is added, and the mixture is heated to the appearance of sulfur trioxide vapors. After dilution with 50 mℓ of water, the whole procedure is repeated. Copper ores and their concentrates are oxidized under the same conditions.[597] A disadvantage of this procedure is introduction of a considerable amount of the solid oxidant into the reaction mixture (up to 3 g MnO_2 per gram of the sample) and difficult control of the dissolution process.

Most common sulfides are rapidly dissolved in hot, concentrated sulfuric acid. Pyrrhotite, sphalerite, and antimonite are dissolved. Arsenides, sulfoarsenides, antimony, and tin sulfosalts are rapidly destroyed. Löllingite, rammelsbergite, arsenopyrite, proustite, tetrahedrite, and tennantite are also attacked. Sulfidic antimony ores are usually dissolved in the presence of reductants that accelerate the decomposition; a piece of filter paper or tartaric acid suffices for the purpose. These substances are dehydrated in the presence of concentrated sulfuric acid and yield carbonaceous compounds with reducing properties.[598] To 0.2 g of a sulfidic antimony ore in a conical flask, 12 g of potassium disulfate, 0.5 g of tartaric acid, and 15 mℓ of concentrated sulfuric acid are added, and the temperature of the reaction mixture is slowly increased until the tartaric acid is completely oxidized, and the sample dissolved. The residue in the flask containing a small amount of sulfuric acid is diluted with water containing 3 g of tartaric acid and heated until the salts dissolve. This procedure is unsuitable for galena, as sparingly soluble lead(II) sulfate is formed. Pyrite and molybdenite are only minimally attacked by hot, concentrated sulfuric acid. Cinnabar is not dissolved even in boiling acid, but is quantitatively dissolved on addition of solid potassium permanganate.[599,600] To 1 g of the ore 5 mℓ of concentrated sulfuric acid is added, and 0.5 g of solid potassium permanganate is added with continuous stirring. The contents of the flask are stirred for 20 min and then heated to the boiling point. After cooling and dilution of the reaction mixture with water, the Mn oxides formed are reduced by an addition of hydrogen peroxide or hydroxylammonium hydrochloride. A virtually identical procedure has been used for dissolution of cinnabar from heavy-mineral concentrates, where it is accumulated as a resistant, heavy mineral.[601] A mixture of nitric and sulfuric acid has been used in determination of mercury in some sulfidic ores; however, cinnabar is not dissolved without an addition of potassium permanganate.[602] Mercury sulfides are present as microscopic intrusions in soils, sediments, and coal.[603-606] Total dissolution of mercury from these polyphase systems requires an increased caution of the analyst, especially during decomposition. Mercury is contained in these materials not only as a sulfide, but also in the form of volatile organometallic compounds.[607] The dissolution is usually carried out using a mixture of nitric and sulfuric acids, with gradually increased temperature in a vessel with a reflux. Potassium permanganate or vanadium pentoxide added oxidizes not only mercury sulfides, but also organometallic complexes and carbonaceous substances in the sample. The preparation of the solution is especially difficult in analyses of coal, as the potassium permanganate amount added must also suffice for the oxidation of the coal matter.[604,608]

Determination of arsenic in soils and fly ashes has become actual in view of environmental contamination. Arsenic compounds are usually leached from soils by acids, without complete dissolution of the material. Total decomposition by hydrofluoric acid in an acid mixture requires careful maintenance of the reaction conditions in an open system, as arsenic may volatilize in the form of fluorides. Therefore, the test soil is usually only leached with a mixture of sulfuric, nitric, and possibly also perchloric acid. Among ten tested procedures for dissolution, the best results have been obtained[609] with a 1 + 2 mixture of sulfuric acid (1 + 1) and nitric acid. Soils and ashes are also leached with solutions of sulfuric and perchloric acids, with an addition of sodium molybdate for catalytic acceleration of the oxidation of organic substances.[610] A 0.1-g sample is wetted with water and heated in a conical flask with 1.5 mℓ of a 3 + 4 mixture of sulfuric and perchloric acid with 3 mℓ of 5% sodium molybdate. The temperature is increased to 150°C and maintained until the solution clears and the insoluble fraction turns white. The solution is then diluted with water, heated, and can be used for analysis.

Sulfuric acid still remains an indispensable solvent in analyses of monazite and xenotime concentrates. These raw materials, industrially important for the obtaining of the rare earth elements, are very difficult to dissolve. Xenotime, the main raw material for yttrium, especially cannot be dissolved in an open system with any acid except for sulfuric acid. Even after treatment of concentrates with sulfuric acid, an undecomposed residue remains often on the bottom of the vessel. The residue mostly contains heavy resistant minerals, such as cassiterite, zircon, rutile, and tantaloniobates of varying composition. The evaporation residue is dissolved in cool water. The sulfate complexes of the elements present that have been thus formed permit dilution of the solution without reprecipitation of phosphate.[611] In determining rare earth elements in monazite and xenotime by emission plasma spectrometry, it is more advantageous to dissolve the residue in concentrated hydrochloric acid,[612] A 0.2-g concentrate sample is heated with 5 mℓ of concentrated H_2SO_4 for 3 hr at 250 to 300°C. The excess of the acid is evaporated to dryness, and the residue is dissolved in 25

$m\ell$ of concentrated HCl. After dilution with water, the insoluble residue is filtered off. Complications are caused by variable composition of the concentrate, especially the presence of decomposable minerals of zirconium and barites. In the presence of zirconium(IV) ions in the solution, the sparingly soluble Zr-phosphate precipitates after dilution of the reaction mixture. After dissolution of barites and celestine and reprecipitation of sulfates on dilution of the mixture, actinoids and lanthanoids are built into the structure of the sparingly soluble sulfates. This procedure has been used in radiochemical analysis for a very efficient preconcentration of actinoids.[613] Quantitative separation of actinoids is best carried out from the medium of dilute sulfuric acid and potassium sulfate. Potassium salts are usually introduced into the solution by preceding fusion with potassium disulfate. Potassium sulfate analogously affects the trapping of the lanthanoid ions in precipitates of strontium and barium sulfates.[614,615] Cocrystallization of these elements is governed by various mechanisms. In the presence of potassium ions, lanthanoids enter the sulfate structure through a heterovalent substitution,

$$2\ Me^{2+} \rightleftharpoons Me^{3+} + Me^{+} \tag{22}$$

Solid solutions may also be formed in the structure of the binary sulfate $(K_2Sr(SO_4)_2$. This compound is a very effective collector of both the light and heavy lanthanoid ions.

These phenomena have serious consequences in decomposition of ores of lanthanoids containing barite and celestine in addition to the economically important minerals. These minerals commonly accompany parasite-bastnesite ores. The ore decomposition cannot be carried out with concentrated sulfuric acid, as the reprecipitated sulfates contain a substantially larger fraction of lanthanoids than the original barites.[616] Analogously, it is not recommended to decompose barites or celestine by fusion with potassium disulfate and dissolve the melt in a solution containing lanthanoids. Light rare earth element ions are primarily trapped in the structure of an insoluble sulfate, especially Ce^{3+} and La^{3+}.

VII. DECOMPOSITION BY PHOSPHORIC ACID

Phosphoric acid considerably differs from other strong inorganic acids in its effects as a solvent for solid substances. It exhibits very low volatility and displaces other acids from the reaction mixture on heating. Excess phosphoric acid cannot be removed by evaporating the solution, as the acid is dehydrated on heating with the formation of a mixture of polyphosphoric acids[617,618] with increasing content of phosphorus pentoxide. The phosphoric anhydride — theoretically the final product of dehydration of the acid — sublimates at 358.9°C at a pressure of 1.01×10^5 Pa. From the point of view of the dissociation of the hydroxonium cation (H_3O^+), solutions of the polyphosphoric acids formed are stronger acids than the original orthophosphoric acid.

Phosphoric acid forms complex compounds with many ions, and some of them can be used analytically. Application of phosphoric acid as a solvent requires the use of special procedures to prevent interference from excess phosphates. Sparingly soluble compounds are formed by phosphates with some cations (especially Zr, Hf, Ti, Sn, Bi, Th, and the cations of the rare earth elements (REE), even in strongly acidic solutions. Excess phosphate is difficult to remove; the separation of cations and phosphates is complicated by the formation of positively charged complex ions (especially with Fe^{3+}) that are sorbed by cation exchangers.[619] Leaching of melts of samples with alkali hydroxides and carbonates by water is not sufficient for complete separation of phosphates from the sparingly soluble hydroxides formed.

In AAS, phosphoric acid exhibits a strong depressive effect, especially in the determination of the alkaline earth metals, and very hot flames are necessary. Even then the interference from phosphate is not completely eliminated (when it is present at an extreme excess, i.e., when the substance has been dissolved in phosphoric acid); moreover, problems are encountered in transporting the solution into the flame, which is very difficult to eliminate by modeling the calibration solutions. Similar difficulties have also been observed[620-622] when using OES-ICP, e.g., in analyses of the raw materials for the production of optical fibers[620] (e.g., $POCl_3$). Phosphoric acid liberated by hydrolysis in the preparation of a test substance solution (*circa* 0.1% H_3PO_4) causes an appreciable decrease in the intensities of some spectral lines. Low concentrations of the acid do not affect the solution viscosity or the transport of the liquid into the plasma, but complicate the desolvation process.[620]

During dissolution of solids, phosphoric acid undergoes very little redox changes compared with other strong inorganic acids; in concentrated solutions it reacts with the redox couple components with formation of variously strong complexes. The redox potentials of some systems [e.g., Fe(III)/Fe(II), Mn(III)/Mn(II), U(VI)/U(IV), Mo(VI)/Mo(V), V(IV)/V(III)] are thus considerably shifted, which permits, e.g., titrations of molybdate, uranyl, and vanadyl ions with a ferrous salt solution[623,625] in a medium of 10 to 13 *M* H_3PO_4. For this reason phosphoric acid is also suitable for the preparation of some less common titrants, such as solutions of vanadium(III) salts[626] that are sufficiently stable and are used for determining iron and manganese in iron ores.

Phosphoric acid reacts with some elements from the V and VI group of the periodical system (especially with vanadates, arsenates, molybdates, and tungstates) with the formation of heteropolyacids that are suitable in analytical practice for spectrophotometric determinations and for extraction separations of complex mixtures. For example, the complex compound, $H_3/SbO(HPO_4)_3/$, is formed with pentavalent antimony in strongly acidic solution, which is difficult to reduce by ferrous and chromium(II) salts; this reaction is used in a coulometric determination of arsenic in the presence of excess antimony.[627]

In strongly acidic solutions containing hydrochloric acid, phosphoric acid decreases the volatility of tin(IV) chloride because of the formation of complexes and permits distillation separation of $SbCl_3$, even in the presence of tin. Tin distillates from this medium in the form of $SnBr_4$, on an addition of hydrobromic acid.

Compounds of unusual composition and properties are formed in the presence of concentrated phosphoric acid; for example, boric acid reacts with the formation of a compound[628] with a composition close to BPO_4, which prevents the formation of volatile boron trifluoride on heating of the reaction mixture with hydrofluoric acid.[629] This phenomenon makes it possible to determine traces of boric acid in HF solutions and in many other inorganic materials, especially in quartz. The procedure is also useful in the preparation of pure nitric acid, without traces of boric acid.

The formation of similar compounds is apparently responsible for a decrease in the loss of germanium that is sometimes observed when dissolving silicates and sulfides in mixtures of hydrofluoric, sulfuric, and nitric acids. This loss is usually ascribed to the formation of volatile germanium tetrafluoride or volatile Ge(II) compounds. However, the problem of increasing the stability of germanium(IV) salts by adding phosphoric acid or an alkali phosphate has not yet been unambiguously solved.[630,631]

Phosphoric acid strongly decreases the vapor pressure of some compounds of rhenium(VII) (especially Re_2O_7 or $HReO_4$) and thus suppresses their volatilization in evaporation of solutions of aqua regia and hydrofluoric acid.[631a] The loss of selenium(IV) was determined using the ^{75}Se nuclide; complete volatilization occurred within 4 hr of heating with H_3PO_4 on a sand bath.[632] Tellurium(IV) does not form volatile compounds under similar conditions.[633]

Phosphoric acid influences the course of analytical separations and the choice of a suitable analytical method substantially more than other strong inorganic acids. However, its ability to dissolve inorganic solids is so pronounced that it is progressively more often used for decomposition of resistant complex oxides (especially single crystals), silicates, and for evaluation of nuclear fuels, despite the drawbacks discussed above. The complications in analytical procedures are compensated by the specific effects of phosphoric acid as a solvent.

Phosphoric acid is usually supplied as a 70 to 85% solution (d = 1.58 − 1.65, 12 − 14.7 *M*), which, however, is insufficiently pure for demanding analyses. Common impurities involve sodium ions leached from glass vessels, calcium ions, fluorides, and some heavy metals introduced during the production of the acid. Weakly reducing effects of some preparations are ascribed to the presence of a small amount of phosphorous acid. Because of a low vapor pressure and condensation reactions, the acid cannot be purified by distillation. To prepare the highly pure acid[634,635] (suitable, e.g., for analyses of lunar rocks from the Apollo 11 expedition[634]), H_3PO_4 freed of lead on a cation exchanger was used. In order to obtain the ultrapure acid, phosphorus pentoxide must be used that has been purified by triple sublimation.[635] In this way a preparation was obtained with a lead content below 1 ng/g; commercial acids contain substantially more lead (*circa* 1 to 10 μg/g). In contrast to the acid, the oxide can be stored in carefully sealed ampules of borosilicate glass.

The selection of vessels for decomposition of solids with phosphoric acid is rather difficult.[636] Platinum vessels are mostly used, although platinum is corroded at higher temperatures.[637,638] Heating with 85% H_3PO_4 during a decomposition procedure leads to a loss of several micrograms of platinum (8 to 9 μg according to Reference 637). On evaporating to dryness, this amount is roughly doubled.[636-639] Quartz vessels are corroded more than with other strong inorganic acids.[636-639] Glassy carbon is not perceptibly attacked by dilute phosphoric acid; no detailed data are available on the effects of condensed phosphoric acids.[636] Borosilicate glass is corroded by boiling concentrated phosphoric acid and is unsuitable for prolonged use. The surface layer of vessels is badly damaged and covered by a film of solid reaction products. The Vycor glass is substantially more resistant. Porcelain dishes are unsuitable for dissolution of substances in phosphoric acid. Among metallic vessels, thick-wall nickel dishes and crucibles have been satisfactory, provided that the reaction solutions contain no strong oxidants. The degree of corrosion is also affected by the presence of some salts (e.g., $FeCl_3$).

The selection of vessels for decompositions with dilute, concentrated, or condensed phosphoric acid mainly depends on the purpose of the analysis (i.e., the elements to be determined), on the concentration and temperature of the acid, and the reaction time that is determined by the resistivity of the test material. For determinations of common components (e.g., the alkali metals, Al, Fe, etc.), the test substance must be dissolved in quartz or platinum vessels. The effective component (polyphosphoric acids) is formed directly in the reaction mixture. In analyses for anionic components (e.g., Cl, S, Se, and As), the decomposition reagents are usually prepared beforehand, and the vessels for these preparations are selected analogously.

The following methods predominate in analytical practice for dissolution of solids in phosphoric acid:

1. Decompositions by "concentrated" H_3PO_4 prepared beforehand or formed by dehydration in the reaction mixture. By "concentrated" H_3PO_4 we understand the acid with a minimal amount of water dependent on the method of preparation of the decomposition reagent or on the temperature course of the decomposition process. Dehydration at 250 to 300°C produces "strong phosphoric acid" (SPA), also called "condensed phosphoric acid" (CPA). The preparation contains variable amounts of polyphosphoric acids, depending on the method of preparation.
2. Dissolution of test substances by phosphoric acid mixed with oxidizing acids, most often with nitric and perchloric acid; excess oxidant is eliminated by heating the solution formed. Sulfuric acid also exhibits oxidizing effects in the presence of polyphosphoric acids.
3. Dissolution of substances in CPA in the presence of solid oxidants (or of their solu-

tions), e.g., $K_2Cr_2O_7$, $Ce(SO_4)_2$, KIO_3, or an alkali vanadate. Excess oxidant cannot be removed from the reaction mixture by heating.

4. Decompositions by CPA in the presence of reductants, especially tin(II) salts. The reagent is usually prepared beforehand. As reductants, tin metal, hypophosphorous acid, red phosphorus, hydroiodic acid, and other substances are also used.
5. Dissolution in CPA in the presence of the alkali (ammonium) halides, chlorides, and bromides being used most often. Volatile halides are separated.
6. Dissolution of substances in dilute phosphoric acid. A decisive factor for the selection of this type of decomposition procedure is usually a low volatility of acids and a limited extent of undesirable (especially oxidative) side reactions.

The selection of the classification criteria, based primarily on the type of chemical reaction involved, is rather subjective. Other criteria can be employed, e.g., the chemical composition or the crystal structure of the test substances, or the kind of analytical method employed for the determination of the required component in the solution obtained. However, the classification scheme based on various types of decomposition reactions permits the best characterization of phosphoric acid as an effective solvent for solids.

The effectiveness of phosphoric acid as a solvent depends mainly on the formation of polyphosphoric acids (CPA) that are either formed during the decomposition process, or are produced beforehand (in quartz or glass vessels). These acids are stronger in solution than the initial orthophosphoric acid from which they are produced by condensation reactions with gradual liberation of water molecules:

$$\begin{array}{c} \text{OH} \\ | \\ \text{HO–P–OH} \\ \| \\ \text{O} \end{array} \rightarrow \begin{array}{c} \text{OH}\quad\text{OH} \\ |\qquad| \\ \text{HO–P–O–P–OH} \\ \|\qquad\| \\ \text{O}\quad\;\;\text{O} \end{array} \rightarrow \begin{array}{c} \text{OH}\quad\text{OH}\quad\text{OH} \\ |\qquad|\qquad| \\ \text{HO–P–O–P–O–P–OH} \\ \|\qquad\|\qquad\| \\ \text{O}\quad\;\;\text{O}\quad\;\;\text{O} \end{array} \tag{23}$$

The conversion into polyphosphoric acids (with the general formula, $H_{n+2}P_nO_{3n+1}$) is never complete; in addition to CPA, the reaction product also contains a substantial amount of orthophosphoric acid. The condensation of H_3PO_4 begins only after complete dehydration of the acid[640-643] (Figure 3) at temperatures of 170 to 180°C. The reagent for analytical purposes is prepared by heating to 250 to 300°C. The following procedure[644] is usually used for the preparation: An amount of 300 to 400 g H_3PO_4 is heated in a conical glass flask, until a thermometer immersed in the liquid reads 300°C. Water vapors are aspirated during the heating through a glass tube placed immediately above the level of the liquid in order to accelerate the condensation reactions. The liquid is heated until the development of the vapors ceases. The syrupy liquid obtained is stored in closed glass or quartz vessels. Polyphosphoric acids form linear chains with the P–O–P bonds (see the above scheme). They are viscous liquids, highly reactive at elevated temperatures and are usually slightly turbid, depending on the method of preparation. They react exothermically on dilution and hydrolyze in aqueous solutions with formation of the original orthophosphoric acid. The reagents for analytical purposes mainly contain di-, tri-, and tetraphosphoric acids. A preparation, obtained by dehydration at 300°C and stored for a long time in laboratory, had the following composition determined chromatographically:[642] 44.4% orthophosphoric acid, 39.8% diphosphoric acid, 12.3% triphosphoric acid, and 3.6% tetraphosphoric acid. Fresh CPA has a different composition depending on the initial amount of H_3PO_4, the temperature program, and the reaction time. Longer reaction times lead to condensation products with higher contents of polyphosphoric acids, even at lower temperatures.[642,643]

The presence of other acids also affects the course of the condensation reactions.[641] Nitric acid hastens the condensation at certain concentrations, whereas sulfuric acid decelerates

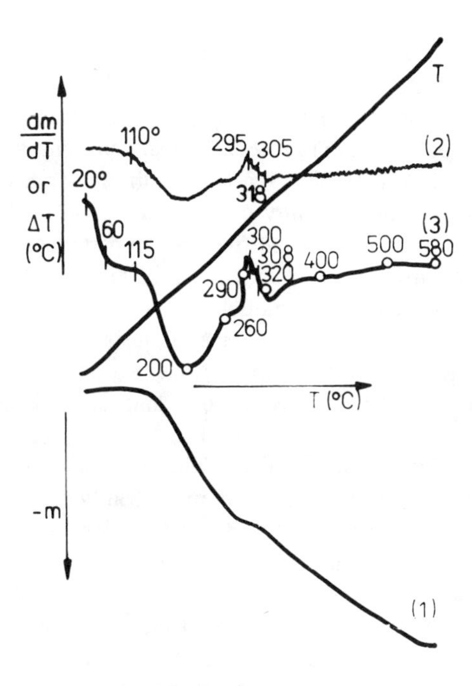

FIGURE 3. Dehydration of phosphoric acid.[25] Curve (1) thermogravimetry (TG); (2) derivative thermogravimetry (DTG); and (3) differential thermal analysis (DTA).

the reaction, even during short heating periods, apparently by retaining water in the reaction mixture. The effect of sulfuric acid depends on time; on prolonged heating (90 min at 270°C) the acid exerts no influence, although sulfates remain in the final product.

Modern chromatographic methods have made it possible to determine the contents of various polyphosphoric acids with a better precision. It has been demonstrated that the hydrolysis (i.e., the back reaction) is not complete and a certain amount of the condensation products (mainly diphosphoric acid) always remains in the solution and forms stable complexes with metal ions [e.g., the formation of Fe(III) diphosphate in materials containing Fe(III)]. This fact must be borne in mind in the subsequent procedures, as diphosphates exhibit strong masking effects in determinations of some elements.[641]

One of the most common applications of phosphoric acid as a solvent in inorganic analysis is the determination of the quartz content in the presence of silicates in samples for evaluation of industrial environment or in petrochemical studies. Methods based on direct determination of mineral phases (petrographic methods or X-ray diffraction) are often insufficiently precise for this purpose; therefore, chemical methods, based on different reactivities of quartz and the accompanying minerals, are still used. The procedures developed by various authors (see, e.g., References 645 to 650) differ considerably. However, all these methods have certain limitations caused by complex kinetics of the actual dissolution and a great variability in the phase composition of the test substances. The reaction is based on a high resistivity of quartz toward hot phosphoric acid under conditions when the crystal lattices of most silicates are destroyed. The reaction time is usually 10 to 20 min at 230 to 250°C. The reaction products are complex compounds of silicon and have been reported to be soluble in water. However, insoluble compounds are also sometimes formed (especially on exceeding a temperature of 250°C), or are produced by hydrolysis on dilution of the reaction mixture with water. Hence, the isolation of insoluble quartz from the reaction products is difficult, and, thus, fluoroboric acid is added to the solution at the end of the decomposition procedure, as it dissolves these products (e.g., the gelatinous precipitate of SiO_2) without attacking the quartz. This procedure has yielded good results, especially in analyses of rocks and industrial dusts consisting of rock-forming minerals.

Diphosphoric acid produced directly in the reaction mixture is considered to be the effective component in the decomposition process.[645-649] Higher polyphosphoric acids and the initial orthophosphoric acid that is present in a great excess, depending on the reaction conditions, in the final stage of the decomposition procedure, also contribute to the dissolution effect.

Various polymorphous forms of SiO_2 react with different rates. The α- and β-quartz exhibit the lowest solubility; the solubility of tridymite is substantially higher, and up to 66% of the cristoballite present is dissolved (with 5-μm particles). The solubility of quartz increases with decreasing particle size.[645-647] It amounts to up to 6% for 5-μm particles and decreases below 1% on an increase in the particle size to 30 μm. For this reason an empirical correction factor is used in analyses of fine industrial dusts. The solubility of SiO_2 obtained by igniting a silicic acid gel exceeds 98%, owing to an extremely large specific surface area of the particles formed predominantly by the tridymite and cristoballite modification of silicon dioxide.

The principal rock-forming silicates exhibit a different behavior under these conditions. Among micas, muscovite, biotite and sericite are most readily degraded. Some silicates with crystal lattices consisting of independent SiO_4^{4-} tetrahedra, such as olivine and garnets (spessartine, almandine, and grossular), are dissolved, whereas zircon, staurolite, and disthene, as well as andalusite, are attacked very little.[645-647] Heavy metal silicates and zeolites are dissolved.[651] Minerals from the group of pyroxene and amphibole, such as diopside, tremolite, and actinolite, are substantially more resistive than feldspars. However, great differences have been observed in the behavior of various minerals, both with plagioclases and alkaline feldspars. Clay minerals with sheet structures from the group of kaolinite and montmorillonite, as well as chlorites, are decomposed in the reaction mixture and the procedure is often used for the separation of quartz from the clay component. Samples of glass, slags, and glazes can be analyzed without great difficulties. After the decomposition, a clear, viscous solution should be obtained that is stable on dilution and from which quartz can readily be separated, e.g., by filtration through a paper filter. Topaz, sillimanite, mullite, beryl, and some kinds of tourmaline are only negligibly attacked by hot phosphoric acid.[645-647]

Among resistive oxides, corundum and rutil are not attacked, whereas the behavior of spinels depends on their chemical composition. These minerals are mostly highly resistant during a short-time exposure to hot phosphoric acid. Some sulfides (pyrite) are only slightly dissolved, and thus they are first preoxidized with a small amount of nitric acid. On the other hand, sphalerite and galenite can be decomposed under these conditions. Fluorite interferes, as the hydrofluoric acid liberated reacts with quartz and silicates.[648] These differences in the behavior of various minerals stem from the variability of the solid solutions in the solid phases analyzed; partly, they can also be ascribed to an imprecision in the determination of the composition of the initial solid or to an insufficient purity of the separated mineral fractions.

The determination of quartz in the presence of silicates is usually carried out in conical flasks made of resistant borosilicate glass. The gelatinous silicic acid produced by corrosion of the vessel is dissolved in the reaction mixture and does not affect the results obtained. The insoluble residue after the decomposition is filtered off and ignited. The quartz content in the residue is determined gravimetrically from the weight difference after removal of the quartz by heating with hydrofluoric acid. In analyses of rocks or analogous polyphase materials, the composition of the insoluble residue should be checked systematically (in parallel experiments) immediately after its isolation. In the determination of quartz, the gel of silicic acid causes positive errors, as well as the residues of undecomposed silicates that are destroyed on subsequent dissolution in hydrofluoric acid. The precision and accuracy of these analyses (i.e., of the determination of individual mineral phases) are substantially poorer than those in the determination of total SiO_2.

The complexity of this problem is demonstrated, e.g., on the results of experiments in which samples of granite and basalt (Japanese rock standards JG 1 and JB 1 with known phase compositions) were decomposed[652] by a mixture of CPA prepared beforehand. The decomposition was carried out in order to determine chlorides in the rocks, and the reaction conditions used for the dissolution did not differ significantly from those recommended for the determination of quartz in the presence of silicates. Sample amounts of 0.5 to 1.0 g were decomposed by 30 g CPA at 250°C. A comparison of the X-ray diffractograms of the original samples and of the solid residues after the decomposition has shown that feldspar and quartz were virtually completely dissolved. The diffraction patterns characteristic of these minerals disappeared and were replaced by lines corresponding to newly formed crystalline phases; they were assigned to the product, $Si_3(PO_4)_4$. Thus, quartz was converted by the decomposition procedure into crystalline compounds of completely different chemical composition and structure. The high solubility of quartz was apparently caused by a different composition and excess of the solvent (i.e., a different ratio of the polyphosphoric acids) and a different thermal program of the decomposition.

Decomposition by phosphoric acid or by its mixtures with sulfuric acid has been useful in the determination of other components in minerals, e.g., in the oxidimetric determination of iron in some Fe chlorites that are difficult to decompose in strong mineral acids. These minerals are usually dissolved by fusion in analytical practice, which leads to alloying of platinum vessels used by the reduced iron and to corrosion of platinum crucibles by the melt. This causes serious complications in the oxidimetric titration of ferrous ions (see Chapter 3). In our experience (but in variance with the data of some authors[646]), some tourmalines are rapidly destroyed by this reagent (e.g., those of the schorl-dravite series) which are otherwise resistant toward hydrofluoric acid. An amount of 0.1 to 0.25 g of the mineral is dissolved by boiling with 20 mℓ of a mixture of H_3PO_4 and H_2SO_4 (2 + 1) within 10 to 15 min. A reaction time of 5 to 7 min suffices for some garnets (almandine). As in solutions of concentrated phosphoric acid (>10 *M*) the redox potential of the Fe^{3+}/Fe^{2+} system is considerably decreased, the Fe^{2+} ions formed are readily oxidized to Fe^{3+}. On the other hand, the subsequent reduction by stannous chloride is substantially slower in the presence of phosphoric acid, and it is necessary to increase the acidity of the medium and the formation of reducible chloride complexes of Fe(III) by adding hydrochloric acid to the test solution.

Easy degradation of some micas (muscovite, lepidolite, biotite) by diphosphoric acid has permitted isolation and determination[653,654] of radiogenic Ca^{2+}. Some resistant mineral phases that do not dissolve in hydrofluoric acid can be degraded (except for sillimanite) by treating the HF-insoluble fraction with polyphosphoric acids.[655-657] The insoluble residue is filtered off and is decomposed again by heating with a mixture of H_3PO_4 and $HClO_4$, until a viscous product is obtained that is suitable for further treatment. This procedure has been recommended, e.g., for the determination of traces of thorium and lithium in rocks. It is, however, unsuitable for the determination of zirconium, as a nonreactive, colloidal zirconium phosphate is formed. Decomposition of coal ash by phosphoric acid is more effective than procedures involving other strong inorganic acids; in the determination of germanium, the use of hydrofluoric acid can thus be avoided.

Dissolution in polyphosphoric acids has found a wide application in analyses of all the crystal modifications of aluminum oxide and of the raw materials for the manufacture of aluminum compounds (bauxite). The alpha modification of aluminum oxide (with defined admixtures of other oxides) of the corundum structure has gained importance because of its favorable physical properties and possibilities of further treatment, in the form of single crystals, polycrystalline phases, or synthetic and natural precious stones (ruby, sapphire). This high-temperature corundum modification is exceptionally resistant toward all strong mineral acids. Great difficulties are also encountered when dissolving natural corundum in a closed system (see Chapter 5), as the necessary reaction temperature lies at the limit

Table 3
CONDITIONS FOR THE DISSOLUTION OF Al_2O_3 SINGLE CRYSTALS[658]

Single crystals	Sample weight (g)	Temperature of the reaction mixture (°C)	Decomposition time (min)	Particle size (μm)
Leukosapphire-α Al_2O_3 dopped with 0.001—0.3% Ca	0, 15	320—330	20	60
β Al_2O_3 dopped with 0.2—10% Na $mAl_2O_3 \cdot nNa_2O$	0, 10	320—330	15	60
Spinel $MgO \cdot n\ Al_2O_3$	0, 15	320—330	15	60
The melting mixture for preparation of the single crystals	0, 30	290—300	5—10	100
Ruby-αAl_2O_3 dopped with 0.005—5% Cr	~0, 10	290—300	30	60
Garnet, $mAl_2O_3 \cdot nY_2O_3 \cdot kNd_2O_3$	0, 30	320—330	15	60

Note: Dissolution is in 10 mℓ 75% H_3PO_4 in the quartz flask.

permissible for the PTFE lining of the pressure vessels (240 to 250°C). Phosphoric acid is an exception in open systems; a mixture of polyphosphoric acids destroys even the resistant crystal structure of alpha Al_2O_3. If the decomposition is carried out in a quartz vessel, then even Li^+, Na^+, and K^+ can be determined in the solution obtained.

The conditions for decomposition of some crystal structures of aluminum oxide have recently been studied again.[658] The main reason for revision of the original procedures has been an increase in demands on the analytical precision and the determination limits of various admixtures. As a considerable difference has been observed in the reactivity of polyphosphoric acid preparations obtained before the decomposition procedure and those produced *in situ* in the reaction mixture, the test samples were dissolved in dilute, 75% H_3PO_4, with the formation of the effective components by dehydration directly in the quartz vessel used. The conditions for dissolution of various single crystals and variously doped aluminum oxides are given in Table 3. The mixture is stirred during the dissolution to prevent fine particles of the test substance from adhering to the bottom of the quartz flask and to avoid local overheating, which lead to the formation of sparingly soluble gel-like compounds. The reaction product is soluble in water, provided that the decomposition is carried out perfectly. The decomposition time depends on the crystal structure of the test material and on the size of the solid particles. Pure leukosapphires (generally alpha Al_2O_3) are decomposed substantially more slowly than other structural modifications of aluminum oxide (doped by other elements). Ruby single crystals and the initial mixture for their production could not be dissolved, even if the reaction time was increased to 30 min, and a greater amount of the acid was used. The fact that even boron can be determined after this decomposition indicates a strong interaction between H_3BO_3 and H_3PO_4; boric acid is not volatilized even at temperatures of 250 to 300°C.

The beta modification of aluminum oxide is decomposed more easily than corundum. Lower temperatures suffice (185°C), at which only a small amount of diphosphoric acid is formed in the solution.[659] If the test material contains chrysoberyl and corundum (electroceramics with 25% of these components), the reaction mixture temperature must be increased up to 300 ± 15°C to dissolve these resistant phases.[660] Quartz vessels are mostly used (or, sometimes, platinum vessels) and the decomposition is followed by an AAS determination. Such a determination of beryllium is not affected by an excess of the decomposition reagent. If magnesium is to be determined as well, the excess of phosphoric acid must be minimized.[659]

In addition to phosphoric acid alone, its mixtures with sulfuric acid are also used to dissolve oxides of aluminum; sulfuric acid limits the formation of higher polyphosphoric acids.[661-665] The procedure has yielded good results in analyses of variously dehydrated aluminum hydroxides, some modifications of aluminum oxides, aluminum catalysts based on aluminum oxide and red mud[662] (a waste product from the treatment of bauxite). The behavior of bauxites depends on their mineralogical composition, characteristic of various exploited deposits. The main component consists of crystalline and amorphous aluminum hydroxides and hydrated oxides of the groups of gibbsite, diaspore, and böhmite, together with hydrated iron oxides, especially "hydroxyhematite" and gethite. Lateritic bauxites and laterites contain an admixture of undecomposed mineral phase originating mainly from the original parent rock that underwent fossil or recent weathering (e.g., some minerals of titanium and niobium, chromspinellides, etc.). The presence of these accompanying minerals affects the behavior of bauxite raw materials during dissolution much more than the amount and ratios of the main aluminum and iron oxide minerals in the test substance.

Differences in the solubility of various bauxites are also reflected in the resistance of these materials toward CPA. The effect of the degree of dehydration of bauxite[666] and of a CPA preparation prepared previously was studied, and an optimal decomposition procedure was proposed for the determination of aluminum, iron, and titanium. To 0.1 g bauxite (preignited in a quartz beaker at 700°C), 10 g CPA (dehydrated at 280°C) is added. The vessel is heated in an electric oven at 300°C. The mixture is stirred by escaping water vapor during a further dehydration of CPA, and the decomposition is completed within 30 min. The solution for the analysis is then obtained by dissolving the residue in the beaker in dilute hydrochloric acid. Titanium minerals are decomposed more slowly than hydrated oxides, but the above time suffices for the dissolution of 99.9% of the TiO_2 present.

Many decomposition procedures have been proposed for analyses of iron ores, the most important raw materials of metallurgy, in order to obtain solutions permitting rapid and reliable determination of the main useful component (iron) and possibly also of accompanying elements. Phosphoric acid is also used as an effective decomposition agent, most often mixed[665,667,668] with sulfuric or perchloric acid (at a 1 + 1 ratio); the ore is completely decomposed in a boiling mixture within 7 to 10 min. Iron silicates (chamosite, thuringite) are also attacked (they are common in some sedimentary iron ores). Condensed polyphosphate acids (CPA, 300°C) prepared beforehand have no advantages for this purpose.[669] Limonite is most rapidly decomposed, and the recommended reaction time varies from 5 to 30 min. In a medium of 9 *M* H_3PO_4 the decomposition rate depends on the Fe(III)/Fe(II) ratio in the test ore. An addition of magnetite substantially accelerates the dissolution of hematite.[670] This phenomenon is connected with the formation of Fe^{2+} in the reaction mixture and has also been observed in other systems, e.g., in the dissolution of iron ores in hydrochloric acid solutions.

The problems of decomposition of chromites and chromium ores have been treated by many authors.[671-683] The reasons for this interest involve mainly:

1. Technological importance of chromites, as the only mineral suitable for industrial production of chromium and some heat-resistant materials
2. The resistance of the mineral toward acids and most fusion agents; changes in the chemical composition of this solid phase owing to extensive isomorphous substitution, connected with changes in the properties of the minerals
3. Application of new analytical methods (OES-ICP, AAS) in which an excess of alkali salts from the fusion agents is undesirable

Chromite ($FeCr_2O_4$) belongs among spinels; most members of this group of minerals (either synthetic or natural) are difficult to dissolve in acids. Similar to corundum, this material can only be dissolved in a finely pulverized form (<325 mesh) in a hot solution of polyphosphoric acids, which enables spectrophotometric determinations of many elements. The temperature interval recommended by various authors[671-674,676,680-682] for the decomposition of chromite is very wide (200 to 350°C) and depends on the compositions of the test material and the initial mixture of acids and on the method of measurement of the temperature in the strongly corrosive environment. Either phosphoric acid alone or its mixtures with sulfuric and perchloric acids are used. In a mixture of chromite, ilmenite, and magnetite, the latter is decomposed most rapidly, while the first dissolves most slowly.[683] The amount of the chromite dissolved is minimal until all the water is evaporated from the solution.[676] After dehydration, the temperature is increased to 310 to 320°C, and the contents of a covered crucible is maintained at this temperature for 1 to 2 hr until all of the sample is dissolved. After diluting with water, the solution is suitable for an AAS determination of major and minor components of the ore; no chemical changes were observed in it during 1 year.[677]

Phosphoric acid reacts with tungstates with formation of phosphotungstic acid (most often formulated as $H_3[P(W_3O_{10})]$), which prevents precipitation of tungstic acid and facilitates dissolution of tungsten ores (with wolframite and scheelite), concentrates,[684-686] and alloys.[687] The materials are usually decomposed by mixtures of acids (especially HNO_3, HCl, HF, and $HClO_4$, sometimes also H_2SO_4) with phosphoric acid. The behavior of wolframite [a solid solution of Fe(II) and Mn(II) tungstates] depends on its composition.[684] Minerals from some localities require several hours of heating, or an undissolved residue must be decomposed by fusion.[684,685]

Niobium-tantalum ores with pyrochlor and calcite are also decomposed by polyphosphoric acids.[685,688-691] This procedure is considered by many authors[685,688,689] more advantageous than fusion of the ore material with sodium or potassium disulfate. The reaction mixture is heated at 300°C and dissolved in dilute hydrochloric acid with hydrogen peroxide, in which niobium forms soluble peroxo-compounds.[689] If the residue is completely dried, then it is insoluble in tartaric acid solutions. Most accompanying sulfides are then not attacked (especially molybdenite and partly pyrite), which simplifies the final spectrophotometric determination of niobium using the thiocyanate complex.[685]

On evaporation of acids containing H_3PO_4 until a strongly viscous liquid (paste) is obtained, volatile components are displaced at temperatures from 200 to 300°C. This procedure is most often used in analyses of sulfidic ores and rocks to remove fluorides, chlorides, and nitrates (or also perchlorates) from the reaction mixture. Sulfates cannot be completely eliminated in this way.[641] This principle has found use in a modified distillation separation of hydrofluoric acid (or H_2SiF_6) according to Williard and Winter.[692-699] The main reasons for application of phosphoric acid as the distillation medium have been reviewed by Maxwell.[647] The temperature of the vapors in the distillation flask should be 130 to 150°C (a mixture of H_3PO_4 and H_2SO_4). In analyses of aluminum materials[697] the distillation temperature is substantially higher (up to 250°C for H_3PO_4 alone). Phosphoric acid should compensate the adverse effect of aluminum salts on the distillation, due to competitive reactions. The mechanism of this interference is unclear and is ascribed to a decrease of the HF partial pressure in the gaseous phase, caused by the formation of fluoride complexes of aluminum in the strongly acidic liquid phase. However, some authors have not observed the positive effect of phosphoric acid.[647] In a simple arrangement (distillation from a quartz or glass test tube heated over a burner) this method has been used to separate fluorides from, e.g., biotite, phosphate, and silicate rocks.[693,694] The completeness of the separation of fluoride was checked by an analysis of rock standards (SRM) and recently also by ion chromatography (in the high-performance form).[694] A similar procedure has given good results in chemical evaluation of fluorite raw materials,[696] claystones for manufacture of bricks, aluminum catalysts,[697] and cryolite.[695]

Monazite, a natural thorium phosphate containing rare earth phosphates and one of the main raw materials for obtaining these components, is very resistant toward acids. It is decomposed more easily in solutions of polyphosphoric acids[700-702] at 270 to 330°C than by sulfuric acid. Reprecipitation of dissolved phosphates has not been observed even in very dilute solutions (0.65 M H_3PO_4).[701] This decomposition procedure has been recommended for the determination of thorium, radium, and lead in this mineral. Phosphoric acid is equally efficient in decomposition of accessory minerals (containing more than 20% TiO_2), isolated in the technological treatment of apatites.

The method of dissolution of sulfidic minerals depends on the purpose of the analysis and on the phase composition of the test ore. In determinations of the individual metallic components, mixtures of acids (H_3PO_4 with HNO_3 or $HClO_4$) are usually employed for the decomposition, during which the sulfidic sulfur is oxidized to sulfate and the compounds of the other elements [Fe(II), Cu(I), Ge(II), etc.] are oxidized to a higher stable valency. Residues of the oxidizing acids (together with hydrofluoric acid, if its addition was necessitated by the need to liberate a metal from the silicate matrix), which might interfere in the subsequent procedure, are expelled by evaporation with phosphoric acid.[631,703,704] Chalcopyrite concentrates are rapidly dissolved by boiling with a mixture of 85% H_3PO_4 and 70% $HClO_4$ (within 3 to 5 min) without a danger of explosive oxidation during heating. Polyphosphates (apparently mainly diphosphate) are formed in the solution, as ferric hydroxide is not precipitated on neutralization of the solution to pH 5 prior to the iodometric determination of copper.[705] Decomposition with a mixture of phosphoric and nitric acids is also suitable for the determination of germanium in sulfidic ores (sphalerites).[631,704,705]

There are some differences in the behavior of sulfidic minerals during treatment with orthophosphoric acid and its condensation products, caused apparently by various reaction conditions employed in the dissolution of a solid phase or by insufficient purity of the sulfides separated. Pyrite is not decomposed on treatment with hydrochloric and phosphoric acids, even if the solution is evaporated to form a dense paste. A mixture of polyphosphoric acids prepared beforehand liberates, at an elevated temperature (280 to 300°C), hydrogen sulfide from pyrite, chalcopyrite, and bornite; this reaction has been used to determine sulfidic sulfur in rocks.[706] However, experiments have indicated[707] that pyrite reacts in this way only under strongly reducing conditions; elemental sulfur is otherwise separated, due to the reaction

$$FeS_2 \rightarrow FeS + S \tag{24}$$

leading to the formation of pyrrhotite. In contrast to the data in Reference 706, Mizoguchi[707] does not recommend this procedure for the determination of sulfidic sulfur in chalcopyrite. Galena, sphalerite, and pyrrhotite release hydrogen sulfide on treatment with condensed phosphoric acid; molybdenite, however, is not decomposed.

Sulfides are substantially less decomposed by 10 M H_3PO_4, compared with the effect of CPA. Conditions can even be found under which only the oxidation products of the sulfides are decomposed in an inert atmosphere, whereas the original sulfidic material is negligibly damaged.[708] In this way it is possible to determine ferrous and ferric oxides in the oxidized samples of pyrite, pyrrhotite, and chalcopyrite; pyrrhotite is least resistant among these mineral phases.

Phosphoric acid, especially its mixtures with strong inorganic acids, have been applied to decompositions of steels and alloys.[709-714] The presence of phosphoric acid substantially accelerates the decomposition procedure. A mixture of H_3PO_4 and H_2SO_4 (9 + 1) permits complete dissolution of boron nitride, which is highly resistant toward sulfuric acid alone.[710] In our experience, the ease of decomposition of boron nitride depends on the sample thermal history and the crystal modification. Under the conditions described in Reference 710, only

48% of the hexagonal and only 12% of the cubic modification of boron nitride were dissolved during 5 hr heating at 270 to 280°C. Silicon nitride, Si_3N_4, is not destroyed by the above aid mixture, but the compounds SiN and AlN are attacked.[712] Uranium nitride is also dissolved.[713] Ammonium ions are produced and are isolated from the test solution by distillation.

Undesirable admixtures of boron salts can be removed from 85% H_3PO_4 by distillation with methanol. The acid purified in this way is a suitable solvent[714] for determination of traces of boron in metallic thorium and thorium dioxide. Dissolution of ThO_2 in CPA is slow (1 to 2 hr), whereas metallic thorium is quantitatively dissolved within 15 to 20 min at 200 to 250°C.

Phosphoric acid facilitates analytical use of many reactions that do not occur in other media or are difficult to carry out. These are especially some redox titrations, electrochemical determinations, spectrophotometry, etc. These reactions only occur in hot concentrated solutions of H_3PO_4 (more exactly, its mixtures with polyphosphoric acids) and its mixtures with oxidizing (or reducing) agents. They are primarily used in determinations of substances in the same valence state as that in the original solid to be analyzed [e.g., Fe(II), Mn(III), U(IV), Co(III), etc.]. Most published papers describe decompositions of complex, resistant oxide materials, where phosphoric acid not only creates a suitable complexing medium, but mainly acts as an efficient decomposition agent. The experimental conditions required for this kind of determination are incomparably more demanding than those for the determination of the total contents of the individual elements, as the solid to be analyzed must be dissolved without a change in the valence of the compound to be determined. In view of the danger of redox changes during the decomposition, the test material must be dissolved as rapidly as possible. It is further necessary to prevent the contact of the reaction mixture with atmospheric oxygen and suppress undesirable interactions among the substances being dissolved that would lead to changes in the valence state of the substances.

The above dissolution procedure has been primarily used in determinations of ferrous oxide in resistant oxides, insoluble in solutions of sulfuric, hydrochloric, and hydrofluoric acids. For total destruction of the test substance, mixtures of phosphoric acid with sulfuric acid are mostly used (at ratios of 4 + 1 to 2 + 1). The decomposition procedure has been successfully applied[669] to analyses of oxide ores containing magnetite (0.1-g samples are decomposed at 300°C by 10 g of a CPA mixture prepared beforehand, in a nitrogen atmosphere), but it has found widest use in analyses of isolated minerals from the group of chrome spinellides, chromium ores, and heat-resistant materials based on chromite. An accurate determination of the FeO content in these materials presents numerous problems. As a result of extensive isomorphous substitutions among the individual oxides in the minerals, chromites, [Fe(II), Mg O·(Cr(III), Al, Fe(III)]$_2$O$_3$, isolated from various chromium ores in a medium of the above acids do not react at the same rates. Ferric ions form much more stable complexes with phosphoric acid and polyphosphoric acids than ferrous ions. Therefore, Fe(II) compounds behave as strong reductants in concentrated solutions of phosphoric acid and are readily oxidized, not only by atmospheric oxygen, but also by other ore components and by the solvent. The problems of the determination of FeO in resistant solids have been reviewed by Schafer[715] and Dinnin.[716] The determination of FeO after decomposition of chromite in a mixture of phosphoric and sulfuric acids includes:

1. Heating in an inert atmosphere, followed by an oxidimetric titration after diluting the reaction mixture or isolating a gaseous reaction product
2. Heating with excess oxidant, e.g., vanadium pentoxide, ceric sulfate, or a manganese(III) salt

For successful determination, the test material must be very finely pulverized (below 325 mesh) in order to accelerate the decomposition. However, with chromite and with other minerals as well (garnet, staurolite, etc.), the time of pulverizing should not be too long to prevent partial oxidation of FeO. For this reason it is recommended to pulverize the minerals in an acetone suspension.

An oxidimetric titration with $KMnO_4$, carried out immediately after the decomposition of the ore by a mixture of H_3PO_4 and H_2SO_4 (4 + 1) at 320°C, is subject to a negative error, as partial oxidation of the ferrous salt cannot be prevented under these conditions.[717] To eliminate the reaction of Fe(II) with atmospheric oxygen, Shein[718] proposed the use of the reaction of Fe^{2+} with excess vanadium pentoxide. This principle has been modified many times[715,716,718-726] and used not only in analyses of chromites, but also for heat-resistant materials (chrome magnesites) and other resistant minerals, e.g., tourmaline.[722] The reaction mixture temperature varies from 360 to 380°C. The reaction products (vanadyl salts) are titrated with permanganate, or the unreacted vanadium pentoxide is titrated reductometrically with a ferrous salt. The procedure verified experimentally by Dinnin[716] involves the decomposition of the material with a (2 + 1) mixture of H_3PO_4 and H_2SO_4 containing excess V_2O_5 at 220°C, the reaction time being extended to 12 hr. An empirical factor is employed in the evaluation of the result of the titration. A chromite sample, in which the FeO content has been determined by several independent analytical methods, is analyzed in parallel. The Shein method was criticized by Goswami,[727] who found that the vanadium pentoxide added is reduced to a greater extent (by 16 to 20%) than required by the assumed reaction stoichiometry. An insufficient stability of polyvanadic acids in hot mixtures of phosphoric and sulfuric acids has also been confirmed by other authors.[728,729] The blank value fluctuates and in the absence of sulfuric acid some kinds of chromium ores are not decomposed, or sparingly soluble phosphates separate.[720] Goswami[727] recommended the use of a defined excess of ceric sulfate instead of vanadium pentoxide. The ceric salt is also reduced to a small extent during decomposition by H_3PO_4 and H_2SO_4 (4 + 1) at 290 to 300°C, but the error thus caused can be corrected for by carrying out the blank experiment. A comparison study of Dixon et al.[730] demonstrated that this method is reliable. Volumes of 100 mℓ H_2SO_4 (d = 1.84) and 400 mℓ H_3PO_4 (d = 1.75) are mixed with 0.45 g $Ce(SO_4)_2{\cdot}4H_2O$ and the mixture is heated to 100 to 110°C. To 0.15 to 0.25 g of the sample, 25 mℓ of this mixture is added. The temperature is slowly increased so that 220°C is attained within 30 min. The decomposition is terminated after 90 min of heating with occasional stirring. The mixture is cooled, and the products are dissolved in 50 mℓ of water. The unreacted ceric salt is determined by titration with Fe^{2+} ions.

Other authors[728,729,731] also prefer the reagent containing excess ceric salt in the determination of FeO in chromium ores. In order to attain a better precision they add the excess oxidant in the form of a solution, which is more precise than the addition of the solid salt.

In view of the high redox potential of the Mn^{3+}/Mn^{2+} system (1.3 V in H_3PO_4), manganese(III) salts are used for the same purpose as vanadium pentoxide and ceric salts.[732] A manganese(III) salt is usually prepared by oxidation of Mn^{2+} with permanganate solutions in a medium of phosphoric and sulfuric acids (2 + 1). The test chromium ores, chromites,[733] and spinels[734] are dissolved in the mixture of acids at temperatures from 280 to 300°C or in H_3PO_4 alone with NaH_2PO_4 added (the Vycor glass is recommended).[733]

This method of determination of FeO in chromites, with the decomposition of the minerals in solutions of phosphoric and sulfuric acids with excess oxidant, exhibits also some drawbacks, in addition to clear advantages (dissolution without the necessity of maintaining an inert atmosphere):

1. Sulfides and organic substances interfere. Some volatile reaction intermediates cannot escape from the reaction mixture and are further oxidized.

2. In the decomposition procedure, the conditions for the complete dissolution of the minerals must be satisfied, simultaneously taking into account a limited thermal stability of the oxidant. The precision of the method depends on the reproducibility of the blank.
3. Substances that may reduce the oxidant added at the given temperature, and the concentrations of the acids must be removed from the acids used. The character of these substances is often not known. The preliminary oxidation of mixtures of acids is usually carried out repeatedly, with small amounts of an oxidant (e.g., $KMnO_4$) and with heating.
4. The reaction mechanism has not been sufficiently clarified. It has not been verified whether there occurs a direct reaction between Fe^{2+} and the oxidant, or whether an intermediate is formed by the action of ferrous salts that is further oxidized by excess Ce^{4+}, Mn^{3+}, or V^{5+}. With vanadium(V) compounds the reduction by Fe^{2+} may lead as far as to the formation of a vanadium(III) salt.[625,626] There is no detailed information on the behavior of manganese(II) ions in the reaction mixture.

Seil[735] has found that during the decomposition of chromium ores with phosphoric and sulfuric acids sulfur dioxide is liberated from the reaction mixture. In hot solutions of polyphosphoric acids, ferrous salts act as strong reductants and reduce the sulfuric acid present,

$$2\ FeO + 4\ H_2SO_4 \rightarrow Fe_2(SO_4)_3 + SO_2 + 4\ H_2O \tag{25}$$

This reaction has been used for the determination of FeO in poorly soluble compounds, especially in chromites. The gaseous reaction product is trapped in a standard solution of dichromate and the excess of dichromate is determined by reductometric titration with a ferrous salt.[716,735-738] As sulfur dioxide is easy to isolate from the reaction mixture, the final titrimetric determination is substantially simplified. The decomposition of chromite at 360 to 380°C in a stream of carbon dioxide is complete within 30 to 90 min. At a lower temperature (220°C), the reaction time increases to 3 to 4 hr.[716]

This procedure, modified for automatic analyzers, has facilitated the determination of FeO not only in ultrabasic rocks containing chromite and chrome diopside, but also in granite rocks and in some rock-forming minerals.[739] The decomposition is carried out in a quartz vessel in a stream of carbon dioxide or nitrogen, at 430°C, and is complete within only 15 min. Organic substances and sulfides interfere. Pyrite is decomposed with liberation of sulfur that reduces sulfuric acid present to sulfur dioxide, which causes a positive error of the FeO determination. However, the interaction between FeO and H_2SO_4 has not yet been unambiguously clarified. As the reaction products, SO_2 and PH_3 are considered.[737] If the decomposition takes place in a stream of carbon dioxide, carbon monoxide, formed by the reaction

$$CO_2 + 2\ FeO \rightarrow Fe_2O_3 + CO \tag{26}$$

is among the reaction products.[738] Carbon monoxide reduces bichromate solutions similar to SO_2

$$3\ CO + Cr_2O_7^{2-} + 8\ H^+ \rightarrow 3\ CO_2 + 2\ Cr^{3+} + 4\ H_2O \tag{27}$$

Using the complexing properties of phosphoric and polyphosphoric acids, some complex problems in the analysis of natural and synthetic spinels (ferrites) can be solved, especially the determination of FeO in the presence of higher oxides of manganese and cobalt.[733,740-742] Compounds of Mn(III) and Co(III) are usually reduced directly during the decomposition of a sample (with a composition of n $Me_2O_3 \cdot$ m FeO), where Me is Co(III) or Mn(III) in H_3PO_4 (d = 1.69) and oxalic acid,[741,742] with refluxing in a carbon dioxide atmosphere. During the dissolution, phosphate and diphosphate complexes of Mn(III) and Co(III) are formed and are selectively reduced by oxalic acid. The Fe^{2+} content in the solution is determined oxidimetrically with a ceric salt, after thermal decomposition of the excess oxalic acid.

The determination of the FeO content in resistant silicate minerals that cannot be decomposed by hydrofluoric acid (e.g., tourmaline, staurolite, some amphiboles, and pyroxenes) is very difficult. For decomposition of tourmaline a mixture of phosphoric acid with ammonium and potassium fluorides has been proposed[743] (using a CO_2 atmosphere). However, it is very difficult to control the critical temperature (210°C) of the reaction mixture. Dissolution in excess vanadium salt has also been tested.[722] The values found are probably lower than the actual FeO content. A detailed comparison study with the data obtained by Mössbauer spectrometry has so far been missing. The importance of an accurate determination of FeO from the point of view of the structural parameters of tourmaline has recently been demonstrated by Povondra.[744]

In a mixture of phosphoric and polyphosphoric acids the complex of manganic ions is sufficiently stable (in contrast to solutions of sulphuric and perchloric acids) and its thermal destruction at 300°C is minimal.[733] The complex is formed by the oxidation of Mn^{2+} with, e.g., nitric or perchloric acid, permanganate, bromate, or bichromate in a medium of polyphosphoric acids. In this solution iodate reacts slowly with Mn^{2+}, which permits a determination of Mn(III) in spinels containing FeO, Fe_2O_3, MnO, and Mn_2O_3. Dissolved Fe^{2+} ions are preferentially oxidized by iodate added,[745] and thus an interaction between Mn^{3+} and Fe^{2+} is prevented. The purple complex (with an absorbance maximum at 500 to 515 nm) is ascribed[750] a composition of $[Mn(III)\,(H_2P_2O_7)_3]^{3-}$; the redox potential of the $[Mn(III)(H_2P_2O_7)_3]^{3-}/[Mn(II)(H_2P_2O_7)_3]^{4-}$ system is 1.2 V normal hydrogen electrode (NHE).

The Mn(III) complex with diphosphate also plays a role in analyses of ferromanganese, manganese alloys, manganese ores, and minerals.[746-750] The material is either directly dissolved in a mixture of phosphoric and perchloric acids, or in solutions of nitric and hydrochloric acids followed by the oxidation by heating the evaporation residue with phosphoric and perchloric acids. The oxidation temperature must be carefully controlled (290 to 300°C), otherwise the oxidation is incomplete.[747] Compounds of Mn(III) are more easily formed by treatment with bromate. Excess bromate is decomposed by heating to 160°C; the bromine formed is simultaneously volatilized. The reaction product is stabilized by an addition of a potassium salt; its composition has been given[747] as $KH_2Mn(III)\,(PO_4)_2$, in contrast to the above composition proposed in Reference 750. The optimal temperature intervals in which oxidants suitable for the oxidation of Mn^{2+} to Mn^{3+} are stable in polyphosphoric acids have been found (the oxidants involve $NaBiO_3$, $KBrO_3$, $K_2Cr_2O_7$ and KIO_3).[751]

Mixtures of phosphoric and perchloric acids oxidize vanadium compounds to vanadate, in addition to the oxidation of Mn^{2+}, during decompositions of iron ores. After selective reduction of Mn(III) by *N*-benzoylphenylhydroxamic acid, the vanadate is titrated by a ferrous salt solution.[752,753]

Changes in the redox potentials of the UO_2^{2+}/U^{4+} and Fe^{3+}/Fe^{2+} systems depending on the phosphoric acid concentration have made it possible to develop new procedures for decomposition and analysis of uranium ores[754-756] and uranium-based nuclear fuels.[757-762] In concentrated solutions of phosphoric acid, ferrous ions behave as a strong reductant.[623,624] Uranium(IV) ions have a lower tendency toward oxidation in this medium than in hydrochloric acid in which they are even oxidized by ferric ions. If the phosphoric acid concentration increases above 10 *M*, uranyl and ferrous ions interact,

$$2\,Fe^{2+} + UO_2^{2+} + 4\,H^+ \rightleftharpoons U^{4+} + 2\,Fe^{3+} + 2\,H_2O \qquad (28)$$

These reactions are utilized when determining various valence forms of uranium. The most common applications[757-762] involve the determination of the U/O ratio (the so-called oxygen stoichiometry) in mixtures of the oxides, UO_2 and UO_3.

Uranium metal is dissolved only slowly in concentrated phosphoric acid.[762] Among uranium oxides, UO_3 is most readily dissolved. In analyses of resistant UO_2 the decomposition is hastened by adding an oxidant. In the determination of the U(VI)/U(IV) ratio in the mixed oxides, UO_{2+x} and U_3O_8, the sample is dissolved in 85% H_3PO_4 in an argon atmosphere and at 180 to 210°C. The decomposition reagent must not contain reductants.[762] To evaluate the oxidation state in uranite ores, interactions among compounds of U(IV) and Fe(III) must be eliminated by a suitable selection of the dissolution conditions.[754-756] Even ignited oxides of other actinoids (ThO_2, Pa_2O_5) are rapidly dissolved on heating with orthophosphoric or diphosphoric acid.[763]

An ideal decomposition agent for the determination of carbon dioxide content in inorganic solids should exhibit the following properties:

1. Carbon dioxide should be quantitatively liberated from the test solid in a shortest possible time. Rapid decomposition facilitates transport of the CO_2 in a stream of an inert gas into the measuring or absorption system. The blank value is decreased and contamination with the atmospheric carbon dioxide suppressed.
2. The decomposition agents should have a minimum volatility. Transport of acid vapors into the measuring or absorption system is then avoided, even without using an acid vapor trap.
3. The agent should exhibit minimal oxidizing properties, even at elevated temperatures. It is then possible to determine carbonates in materials containing organic substances, such as organogenic limestones, slates, soils, and sediments.

Among common inorganic acids, only phosphoric acid meets the above requirements.[764-769] All carbonates, including magnesite, siderite, scapolite,[768] and natural fluorocarbonates of the rare earths (bastnesite and parisite) are dissolved within 10 to 15 min in boiling 85% H_3PO_4. If the acid concentration decreases below 50%, the rate of the decomposition of carbonates is substantially decreased,[764,766,767] and the reaction time must be increased to 30 to 90 min. In dilute phosphoric acid, calcite is dissolved most rapidly, magnesite and siderite react slowly, and scapolite is only partially attacked.[764,766] The reaction

$$MeCO_3 + 2\,H^+ \rightarrow Me^{2+} + CO_2 + H_2O \tag{29}$$

has been used to isolate and determine the CO_2 content in rocks, soils, and minerals, employing mostly a conductometric or coulometric titration or gas chromatography. The determination of paleotemperatures of the formation of carbonate minerals by mass spectrometry, measuring the $^{18}O/^{16}O$ ratio (e.g., after decomposition[769] in 100% H_3PO_4 at 95°C) in the carbon dioxide liberated by reaction,[29] suffers from certain problems. Two oxygen atoms from the solid carbonate are transferred into the gaseous phase as carbon dioxide, and the third oxygen atom yields a water molecule by reaction with H^+ ions. The isotopic composition of the oxygen in CO_2 does not correspond to the ratio of the oxygen isotopes in the original material. Because of a departure from equilibrium the δ value is shifted by as much as 10%. This shift is different for different carbonate minerals and also depends on the decomposition conditions.

The determination of carbonate may be immediately followed by a determination of noncarbonate forms of carbon, which are inaccurately denoted as the "organic" carbon. One of the possible solutions of this problem is a conversion of carbonaceous substances into carbon dioxide by chromic acid in a medium of phosphoric acid.[647,770-775] Carbonates are first removed by treatment with phosphoric acid from rock or soil samples; the carbonaceous substances remain unchanged. In the second stage, excess CrO_3 is added, and the CO_2 produced by the oxidation is determined in the same apparatus and by the same procedure as those used in the preceding determination of carbonate. The mixture of phosphoric acid and chromium trioxide is often called the Dixon mixture.[647,770] The remaining problem concerns the rate of the oxidation that depends on the chemical composition of the carbonaceous compound present, on the phosphoric acid concentration, the reaction mixture temperature, and on the kind of the catalyst, if present. Most common carbonaceous substances in geological materials are bituminous compounds and graphite (eruptive and metamorphosed rocks); in sedimentary rocks kerogen is usually present (highly condensed polyaromatics), or amorphous carbon. This kind of decomposition of rocks containing crystalline graphite is subject to a negative error caused, either by incomplete oxidation of graphite or by migration of a graphite film on the surface of vessels to the upper part of the apparatus, away from the liquid oxidizing mixture.[771] The oxidation with potassium bichromate in a mixture of phosphoric and sulfuric acids has yielded good results in analyses of river sediments[774] and soils.[775,776] With a sufficiently long reaction time (more than 12 hr) the oxidation occurs even at 25°C.

In contrast to aqueous solutions, compounds of Cr(VI) are stronger oxidants than ceric salts in anhydrous polyphosphoric acids (CPA) at temperatures higher than 200°C, which can be explained by the formation of very stable Cr(III)-CPA complexes. The behavior of CPA is in this respect comparable with the properties of melts.[644] The role of the acceptor of O_2^- anions affecting the value of the formal redox potential of the Cr(VI)/Cr(III) system is played here by the polyphosphoric acids present. By a mixture of solid ceric sulfate with CPA, various forms of osmium are converted into volatile OsO_4 that is separated by distillation from the reaction mixture. As demonstrated by experiments with ^{106}Ru, ruthenium is not oxidized to RuO_4 under these conditions. However, volatile tetraoxide is readily formed by the oxidation of ruthenium metal, Ru_2S_3 or $RuCl_3$ with solid potassium bichromate in anhydrous CPA. As polyphosphoric acids function as a highly aggressive decomposition agent at high temperatures, this procedure permits a combination of the difficult operation of dissolution of the solid phase with the subsequent separation of the two components, which substantially shortens the time required for the obtaining the distillate for the final determination. This principle has been used to determine ^{106}Ru in sea sediments and natural ruthenium in igneous rocks.[644,777] For the determination of ruthenium in geological samples, the material decomposition and the sample distillation are carried out after the activation of the material by thermal neutrons in a reactor. If the sample contains bromides (sea sediments), part of the selenium, mercury, and arsenic present is transferred to the distillate. The photopeaks of ^{75}Se in the γ-spectrum interfere in the determination of the ^{191}Os content from the 129.4- and 139.0-keV lines. Therefore, the determination of osmium is preceded by a separation of $SeBr_4$. Complete isolation of osmium and ruthenium can be attained in a simple apparatus, even from platinum ores with a complicated composition.[778] The samples need not be fused, and a spectrophotometric determination of isolated osmium and ruthenium can be readily performed in the distillates obtained.

Polyphosphoric acids displace highly reactive hydrogen halides from the alkali halides, and the hydrogen halides immediately react with the dissolved substances with formation of volatile halides that are readily separated from the reaction mixture by distillation. Most applications have been directed toward determinations of those elements with which an unsuitable decomposition technique (incomplete transfer of the test components into a solution and losses through volatilization) is the main source of error. Therefore, the primary application is to selenium; solid sodium or ammonium bromide is used as the source of the hydrogen halide.[777,779-783] Samples of 0.1 to 1.0 g of a rock (sediment) are dissolved in 30 to 40 g of CPA prepared beforehand and containing 0.1 to 0.2 g of ammonium bromide, in a glass or quartz distillation apparatus through which air is passed.[777] The decomposition is complete when the reaction mixture temperature attains a value of 250°C (within 30 to 50 min). Selenium forms volatile $SeBr_4$ and compounds of Se(VI) are reduced by the hydrogen bromide formed to Se(IV). CPA does not exhibit oxidizing effects even at high temperatures, and thus selenides and elemental selenium do not react under these conditions and are oxidized by iodate.[779] Galena and chalcopyrite are decomposed only partially, and pyrite is degraded with the formation of elemental sulfur which is partly transferred into the distillate.[779,782] Part of the $SeBr_4$ decomposes in the gaseous phase according to the reaction

$$2\ SeBr_4 \rightleftharpoons Se_2Br_2 + 3\ Br_2 \qquad (30)$$

The instable selenium(I) bromide is converted into selenium metal that condenses in the upper part of the distillation apparatus. To suppress Reaction 30, excess bromide is used, and the upper part of the distillation apparatus is heated in order to transport all the selenium into the absorption solutions.[779] When decomposing rocks and sediments for a determination of arsenic, sodium chloride is added to the reaction mixture;[784] As_2S_3 and As_2O_5 react only after an addition of sodium bromide.[783] The samples are first irradiated by thermal neutrons in a reactor and then decomposed in the presence of a carrier. Compromise conditions are selected when determining selenium, arsenic, and antimony simultaneously in a single sample.[783] The distillation is carried out from a mixture of CPA and NaBr at 180°C. The same system has also found use in the isolation of mercury from granite and basalt.[785] To compensate for the effects of the phase composition, the dissolution is carried out in the presence of vanadate. A decomposition with immediate separation of $GeBr_4$ has not yet been applied in rock analysis, although some germanium compounds (GeS_2 and some crystal modifications of GeO_2) are quantitatively transferred into the distillate in the form of the bromides.[786] Sulfides, coal, and soils are oxidized directly in a closed distillation apparatus by potassium permanganate in phosphoric acid. On an addition of hydrochloric acid to the reaction mixture the $GeCl_4$ formed is distilled off. The chlorine formed has a favorable effect on the oxidative dissolution of the test substances.[787]

An important application in silicate analysis is the determination of chloride in rocks and soils.[652] Hydrogen chloride is quantitatively liberated by CPA at 250°C and is trapped in an absorption vessel containing 0.05 *M* NaOH.

The decomposition involving CPA and halides permits differential determinations of elements that are present in various forms in the solid phase. The sample destruction is accompanied by rapid separation of a test element from a complex mixture, and this procedure has yielded good results in γ spectrometric determinations of some components in the separated distillates. The principal condition for the obtaining reliable results is detailed knowledge of the phase composition of the test material, which is a great problem with minor and trace elements.

Polyphosphoric acids have found the widest application in combination with reductants. The decomposition of the material is combined with a separation of the test component;[644,788-802] tin(II) salts, tin metal, red phosphorus, titanium(III) salts, hydrogen iodide, and hypophosphorous acid are mostly used as reductants. Effective reagents are most often obtained by heating an anhydrous mixture of polyphosphoric acids (CPA, for the preparation see Page 80) with $SnCl_2{\cdot}2H_2O$ (Sn(II)-CPA) at 250 to 300°C in a glass or quartz apparatus.[788-791,794-796] Some authors prefer preparations obtained from orthophosphoric acid and stannous chloride or reagents prepared from the same compounds *in situ*, during the sample decomposition.[802] The reagent is a highly viscous liquid, sometimes slightly turbid, depending on the preparation method. It is stored in closed glass or quartz vessels and is stable at a low temperature; at 30°C about 10% of the reducing power is lost within 1 month.[788] The reagent exhibits no reducing properties at temperatures below 120°C, but sulfate is reduced to hydrogen sulfide at 250 to 300°C. The latter is absorbed and determined by a iodometric titration, spectrophotometrically, or with an ion-selective electrode. This decomposition procedure makes it possible to determine sulfate, even in barites and precipitated barium sulfate, with which other procedures are complicated by a high chemical resistance of the materials.[706,707,788,791] A mixture of CPA with sodium hypophosphite, tin metal, or potassium iodide[707,795] is more efficient than Sn(II)-CPA if barites and anglesite, gypsum, and alunite are to be completely destroyed. The decomposition of barium sulfate with a mixture of CPA and Ti(III) (obtained from titanium metal) with $SnCl_2$ is complete within only 15 to 20 min at 220 to 260°C.[793] The Sn(II)-CPA reagent also reduces sulfates in coal,[792] rocks,[706] and roasted pyrite.[794]

Sulfides are decomposed together with sulfates. On treatment with Sn(II)-CPA, pyrite, galena, sphalerite, pyrrhotite, and pentlandite release hydrogen sulfide.[706,707,788,790,791] According to Nagashima,[706] copper sulfides, such as chalcopyrite, bornite, covellite, cubanite, and digenite, are also easy to decompose. Molybdenite is very little attacked under these conditions.[707] However, copper-molybdenum ores are completely decomposed by Sn-CPA containing a titanium(III) salt.[793] Mizobuchi[707] reports certain difficulties in the decomposition of chalcopyrite. In a reaction mixture of Sn(II)-CPA (or CPA with tin metal and sodium hypophosphite) a black precipitate of cuprous sulfide is formed that is resistant toward the reductant. To release hydrogen sulfide quantitatively, a small amount of potassium or lead iodide must be added to the reaction mixture.

The above reaction underlie a difference determination of sulfide and sulfate sulfur, e.g., in rocks[706] and coal.[792] The total sulfur is determined first by measuring the hydrogen sulfide liberated during the decomposition of the substance by CPA with a reductant [usually Sn(II)-CPA]. In a parallel sample, the content of sulfide is determined after destruction of the material by CPA. However, a different behavior has been observed with pyrites, which may distort the results.[706,707] To differentiate between the contents of pyrite and sulfate in coal, the reduction with the Sn(II)-CPA reagent was carried out at a gradually increased temperature. According to Purnell,[792] sulfate is reduced at 210 to 220°C, whereas pyrite is destroyed at 265 to 270°C. However, the precision of the determination of the individual forms of sulfur in the presence of one another is substantially poorer than that of the determination of total sulfur. The accuracy of the results cannot be evaluated by analyzing reference materials (e.g., rocks), as only the values of total sulfur have been determined. The methods based on the reaction of the test substance with Sn(II)-CPA thus have found main use in the determination of total sulfur in rocks, airborne ashes, roasted pyrites, bauxites, chromium and iron ores, and aluminum, thorium, and zirconium oxides.[796,797]

The Sn(II)-CPA reagent releases hydrogen selenide from pyrites,[801] similar to the liberation of hydrogen sulfide; hydrogen selenide is readily separated from the reaction mixture in a stream of hydrogen. Under the same conditions, nitrides of titanium, zirconium, hafnium, aluminum, and molybdenum are also decomposed. No nitrogen is lost during the decomposition procedure. Ammonia is separated from the ammonium salts formed by distillation form a medium of an alkali hydroxide.[802]

In spite of certain complications involved in the use of phosphoric acid as a solvent in the analytical process, the acid is finding progressively greater application in decompositions of resistant inorganic substances. It is preferred for decomposition of minerals, ores, and natural and synthetic oxides that would otherwise have to be fused to be dissolved. The modern techniques of spectral analysis (especially OES-ICP and AAS) are also capable to deal with the depressive effects of phosphoric acid in determinations of some elements; e.g., the recent work of Hannaker et al.[803] describes the successful use of a polyphosphoric acid medium in a determination of major components in various minerals, soils, and industrial products. At a temperature of 290°C, the time required for the decomposition does not exceed 45 min for most minerals (Table 4). Sulfides are difficult to dissolve.

Table 4
EFFECT OF H_3PO_4 AND $H_3PO_4/HClO_4$ ON THE DISSOLUTION TIME FOR 150 mg OF 300-MESH NATURAL MINERALS[803]

Mineral	H_3PO_4	$H_3PO_4/HClO_4$	Mineral	H_3PO_4	$H_3PO_4/HClO_4$	Mineral	H_3PO_4	$H_3PO_4/HClO_4$
Alabandite	45	30	Epidote	45	45	Psilomelane	30	30
Allanite	45	45	Fluorite	15	15	Pyrite	195a	30
Amblygonite	30	30	Franklinite	30	30	Pyrolusite	30	30
Amphibole	30	30	Galena	15	15	Pyroxene	45	45
Anhydrite	15	15	Garnet	30	30	Pyrrhotite	75	30
Apatite	30	30	Garnierite	15	15	Rhodocrosite	15	15
Apophyllite	15	15	Gypsum	15	15	Serpentine	30	30
Azurite	15	15	Hematite	135	45	Siderite	15	15
Bornite	195[a]	30	Hemimorphite	45	45	Smithsonite	15	15
Calcite	15	15	Ilmenite	165	45	Sphalerite	15	15
Cerussite	15	15	Kaolinite	30	30	Spinel	30	30
Chalcanthite	15	15	Labradorite	75	75	Staurolite	120	120
Chalcopyrite	195[a]	30	Lepidolite	60	60	Stilbite	15	15
Clinochlore	30	30	Limonite	15	15	Talc	90	90
Chromite	150	30	Malachite	15	15	Tetrahedrite	195	15
Chrysolite	45	45	Magnesite	15	15	Tourmaline	150	150
Colemanite	30	30	Magnetite	15	15	Wavellite	30	30
Corundum	180	180	Manganite	15	15	Wernerite	30	30
Cryolite	30	30	Millerite	105	30	Wolframite	30	30
Cuprite	15	15	Molybdenite	195[a]	45	Wollastonite	30	30
Datolite	15	15	Nepheline	105	105	Zincite	15	15
Embolite	—	15	Orthoclase	30	30	Zircon	—	150
Enstatite	45	45	Pectolite	30	30			

[a] Incomplete dissolution.

In contrast to the data given in Reference 706, chalcopyrite and bornite are not decomposed; molybdenite and pyrite react slowly. Among oxide minerals, corundum, haematite, chromite, and ilmenite are least attacked, as are staurolite and some kinds of tourmalines among silicates. On addition of a small amount of perchloric acid (0.5 mℓ added to 10 mℓ H_3PO_4), the decomposition becomes oxidizing in character, and the reaction time required for destruction of sulfidic phases is considerably shortened (at most 45 min for molybdenite) even for chromite and haematite. After 150 min, even zircon is decomposed, which is highly resistant even toward effects of acids in autoclaves, at elevated pressure and temperature.

VIII. DECOMPOSITION BY THE ACTION OF ION EXCHANGERS

Mineral acids and hydroxide solutions, as solvents for solids, can be replaced by ion exchangers that act, depending on the chemical character of their functional groups, as strong, weak, or complexing solvents. They have an advantage in that they react only in equivalent amounts with respect to the solid to be dissolved, are not corrosive, and introduce only minimal amounts of impurities into the system. The decomposition is based on the fact that the ion exchanger removes from a saturated solution of a sparingly soluble substance the cations, or the anions, or both kinds of ion simultaneously when a mixture of the two kinds of ion exchanger is employed. The equilibrium in the solution is thus disturbed and is renewed by a further dissolution of the substance. If the counterion, liberated on the ion exchange, is removed, or if a large excess of the ion exchanger is used, the solid is finally completely dissolved. The procedure is experimentally simple: a sufficient excess of an ion exchanger is added to an aqueous suspension of the substance to be dissolved, and the mixture is stirred at laboratory or an elevated temperature, until the solid is dissolved. The rate and completeness of dissolution primarily depend on the magnitude of the distribution coefficients of the ions between the solution and the solid phase. The amount of a cation exchanger required for complete dissolution of a solid of the B^+A^- type is given by the relationship

$$\overline{Q} \gg Q^o_{BA} \left[1 + \frac{(\alpha\ Q^o_{BA})^2}{L_{pBA} \cdot K^B_H \cdot V^2} \right] \tag{31}$$

where $\overline{Q}$ is the required amount of the ion exchanger, Q^o_{BA} is the amount of substance BA (moles), L_{pBA} is the solubility product of substance BA, K^B_H is the distribution coefficient for the cation exchange, B^{n+}/nH^+, α is the degree of dissociation of the saturated solution of BA, and V is the solution volume in liters. In order to decrease the required amount of the ion exchanger, it is advantageous to combine a cation exchanger in the H^+ form with an anion exchanger in the OH^- form, as the amount required for complete reaction is then given by

$$\overline{Q} > Q^o_{BA} \left[1 + \frac{K_w}{L_{pBA} \cdot K^B_H \cdot K^A_{OH}}\ 1/2 \right] \tag{32}$$

where the symbols are the same as in Equation 34, K^A_{OH} is the distribution coefficient for the anion exchange, A^{n-}/nOH^-, and K_w is the ionic product of water. After the decomposition, the two exchanger types can be separated by a gravity method in a suitable mixture of dichloroethane and carbon tetrachloride, utilizing the different densities of the two solid phases.[804]

This principle of the action of ion exchangers has mainly been used in decomposition of sparingly soluble fluorides, sulfates, and phosphates. Among natural equivalents, the procedure has been successfully applied to decomposition of apatites.[805] A 0.05-g sample of the dried mineral is stirred for 12 hr in 35 mℓ of hot water containing 5 to 10 g of a Zeo-Karb 225 cation exchanger in the H^+ form. The ion exchanger is then filtered off on a plastic Buechner funnel and thoroughly washed with water. The solution obtained is neutralized and diluted to 200 mℓ, taking suitable aliquot parts for determinations of phosphorus and fluorine.

This decomposition technique for apatite has been later compared[806] with acid decomposition and fusion with an alkali hydroxide, determining the fluoride. It has been found that the results obtained after the decomposition by ion exchanger are systematically lower than those after fusion. It seems that aluminum phosphate that might be present is more difficult to decompose by ion exchange. The time of the ion exchanger action, the suspension temperature, and the effect of the ion exchanger excess on the completeness of decomposition of technical calcium phosphate have also been systematically studied. An optimal ratio is 1 + 10—20, when the decomposition is virtually complete within 30 min.[807] To decompose calcium phosphate,[808] the effect of cation and anion exchangers has been combined, according to Reference 805. The ion exchangers of very different grain sizes were employed and separated by sieving after dissolution of the solid. The fractions obtained were then eluted with suitable reagents, obtaining an acidic solution containing Ca^{2+} and an alkaline solution containing all the anions. The ''accessible'' phosphoric acid can be extracted from soil suspensions using an anion exchanger.[809] Segall and Schumacher[810] have described the equilibria among insoluble calcium salts (carbonate, oxalate) and the Dowex A-1 cation exchanger containing iminodiacetic acid as the functional group.

Barium sulfate can be analogously decomposed in a water suspension, using a cation exchanger or its mixture with an anion exchanger. The dissolution proceeds best at 80 to 90°C. The procedure is also suitable for an indirect determination of sulfur in some organic and inorganic compounds. The substances are decomposed in acids or burned to produce sulfate that is precipitated with barium ions. The barium sulfate is then dissolved by agitating with the above mixture of ion exchangers in the H^+ and Cl^- forms. The chloride that has passed into the solution and is equivalent to the sulfate content is determined argentometrically.

A cation exchanger has been used by Govindaraju[811,812] for decomposition of borate glasses prepared by fusion of silicates with lithium metaborate. As the product is difficult to dissolve in water and acids, a cation exchanger in the H^+ form has been used in an aqueous suspension. After the decomposition, the ion exchanger is filtered off, dried, and applied to a spectrochemical determination of many major, minor, and trace elements in the rocks analyzed. The procedure for dissolution of the melt with an ion exchanger is described in more detail in Chapter 8 (Section VIII.E). Among dissolution techniques with ion exchangers, the procedure of Povondra and Hejl,[813] employed in determination of boron in tourmalines and other borosilicates, can also be included. The minerals are decomposed by sintering with a mixture of sodium carbonate and zinc oxide, and the sinter is leached with water. All the ions leached (Na^+, Me^{2+}, CO_3^{2-}) are then removed from this solution by static action of a cation exchanger, according to Equation 33.

$$2\ Na^+ + Me^{n+} + (n + 2)\ RSO_3H + CO_3^{2-} \rightarrow$$

$$2\ RSO_3Na + n(RSO_3)n\ Me + CO_2 + H_2O + nH^+ \qquad (33)$$

The ion exchanger is separated, washed, and a deionized solution is thus obtained containing boric acid, which can be determined by acid-base titration.

REFERENCES

1. **Janoušek, I.,** *Chem. Listy,* 78, 1320, 1984.
2. **Novák, J., Funke, A., and Kleinert, P.,** *Fresenius Z. Anal. Chem.,* 237, 339, 1968.
3. **Duclos, P.,** *Analusis,* 4, 98, 1976.
4. **Hetényi, M. and Varsányi, I.,** *Acta Mineral. Petrogr. Szeged,* 22, 165, 1975.
5. **Brooks, R. R., Holzbecher, J., Ryan, D. E., Zhang, H. F., and Chatterjee, A. K.,** *At. Spectrosc.,* 2, 151, 1981.
6. **Shane, S. Q. H., Macdonald, T. J., and Boyle, J. R.,** *Anal. Chem.,* 57, 1242, 1985.
7. **Russell, G. M.,** *MINTEK Rep.,* No. M289, 1986.
8. **Huang, M.,** *Fenxi Huaxue,* 13, 203, 1985; *Anal. Abstr.,* 47, 12B53, 1985.
9. **Tsimbalist, V. G., Yukhin, Yu. M., and Bukhlova, T. I.,** *Zh. Anal. Khim.,* 38, 993, 1983.
10. **Dietze, U., Braun, J., and Peter, H. J.,** *Fresenius Z. Anal. Chem.,* 322, 17, 1985.
11. **Greaves, M. C.,** *Nature,* 199, 552, 1963.
12. **Donaldson, E. M.,** *Talanta,* 29, 1069, 1982.
13. **Donaldson, E. M., Mark, E., and Leaver, M. E.,** *Talanta,* 31, 89, 1984.
14. **Guo, M.,** *Fenxi Huaxue,* 10, 254, 1982; *Anal. Abstr.,* 43, 6B80, 1982.
15. **Tuttle, M. L., Goldhaber, M. B., and Williamson, D. L.,** *Talanta,* 33, 953, 1986.
16. **Thompson, M. and Liang-Lao,** *Analyst,* 110, 229, 1985.
17. **Nakamura, Y. and Kobayashi, Y.,** *Bunseki Kagaku,* 35, 446, 1986.
18. **Ru-Shi Liu, and Mo-Hsiung Yang,** *Fresenius Z. Anal. Chem.,* 325, 272, 1986.
19. **Varanova, L. L., Berliner, L. D., Kaplan, B. Ya., Malyutina, T. M., Nazarova, M. G., and Razumova, L. S.,** *Zavod. Lab.,* 50(5), 10, 1984.
20. **Olade, M. and Fletcher, K.,** *J. Geochem. Explor.,* 3, 337, 1974.
21. **Viets, J. G.,** *Anal. Chem.,* 50, 1097, 1978.
22. **Motooka, J. M., Mosier, E. L., Sutley, S. J., and Viets, J. G.,** *Appl. Spectrosc.,* 33, 456, 1979.
23. **Chao, T. T. and Sansolone, R. F.,** *J. Res. U.S. Geol. Surv.,* 5, 409, 1977.
24. **Viets, J. G., O'Leary, R. M., and Clark, J. R.,** *Analyst,* 109, 1589, 1984.
25. **O'Leary, R. M. and Viets, J. G.,** *At. Spectrosc.,* 7, 4, 1986.
26. **Murphy, J. M. and Sergeant, G. A.,** *Analyst,* 99, 515, 1974.
27. **Van der Veen, N. G., Keukens, H. J., and Vos, G.,** *Anal. Chim. Acta,* 171, 285, 1985.
28. **Aslin, G. E. M.,** *J. Geochem. Explor.,* 6, 321, 1976.
29. **Forehand, T. J., Dupuy, A. E., Jr. and Tai, H.,** *Anal. Chem.,* 48, 999, 1976.
30. **Peachey, D. and Vickers, B. P.,** *Rep. Inst. Geol. Sci. U.K.,* 80(1), 6, 1980; *Anal. Abstr.,* 42, 3B150, 1982.
31. **Somer, G. and Aydin, H.,** *Analyst,* 110, 631, 1985.
32. **Otsuki, T.,** *Bunseki Kagaku,* 28, T15, 1979.
33. **Jones, E. A.,** *Natl. Inst. Metall. Repub. S. Afr. Rep.,* No. 1869, 1977.
34. **Banerjee, S. and Dutta, R. K.,** *Talanta,* 27, 448, 1980.
35. **Ren, Q. Li. S. and Zhou, Y.,** *Fenxi Huaxue,* 10, 227, 1982; *Anal. Abstr.,* 43, 6B31, 1982.
36. **Tindall, F. M.,** *At. Absorpt. Newsl.,* 16, 37, 1977.
37. **Wemyss, R. B. and Scott, R. H.,** *Anal. Chem.,* 50, 1694, 1978.
38. **Diamantatos, A.,** *Anal. Chim. Acta,* 147, 219, 1983.
39. **Alekseeva, I. I., Latysheva, G. N., Romanovskaya, L. E., and Tikhonova, L. P.,** *Zavod. Lab.,* 50(3), 5, 1984.
40. **Kothny, E. L.,** *J. Geochem. Explor.,* 3, 291, 1974.
41. **Ptushkina, M. N., Lebedeva, L. I., Shcheredina, M. I., and Stolyarov, K. P.,** *Zavod. Lab.,* 49(5), 12, 1983.
42. **Coetze, C. F. B. and Fieberg, M. M.,** *MINTEK Rep.,* No. M93D, 1985.
43. **Šulcek, Z.,** unpublished results, 1984.
44. **Bajo, C., Rybach, L., and Weibel, M.,** *Chem. Geol.,* 39, 281, 1983.
45. **Dolgova, S. I. and Aleksanin, N. I.,** *Zavod. Lab.,* 46, 804, 1980.
46. **Bhuchar, V. M. and Marayan, V. M.,** *Indian J. Technol.,* 20, 283, 1982; *Anal. Abstr.,* 44, 4B34, 1983.
47. **Karpenko, L. I., Fadeeva, L. A., and Shevchenko, L. D.,** *Zh. Anal. Khim.,* 30, 1330, 1975.
48. **Steger, H.,** *Talanta,* 23, 81, 1976.
49. **Ryakchikov, D. I. and Golbraikh, E. K.,** *Analiticheskaya Khimia Toria,* Nauka, Moskva, 1960, 159.
50. **Foster, J. R.,** *Can. Miner. Metal. Bull.,* 66, 85, 1973; *Can. Miner. Metal. Bull. Spec. Vol.,* 11, 554, 1971.
51. **Novák, J.,** unpublished results, 1980.
52. **Wang, X., Lu, P., and Zhang, G.,** *Guangpuxue Yu Guangpu Fenxi,* 5, 63, 1985.
53. **Hlasivcová, N. and Novák, J.,** *Silikáty,* 13, 157, 1969.

54. **Novák, J. and Arend, H.,** *J. Am. Ceram. Soc.*, 47, 530, 1964.
55. **Krtil, J.,** *Jad. Energ.*, 29, 440, 1983; *Jad. Energ.*, 30, 91, 1984.
56. **Milner, G. W. C., Phillips, G., and Fudge, A. J.,** *Talanta*, 15, 1241, 1968.
57. **Sager, M.,** *Microchim. Acta*, II, 381, 1984.
58. **Chan, Y. K. and Riley, J. P.,** *Anal. Chim. Acta*, 33, 36, 1965.
59. **Bock, R., Jacob, D., Fariwar, M., and Frankenfeld, K.,** *Fresenius Z. Anal. Chem.*, 200, 81, 1964.
60. **Bock, R. and Tschöpel, P.,** *Fresenius Z. Anal. Chem.*, 246, 81, 1969.
61. **Beshikdashyan, M. T., Vasileva, M. G., and Korostyleva, L. D.,** *Zh. Anal. Khim.*, 36, 1021, 1981.
62. **Kvaček, M.,** *Chem. Listy*, 58, 305, 1964.
63. **Johnson, J. S. and Kranska, K. A.,** *J. Phys. Chem.*, 63, 440, 1959.
64. **Van Loon, J. C.,** *Trends Anal. Chem.*, 3, 272, 1984.
65. **Qu, Y., Zhao, D., and Wang, J.,** *Fenxi Huaxue*, 12, 623, 1984; *Anal. Abstr.*, 47, 4B155, 1984.
66. **Fishkova, N. L. and Talypina, O. P.,** *Zh. Anal. Khim.*, 38, 452, 1983.
67. **Raoot, S., Athavale, S. V., and Rao, T. H.,** *Analyst*, 111, 115, 1986.
68. **Abele, C., Weichbrodt, G., and Wichmann, K. H.,** *Fresenius Z. Anal. Chem.*, 322, 11, 1985.
69. **Korotaeva, I. Ya., and Zankhabaeva, V. Z.,** *Zavod. Lab.*, 45, 403, 1979.
70. **Galanova, A. P., Pronin, V. A., Yudelevich, I. G., and Gilbert, E. N.,** *Zavod. Lab.*, 38, 646, 1972.
71. **Fishkova, N. L., Falkova, O. B., and Meshalkina, R. D.,** *Zh. Anal. Khim.*, 27, 1916, 1972.
72. **Mirova, L. F., Tolkacheva, L. F., and Malkov, E. M., and Talipov, S. T.,** *Zavod. Lab.*, 38, 645, 1972.
73. **Blyum, I. A., Pavlova, N. I., and Kalupina, F. P.,** *Zh. Anal. Khim.*, 26, 55, 1971.
74. **Strong, B. and Murray-Smith, R.,** *Talanta*, 21, 1253, 1974.
75. **Kulikova, A. B., Pogrebnyak, Yu. F., and Tatyankina, E. M.,** *Zavod. Lab.*, 46, 127, 1980.
76. **Larkins, P. L.,** *Anal. Chim. Acta*, 173, 77, 1985.
77. **Terenteva, L. A., Klimovich, T. S., and Torgov, V. G.,** *Zavod. Lab.*, 52(2), 5, 1986.
78. **Kulikova, A. B. and Pogrebnyak, Yu. F.,** *Zh. Anal. Khim.*, 35, 793, 1980.
79. **Kulikov, A. A. and Kulikova, A. B.,** *Zh. Anal. Khim.*, 38, 642, 1983.
80. **Vall, G. A., Polubnaya, L. P., Yudelevich, I. G., and Zolotov, Yu. A.,** *Zh. Anal. Khim.*, 34, 885, 1979.
81. **Pogrebnyak, Yu. F.,** *Zh. Anal. Khim.*, 34, 91, 1979.
82. **Farwell, S. O. and Kagel, C. T.,** *Anal. Chim. Acta*, 178, 325, 1985.
83. **Rubeška, I., Korečková, J., and Weiss, D.,** *At. Absorpt. Newsl.*, 16, 1, 1977.
84. **Shvartzman, S. I., Falkova, O. B., Kurskii, A. N., and Popova, N. E.,** *Zh. Anal. Khim.*, 39, 2113, 1984.
85. **Fryer, B. J. and Kerich, R.,** *At. Absorpt. Newsl.*, 17, 4, 1978.
86. **Zhao Shi-shan.,** Fresenius Z. *Anal. Chem.*, 321, 376, 1985.
87. **Fernandez Sanchez, M. L., Garcia Ortis, C., Arribas, Jimeno, S., and Sans Mendel, A.,** *At. Spectrosc.*, 5, 197, 1984.
88. **Bazhov, A. S. and Sokolova, E. A.,** *Zh. Anal. Khim.*, 32, 65, 1977.
89. **Terenteeva, L. A., Afanaseva, L. O., Chalsora, G. K., Vanifatova, N. G., Torgov, V. G., and Zolotov, Yu, A.,** *Zavod. Lab.*, 49(8), 25, 1983.
90. **Grimaldi, F. S. and Schnepfe, M. M.,** *U.S. Geol. Surv. Prof. Pap.*, 575C, C141, 1967; *U.S. Geol. Surv. Prof. Pap.*, 600B, B99, 1968.
91. **Cruickshank, Z. and Munro, H. C.,** *Analyst*, 104, 1050, 1979.
92. **Coombes, R. J. and Chow, A.,** *Talanta*, 26, 991, 1979.
93. **Young, R. S.,** *Talanta*, 28, 25, 1981.
94. **Brown, R. J. and Biggs, W. R.,** *Anal. Chem.*, 56, 646, 1984.
95. **Palmer, I., Streichert, G., and Wilson, A.,** *Natl. Inst. Metall. Repub. S. Afr. Rep.*, No. 1218, 1971.
96. **Parkes, A. and Murray Smith, R.,** *At. Absorpt. Newsl.*, 18, 57, 1979.
97. **Wang, J., Han, P., and Fan, B.,** *Fenxi Huaxue*, 13, 101, 1985; *Anal. Abstr.*, 48, 1B61, 1986.
98. **Kuldvare, A.,** *Analyst*, 110, 1487, 1985.
99. **Bye, R.,** *Fresenius Z. Anal. Chem.*, 317, 27, 1984.
100. **Jordanov, N. and Futekov, L.,** *Talanta*, 15, 850, 1968.
101. **Kuznetsov, V. V. and Korchagina, O. A.,** *Zavod. Lab.*, 51(8), 17, 1985.
102. **Ferri, D., Zignani, F., and Buldini, P. L.,** *Fresenius Z. Anal. Chem.*, 313, 539, 1982.
103. **Lanza, P. and Zappoli, S.,** *Anal. Chim. Acta*, 185, 219, 1986.
104. **Sharma, K. D.,** *Talanta*, 30, 493, 1983.
105. **Steele, K. F. and Wagner, G. H.,** *J. Sediment. Petrol.*, 45, 310, 1975.
106. **Christiensen, T. T., Pedersen, L. R., and Tjell, J. C.,** *Environ. Anal. Chem.*, 12, 41, 1982.
107. **Griepink, B., Muntau, H., and Colinet, E.,** *Fresenius Z. Anal. Chem.*, 318, 490, 1984.
108. **Schlösser, W. and Schwedt, G.,** *Fresenius Z. Anal. Chem.*, 321, 136, 1985.

109. **Götz, A. and Heumann, K. G.,** *Fresenius Z. Anal. Chem.*, 325, 24, 1986.
110. **Sinex, A. S., Castillo, A. Y., and Hetz, G. R.,** *Anal. Chem.*, 52, 2342, 1980.
111. **Berrow, M. L. and Stein, W. M.,** *Analyst,* 108, 277, 1983.
112. **Fletcher, W. K.,** *Analytical Methods in Geochemical Prospecting,* Elsevier, Amsterdam, 1981, 62.
113. **Haring, B. J., van Delft, W., and Bom, C. B.,** *Fresenius Z. Anal. Chem.*, 310, 217, 1982.
114. **Rubeška, I. and Hlavinková, V.,** *At. Absorpt. Newsl.*, 18, 5, 1979.
115. **Sansolone, F. R. and Chao, T. T.,** *Analyst,* 108, 58, 1983.
116. **Suo, Y., Sui, Y., and Wang, L.,** *Guangpuxue Yu Guangpu Fenxi,* 5, 36, 1985; *Anal. Abstr.*, 48, 5B61, 1986.
117. **Janáček, J.,** unpublished results, 1984.
118. **Cook, E. B. T.,** *Natl. Inst. Metall. S. Afr. Lab. Method,* No. 42/3, 1969.
119. **Berndt, H., Masserschnidt, J., Alt, F., and Sommer, D.,** *Fresenius Z. Anal. Chem.*, 306, 385, 1981.
120. **Ruseva, E., Havezov, I., Spivakov, B. Ya, and Shkinev, V. M.,** *Fresenius Z. Anal. Chem.*, 315, 499, 1983.
121. **Young, R. S.,** *Talanta,* 33, 561, 1986.
122. **Parker, G. A.,** *Analytical Chemistry of Molybdenum,* Springer-Verlag, Berlin, 1983, 133.
123. **Diamantatos, A.,** *Anal. Chim. Acta,* 165, 263, 1984.
124. **Beamish, F. E. and Seath, J.,** *Ind. Eng. Chem. Anal. Ed.*, 10, 639, 1938.
125. **Samadi, A. A., Grynszpan, R., and Fedoroff, M.,** *Talanta,* 23, 829, 1976.
126. **Borokha, A. E., Demidová, A. I., Polinskaya, M. B., and Knyazeva, N. F.,** *Zavod. Lab.*, 52(2), 17, 1986.
127. **Thévenot, F., Goeuriot, P., and Gilbert, R.,** *Analusis,* 6, 359, 1976.
128. **Lengauer, W.,** *Fresenius Z. Anal. Chem.*, 322, 23, 1985.
129. **Young, R. S.,** *Analyst,* 107, 721, 1982.
130. **Green, L. W., Knight, C. H., Longhurst, T. M., and Cassidy, R. M.,** *Anal. Chem.*, 56, 696, 1984.
131. **Green, L. W., Elliot, N. L., and Longhurst, T. H.,** *Anal. Chem.*, 55, 2394, 1983.
132. **Drummond, J. L.,** *Talanta,* 13, 477, 1966.
133. **Chong, C. H. H., Crockett, T. W., and Doty, J. W.,** *J. Inorg. Chem.*, 31, 81, 1969.
134. **Egorova, V. A., Kirillova, Z. P., Kocherba, L. V., and Merisov, Yu. I.,** *Zavod. Lab.*, 47(4), 31, 1981.
135. **Popova, M. I.,** *Zh. Anal. Khim.*, 41, 1590, 1986.
136. **Bajo, S.,** *Anal. Chem.*, 50, 649, 1978.
137. **Janghornbarni, M., King, B. T. G., Nahapetian, A., and Young, V. R.,** *Anal. Chem.*, 54, 1188, 1982.
138. **Rybakov, A. A. and Ostroumov, E. A.,** *Zh. Anal. Khim.*, 39, 2168, 1984.
139. **Shawky, M. and White, C. L.,** *Anal. Chem.*, 48, 1484, 1976.
140. **Simeonova, S., Bekyarov, G., and Futekov, L.,** *Fresenius Z. Anal. Chem.*, 325, 478, 1986.
141. **Elson, O. M. and Macdonald, A. S.,** *Anal. Chim. Acta,* 110, 153, 1979.
142. **Fishkova, N. L., Nazarenko, I. I., Vilenkin, V. A., and Petrakova, Z. A.,** *Zh. Anal. Khim.*, 36, 115, 1981.
143. **Ivankova, A. I. and Blyum, I. A.,** *Zavod. Lab.*, 27, 371, 1961.
144. **Nurtaeva, A. K., Kabdulkarimova, K. K., Ilyukevich, Yu, A., Gladyshev, V. P., and Kucheryavenko, V. V.,** *Zavod. Lab.*, 51(2), 18, 1985.
145. **Baranova, L. L., Kaplan, B. Ya., Nazarova, M. G., and Razumova, L. S.,** *Zavod. Lab.*, 52(4), 9, 1986.
146. **Watterson, J. R. and Neuerburg, G. J.,** *J. Res. U.S. Geol. Surv.*, 3, 191, 1975.
147. **Corbett, J. A. and Godbeer, W. C.,** *Anal. Chim. Acta,* 91, 211, 1977.
148. **Brown, R. M., Fry, R. C., Moyers, J. L., Northway, S. J., Denton, M. B., and Wilson, G. S.,** *Anal. Chem.*, 53, 1560, 1981.
149. **Cutter, G. A.,** *Anal. Chim. Acta,* 149, 391, 1983.
150. **Lazareva, V. I., Lazarev, A. I., and Kurkina, E. A.,** *Zavod. Lab.*, 47(11), 24, 1981.
151. **Donaldson, E. M.,** *Talanta,* 24, 105, 1977.
152. **Donaldson, E. M. and Wang, M.,** *Talanta,* 33, 233, 1986.
153. **Thompson, A. J. and Thoresby, P. A.,** *Analyst,* 102, 9, 1977.
154. **Elkhatib, E. A., Bennett, O. L., and Wright, R. J.,** *Soil Sci. Soc. Am. J.*, 47, 836, 1983; *Anal. Abstr.*, 46, 866, 1984.
155. **Rubeška, I., Šulcek, Z., and Moldan, B.,** *Anal. Chim. Acta,* 37, 27, 1967.
156. **Streško, V. and Martiny, E.,** *At. Absorpt. Newsl.*, 11, 4, 1972.
157. **Maren, T. A.,** *Anal. Chem.*, 19, 487, 1947.
158. **Donaldson, E. M.,** *Talanta,* 26, 999, 1979.
159. **Bibinov, S. A., Gladyshev, V. P., Yarmolik, A. S., Kim, A. Ch., and Sokup, N. P.,** *Zavod. Lab.*, 50(2), 12, 1984.
160. **Grossmann, O.,** *Fresenius Z. Anal. Chem.*, 321, 442, 1985.

161. **Donaldson, E. M.,** *Talanta,* 27, 79, 1980.
162. **Agemian, H. and Chau, A. S. Y.,** *Analyst,* 101, 91, 1976.
163. **Chiu, Ch. H. and Hilborn, J. C.,** *Analyst,* 104, 1159, 1979.
164. **Ovrutskii, M. I., Kozachuk, N. S., and Freger, S. V.,** *Gig. Sanit.,* 55(3), 1981; *Anal. Abstr.,* 42, 4613, 1982.
165. **Knechtel, J. R. and Fraser, J. L.,** *Anal. Chem.,* 51, 315, 1979.
166. **McQuaker, N. R., Brown, D. F., and Kluckner, P. D.,** *Anal. Chem.,* 51, 1082, 1979.
167. **Stendal, H.,** *Chem. Erde,* 39, 276, 1980.
168. **Staiger, K., Machelett, B., and Podlesak, W.,** *Bodenkultur,* 36, 99, 1985; *Anal. Abstr.,* 48, 2615, 1986.
169. **Clemency, C. V. and Hagner, A. F.,** *Anal. Chem.,* 33, 888, 1961.
170. **Grekova, I. M., Golik, N. N., and Serbinovich, V. V.,** *Zh. Anal. Khim.,* 38, 443, 1983.
171. **Weiss, D.,** *Chem. Listy,* 79, 205, 1985.
172. **Haskell, R. J. and Wright, J. C.,** *Anal. Chem.,* 59, 427, 1987.
173. **Donaldson, E. M.,** *Methods for the Analysis of Ores, Rock, and Related Materials,* Mines Branch Monogr. 881, Ottawa, 1974, 81.
174. **Peták, P. and Koubová, V.,** *Analyst,* 103, 179, 1978.
175. **Sill, C. W. and Willis, C. P.,** *Anal. Chem.,* 49, 302, 1977.
176. **Chao, T. T., Sansolone, R. F., and Hubert, A. E.,** *Anal. Chim. Acta,* 96, 251, 1978.
177. **Hubert, A. E.,** *U.S. Geol. Surv. Prof. Pap.,* 750B, B188, 1971.
178. **Ward, F. N., Nakagawa, H. M., Harms, T. F., and Van Sickle, G. H.,** *U.S. Geol. Surv. Bull.,* 1289, 35, 1969.
179. **Thompson, C. E., Nakagawa, H. M., and Van Sickle, G. H.,** *U.S. Geol. Surv. Prof. Pap.,* 600B, B130, 1968.
180. **Lutrin, A. and Šikl, J.,** unpublished results, 1980.
181. **Meier, A. L.,** *J. Geochem. Explor.,* 13, 77, 1980.
182. **Fletcher, W. K.,** *Analytical Methods in Geochemical Prospecting,* Elsevier, Amsterdam, 1981, 62.
183. **Thompson, C. E.,** *U.S. Geol. Surv. Prof. Pap.,* 575D, D236, 1969.
184. **Berecki-Biedermann, C.,** *Ark. Kemi,* 26, 391, 1967.
185. **Šmejkal, V.,** private communication, 1981.
186. **Vandael, C.,** *Chim. Anal. (Paris),* 44, 295, 1962.
187. **Murthy, A. R. V., Narayana, V. A., and Rao, M. R. A.,** *Analyst,* 81, 373, 1956.
188. **Murthy, A. R. V. and Sharada, K.,** *Analyst,* 85, 299, 1960.
189. **Sorensen, D. L., Kneib, W. A., and Porcella, D.,** *Anal. Chem.,* 51, 1870, 1979.
190. **Varavko, T. N. and Kaplan, B. Ya.,** *Zavod. Lab.,* 50(5), 7, 1984.
191. **Agterdenbos, J. and Vlogtman, J.,** *Talanta,* 19, 1295, 1972.
192. **Sillen, L. G. and Martell, A. E.,** *Stability Constants of Metal Ion Complexes,* Burlington House, London, 1964, 256.
193. **Headridge, J. B.,** *CRC Crit. Rev. Anal. Chem.,* 1, 461, 1972.
194. **Fahey, J. J.,** *Am. Miner.,* 56, 2145, 1971.
195. **Masaytis, V. L., Raykhlin, A. I., Reshetnyak, N. B., Selinovskaya, T. V., and Shitov, V. A.,** *Zap. Vses. Miner. Ova,* 103, 122, 1974.
196. **Tso, S. T. and Pask, J. A.,** *J. Am. Ceram. Soc.,* 65, 360, 1982.
197. **Palmer, W. G.,** *J. Chem. Soc.,* 2, 1657, 1930.
198. **Blumberg, A. A.,** *J. Phys. Chem.,* 63, 1129, 1959.
199. **Judge, S.,** *J. Electrochem. Soc.,* 118, 1772, 1971.
200. **Langmyhr, F. J. and Graff, P. R.,** *Anal. Chim. Acta,* 21, 334, 1959.
201. **Muster, P. A., Aepli, O. T., and Kossatz, R. A.,** *Ind. Eng. Chem.,* 39, 427, 1947.
202. **Bastius, H.,** *Fresenius Z. Anal. Chem.,* 288, 344, 1977.
203. **Lin, B.,** *Fenxi Huaxue,* 12, 794, 1984; *Anal. Abstr.,* 47, 5B72, 1985.
204. **Kilroy, W. P. and Moynihan, C. T.,** *Anal. Chim. Acta,* 83, 389, 1976.
205. **Piryutko, M. M. and Makarova, T. M.,** *Zavod. Lab.,* 41, 393, 1976.
206. **Markušić, T. and Vranešević, M.,** *Kem. Ind.,* 13, 82, 1964; *Anal. Abstr.,* 12, 2733, 1965.
207. **Glasö, Ö. and Patzauer, G.,** *Anal. Chim. Acta,* 25, 189, 1961.
208. **Fox, E. and Jackson, W. A.,** *Anal. Chem.,* 31, 1656, 1960.
209. **Geilmann, W. and Tölg, G.,** *Glastechn. Ber.,* 33, 245, 1960.
210. **Kaloczai, G. I. Z. and Hockley, J. J.,** *Mineral. Mag.,* 38, 618, 1972.
211. **Bosch, R. F., Hernandis, M. V., and Anton, S. G.,** *An. Quim. Ser. B,* 78, 317, 1982; *Anal. Abstr.,* 44, 6B98, 1983.
212. **Huka, M.,** unpublished results, 1986.
213. **Voigt, A.,** *Fresenius Z. Anal. Chem.,* 133, 44, 1951.
214. **Abbey, S. and Maxwell, J. A.,** *Chem. Can.,* 12, 37, 1960.

215. **Ingamells, C. O.,** *Talanta,* 9, 781, 1962.
216. **Easton, A. J. and Lovering, J. F.,** *Anal. Chim. Acta,* 30, 543, 1964.
217. **Goldich, S. S. and Oslund, E. H.,** *Bull. Geol. Soc. Am.,* 67, 811, 1956.
218. **Lebedev, V. I.,** *Zh. Anal. Khim.,* 14, 283, 1959.
219. **Jakob, J.,** *Chemische Analyse der Gesteine und Silikatischen Mineralien,* Birkhäuser, Basel, 1952, 11.
220. **Jeffery, P. G. and Hutchison, D.,** *Chemical Methods of Rock Analysis,* 3rd ed., Pergamon Press, Oxford, 1981, 19.
221. **Johnson, W. H. and Maxwell, J. A.,** *Rock and Mineral Analysis,* John Wiley & Sons, New York, 1981, 77.
222. **Langmyhr, F. J. and Sveen, S.,** *Anal. Chim. Acta,* 32, 1, 1965.
223. **Scharm, B., Čadek, J., Čadková, Z. Hájková, H., Kühn, P., Lepka, F., Obr, F., Baloun, S., and Parobek, P.,** *Sb. Ústřed. Ústavu Geol., Ložisk. Geol. Mineral.,* 26, 9, 1984.
224. **Howling, H. L. and Landolt, P. E.,** *Anal. Chem.,* 31, 1818, 1959.
225. **Sykes, P. W.,** *Analyst,* 81, 283, 1956.
226. **Donaldson, E. M.,** *Methods for the Analysis of Ores, Rocks, and Related Materials, Energy, Mines and Resources,* Canada Mines Branch Monogr. 881, Ottawa, 1974, 33.
227. **Tewari, D. N.,** *Talanta,* 25, 269, 1978.
228. **Antweiler, J. C.,** *U.S. Geol. Surv. Prof. Pap.,* 424B, B322, 1961.
229. **Israel, Y. and Paschkes, B.,** *Micr. Acta,* II, 69, 1981.
230. **Hoops, G. K.,** *Geochim. Cosmochim. Acta,* 28, 405, 1964.
231. **Gedeon, A. Z., Butt, C. R. M., Gardner, K. A., and Hart, M. K.,** *J. Geochem. Explor.,* 8, 283, 1977.
232. **Povondra, P.,** *Acta Univ. Carol. Geol.,* No. 3, 223, 1981.
233. **Rice, T. D.,** *Talanta,* 23, 359, 1976.
234. **Rice, T. D.,** *Anal. Chim. Acta,* 91, 221, 1977.
235. **Habrman, Z., Křesťan, V., and Paleček, M.,** unpublished results, 1985.
236. **Bond, A. M., O'Donnell, T. A., Waugh, A. B., and McLaughlin, R. J. W.,** *Anal. Chem.,* 42, 1168, 1970.
237. **Yamashige, T., Yamamoto, M., Shigetomi, Y., and Yamamoto, Y.,** *Bunseki Kagaku,* 33, 221, 1984; *Anal. Abstr.,* 47, 1B212, 1985.
238. **Yamashige, T., Ida, H., Yamamoto, M., Shigetomi, Y., and Yamamoto, Y.,** *Bunseki Kagaku,* 32, 169, 1983; *Anal. Abstr.,* 45, 4H8, 1983.
239. **Silberman, D. and Fisher, G. L.,** *Anal. Chim. Acta,* 106, 59, 1979.
240. **Uchida, H., Uchida, T., and Iida, C.,** *Anal. Chim. Acta,* 108, 87, 1979; *Anal. Chim. Acta,* 116, 433, 1980.
241. **Schinkel, H.,** *Fresenius Z. Anal. Chem.,* 317, 10, 1984.
242. **Bosch, A. F., Bosch, R. F., and Hernandez, M. V.,** *Quim. Anal.,* 30, 399, 1976; *Anal. Abstr.,* 33, 3B180, 1977.
243. **Brooks, R. R.,** *Nature,* 185, 837, 1960.
244. **Lee, A. F. and Steele, T. W.,** *Natl. Inst. Metall. Repub. S. Afr. Rep.,* No. 2083, 1980.
245. **Sanzolone, R. F. and Chao, T. T.,** *Anal. Chim. Acta,* 86, 163, 1976.
246. **Janáček, J.,** unpublished data, 1985.
247. **Harrison, R. M. and Laxen, D. P. H.,** *Water Air Soil Pollut.,* 8, 387, 1977.
248. **Wlotzka, F.,** *Geochim. Cosmochim. Acta,* 24, 106, 1961.
249. **Aruscavage, P. J. and Campbell, E. Y.,** *Talanta,* 30, 745, 1983.
250. **Bastius, H. and Kroenert, W.,** *Tonind. Ztg.,* 102, 463, 1978.
251. **Sixta, V.,** *Fresenius Z. Anal. Chem.,* 285, 369, 1977.
252. **Crossley, H. E.,** *J. Inst. Fuel.,* 25, 221, 1952.
253. **Langmyhr, F. J. and Graff, P. R.,** *A Contribution to the Analytical Chemistry of Silicate Rocks,* No. 230, Universitetforlaget, Oslo, 1965, 18.
254. **Langmyhr, F. J. and Kringstad, K.,** *Anal. Chim. Acta,* 35, 161, 1966.
255. **Tarutani, T. and Toshikazu, T.,** *Nippon Kagaku Zashi,* 77, 1292, 1956; *Chem. Abstr.,* 52, 974, 1958.
256. **Langmyhr, F. J. and Wendelborg, R.,** *Anal. Chim. Acta,* 45, 171, 1969.
257. **Croudace, I. W.,** *Chem. Geol.,* 31, 153, 1980.
258. **Kotrba, Z. and Šulcek, Z.,** unpublished data, 1976.
259. **Wilson, A. D.,** *Analyst,* 85, 823, 1960.
260. **Pearce, W. C., Thornewill, D., and Marston, J. H.,** *Analyst,* 110, 625, 1985.
261. **Yaschenko, M. L. and Varshavskaya, E. S.,** *Zavod. Lab.,* 26, 275, 1960.
262. **Jeczalik, A. and Lis, B.,** *Chem. Anal. (Warsaw),* 12, 1251, 1967.
263. **Saxby, J. D.,** *Chem. Geol.,* 6, 173, 1970.
264. **Riley, K. W.,** *Analyst,* 109, 181, 1984.

265. **Koenig, E. W.,** *Ind. Eng. Chem. Anal. Ed.*, 7, 314, 1935.
266. **Selch, E.,** *Fresenius Z. Anal. Chem.*, 54, 395, 1915.
267. **Langmyhr, F. J.,** *Anal. Chim. Acta,* 39, 516, 1967.
268. **Schmidt, W., Konopicky, K., and Kostyra, J.,** *Fresenius Z. Anal. Chem.*, 206, 174, 1964.
269. **Su, Y.-S. and Campbell, D. E.,** *Anal. Chim. Acta,* 55, 265, 1971.
270. **Su, Y.-S., Strzegowski, W. R., Kacyon, A. R., and Lichtenstein, I. E.,** *Anal. Chim. Acta,* 81, 167, 1976.
271. **Huka, M. and Šulcek, Z.,** unpublished data, 1985.
272. **Wise, W. M., Burdo, R. A., and Sterlace, J. S.,** *Prog. Anal. At. Spectrosc.*, 1, 201, 1978.
273. **Unruh, D. M., Stille, P., Patchett, P. J., and Tatsumoto, M.,** *J. Geophys. Res.*, 89 (Suppl.), B459, 1984.
274. **Puffer, J. H. and Cohen, R. S.,** *Chem. Geol.*, 15, 217, 1975.
275. **Cohen, R. S., Hemmes, P., and Puffer, J. H.,** *Chem. Geol.*, 16, 307, 1975.
276. **Luginin, V. A. and Cherkovnitskaya, I. A.,** *Zh. Anal. Khim.*, 26, 1593, 1971.
277. **Bancroft, G. M., Sham, T. K., Riddle, C. H., Smith, T. E., and Turek, A.,** *Chem. Geol.*, 19, 277, 1977.
278. **Whipple, E. R.,** *Chem. Geol.*, 14, 223, 1974.
279. **Schafer, H. N. S.,** *Analyst,* 91, 755, 1966.
280. **Kiss, E.,** *Anal. Chim. Acta,* 89, 303, 1977.
281. **Peck, C. L.,** *U.S. Geol. Surv. Bull.*, 1170, 39, 1964.
282. **Heffernan, B. J.,** *Lab. Pract.*, 23, 427, 1974.
283. **French, W. J. and Adams, S. J.,** *Analyst,* 97, 828, 1972.
284. **Peters, A.,** *N. Jb. Mineral. Mh.*, 119, 1968.
285. **Wilson, A. D.,** *Bull. Geol. Surv. G.B.*, 9, 56, 1955.
286. **Abramov, V. V., Chesnokova, S. M., Andreev, P. A., and Sapalova, O. I.,** *Zh. Anal. Khim.*, 37, 2031, 1982.
287. **Carmichael, I. S. E., Hampel, J., and Jack, R. N.,** *Chem. Geol.*, 3, 59, 1968.
288. **Ishibashi, I. M. and Kusaka, Y.,** *J. Chem. Soc. Jpn.*, 71, 160, 1950.
289. **Babčan, J.,** unpublished results, 1958.
290. **Whitehead, D. and Malik, S. A.,** *Anal. Chem.*, 47, 554, 1975.
291. **Vieux, A. S. and Kabwe, C.,** *Chim. Anal.*, 52, 866, 1970.
292. **Popov, N. P. and Stolyarova, I. A.,** *Khimicheskii, Analiz Gornykh Porod i Mineralov,* Nedra, Moskva, 1974, 18.
293. **Jackson, P. J.,** *Appl. Chem.*, 7, 605, 1957.
294. **Povondra, P.,** unpublished data, 1985.
295. **Spivakovskii, V. B. and Zimina, V. A.,** *Zavod. Lab.*, 28, 290, 1962.
296. **Reichen, L. E. and Fahey, J. J.,** *U.S. Geol. Surv. Bull.*, 1144B, B3, 1962.
297. **Meyrowitz, R.,** *Am. Miner.*, 48, 340, 1963.
298. **Meyrowitz, R.,** *Anal. Chem.*, 42, 1110, 1970.
299. **Abed, U.,** *Anal. Chim. Acta,* 47, 495, 1969.
300. **Pamnani, K. and Agnihotri, K. S.,** *Lab. Pract.*, 15, 867, 1966.
301. **Ungethüm, H.,** *Z. Angew. Geol.*, 11, 500, 1965.
302. **Hetman, J. S.,** *Bull. Centr. Rech. Pau.*, 8, 153, 1974.
303. **Van Loon, J. C.,** *Talanta,* 12, 599, 1965.
304. **Hey, M. H.,** *Mineral. Mag.*, 26, 116, 1941.
305. **Nicholls, G. D.,** *J. Sediment. Petrol.*, 30, 603, 1960.
306. **Banerjee, S.,** *Anal. Chem.*, 46, 782, 1974.
307. **Das Gupta, H. N. and Mitra, N. K.,** *J. Proc. Inst. Chem. India,* 39, 250, 1967; *Anal. Abstr.*, 17, 110, 1969.
308. **Shapiro, S.,** *U.S. Geol. Surv. Prof. Pap.*, 400B, B496, 1960.
309. **Riley, J. P. and Williams, H. P.,** *Micr. Acta,* p. 516, 1959.
310. **Walker, J. L. and Sherman, G. D.,** *Soil Sci.*, 93, 325, 1962.
311. **Pruden, G. and Bloomfield, C.,** *Analyst,* 94, 688, 1969.
312. **Beghein, L. T.,** *Analyst,* 104, 1055, 1979.
313. **Kiss, E.,** *Anal. Chim. Acta,* 72, 127, 1974.
314. **Kiss, E.,** *Anal. Chim. Acta,* 161, 231, 1984.
315. **Vydra, F. and Vorlíček, J.,** *Hutn. Listy,* 18, 733, 1963; *Chemist Analyst,* 53, 103, 1964.
316. **Afanaseva, L. I. and Bugrova, V. D.,** *Zh. Anal. Khim.*, 30, 627, 1975.
317. **Murphy, J. M., Read, J. I., and Sergeant, G. A.,** *Analyst,* 99, 273, 1974.
318. **Clemency, C. V. and Hagner, A. F.,** *Anal. Chem.*, 33, 888, 1961.
319. **Lo-Sun, J.,** *Anal. Chim. Acta,* 66, 315, 1973.

320. **Ritchie, J. A.,** *Geochim. Cosmochim. Acta,* 32, 1363, 1968.
321. **MacCardle, L. E. and Scheffer, E. R.,** *Anal. Chem.,* 23, 1169, 1951.
322. **Grimaldi, F. S., May, I., Fletcher, M. H., and Titcomb, J.,** *U.S. Geol. Surv. Bull.,* 1006, 11, 1954.
323. **Florence, T. M. and Farrar, Y. J.,** *Anal. Chem.,* 42, 271, 1970.
324. **Cook, E. B. T. and Gereghty, A.,** *Natl. Inst. Metall. Repub. S. Afr. Rep.,* No. 1145, 1971.
325. **Feng, Y. and Yin, Z.,** *Fenxi Huaxue,* 13, 37, 1985; *Anal. Abstr.,* 47, 9B58, 1985.
326. **Bajo, C., Rybach, L., and Weibel, M.,** *Chem. Geol.,* 39, 281, 1983.
327. **May, I. and Jenkins, L. B.,** *U.S. Geol. Surv. Prof. Pap.,* 525D, D192, 1965.
328. **Sill, C. W.,** *Anal. Chem.,* 33, 1684, 1961.
329. **Sill, C. W.,** *Health Phys.,* 33, 393, 1977.
330. **Sill, C. W.,** *Anal. Chem.,* 49, 618, 1977.
331. **Sill, C. W.,** *Anal. Chem.,* 49, 302, 1977.
332. **Rosner, G., Bunzl, K., Hoetzl, H., and Winkler, R.,** *Nucl. Instrum. Methods Phys. Res.,* 223, 585, 1984; *Anal. Abstr.,* 47, 8G10, 1985.
333. **Anderson, R. F. and Fleer, A. P.,** *Anal. Chem.,* 59, 1142, 1982.
334. **Sill, C. W.,** *Health Phys.,* 29, 619, 1975.
335. **Veselsky, J. C.,** *Österr. Chem. Z.,* 77, 2, 1976.
336. **Veselsky, J. C.,** *Anal. Chim. Acta,* 90, 1, 1977.
337. **Hiatt, M. H. and Hahn, P. B.,** *Anal. Chem.,* 51, 295, 1979.
338. **Knab, D.,** *Anal. Chem.,* 51, 1095, 1979.
339. **Sill, C. W., Hindman, F. D., and Anderson, J. I.,** *Anal. Chem.,* 51, 1307, 1979.
340. **Chu, N. Y.,** *Anal. Chem.,* 43, 449, 1971.
341. **Ryabchikov, D. I. and Golbraikh, E. K.,** *Analiticheskaya Khimia Thoria,* Akad. Nauk SSSR, Moskva, 1960, 9.
342. **Green, L. W., Elliott, N. L., and Longhurst, T. H.,** *Anal. Chem.,* 55, 2394, 1983.
343. **Green, L. W., Knight, C. H., Longhurst, T. H., and Cassidy, R. M.,** *Anal. Chem.,* 56, 696, 1984.
344. **Krtil, J.,** *Jad. Energ.,* 29, 441, 1983.
345. **Savvin, S. B. and Bagrev, V. V.,** *Zavod. Lab.,* 26, 412, 1960.
346. **Udaltsova, N. I.,** *Izv. Sib. Otdel. AN SSSR, Ser. Khim. Nauk,* 5, 53, 1968; *Anal. Abstr.,* 18, 1558, 1968.
347. **Barnabee, R. P.,** *Health Phys.,* 44, 688, 1983.
348. **Vladimirova, V. M. and Davidovich, N. K.,** *Zavod. Lab.,* 26, 1210, 1960.
349. **Husler, J.,** *Talanta,* 19, 863, 1972.
350. **Faye, G. H.,** *Chem. Can.,* 10, 90, 1958.
351. **Bakes, J. M., Gregory, G. R. E. C., and Jeffery, P. G.,** *Anal. Chim. Acta,* 27, 540, 1962.
352. **Kallman, S., Oberthin, H., and Liu, R.,** *Anal. Chem.,* 34, 609, 1962.
353. **Steger, H.,** *Talanta,* 23, 81, 1976.
354. **Nagashima, K., Yano, Y., and Hamada, M.,** *Bunseki Kagaku,* 33, T91, 1984.
355. **Pilai, C. K., Natarajan, S., and Venkateswarlu, C.,** *At. Spectrosc.,* 6, 53, 1985.
356. **Strelow, F. W. E.,** *Anal. Chem.,* 39, 1454, 1967.
357. **Pollock, J. B.,** *Analyst,* 93, 93, 1968.
358. **Muir, C. W. A. and Lloyd, P. A.,** *Natl. Inst. Metall. Repub. S. Afr. Rep.,* No. 1178, 1970; No. 1010, 1970.
359. **Majmudar, A. A., Gokhale, Y. W., and Iyer, R. K.,** *Indian J. Technol.,* 21, 532, 1983; *Anal. Abstr.,* 47, 5B110, 1985.
360. **Ryabchikov, D. I. and Ryabukhin, V. A.,** *Analiticheskaya Chimia Redkozemelnykh Elementov i Ittrya,* Nauka, Moskva, 1966, 218.
361. **Marzys, A. E. O.,** *Analyst,* 79, 327, 1954.
362. **Childress, A. E. and Greenland, L. P.,** *Anal. Chim. Acta,* 116, 185, 1980.
363. **Allen, R. O. and Steinnes, E.,** *Anal. Chem.,* 50, 903, 1978.
364. **Milner, G. W. C. and Smales, A. A.,** *Analyst,* 79, 315, 1954.
365. **Esson, J. E.,** *Analyst,* 90, 489, 1965.
366. **Grimaldi, F. S.,** *Anal. Chem.,* 32, 119, 1960.
367. **Greenland, L. P. and Campbell, E. Y.,** *J. Res. U.S. Geol. Surv.,* 2, 353, 1974.
368. **Caletka, R. and Krivan, V.,** *Fresenius Z. Anal. Chem.,* 313, 125, 1982.
369. **Caletka, R. and Krivan, V.,** *Talanta,* 30, 465, 1983.
370. **Sen Gupta, J. G.,** *Talanta,* 23, 343, 1976.
371. **Dolgorev, A. V. and Lysak, Ya. G.,** *Zh. Anal. Khim.,* 30, 1951, 1975.
372. **Sen Gupta, J. G.,** *Anal. Chim. Acta,* 138, 295, 1982.
373. **Šulcek, Z.,** unpublished data, 1985.
374. **Crock, J. G., Lichte, F. E., and Wildeman, T. R.,** *Chem. Geol.,* 45, 149, 1984.
375. **Crock, J. G. and Severson, R. C.,** *U.S. Geol. Surv. Circ.,* No. 841, 1981.

376. **Yoshida, K. and Haraguchi, H.,** *Anal. Chem.,* 56, 2580, 1984.
377. **Brenner, I. B., Watson, A. E., Jones, E. A., and Goneales, M.,** *Spectrochim. Acta,* 36B, 785, 1981.
378. **Povondra, P.,** unpublished data, 1986.
379. **Novák, J.,** unpublished data, 1986.
380. **Milner, G. W. C.,** *Analyst,* 87, 125, 1962.
381. **Lunina, G. E. and Romanenko, E. G.,** *Zavod. Lab.,* 34, 538, 1968.
382. **Sarkar, A. K. and Das, J.,** *Anal. Chem.,* 39, 1608, 1967.
383. **Neirinckx, R., Adams, F., and Hoste, J.,** *Anal. Chim. Acta,* 48, 1, 1969.
384. **Campbell, A. S.,** *Clay Miner.,* 10, 57, 1973.
385. **Meyrowitz, R.,** *U.S. Geol. Surv. Prof. Pap.,* 750B, B165, 1971.
386. **Shlenskaya, V. I. and Khvostova, V. P.,** *Zh. Anal. Khim.,* 29, 314, 1974.
387. **Khvostova, V. P. and Golovnaya, S. V.,** *Zavod. Lab.,* 48(7), 3, 1982.
388. **Simonsen, A.,** *Anal. Chim. Acta,* 49, 368, 1970.
389. **Rubeška, I., Korečková, J., and Weiss, D.,** *At. Absorpt. Newsl.,* 16, 1, 1977.
390. **Korečková, J., Mrázek, P., and Weiss, D.,** unpublished data, 1986.
391. **Hu, Q. and Huang, X.,** *Fen Hsi Hua Hsueh,* 9, 66, 1981; *Anal. Abstr.,* 41, 3B50, 1981.
392. **Bazhov, A. S. and Sokolova, E. A.,** *Zh. Anal. Khim.,* 32, 65, 1977.
393. **Kuznetsov, R. A.,** *Zh. Anal. Khim.,* 33, 294, 1978.
394. **Hubert, A. E. and Chao, T. T.,** *Talanta,* 32, 568, 1985.
395. **Headridge, J. B. and Taylor, M. S.,** *Analyst,* 88, 590, 1963.
396. **Quin, B. F. and Brooks, R. R.,** *Anal. Chim. Acta,* 65, 206, 1973.
397. **Chan, K. M. and Riley, J. P.,** *Anal. Chim. Acta,* 39, 103, 1967.
398. **Hase, U., Yoshimura, K., and Tarutani, T.,** *Anal. Chem.,* 57, 1416, 1985.
399. **Stepanova, N. A. and Yakumina, G. A.,** *Zh. Anal. Khim.,* 17, 858, 1962.
400. **Parker, G. A.,** *Analytical Chemistry of Molybdenum,* Springer-Verlag, Berlin, 1983, 124.
401. **Donaldson, E. A.,** *Talanta,* 27, 79, 1980.
402. **Rader, L. F. and Grimaldi, F. S.,** *U.S. Geol. Surv. Prof. Pap.,* 391A, A23, 1961.
403. **Lillie, E. G. and Greenland, L. P.,** *Anal. Chim. Acta,* 69, 313, 1974.
404. **Kim, C. H., Owens, C. M., and Smythe, L. E.,** *Talanta,* 21, 445, 1974.
405. **Willard, H. H. and Rulfs, C. L.,** *Treatise on Analytical Chemistry,* Vol. 2 (Part 1), Kolthoff, I. M. and Elving, P. J., Eds., Interscience, New York, 1961, 1040.
406. **Lengauer, W.,** *Fresenius Z. Anal. Chem.,* 322, 23, 1985.
407. **Mikhailova, M.,** *Khim. Ind.,* (Sofia), p. 111, 1984; *Anal. Abstr.,* 46, 11B39, 1984.
408. **Malyutina, T. M., Namvrina, E. G., and Shiryaeva, O. A.,** *Zavod. Lab.,* 47(9), 8, 1981.
409. **Barkovskii, V. F. and Radovskaya, T. L.,** *Zavod. Lab.,* 35, 160, 1969.
410. **Henrion, G. and Lippert, H. H.,** *Z. Chem.,* 20, 108, 1980.
411. **Pyatnitskii, I. V. and Simonenko, V. I.,** *Zavod. Lab.,* 49(4), 24, 1983.
412. **Kerrich, R. and Starkey, J.,** *Am. Mineral.,* 64, 452, 1979.
413. **Chapman, S. L., Syers, J., and Jackson, M. L.,** *Soil Sci.,* 107, 348, 1969.
414. **Hlasivcová, N. and Novák, J.,** *Chem. Listy,* 63, 129, 1969.
415. **Strock, L. W.,** *Fresenius Z. Anal. Chem.,* 99, 321, 1934.
416. **Portmann, J. E. and Riley, J. P.,** *Anal. Chim. Acta,* 31, 509, 1964.
417. **Lounamaa, K.,** *Fresenius Z. Anal. Chem.,* 146, 422, 1955.
418. **Close, P., Shepherd, H. M., and Drummond, C. H.,** *J. Am. Ceram. Soc.,* 41, 455, 1958.
419. **Chapman, F. W., Marvin, G. G., and Tyree, S. Y.,** *Anal. Chem.,* 21, 700, 1949.
420. **Ploum, H.,** *Arch. Eisenhüttenw.,* 27, 761, 1956.
421. **Tölg, G.,** *Fresenius Z. Anal. Chem.,* 190, 161, 1962.
422. **Maher, W. A.,** *Chem. Geol.,* 45, 173, 1984.
423. **Fainberg, S. Yu.,** *Analiz Rud Tsvetnykh Metallov,* Izd. Liter. po Chernoy i Tsvetnoy Metallurgii, 3rd ed., Moskva, 1963, 263.
424. **Bajo, S.,** *Anal. Chem.,* 50, 649, 1978.
425. **Agemian, H. and Bedek, E.,** *Anal. Chim. Acta,* 119, 323, 1980.
426. **Hetman, J., Hauric, M., and Puyo, M.,** *Bull. Centr. Rech. Pau.,* 6, 215, 1972.
427. **Terashima, S.,** *Anal. Chim. Acta,* 86, 43, 1976.
428. **Terashima, S.,** *Jpn. Analyst,* 23, 1331, 1974.
429. **Aslin, G. E. M.,** *J. Geochem. Explor.,* 6, 321, 1976.
430. **Šulcek, Z.,** unpublished data, 1986.
431. **Xiao-Quan, S., Zhe-Ming, N., and Zhang, L.,** *Anal. Chim. Acta,* 151, 179, 1983.
432. **Marshall, N. J.,** *J. Geochem. Explor.,* 10, 307, 1978.
433. **Kiperman, M. G. and Erofeeva, L. E.,** *Zavod. Lab.,* 43, 418, 1977.
434. **Marin, L. and Vernet, M.,** *Analusis,* 7, 33, 1979.

435. **Wise, W. M. and Williams, J. P.**, *Anal. Chem.*, 36, 19, 1964.
436. **Van der Veen, N. G., Keukens, H. J., and Vos, G.**, *Anal. Chim. Acta*, 171, 285, 1985.
437. **Sansolone, R. F., Chao, T. T., and Welsch, E. P.**, *Anal. Chim. Acta*, 108, 357, 1979.
438. **Jones, E. A., Russell, B. G., and Steele, T. W.**, *Natl. Inst. Metall. Repub. S. Afr. Rep.*, No. 151, 1967.
439. **Janáček, J. and Hajzler, Z.**, unpublished data, 1985.
440. **Abu Hilal, A. H. and Riley, J. P.**, *Anal. Chim. Acta*, 131, 175, 1981.
441. **Sandell, E. B.**, *Colorimetric Determination of Traces of Metals*, 2nd ed., Interscience, New York, 1959, 271.
442. **Onishi, H. and Sandell, E. B.**, *Anal. Chim. Acta*, 11, 444, 1954.
443. **Perezhogin, G. A. and Gavrilova, L. I.**, *Zh. Anal. Khim.*, 30, 725, 1975.
444. **De Doncker, K., Dumarey, R., Dams, R., and Hoste, J.**, *Anal. Chim. Acta*, 153, 33, 1980.
445. **Zemanová, D. and Knotková, J.**, unpublished data, 1971.
446. **Bock, R. and Jacob, D.**, *Fresenius Z. Anal. Chem.*, 200, 81, 1964.
447. **Chan, C. C. Y.**, *Anal. Chem.*, 57, 1482, 1985.
448. **Maher, W. A.**, *Anal. Lett.*, 16A, 491, 1983.
449. **Chau, Y. K. and Riley, J. P.**, *Anal. Chim. Acta*, 33, 36, 1965.
450. **Golembeski, T.**, *Talanta*, 22, 547, 1975.
451. **Sanzolone, R. F. and Chao, T. T.**, *Analyst*, 106, 647, 1981.
452. **Imai, M., Terashima, S., and Ando, A.**, *Bunseki Kagaku*, 33, 288, 1984.
453. **Greenland, L. P. and Campbell, E. Y.**, *J. Res. U.S. Geol. Surv.*, 5, 403, 1977.
454. **Knab, D. and Gladney, E. S.**, *Anal. Chem.*, 52, 825, 1980.
455. **Nazarenko, I. I. and Kislova, I. V.**, *Zavod. Lab.*, 37, 414, 1971.
456. **Tamari, I.**, *Bunseki Kagaku*, 33, E115, 1984.
457. **Bock, R. and Tschöpel, P.**, *Fresenius Z. Anal. Chem.*, 246, 81, 1969.
458. **Beaty, R. D.**, *At. Absorpt. Newsl.*, 13, 38, 1974.
459. **Weiss, D.**, unpublished data, 1979.
460. **Greenland, L. P. and Campbell, E. Y.**, *Anal. Chim. Acta*, 87, 323, 1976.
461. **Sighinolfi, G. P., Santos, A. M., and Martinelli, G.**, *Talanta*, 26, 143, 1979.
462. **Corbett, J. A. and Godbeer, W. C.**, *Anal. Chim. Acta*, 91, 211, 1977.
463. **Volf, J. and Houzim, V.**, unpublished data, 1966.
464. **Cheng, K. L. and Goydish, B. L.**, *Anal. Chem.*, 35, 1273, 1963.
465. **Vrbský, J. and Tymáň, V.**, *Sb. Vys. Sk. Chem. Technol. Praze*, H1, 77, 1967.
466. **Schneider, W. A. and Sandell, E. B.**, *Micr. Acta*, 263, 1954.
467. **Yermakov, A. A., Makan, S. Y., and Perfil, A. I.**, *Zh. Neorg. Chem.*, 22, 577, 1957.
468. **Donaldson, E. M.**, *Talanta*, 31, 997, 1984.
469. **Hybinette, A. G. and Sandell, E. B.**, *Ind. Eng. Chem. Anal. Ed.*, 14, 715, 1942.
470. **Vozisova, V. F. and Podchainova, V. N.**, *Tr. Ural. Politekh. Inst.*, 163, 5, 1967; *Anal. Abstr.*, 16, 2412, 1969.
471. **Halicz, L.**, *Analyst*, 110, 943, 1985.
472. **Nazarenko, V. A., Lebedeva, N. V., and Vinarova, L. I.**, *Zh. Anal. Khim.*, 27, 128, 1972.
473. **Nazarenko, V. A., Lebedeva, N. V., and Ravitskaya, R. V.**, *Zabod. Lab.*, 24, 9, 1958.
474. **Stricland, E. H.**, *Analyst*, 80, 548, 1955.
475. **Lukin, A. M., Efremenko, O. A., and Podolskaya, B. L.**, *Zh. Anal. Khim.*, 21, 970, 1966.
476. **Lebedeva, N. V. and Lyakh, R. A.**, *Zavod. Lab.*, 32, 1334, 1966.
477. **Basinska, M. and Rutkowski, W.**, *Chem. Anal. (Warsaw)*, 8, 353, 1963.
478. **Nemirovskaya, E. M. and Kolosova, G. M.**, *Zavod. Lab.*, 45, 311, 1979.
479. **Masalovich, V. M., Menshenina, G. S., and Moshareva, G. S.**, *Tr. Ural. Nauchno. Issled. Khim. Inst.*, 42, 56, 1977; *Anal. Abstr.*, 36, 2B 146, 1979.
480. **Lanza, P. and Buldini, P. L.**, *Anal. Chim. Acta*, 70, 341, 1974.
481. **Takahari, T. and Kosaka, M.**, *Bunseki Kagaku*, 25, 192, 1976.
482. **Grallath, E., Tschöpel, P., Kölblin, G., Stix, U., and Tölg, G.**, *Fresenius Z. Anal. Chem.*, 302, 40, 1980.
483. **Bhargava, O. P. and Grant Hines, W.**, *Talanta*, 17, 61, 1970.
484. **Čižek, Z. and Študlarová, V.**, *Talanta*, 31, 547, 1984.
485. **Karalova, Z. K. and Nemodruk, A. A.**, *Zh. Anal. Khim.*, 17, 985, 1962.
486. **Stanton, R. E. and McDonald, A. J.**, *Analyst*, 91, 775, 1966.
487. **Tölg, G.**, *Pure Appl. Chem.*, 44, 645, 1975.
488. **Hulmston, P.**, *Anal. Chim. Acta*, 155, 247, 1983.
489. **Bykova, V. S. and Skrizhinskaya, V. I.**, *Khimicheskie Analizy i Formuly Mineralov*, Nauka, Moskva, 1969, 6.
490. **Grazulene, S. S., Grossman, O. V., Kyuncher, K. K., Maligina, L. I., Myuller, E. N., and Telegin, G. F.**, *Zh. Anal. Khim.*, 40, 674, 1985.

491. **Kluger, F. and Koeberl, Ch.,** *Anal. Chim. Acta,* 175, 127, 1985.
492. **Kochen, R. L.,** *Anal. Chim. Acta,* 71, 451, 1974.
493. **Kilroy, W. P. and Moynihan, C. T.,** *Anal. Chim. Acta,* 83, 389, 1976.
494. **Burdo, R. A. and Snyder, M. L.,** *Anal. Chem.,* 51, 1502, 1979.
495. **Vasilevskaya, L. S., Kondrashina, A. I., and Shifrina, G. G.,** *Zavod. Lab.,* 28, 674, 1962.
496. **Shafran, I. G.,** *Sb. Kachestvo Materialov dlya Poluprovodnikovoi Tekhn.,* Metallurgizdat, Moskva, 1959, 13.
497. **Semov, M. P.,** *Zavod. Lab.,* 29, 1450, 1963.
498. **Piryutko, M. M., and Gileva, K. G.,** *Zavod. Lab.,* 44, 538, 1978.
499. **Pritchard, M. W. and Lee, J.,** *Anal. Chim. Acta,* 157, 313, 1984.
500. **Dobeš, I.,** *Sklář Keram.,* 11, 14, 1961.
501. **Kuznetsov, V. I. and Myasoedova, G. V.,** *Zh. Prikl. Khim.,* 29, 1875, 1956.
502. **Calderoni, G. and Ferri, T.,** *Talanta,* 29, 371, 1982.
503. **Eckmann, W. D. and Huizenga, J. R.,** *Geochim. Cosmochim. Acta,* 17, 125, 1959.
504. **Fornaseri, A. and Penta, A.,** *Metallurgy,* 55, 437, 1963; *Anal. Abstr.,* 11, 5391, 1964.
505. **Matthews, A. D. and Riley, J. P.,** *Anal. Chim. Acta,* 48, 25, 1969.
506. **Schnepfe, M. M.,** *Anal. Chim. Acta,* 79, 101, 1975.
507. **Ikramuddin, M.,** *At. Spectrosc.,* 4, 101, 1983.
508. **Brooks, R.,** *Anal. Chim. Acta,* 24, 456, 1961.
509. **Fratta, M.,** *Can. J. Spectrosc.,* 19, 33, 1974.
510. **Šulcek, Z.,** unpublished data, 1979.
511. **Elson, C. M. and Albuquerque, C. A. R.,** *Anal. Chim. Acta,* 134, 393, 1982.
512. **Nadkarni, R. A. and Haldar, B. C.,** *Radiochem. Anal. Lett.,* 11, 367, 1972.
513. **Gorbauch, H., Rumpf, H. H., Alter, G., and Schmitt-Henco, C. H.,** *Fresenius Z. Anal. Chem.,* 317, 236, 1984.
514. **Calderoni, G., Ferrini, V., and Masi, V.,** *Chem. Geol.,* 51, 29, 1985.
515. **Blyum, I. A. and Khuvileva, A. I.,** *Zh. Anal. Khim.,* 25, 18, 1970.
516. **Cuttitta, F.,** *U.S. Geol. Surv. Prof. Pap.,* 424C, 384, 1961.
517. **Beyermann, K.,** *Fresenius Z. Anal. Chem.,* 183, 91, 1961.
518. **Vilczek, E. and Lohman, G.,** *Fresenius Z. Anal. Chem.,* 304, 395, 1980.
519. **Findlay, W. J., Zdrojewski, A., and Quikert, N.,** *Spectrosc. Lett.,* 7, 355, 1974.
520. **Maketon, S. and Tarter, J. G.,** *Anal. Lett.,* 18(A2), 181, 1985.
521. **Smith, G. F.,** *The Wet Chemical Oxidation of Organic Compositions Employing Perchloric Acid With or Without Added* HNO_3-H_5IO_6-H_2SO_4, G. F. Smith Chemical Company, Columbus, Ohio, 1965, 1.
522. **Vrbský, J. and Tymáň, V.,** *Sb. Vys. Skoly Chem. Technol. Praze,* H1, 77, 1967.
523. **Spieholz, G. I. and Diehl, H.,** *Talanta,* 13, 991, 1966.
524. **Buzzeli, G. and Mosen, A. W.,** *Talanta,* 24, 383, 1977.
525. **Aruscavage, P.,** *J. Res. U.S. Geol. Surv.,* 5, 405, 1977.
526. **May, K. and Stoeppler, M.,** *Fresenius Z. Anal. Chem.,* 293, 127, 1978.
527. **Shvartsman, S. I., Falkova, O. B., Kurskii, A. N., and Popova, N. E.,** *Zh. Anal. Khim.,* 39, 1213, 1984.
528. **Analytical Methods Committee,** *Analyst,* 84, 214, 1959.
529. **Birkhahn, W.,** *GIT Fachz. Lab.,* 25, 35, 1981; *Anal. Abstr.,* 41, 2A 34, 1981.
530. **Michelotti, F. W.,** *Anal. Chem.,* 51, 441 A, 1979.
531. **Levens, E.,** *Perchlorates, Their Properties, Manufacture and Uses,* Schuhmacher, J. C., Ed., Am. Chem. Soc. Monogr. No. 116, Reinhold, New York, 1960, 195.
532. **Steere, N. V.,** *CRC Handbook of Laboratory Safety,* 2nd ed., CRC Press, Boca Raton, Fla., 1985.
533. **Cook, R. E. and Robinson, P. J.,** *Chem. Br.,* 18, 859, 1982; *Anal. Abstr.,* 44, 6A 21, 1983.
534. **Robinson, P. J.,** *Chem. Br.,* 17, 560, 1981; *Anal. Abstr.,* 42, 6A 30, 1982.
535. **Muir, G. D.,** *Hazards in the Chemical Laboratory,* 2nd ed., Chemical Society, London, 1977.
536. **Dymova, M. S., Kozina, G. V., and Titova, T. V.,** *Zavod. Lab.,* 50(6), 18, 1984.
537. **Trofimov, I. V. and Busev, A. I.,** *Zavod. Lab.,* 49(3), 5, 1983.
538. **Smith, G. F. and Taylor, W. H.,** *Talanta,* 10, 1107, 1963.
539. **Šulcek, Z., Povondra, P., and Kratochvíl, V.,** *Collect. Czech. Chem. Commun.,* 34, 3711, 1969.
540. **Bounsall, E. J. and McBryde, W. A. E.,** *Can. J. Chem.,* 38, 1488, 1960.
541. **Goetz, C. A. and Debbrecht, F. J.,** *Anal. Chem.,* 27, 1972, 1955.
542. **Rüssel, H.,** *Fresenius Z. Anal. Chem.,* 189, 256, 1962.
543. **Fan, Z.,** *Fenxi Shiyanshi,* 5, 64, 1986; *Anal. Abstr.,* 48, 11B44, 1986.
544. **Gallorini, M., Greenberg, R. R., and Gills, T. E.,** *Anal. Chem.,* 50, 1479, 1978.
545. **Bajo, S. and Suter, U.,** *Anal. Chem.,* 54, 49, 1982.
546. **Merry, R. H. and Zarcinas, B. A.,** *Analyst,* 105, 558, 1980.

547. **Crock, J. G. and Lichte, F. E.,** *Anal. Chim. Acta,* 144, 223, 1983.
548. **Donaldson, E. M.,** *Talanta,* 26, 999, 1979.
549. **Sansolone, R. F., Chao, T. T., and Welsch, E.,** *Anal. Chim. Acta,* 108, 357, 1979.
550. **Bajo, S.,** *Anal. Chem.,* 50, 649, 1978.
551. **Subramanian, K. S.,** *Fresenius Z. Anal. Chem.,* 305, 382, 1981.
552. **Itoh, K., Nakayama, M., Chikuma, M., and Tanaka, H.,** *Fresenius Z. Anal. Chem.,* 325, 539, 1986.
553. **Rybakov, A. A. and Ostroumov, E. A.,** *Zh. Anal. Khim.,* 39, 2168, 1984.
554. **Robberecht, H. J., Van Grieken, R. E., Van den Bosch, P. A., Deelstra, H., and Vanden Berge, D.,** *Talanta,* 29, 1025, 1982.
555. **Akiba, M., Shimioshi, Y., and Toei, K.,** *Analyst,* 100, 648, 1975.
556. **Ooghe, W. and Verbeek, F.,** *Anal. Chim. Acta,* 73, 87, 1974.
557. **Pal, B. K., Toneguzzo, F., Corsini, A., and Ryan, D. E.,** *Anal. Chim. Acta,* 88, 353, 1977.
558. **Lee, C. S., Chang, F. S., and Yeh, Y. H.,** *J. Chin. Chem. Soc. Taipei,* 29, 81, 1982; *Anal. Abstr.,* 44, 2B94, 1983.
559. **Iwasaki, K., Fuwa, K., and Haraguchi, H.,** *Anal. Chim. Acta,* 183, 239, 1986.
560. **Malhotra, P. D. and Prasada Rao, G. H. S. V.,** *Rec. Geol. Surv. India,* 93, 215, 1966.
561. **Khvostova, V. P. and Golovnya, S. V.,** *Zavod. Lab.,* 48(7), 3, 1982.
562. **Shlenskaya, V. I., Khvostova, V. P., and Bulakova, V. I.,** *Zh. Anal. Khim.,* 29, 314, 1974.
563. **Sill, C. W.,** *Anal. Chem.,* 50, 1559, 1978; **Sill, C. W.** *Anal. Chem.,* 52, 1452, 1980.
564. **Sill, C. W.,** *Health Phys.,* 33, 393, 1977.
565. **Sill, C. W., Hindman, F. D., and Anderson, J. I.,** *Anal. Chem.,* 51, 1307, 1979.
566. **Sill, C. W. and Williams, R. L.,** *Anal. Chem.,* 53, 412, 1981.
567. **Barnabee, R. P., Percival, D. R., and Hindman, F. D.,** *Anal. Chem.,* 52, 2351, 1980.
568. **Hsu, C. G. and Locke, D. C.,** *Anal. Chim. Acta,* 153, 313, 1983.
569. **Scott, K.,** *Analyst,* 103, 754, 1978.
570. **Chao, S. S. and Pickett, E. E.,** *Anal. Chem.,* 52, 335, 1980.
571. **Xu Li-quiang and Rao Zhu,** *Fresenius Z. Anal. Chem.,* 325, 534, 1986.
572. **Diamantatos, A.,** *Analyst,* 111, 213, 1986.
573. **Diamantatos, A.,** *Anal. Chim. Acta,* 165, 263, 1984.
574. **Schlieckmann, F. and Umland, F.,** *Fresenius Z. Anal. Chem.,* 318, 495, 1984.
575. **Chernov, R. V., Lyubova, L. D., and Ivanova, E. G.,** *Zavod. Lab.,* 50(9), 18, 1984.
576. **Trofimov, I. V. and Busev, A. I.,** *Zavod. Lab.,* 49(3), 5, 1983.
577. **Donaldson, E. M.,** *Talanta,* 26, 999, 1979.
578. **Molchanova, N. G. and Strekalovskii, V. N.,** *Zavod. Lab.,* 50(10), 10, 1984.
579. **Donaldson, E. M.,** *Talanta,* 28, 825, 1981.
580. **Rigin, V. I.,** *Zavod. Lab.,* 40, 1195, 1974.
581. **Holen, B., Bye, R., and Lund, W.,** *Anal. Chim. Acta,* 131, 37, 1981.
582. **Imahashi, M. and Takamatsu, N.,** *Bull. Soc. Chem. Jpn.,* 49, 1549, 1976.
583. **Neirinx, R., Adams, F., and Hoste, J.,** *Anal. Chim. Acta,* 48, 1, 1969.
584. **Stabryn, J.,** *Hutn. Listy,* 14, 515, 1959.
585. **Shlenskaya, V. I., Khvostova, V. P., and Bulakova, V. I.,** *Zh. Anal. Khim.,* 29, 314, 1974.
586. **Khvostova, V. P. and Golovnya, S. V.,** *Zavod. Lab.,* 48(7), 3, 1982.
587. **Pilipenko, A. T., Vasilchuk, T. A., and Volkova, A. I.,** *Zavod. Lab.,* 50(3), 7, 1984.
588. **Kushparenko, Yu. S.,** *Zavod. Lab.,* 45, 302, 1979.
589. **Angeletti, L. M. and Bartscher, W. J.,** *Anal. Chim. Acta,* 60, 238, 1972.
590. **Krtil, J.,** *Jad. Energ.,* 29, 440, 1983.
591. **Foner, H. A.,** *Analyst,* 109, 1469, 1984.
592. **Novoselova, I. M.,** *Zavod. Lab.,* 45, 884, 1979.
593. **Ginrburg, S. I., Ezerskaya, N. A., Prokofeva, I. V., Fedorenko, N. V., Shlenskaya, V. I., and Belskii, N. K.,** *Analiticheskaya Khimia Platinovykh Metallov,* Nauka, Moskva, 1972, 466.
594. **Shifris, B. S. and Konpakova, N. A.,** *Zh. Anal. Khim.,* 37, 2217, 1982.
595. **Mandal, S. K., Rao, S. B., and Sant, B. R.,** *Talanta,* 26, 135, 1979.
596. **Mandal, S. K., Rao, S. B., and Sant, B. R.,** *Talanta,* 28, 771, 1981.
597. **Mandal, S. K., Rao, S. B., and Sant, B. R.,** *Talanta,* 28, 121, 1981.
598. **Donaldson, E. M.,** *Methods for the Analysis of Ores, Rocks and Related Materials,* Energy, Mines and Resources Canada, Mines Branch Monogr. 881, Ottawa, 1974, 227.
599. **Kryukova, L. V. and Usatenko, Yu. I.,** *Zavod. Lab.,* 41, 387, 1975.
600. **Bogatyreva, L. P. and Afanaseva, G. V.,** *Zavod. Lab.,* 36, 663, 1970.
601. **Jeczalik, A.,** *Chem. Anal.* (Warsaw), 16, 1271, 1971.
602. **Ure, A. M.,** *Analyst,* 102, 50, 1977.
603. **Iskandar, I. K., Syers, J. K., Jacobs, L. W., Kenney, D. R., and Gilmour, J. T.,** *Analyst,* 97, 388, 1972.

604. **Huffman, C., Rahill, R. L., Van Shaw, E., and Norton, D. R.,** *U.S. Geol. Surv. Prof. Pap.*, 800C, C203, 1972.
605. **Ovrutskii, M. I., Kozachuk, N. S., and Freger, S. V.,** *Gig. Sanit.*, 55(3), 1981; *Anal. Abstr.*, 42, 4G63, 1982.
606. **Anon.,** *Fen Hsi Hua Hsueh*, 7, 46, 1979; *Anal. Abstr.*, 39, 5G8, 1980.
607. **Head, P. C. and Nicholson, R. A.,** *Analyst*, 98, 53, 1972.
608. **Píša, J.,** personal communication, 1985.
609. **Van der Veen, N. G., Keukens, H. J., and Vos, G.,** *Anal. Chim. Acta*, 171, 285, 1985.
610. **Xiao-Quon, S., Zhe-Ming, N., and Zhang-Li,** *Anal. Chim. Acta*, 151, 179, 1983.
611. **Hughes, K. C. and Carswell, D. J.,** *Analyst*, 95, 302, 1970.
612. **Iwasaki, K., Fuwa, K., and Haraguchi, H.,** *Anal. Chim. Acta*, 183, 239, 1986.
613. **Sill, C. W. and Williams, R. L.,** *Anal. Chem.*, 53, 412, 1981.
614. **Tsizin, G. I., Malofeeva, G. I., Tobelko, K. I., Urutov, V. S., Kalinichenko, N. B., Marov, N. I., and Zolotov, Yu A.,** *Zh. Anal. Khim.*, 39, 389, 1984.
615. **Tsizin, G. I., Malofeeva, G. I., Tobelko, K. I., Zolotov, Yu. A., Urusov, V. S., Kalinichen, N. B., and Marov, I. N.,** *Zh. Anal. Khim.*, 38, 1027, 1983.
616. **Weiss, D.,** *Chem. Listy*, 79, 205, 1985.
617. **Remy, H.,** *Anorganická Chemie* (transl.), Part I, SNTL, Praha, 1961, 622.
618. **Cotton, F. A. and Wilkinson, G.,** *Anorganická Chemie* (transl.), Academia, Praha, 1973, 502.
619. **Povondra, P. and Roubalová, D.,** *Collect. Czech. Chem. Commun.*, 25, 1890, 1960.
620. **Shen, X. E. and Chen, Q. L.,** *Spectrochim. Acta*, B-38, 115, 1983.
621. **Greenfield, S., McGeachin, M. McD., and Smith, P. B.,** *Anal. Chim. Acta*, 84, 67, 1976.
622. **Dahlquist, R. L. and Knoll, J. W.,** *Appl. Spectrosc.*, 32, 1, 1978.
623. **Rao, G. G. and Sagi, S. R.,** *Talanta*, 9, 715, 1962.
624. **Rao, G. G. and Sagi, S. R.,** *Talanta*, 10, 169, 1963.
625. **Rao, G. G. and Dikshitulu, L. S. A.,** *Talanta*, 10, 295, 1963.
626. **Mandal, S. K.,** *Talanta*, 26, 133, 1979.
627. **Bulycheva, I. B., Masalovich, V. M., and Agasyan, P. K.,** *Zh. Anal. Khim.*, 32, 283, 1977.
628. **Vláčil, F. and Drábal, K.,** *Chem. Listy*, 62, 1371, 1968.
629. **Dobeš, I.,** *Sklář. Keram.*, 9, 114, 1961.
630. **Vrbský, J. and Tymáň, V.,** *Sb. Vys. Sk. Chem. Technol. Praze*, Hl, 77, 1967.
631. **Strickland, E. H.,** *Analyst*, 80, 548, 1955.
631a. **Kallman, S. and Oberthin, H. K.,** *Anal. Chem.*, 37, 280, 1965.
632. **Bock, R., Jacob, D., Fariwar, M., and Frankenfeld, K.,** *Fresenius Z. Anal. Chem.*, 200, 81, 1964.
633. **Bock, R. and Tschöpel, P.,** *Fresenius Z. Anal. Chem.*, 246, 81, 1969.
634. **Tatsumoto, M. and Rosholt, J. N.,** *Science*, 167, 461, 1970.
635. **Zief, M. and Mitchell, J. W.,** *Contamination Control in Trace Element Analysis*, John Wiley & Sons, New York, 1976, 135.
636. **Trofimov, I. V. and Busev, A. I.,** *Zavod. Lab.*, 49(3), 5, 1983.
637. **Oelschläger, W.,** *Fresenius Z. Anal. Chem.*, 246, 376, 1969.
638. **Oelschläger, W.,** *Landwirtsch. Forsch.*, 22, 218, 1969.
639. **Kleinteich, R.,** *GIT Fachz. Lab.*, 5, 334, 1961.
640. **Shchegrov, L. N., Pechkovskii, V. V., and Eshchenko, L. S.,** *Zh. Prikl. Khim.*, 43, 990, 1970.
641. **Jordanov, N., Nikolova, B., and Havezov, I.,** *Talanta*, 25, 275, 1978.
642. **Ohashi, S. and Sugatami, H.,** *Bull. Chem. Soc. Jpn.*, 30, 864, 1957.
643. **Higgins, C. E. and Balwin, W. H.,** *Anal. Chem.*, 27, 1780, 1955.
644. **Kiba, T., Terada, T., and Suzuki, K.,** *Talanta*, 19, 451, 1972.
645. **Talvitie, N. A.,** *Anal. Chem.*, 23, 623, 1951.
646. **Popov, N. P. and Stolyarova, I. A.,** *Khimicheskii analiz gornykh porod i mineralov*, Nedra, Moskva, 1974, 61.
647. **Maxwell, J. A.,** *Rock and Mineral Analysis*, John Wiley & Sons, New York, 1968, 95.
648. **Jarzebowska, J.,** *Chem. Anal. (Warsaw)*, 12, 835, 1967.
649. **Stognii, N. J.,** *Zavod. Lab.*, 25, 420, 1959.
650. **Schmidt, K. G.,** *Staub*, 20, 404, 1960.
651. **Haftka, F. J.,** *Fresenius Z. Anal. Chem.*, 231, 321, 1969.
652. **Terada, K., Hirakawa, S., and Kiba, T.,** *Bull. Chem. Soc. Jpn.*, 50, 396, 1977.
653. **Baranov, B. I. and Chen, Y. V.,** *Zh. Anal. Khim.*, 15, 162, 1960.
654. **Sinyakova, S. YI. and Chen, Ya. V.,** *Zh. Anal. Khim.*, 15, 277, 1960.
655. **Strelow, F. W. E., Weinert, C. H. S. W., and van der Walt, T. N.,** *Anal. Chim. Acta*, 71, 123, 1974.
656. **Strelow, F. W. E., Liebenberg, C. J., and Victor, A. H.,** *Anal. Chim. Acta*, 46, 1409, 1974.
657. **Victor, A. H. and Strelow, F. W. E.,** *Anal. Chim. Acta*, 138, 285, 1982.

658. **Zolotovitskaya, E. S. and Potapova, V. G.,** *Zh. Anal. Khim.*, 39, 1781, 1984.
659. **Tamnev, B., Havezov, I., and Jotova, L. K.,** *Fresenius Z. Anal. Chem.*, 271, 349, 1974.
660. **Havezov, I. and Tamnev, B.,** *Fresenius Z. Anal. Chem.*, 290, 299, 1978.
661. **Nikolova, B. and Jordanov, N.,** *Talanta,* 29, 861, 1982.
662. **Palmer, T. A. and Winkler, J. M.,** *Anal. Chim. Acta,* 113, 301, 1980.
663. **Mendlina, N. G., Novoselova, A. A., and Rychkov, R. S.,** *Zavod. Lab.*, 25, 1293, 1959.
664. **Yamauchi, F. and Otaka, T.,** *Bunseki Kagaku,* 17, 1384, 1968.
665. **Williard, H. H. and Thompson, J. J.,** *Ind. Eng. Chem. Anal. Ed.*, 3, 399, 1931.
666. **Mizoguchi, T. and Ishii, H.,** *Talanta,* 26, 33, 1979.
667. **Goetz, C. A. and Wadsworth, E. P.,** *Anal. Chem.*, 28, 375, 1956.
668. **Zhao, Y.,** *Fen Hsi Hua Hsueh,* 9, 306, 1981; *Anal. Abstr.*, 42, 2B161, 1982.
669. **Mizoguchi, T. and Ishii, H.,** *Talanta,* 25, 311, 1978.
670. **Popova, I. M.,** *Zavod. Lab.*, 50(11), 16, 1984.
671. **Doležal, J., Povondra, P., and Šulcek, Z.,** *Decomposition Procedures in Inorganic Analysis,* Iliffe, London, 1968, 74.
672. **Bock, R.,** *Aufschlussmethoden der anorganischen und organischen chemie,* Weinheim, West Germany, 1972, 63.
673. **Šulcek, Z., Povondra, P., and Doležal, J.,** *CRC Crit. Rev. Anal. Chem.*, 6, 255, 1977.
674. **Bock, R.,** *Decomposition Methods in Analytical Chemistry,* International Textbook Company, Glasgow, 1979, 82.
675. **Lucas, R. P. and Ruprecht, B. C.,** *Anal. Chem.*, 43, 1013, 1971.
676. **Stoch, H. and Dixon, K.,** *A Manual of Analytical Methods Used at Mintek,* Spec. Publ. No. 4, Mintek, Randburg, S. Africa, 1983, 124.
677. **Hofton, M. E. and Baines, M.,** *Br. Steel Corp. Open (Rep.),* GS/EX/43/72/C, 1975; *Chem. Abstr.*, 83, 201523 k, 1975.
678. **Hofton, M. E. and Baines, M.,** U.S. NTIS, PB Rep. 1975; *Chem. Abstr.*, 83, 125704c, 1975.
679. **Kreingold, S. V., Shigina, E. D., Vzorova, I. F., and Sosenko, L. J.,** *Tr. Vses. Nauchno Issled. Inst. Khim. Reakt. Osobo. Chist Khim. Veshchestv,* No. 46, 169, 1984; *Anal. Abstr.*, 46, 6B6, 1985.
680. **Balyuk, S. T. and Zilberg, E. S.,** *Ogneupory,* p. 378, 1963; *Anal. Abstr.*, 11, 5451, 1964.
681. **Rao, J. R. and Pradhan, D.,** *Indian J. Technol.*, 13, 575, 1975; *Anal. Abstr.*, 31, 3B131, 1976.
682. **Smith, G. F. and Getz, C. A.,** *Ind. Eng. Chem. Anal. Ed.*, 9, 518, 1937.
683. **Dinnin, J. I. and Williams, E. G.,** *U.S. Geol. Surv. Prof. Pap.*, No. 424D, D394, 1961.
684. **Wünsch, G., and Czech, N.,** *Fresenius Z. Anal. Chem.*, 317, 5, 1984.
685. **Donaldson, E. M.,** *Methods for the Analysis of Ores, Rocks and Related Materials,* Energy, Mines and Resources Canada, Mines Branch Monogr. 881, Ottawa, 1974, 57.
686. **Nishida, H.,** *Jpn. Analyst,* 6, 299, 1957.
687. **Ohls, K., Sebastiani, E., and Riemer, G.,** *Fresenius Z. Anal. Chem.*, 281, 142, 1976.
688. **Faye, G. H.,** *Chem. Can.*, 10, 90, 1958.
689. **Medel, A. S. and Diaz-Garcia, E. M.,** *Analyst,* 106, 1268, 1981.
690. **Xu, S., Chen, Y., and Huang, X.,** *Fenxi Huaxue,* 12, 78, 1984; *Anal. Abstr.*, 46, 12B116, 1984.
691. **Ryabchikov, D. I., Gokhshtein, Ya. P., and Kao, T. Sh.,** *Zh. Anal. Khim.*, 16, 709, 1961.
692. **Williard, H. H. and Winter, O. B.,** *Ind. Eng. Chem. Anal. Ed.*, 5, 7, 1933.
693. **Shapiro, L.,** *U.S. Geol. Surv. Prof. Pap.*, 575D, D233, 1967.
694. **Kennedy, W. T., Hubbard, W. B., and Tarter, G. J.,** *Anal. Lett.*, 16, 1133, 1983.
695. **Shiraishi, N., Morishige, T., and Hagesawa, T.,** *Jpn. Analyst,* 18, 258, 1969.
696. **Zönnchen, W., Mehlhorn, M., and Volke, K.,** *Freiberg. Forschungsh. A,* 389, 35, 1966.
697. **Shimon, A. and Papp, L.,** *Zavod. Lab.*, 44, 936, 1978.
698. **Angot, J. and Mevel, N.,** *Chim. Anal.*, 45, 111, 1963.
699. **Fässler, A.,** *Erzmetall,* 22, 175, 1969.
700. **Milner, G. W. C. and Slee, L. J.,** *Analyst,* 82, 139, 1957.
701. **Ivanova, K. S.,** *Radiokhimiya,* 3, 348, 1961.
702. **Marchand, B.,** *Atomkernenergie,* 7, 371, 1962.
703. **Shcherbov, D. P., Plotnikova, P. N., and Astafeva, I. N.,** *Zavod. Lab.*, 36, 528, 1970.
704. **Volkova, G. A. and Sochevanov, V. G.,** *Zavod. Lab.*, 31, 541, 1965.
705. **Hoyle, W. C. and Diehl, H.,** *Talanta,* 18, 1072, 1971.
706. **Nagashima, S., Yoshida, M., and Ozawa, T.,** *Bull. Chem. Soc. Jpn.*, 45, 3446, 1972.
707. **Mizoguchi, T. and Ishii, H.,** *Talanta,* 27, 525, 1980.
708. **Steger, H. F.,** *Talanta,* 24, 251, 1977.
709. **Akiyoshi, T. and Tsukamoto, T.,** *Bunseki Kagaku,* 27, 85, 1978.
710. **Pietrosz, J. and Czyz, J.,** *Hutn. Listy,* 34, 658, 1979.
711. **Ischenko, A. V., Shvarts, E. M., Stashkova, N. V., Dzene, A. Ya., and Bernane, A. A.,** *Zavod. Lab.*, 51(4), 8, 1985.

712. **Lev, I. E., Lazarev, B. G., and Mitskevich, N. S.,** *Zavod. Lab.*, 49(9), 19, 1985.
713. **Bollman, D. H. and Mortimore, D. M.,** *Anal. Chem.*, 43, 154, 1971.
714. **Onishi, H., Ishiwatari, N., and Nagai, H.,** *Bull. Chem. Soc. Jpn.*, 33, 1687, 1960.
715. **Schafer, H. N. S.,** *Analyst,* 91, 755, 1966.
716. **Dinnin, J. I.,** Rapid analysis of chromite and chrome ores, *U.S. Geol. Survey Bull.*, 1084B, 1959.
717. **Konopicky, K. and Caesar, F.,** *Ber. Deutsch. Keram. Ges.*, 20, 362, 1939.
718. **Shein, A. V.,** *Zavod. Lab.*, 6, 1199, 1937.
719. **Balyuk, S. T. and Mirakyan, V. M.,** *Zavod. Lab.*, 15, 1004, 1949.
720. **Dippel, P.,** *Silikattechnik,* 13, 51, 1963.
721. **Pilnik, R. S. and Ivanova, T. V.,** *Tr. Ural. Nauch. Issled. Khim. Inst.*, 8(11), 1964; *Anal. Abstr.*, 13, 635, 1966.
722. **Gekht, I. I. and Putok, S.,** *Vestn. Akad. Nauk. Kazakh. SSR,* 9, 68, 1960.
723. **Kondrakhina, E. G., Egorova, L. G., and Songina, O. A.,** *Izv. Akad. Nauk. Kazakh. SSR,* 1, 45, 1957.
724. **Nagaoka, T. and Yamazaki, S.,** *Jpn. Analyst,* 3, 408, 1954.
725. **Sasuga, H. and Tida, Y.,** *Jpn. Analyst,* 7, 248, 1958.
726. **Samanta, H. B. and Sen, N. B.,** *Indian Ceram. Soc. Trans.*, 5, 97, 1946.
727. **Goswami, N.,** *Sci. Cult. India,* 22, 398, 1957.
728. **Brežný, B.,** *Hutn. Listy,* 15, 552, 1960.
729. **Radovskaya, T. L. and Borovik, K. G.,** *Zavod. Lab.*, 39, 1450, 1973.
730. **Dixon, K., Cook, E. B. T., and Silverthorne, D. F.,** *Natl. Inst. Metall. Repub. S. Afr. Rep.*, No. 1394, 1972.
731. **Cheng, K. L.,** *Anal. Chem.*, 36, 1666, 1964.
732. **Songina, O. A., Kemeleva, N. G., Egorova, L. G., and Lebedyantseva, A. I.,** *Zavod. Lab.*, 34, 794, 1968.
733. **Ingamells, C. O.,** *Talanta,* 4, 268, 1960.
734. **Sabine, P. A., Sergeant, G. A., and Young, B. R.,** *Mineral. Mag.*, 36, 948, 1967.
735. **Seil, G. E.,** *Ind. Eng. Chem. Anal. Ed.*, 15, 189, 1943.
736. **Malhotra, P. D. and Prasada Rao, G.H.S.V.,** *Rec. Geol. Surv. India,* 93, 215, 1966.
737. **Tikhomirova, O. F., Strebulaeva, E. N., and Sazonova, Z.,** *Sb. Tr. Tsentr Nauchno Issled. Inst. Chern. Metall.*, p. 180, 1963; *Anal. Abstr.*, 11, 1737, 1964.
738. **Bilgrami, S. A. and Ingamells, C. O.,** *Am. Miner.*, 45, 576, 1960.
739. **Bouvier, J. L., Sen Gupta, J. G., and Abbey, S.,** *Geol. Surv. Can. Pap.*, p. 72, 1972.
740. **Varzaru, E.,** *Rev. Chim.*, 20, 764, 1969.
741. **Pozdnyakova, E. A., Komissarova, T. E. and Barsukova, A. M.,** *Zh. Anal. Khim.*, 26, 1128, 1971.
742. **Komissarova, T. E. and Pozdnyakova, E. A.,** *Zh. Anal. Khim.*, 29, 306, 1974.
743. **Weiss, D.,** *Chem. Zvesti,* 23, 671, 1969.
744. **Povondra, P.,** *Acta Univ. Carol. Geol.*, No. 3, 223, 1981.
745. **Marinenko, J.,** *Anal. Lett.*, 9, 755, 1976.
746. **Ingamells, C. O. and Bradshaw, W. S.,** *Chemist Analyst,* 47, 11, 1958.
747. **Kluh, I., Doležal, J., and Zýka, J.,** *Fresenius Z. Anal. Chem.*, 177, 14, 1960.
748. **Doležal, J., Zýka, J., and Donose, G.,** *Anal. Chim. Acta,* 29, 70, 1963.
749. **Ingamells, C. O.,** *Talanta,* 2, 171, 1959.
750. **Knoeck, J. and Diehl, H.,** *Talanta,* 14, 1083, 1967.
751. **Kitagawa, H. and Shibata, N.,** *Bunseki Kagaku,* 9, 597, 1960.
752. **Bykhovtsova, T. T. and Kaukhova, L. V.,** *Zavod. Lab.*, 41, 281, 1975.
753. **Bykhovtsova, T. T.,** *Zavod. Lab.*, 49(9), 22, 1983.
754. **Athavale, V. T. and Murugaiyan, P.,** *Indian J. Chem.*, 1, 67, 1963; *Anal. Abstr.*, 11, 144, 1964.
755. **Luginin, V. A. and Tserkovnitskaya, I. A.,** *Radiokhimiya,* 12, 890, 1970.
756. **Borstel, D. and Halbach, P.,** *Fresenius Z. Anal. Chem.*, 310, 431, 1982.
757. **Stonhill, L. G.,** *Can. J. Chem.*, 36, 1487, 1958.
758. **Zpěváčková, V. and Krumlová, L.,** *Radiochem. Radioanal. Lett.*, 14, 193, 1973.
759. **Buldini, P. L., Ferri, D., Pauluzzi, E., and Zambianchi, M.,** *Analyst,* 109, 225, 1984.
760. **Papež, V., Večerník, J., and Krtil, J.,** *Jad. Energ.*, 29, 410, 1983.
761. **Kihara, S., Yoshida, Z., Muto, H., Aoyagi, H., Baba, Y., and Hashitani, H.,** *Anal. Chem.*, 52, 1601, 1980.
762. **Krtil, J.,** *Jad. Energ.*, 29, 440, 1983.
763. **Sill, C. W.,** *Anal. Chem.*, 50, 1559, 1978.
764. **Jeffery, P. G. and Kipping, P. J.,** *Analyst,* 87, 379, 1972.
765. **Marinenko, J. and May, I.,** *U.S. Geol. Surv. Prof. Pap.*, 700D, D103, 1970.
766. **Jeffery, P. G.,** *Chemical Methods of Rock Analysis,* Pergamon Press, Oxford, 1970, 171.
767. **Read, J. I.,** *Analyst,* 97, 134, 1972.

768. **Sixta, V.**, *Fresenius Z. Anal. Chem.*, 285, 369, 1977.
769. **Cornides, I. and Kusakabe, M.**, *Fresenius Z. Anal. Chem.*, 287, 310, 1977.
770. **Dixon, B. E.**, *Analyst*, 59, 739, 1934.
771. **Jeffery, P. G. and Wilson, A. D.**, *Analyst*, 85, 749, 1960.
772. **Hoefs, J.**, *Geochim. Cosmochim. Acta*, 29, 399, 1965.
773. **Bush, P. R.**, *Chem. Geol.*, 6, 59, 1970.
774. **Suzuki, J., Yokoyama, Y., Unno, Y., and Sazuki, J.**, *Water Res.*, 17, 431, 1983.
775. **Allison, L. E., Bollen, W. K., and Moodie, C. D.**, *Agronomy*, 9, 1346, 1965.
776. **Snyder, J. D. and Trofymov, J. A.**, *Commun. Soil Sci. Plant Anal.*, 15, 587, 1984.
777. **Terada, K., Matsumoto, K., and Kiba, T.**, *Bull. Chem. Soc. Jpn.*, 48, 2567, 1975.
778. **Khvostova, V. P. and Golovnya, S. V.**, *Zavod. Lab.*, 48(7), 3, 1982.
779. **Terada, K., Ooba, T., and Kiba, T.**, *Talanta*, 22, 41, 1975.
780. **Rybakov, A. A. and Ostroumov, E. A.**, *Zh. Anal. Khim.*, 38, 446, 1983.
781. **Rybakov, A. A. and Ostroumov, E. A.**, *Okeanologyia*, 23, 1039, 1983.
782. **Rybakov, A. A. and Ostroumov, E. A.**, *Zh. Anal. Khim.*, 39, 2168, 1984.
783. **Terada, K., Okuda, K., Maeda, K., and Kiba, T.**, *J. Radioanal. Chem.*, 46, 217, 1978.
784. **Terada, K., Okuda, K., and Kiba, T.**, *J. Radioanal. Chem.*, 36, 47, 1977.
785. **Terada, K. and Yoshida, K.**, *Nippon Kagaku Kaishi*, No. 1, 110, 1981.
786. **Terada, K. and Okuda, K.**, *Fresenius Z. Anal. Chem.*, 320, 773, 1985.
787. **Sager, H.**, *Microchim. Acta*, II, 381, 1984.
788. **Kiba, T., Takagi, T., Yoshimura, Y., and Kishi, I.**, *Bull. Chem. Soc. Jpn.*, 28, 641, 1955.
789. **Kiba, T. and Kishi, I.**, *Bull. Chem. Soc. Jpn.*, 30, 44, 1957.
790. **Kiba, T., Akaza, I., and Sugishita, N.**, *Bull. Chem. Soc. Jpn.*, 30, 972, 1957.
791. **Arikawa, Y., Ozawa, T., and Iwasaki, I.**, *Bunseki Kagaku*, 21, 920, 1972.
792. **Purnell, A. L. and Doolan, K. J.**, *Fuel*, 62, 1107, 1983.
793. **Bebeshko, G. I. and Oleshko, O. N.**, *Zh. Anal. Khim.*, 37, 640, 1982.
794. **Kölling, W.**, *Chem. Technol.*, 15, 747, 1963.
795. **Mizoguchi, T., Iwahori, H., and Ishii, H.**, *Talanta*, 27, 519, 1980.
796. **Yamazaki, Y.**, *Bunseki Kagaku*, 19, 187, 1970.
797. **Bajpel, H. N., Iyer, C. S. P., and Das, M. S.**, *Anal. Chim. Acta*, 72, 423, 1974.
798. **van Grondelle, M. C., van der Craats, F., and van der Laarse, J. D.**, *Anal. Chim. Acta*, 92, 267, 1977.
799. **Murphy, J. M. and Sergeant, G. A.**, *Analyst*, 99, 515, 1974.
800. **Johnson, C. M. and Nishita, H.**, *Anal. Chem.*, 24, 736, 1952.
801. **Kiba, T., Akaza, I., and Hachino, H.**, *Bull. Chem. Soc. Jpn.*, 32, 454, 1959.
802. **Bollman, D. H.**, *Anal. Chem.*, 44, 887, 1972.
803. **Hannaker, P. and Hou, Q. L.**, *Talanta*, 31, 1153, 1984.
804. **Bogatyrev, V. L., Bulikh, A. I., and Sokolova, S. I.**, *Zh. Anal. Khim.*, 22, 837, 1967.
805. **Schafer, H. N. S.**, *Anal. Chem.*, 35, 53, 1963.
806. **Evans, L., Hoyle, R. D., and Macaskill, J. B.**, *N.Z. J. Sci.*, 13, 143, 1970; *Chem. Abstr.*, 72, 128385, 1970.
807. **Nikolina, E. S., Frolov, I. A., and Aziev, R. G.**, *Zh. Anal. Khim.*, 31, 1361, 1976.
808. **Taneeva, G. G. and Freze, N. A.**, *Zavod. Lab.*, 36, 665, 1970.
809. **Macháček, V.**, unpublished data.
810. **Segall, E. and Schmuckler, G.**, *Talanta*, 14, 1253, 1967.
811. **Govindaraju, K.**, *Anal. Chem.*, 40, 24, 1968.
812. **Govindaraju, K.**, *Analusis*, 2, 367, 1973.
813. **Povondra, P. and Hejl, V.**, *Collect. Czech. Chem. Comm.*, 41, 1343, 1976.

Chapter 5

DECOMPOSITION IN CLOSED SYSTEMS

I. INTRODUCTION

Progressively increasing demands on chemical analyses have often necessitated studies of the decomposition techniques for solids, in addition to the study of the methods for the final determination. The only methods available earlier for dissolutioin of resistant materials involved fusion with alkaline or acidic agents, which at that time greatly complicated the subsequent analytical operations.

Chemists have tried to increase the effect of acids in decompositions for more than a century. Although many findings useful for this purpose were available as early as at the end of the 19th century (obtained, e.g., in the synthesis of organic and inorganic substances under pressure), the first experiments of Carius,[1] Mitscherlich,[2] and Jannasch[3-5] with decompositions in closed systems aroused little attention, and the development of decomposition techniques further stagnated. Only the pioneering works of the chemists of the National Bureau of Standards of the U.S., dealing with dissolution of the platinum metals and some minerals in sealed glass ampules, paved the way for increasing the effect of acids.[6] Metallic autoclaves with hermetic internal vessels of platinum or gold found use substantially later. In this apparatus, even hydrofluoric acid could be used for decomposition at temperatures exceeding 400°C. However, the apparatus was complicated and costly, and thus, autoclaves of this type have never been marketed. The analytical applications were limited to decompositions of small amounts of resistant substances for special purposes.

Decompositions in closed systems at elevated temperatures started to develop successfully only after introduction of fluorinated hydrocarbon polymers as the materials for the autoclave internal vessels. Polytetrafluoroethylene (PTFE, Teflon®) has still remained the most common material because of its high degree of chemical and thermal resistance. During the last 25 years, various types of vessels and various methods of their sealing have been proposed. Chemical reactivity of polymers with all inorganic acids and leaching of impurities from the vessel walls have been studied. The construction is still being perfected, and the metal mantle materials are still changing. New modifications of PTFE reaction vessels appeared that are sealed by a simple device without a metallic autoclave.

Even these acid decompositions in closed systems have not been able to cope with extremely high requirements of analyses of highly pure substances. Therefore, pressure decomposition methods have been developed in which the test material is placed on a Teflon® dish and dissolved by vapors of acids at temperatures of circa 150 to 220°C. To decrease the blank value, the decomposition may also be performed in a quartz test tube placed in a Teflon® crucible or directly in the autoclave body. A modern modification of the method consists of dissolution in a closed Teflon® vessel provided with a safety valve, with microwave supply of the energy to the reaction medium.

The principal advantages of decomposition at elevated temperature and pressure in a closed system can be summarized as follows:

1. Resistant solids, undecomposable by acids in open systems, can be dissolved.
2. The dissolution time is shortened.
3. Losses of material in the form of volatile compounds are prevented.
4. The volumes of acids (reagents) required for decomposition are small.
5. The reaction mixture is isolated during the dissolution, and contamination from the laboratory atmosphere is prevented.
6. The blank value is substantially decreased.

Table 1
LEACHING OF SEALED GLASS AND SILICA TUBES BY CONCENTRATED HYDROCHLORIC ACID[6,12]

Tube material	Temperature (°C)	Leaching time (hr)	Material leached ($mg/100\ cm^2$ surface)
Silica	350	13	1,6
	350	46	3,5
Pyrex glass	300	24	27

The importance of the above points depends on the given decomposition method. Any system has limitations stemming from the principle of the decomposition technique. The specific features of the individual dissolution techniques are characterized in detail below. This chapter deals with dissolution in closed systems containing aqueous solutions of acids, their vapors, and possibly other reagents. Decompositions in autoclaves in anhydrous media (e.g., pressure fluorination, chlorination, or bromination) are discussed in Chapter 11. Dissolution of solids using microwaves is treated separately, as this method of supplying energy has found use in both open and closed systems.

Instrumentation for pressure decompositions and the analytical use of the method are discussed in the literature.[7-10] Decompositions in closed systems have now become an indispensable laboratory technique. This chapter does not aim at an exhaustive survey of the published data, but attempts general evaluation of the present state and the trends in this field, together with a critical selection of analytical applications.

II. DECOMPOSITIONS IN SEALED AMPULES

Decomposition in sealed glass ampules belongs among the oldest methods of pressure decomposition.[1-5] It is still the cheapest approach to decompositions in closed systems at reaction temperatures between 250 and 350°C. Freely heated glass ampules are very advantageous for routine applications. In a suitable arrangement, the dissolution process can be monitored visually, and the reaction time thus minimized.[7,8] The ampules are mostly made of thick-walled (2 to 2.5 mm) tubes of borosilicate glass, e.g., of the Pyrex® type, or of Vycor type quartz glass. The faultless glass tube is usually formed to obtain a test-tube shape, narrowed in the upper part. The test-tube orifice is widened to accommodate a funnel through which the sample and the decomposition agent are introduced. In some determinations, the air must be expelled from the ampule by a stream of argon, nitrogen, or carbon dioxide. Depending on the character of the test substance and of the kind of analysis, the bottom part of the ampule is efficiently cooled during the sealing in an ice bath, solid carbon dioxide, or liquid nitrogen. On completion of the decomposition reaction, the ampule is cooled again, and the aerosols are allowed to coagulate. If necessary, the ampule is opened in an inert gas atmosphere or in a vacuum.[11]

Pressure decomposition with acids at an elevated temperature leads to corrosion of the ampule internal walls and to contamination of the solution obtained by compounds of silicon, boron, sodium, zinc, etc. Therefore, in demanding analyses, the glass ampules must be replaced by more expensive quartz tubes. The amounts of impurities leached by concentrated hydrochloric acid from the two types of glass are given in Table 1. The walls of quartz ampules are considerably less attacked than glass walls, even at prolonged reaction times and higher temperatures. Whereas virtually only silicon compounds are leached from quartz glass, with Pyrex glass the content of silicon dioxide amounts to less than one half of the sum of the corrosion products.[6,12] As maximal working temperatures, 300 and 350°C are usually considered for glass and quartz ampules, respectively. In decompositions of resistant substances, temperatures of 390 to 420, or even 470°C (Vycor glass), are exceptionally attained.[12,13]

The application of sealed ampules is limited by the mechanical and chemical properties of the material and by the limited strength of the vessel at the working temperature. The pressure inside the ampule is mainly determined by the pressures of the water vapor, the vapors of the acids used, the gaseous reaction products, the volume of the decomposition agents, and the selected working temperature. The calculation of the pressure in real systems is very difficult, and the published data are only valid for simple systems, such as hydrochloric acid.[6-8,14] According to empirical equations, the pressure attains almost 0.7 MPa for 36% hydrochloric acid at temperatures below 200°C. Therefore, safety precautions must be observed in work with sealed ampules.[7,8] Freely heated ampules without any external protection may be used at pressures not exceeding 0.2 MPa. If a higher pressure may develop, the ampule must be inserted into a thick-walled steel tube with a hermetic, screw closing. Volatile substances are added into the protective tube, in order that the pressure of their vapors compensates at higher temperatures the pressure inside the ampule (e.g., gasoline, trichloroethylene, ethers, etc.). A suitable compensating substance is solid carbon dioxide, which exhibits an almost linear dependence of the pressure on temperature within the most common interval of 250 to 300°C.

In an effort to improve the safety of work with sealed ampules, the apparatus has been considerably altered. The sealed quartz ampules are placed in a metallic autoclave located in a box with controllable temperature.[12,15,16] The compensating counterpressure is provided by nitrogen introduced from a pressure cylinder through an automatically controlled valve. The apparatus operates reliably,[12,15] even in long-term experiments at temperatures above 350°C.

Pressure decompositions in sealed ampules first found important analytical application in analyses of the platinum metals and their alloys. In open systems, only platinum and palladium can completely be dissolved in aqua regia, while ruthenium, iridium, and rhodium are only partially dissolved. The preparation of a solution for analysis thus required rather complicated procedures, mostly chlorination with gaseous chlorine in the presence of sodium chloride, under exactly defined reaction conditions. These problems have generally been solved by decomposition with an oxidizing mixture of acids in sealed glass ampules.[6] Hydrochloric acid with oxidizing admixtures, e.g., nitric or perchloric acid, chlorate, or bromine, has yielded the best results.[6,16-23] The dissolution temperature usually does not exceed 300°C, and the reaction time can be shortened to less than 12 hr under suitable conditions. The procedures are useful not only in the determination of impurities in pure metals, but also in analyses of alloys and technological intermediates, such as matte, leached matte, and prills.[17,18,20] In dissolution in sealed ampules, difficulties with the formation of volatile compounds are overcome. The reaction solution is only contaminated with the components leached from the glass.

A mixture of hydrochloric and nitric acids is especially useful for dissolution of resistant prills (with the platinum metals) obtained by reductive fusion of chromite.[17,18,20] The common ampules are 30 mm long, 10 mm in internal diameter, and a wall thickness of to 2 to 2.5 mm. The prill is transferred to the tube, whose bottom is cooled in an ice bath, and 2 mℓ concentrated HCl and 0.04 mℓ fuming HNO_3 are added. The mixture is cooled again, and the ampule is sealed. The tube is heated in a steel cylinder provided with screw closings and placed in an oven at 200°C for 3 hr. The compensating pressure is created by placing a piece of solid carbon dioxide into the steel mantle. After cooling, the ampule is placed in an ice bath and cautiously opened.

Decomposition with brominated hydrochloric acid to which nitric acid is added is suitable for matte and the leached matte residue, but the sample must not contain organic substances, as they readily volatilize by the action of the oxidizing mixture and their pressure prevents perfect sealing of the ampule. A volume of 1.5 mℓ of fuming HNO_3 is added to the ampule, and the latter is cooled in a solid CO_2 bath. Then 4 mℓ of brominated HCl are added (900 mℓ concentrated HCl and 100 mℓ Br_2), and the end of the ampule is again immersed in the cooling bath. A 1-g sample is introduced to the reaction mixture, the contents of the ampule are thoroughly frozen, and the ampule is sealed. The reaction time and temperature are the same as above. If required, silicon dioxide is removed from the solution obtained by evaporation with HF. The undecomposed residue is burned, after separation, in a porcelain crucible and reduced in a stream of hydrogen. In the final stage of decomposition, the reduced portion is dissolved, using the above procedure, in a medium of hydrochloric acid and liquid chlorine.

Decomposition with hydrochloric acid in the presence of liquid chlorine is also efficient for rhodium, iridium, ruthenium and osmium metals, and for reduced metal sponge. The pure metals are reduced by heating in a stream of hydrogen (*circa* 0.5-g samples), to remove passivating surface oxidic films. To the sample in the ampule, 5 mℓ concentrated HCl are added, and the mixture is frozen in a solid CO_2 bath, followed by addition of 1 mℓ of liquid Cl_2, freezing, and sealing of the ampule. The tightness of the seal is checked by placing a litmus paper over it. The tube is inserted in a steel mantle containing a piece of solid CO_2, and the latter is fixed in a mechanical device permitting rotation of the ampules in the slowly heated box. The dissolution takes 12 to 15 hr[17,18,20] at 270°C. Up to 2 g of iridium metal were dissolved in 20 mℓ of 36% HCl with addition of 2 g Cl_2 or 1 g $NaClO_3$, at 300°C and a reaction time prolonged to 24 hr.[6]

Oxidizing dissolution of steel in a mixture of hydrochloric and nitric acids in sealed ampules prevents losses in volatile sulfur compounds during the decomposition.[24] By the action of water and water vapor on selenium metal, the chlorine and bromine compounds present are leached at 300°C in quartz ampules. The heavy metal ions are extracted with 2 *M* hydrochloric acid, whereas iodides are extracted with a 0.5 *M* ammonium hydroxide solution. The selenium is dissolved to a minimal extent, which simplifies the subsequent analytical operations.[25]

The conditions proposed by various authors[7,8,12,26-28] for dissolution of aluminum oxide are difficult to compare. The main cause of the discrepancies is incomplete characterization of the test substance, as the thermal treatment of the sample and its physical state are usually not specified. Single crystals of α-Al_2O_3 are most difficult to dissolve. With the same particle size, calcined oxides are destroyed more easily than fused ones. The selection of acids suitable for the obtaining of a solution is very limited,[12,15] 20 to 36% hydrochloric acid being considered most efficient. Sulfuric acid (25 to 40%) has exceptionally been used. The effect of hydrochloric acid increases up to a temperature of 350°C and then remains the same or decreases. The decreased efficiency of hydrochloric acid in an interval of 380 to 390°C is probably connected with the disappearance of water as the liquid phase that dissolves the aluminum chloride formed at the interface. The reaction conditions required for the obtaining of a clear solution from variously pretreated samples of aluminum oxide are listed in Table 2. When working in a temperature range of 300 to 400°C, the safety of the operator must be ensured. The reaction was carried out in a specially constructed pressure vessel made of a Cr-Mo steel, with a compensation counterpressure created by gaseous nitrogen. The pressure and temperature in the steel vessel was continuously measured, and the whole system was provided with safety valves. The operator was separated from the heated box by a concrete housing.

Leaching of fluorides by water from a corundum (ruby) single crystal is interesting from the point of view of the reaction mechanism. By action of 5 mℓ H_2O, more than 96% of

Table 2
DISSOLUTION OF ALUMINA IN SEALED QUARTZ TUBES BY HYDROCHLORIC ACID[12]

Sample material	Temperature (°C)	Heating period (hr)
Calcined alumina 1600°C, fine powder	260	14
Fused alumina, 30—60 mesh	350	45
Fused alumina, 10—30 mesh	350	72
Sapphire, chips 5 × 3 × 2 mm	340	80
Recrystallized alumina, chips 5 × 5 × 2 mm	325	10
Recrystallized alumina, chips 5 × 5 × 2 mm	277	24

Note: Dissolution is by 8 mℓ 36% HCl per 1 g of sample.

the fluorides are leached from a single crystal 0.5 mm thick with a surface area of 1 cm^2, within 24 hr at 230°C. After the leaching, the surface of the single crystal is preserved for further measurements. As demonstrated by IR spectra, fluorides are replaced by hydroxyl in the solid phase through an ion-exchange process.[8,29]

A specially constructed apparatus has been proposed for decomposition of nuclear fuels and other resistant substances.[16] Quartz tubes are inserted into a metallic autoclave, and the whole system operates with a controllable internal pressure (CO_2). The reaction temperature is automatically maintained at 300°C. A mixture of hydrochloric and nitric acids dissolves even the high-temperature ignited oxides ThO_2, ZrO_2 and V_2O_5 within 5 to 24 hr. In this arrangement, the safety of the operator is substantially improved, and the danger of contamination of the environment is suppressed, which is the principal requirement when decomposing highly toxic radioactive substances, e.g., plutonium dioxide. A mixture of hydrochloric and perchloric acids quantitatively dissolves a 0.5-g sample of this oxide within 2 to 4 hr at 390°C; mixed oxides of Pu(IV) and U(IV) are decomposed more easily. Dissolution of nuclear fuels and alloys employed in nuclear energy production by decomposition in sealed ampules has recently been critically discussed.[30]

Even this energetic decomposition has been found insufficiently efficient for dissolution of single crystals of the rutile form of titanium dioxide.[6,7,31,32] Of a 50-mg sample, only 50% were dissolved in 12 *M* hydrochloric acid after 12 hr at 225°C.[31] As pressure decomposition by sulfuric or nitric acid does not yield a clear solution, the sample must be fused with, e.g., $K_2S_2O_7$ or KHF_2. Attempts to dissolve barium titanate in dilute and concentrated sulfuric acid or the strontium salt in hydrochloric acid have also been unsuccessful. However, pulverized single crystals of an aluminum-yttrium garnet have been successfully dissolved in 6 *N* HCl or 30% H_2SO_4 by heating for 24 hr at 220°C. If the sulfuric acid concentration is increased to 60 to 80%, then hafnium dioxide is also completely dissolved. Cassiterite is decomposed by prolonged action of hydrochloric acid at 300°C.[6-8,32]

A melt of selenium dioxide (mp 390°C with a vapor pressure of 0.9 MPa) dissolves in sealed ampules even such highly resistant oxides as those of niobium (V) and tantalum(V).[33] The bottom part of the tube must be heated up to 420°C, which places high demands on the glass quality. The fusion agent circulates within the reaction space due to a temperature gradient between the top and the bottom of the tube.

A decomposition in a sealed ampule by dilute sulfuric acid was proposed more than 100 years ago for a determination of ferrous oxide in many natural and synthetic substances.[2,34] A completely inert atmosphere can also be readily maintained in a sealed ampule. The redox reactions of the solution with the tube walls are minimal, even during prolonged reactions. The course of dissolution can be observed visually when the experimental arrangement is suitable. As decomposition agents, dilute sulfuric, hydrochloric, or phosphoric acids are primarily employed at various concentration ratios. The stability of the Fe^{2+} and Fe^{3+} ions in these solutions is not completely the same in open and closed systems (even if access of oxygen is completely prevented). The differences observed in the interaction of acids with these ions may partially be ascribed to different reaction temperatures. Another possible cause is the formation of gaseous products that cannot escape from a sealed ampule and affect the redox equilibria.

Although dilute sulfuric acid is a suitable solvent in open systems, it exhibits weakly-reducing effects on prolonged heating in an ampule.[35] The reduction of Fe^{3+} on dissolution of a hematite single crystal is probably caused by a small amount of sulfur dioxide produced by thermal decomposition of sulfuric acid,

$$2\ H_2SO_4 \rightarrow 2\ SO_2 + 2\ H_2O + O_2 \tag{1}$$

A several-hour boiling with 6 *M* hydrochloric acid in an open system also causes reduction of Fe^{3+} to a low, but unambiguously confirmed, extent. However, this phenomenon has not been observed in sealed tubes.[35,36] The chlorine produced by the oxidation cannot escape from the closed space, an equilibrium is rapidly established, and the reduction of ferric ions is minimal.

An analysis of the gaseous phase from an ampule after sample dissolution in 85% H_3PO_4 in an inert atmosphere has indicated the presence of volatile, phosphorus-containing substances that reduce, e.g., a potassium bichromate solution. These results suggest that hot, concentrated phosphoric acid oxidizes a small part of ferrous ions into ferric ions even in the absence of oxygen.

$$2\ Fe_3(PO_4)_2 + 5\ H_3PO_4 \rightarrow 6\ FePO_4 + 3\ H_3PO_3 + 3\ H_2O \tag{2}$$

The phosphorous acid formed is further decomposed with formation of phosphanes or gaseous hydrogen and phosphoric acid. Therefore, concentrated phosphoric acid is not suitable for decomposing substances with small contents of ferrous oxide. Dilute, 40 to 45% phosphoric acid does not exhibit these oxidizing effects and is thus a better solvent for closed systems from this point of view; however, it attacks solids much more slowly than the concentrated acid.

The determination of ferrous oxide based on decomposition of the substance in a sealed ampule with an inert atmosphere is still used in analytical practice.[7,8,37] The selection of the test materials is limited to the substances decomposable by hydrochloric, sulfuric, or phosphoric acid at temperatures of 180 to 250°C. Pyrites and other sulfides yield reducing compounds on interaction with the acids used (H_2S, SO_2, and S) which are highly reactive in a closed system and reduce ferric ions. Most organic compounds and carbonaceous substances have similar effects.[34,37] In spite of these drawbacks, the method is still used, e.g., to determine ferrous oxide in materials with minimal sulfur contents, e.g., hematite or oxidic iron ores.[37] A 0.2- to 0.5-g sample of the crushed (but not pulverized) material is placed in a Pyrex glass tube with a wall thickness of 2 to 2.5 mm. The tube is rinsed with argon, 3 mℓ of concentrated HCl are added, and the tube is sealed and heated in a steel protective mantle to 200 to 250°C until the sample dissolves. The FeO content is determined by oxidimetric titration. With hematite single crystals, spectrophotometry with 1,10-phenanthroline is more suitable for determination than the titration when the content of FeO is small.[35,38]

Resistant complex oxides with the spinel structure can only be completely dissolved when they are heated with acids in a sealed ampule.[7,8,35,38-41] The difficult determination of ferrous oxide in ores containing chromite can be carried out[42] using oxidimetric titration after a 7-

to 10-hr dissolution in dilute sulfuric acid (d = 1.35) at 230 to 250°C. No reducing effect of sulfuric acid was observed, in contrast to Reference 35. High-temperature ignited ferrites cannot be decomposed with acids in open systems, and thus, decomposition in sealed ampules has become an indispensable technique for determination of oxygen stoichiometry of ferrites.[7,8,40] Deviations from the theoretical oxygen content cause defects in their structure due to heterovalent substitution, leading to changes in the physical properties of these substances. In view of the danger of oxidation and contamination, the samples are not pulverized, and 1- to 3-mm lumps are dissolved, which, of course, leads to a substantially longer reaction time.[41] Phosphoric acid is suitable for determination of small contents of ferrous oxide in the compound $NiFe_2^{III}O_4$ or in hematite.[7,8,35] To a 50-mg sample of the pulverized material in an ampule, 3mℓ of 40 to 45% H_3PO_4 is added, and the air is completely expelled from the ampule by a stream of nitrogen. The reaction mixture is heated for 6 hr at 150 to 200°C, and the FeO content is determined spectrophotometrically with 1,10-phenanthroline. With nickel(II) ferrites with higher FeO contents ($Ni,Fe^{II}/Fe_2^{III}O_4$), the reaction time can be shortened down to 30 min when using 85% acid and increasing the temperature to 220°C. Decompositions with hydrochloric acid solutions with ferrous and ferric ions added have found use with nickel(II) ferrites and mixed Mn(III)-Cr(III) oxides.[36,38] The decomposition and interaction between the ferrous and manganic ions are quantitative after 16 to 72 hr at 180 to 260°C. The unconsumed excess of the ferrous salt is determined by cerimetric titration. The solution obtained on decomposition of ferrites and Mg-Cr(VIII) spinels (including chromium ores) is used not only for determination of active oxygen, but also for that of other dissolved components, e.g., chromium, magnesium, lithium, etc.[32,43,44]

Rock-forming silicate minerals can be dissolved up to 99% in concentrated hydrochloric acid at 300°C. Even highly resistant crystal structures are decomposed, such as those of sillimanite and tourmaline. Some kinds of amphiboles, muscovites and biotite, among micas react analogously. Many plagioclases are differentiated during the decomposition. Basic members, such as anorthite and bytownite, are more readily decomposed than the more acidic albite or oligoclase.[6] Among industrially important aluminosilicates, kaolinite could be dissolved, as well as claystone, among rocks.[32] After a 2-hr heating at 180 to 200°C, all the uranium is leached from analyzed sandstones.[45] Compounds of copper, cadmium, manganese, and vanadium are liberated from river sediments by action of nitric acid in a sealed glass tube placed in an autoclave heated to 125°C. The leaching of chromium, iron, and calcium was not complete under these conditions.[46]

This technique has found an important application in the study of the nitrogen content in plutonites and sedimentary rocks.[13,47] The samples are heated with concentrated H_2SO_4 in tubes of borosilicate glass up to 420°C and in Vycor glass tubes up to 470°C. Most of the nitrogen liberated is in the form of ammonium ions. The nitrogen compounds are extracted from rocks and main rock-forming minerals within 60 to 90 min in the original form, without thermal destruction of the ammonium ions.

As boron nitride is an outstanding refractory ceramic material used in industry and space research, a great attention has been paid to its composition.[7,8,48-50] The hexagonal modification can be dissolved in acids in sealed ampules, but the cubic form is not attacked. Complete dissolution of boron nitride in freely placed ampules can be attained within 2 to 4 days using dilute (1 + 1) hydrochloric or sulfuric acid at 170°C. The decomposition time is shortened to 8 to 10 hr[49] on increasing the temperature to 300°C and the sulfuric acid concentration to 80%, but the tubes must be protected by a steel mantle. The dissolution process is accelerated by adding potassium perchlorate.[48] The sample is hydrolyzed during the decomposition,

$$BN + 3\,H_2O + H^+ \rightarrow H_3BO_3 + NH_4^+ \qquad (3)$$

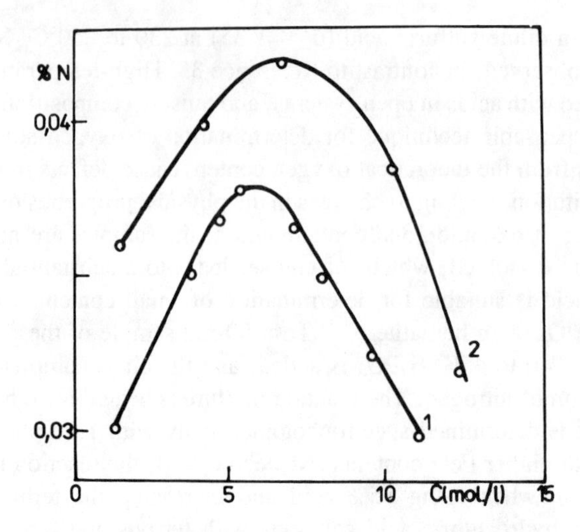

FIGURE 1. Decomposition of silicon nitride in glass sealed tubes.[53] Dissolution of silicon nitride (as % N) leached from silicon carbide. Curve (1) HCl, curve (2) H_2SO_4. Temperature 180°C, sample mass 0.05 g, leaching time 24 hr.

High-temperature pretreated aluminum nitride, AlN, mostly resists action of acids in open systems.[51] In closed tubes, 100-mg samples of the substance dissolve within 12 hr in 7 mℓ of 5 *N* HCl at 230°C, the nitrogen being quantitatively converted into an ammonium salt.[50-52] Sulfuric acid of the same concentration can also be used for the purpose. The internal pressure in the ampule, calculated from Gordon's equations[14] for the HCl-H_2O system at 230°C, amounts to 0.4 MPa.

Silicon nitride present in silicon carbide based abrasives can be leached within 24 hr by the above acids.[53] The optimal conditions for the acid extraction can be seen in Figure 1. Silicon carbide, SiC, is inert toward the acids. For total destruction of boron carbide,[54] a 0.2-g sample must be heated in a quartz tube for 72 hr with 60% perchloric acid at 300°C.

Dissolution of substances in acids in sealed tubes is the simplest technique for attainment of temperatures from 250 to 400°C during decomposition. The only other technique available for prolonged heating in this temperature range involves the use of expensive metallic autoclaves with platinum internal vessels. A combination of an autoclave and the sealed ampule technique has given rise to a new, ''hybrid'' decomposition technique. A mixture of a sample and nitric acid is placed in a quartz test tube with a quartz lid, and the test tube is inserted into a hole in a heated aluminum block located on the bottom of a metallic autoclave.[55-57] A gas is fed to the autoclave through a side inlet and maintains an overpressure of *circa* 10 MPa, which presses the lid to the orifice of the test tube and thus keeps the vessel sealed. In this arrangement the temperature can be increased up to 320°C and the necessary decomposing effect attained with minimal amounts of acids (Figure 2).

III. DECOMPOSITIONS IN AUTOCLAVES

Decompositions of substances in autoclaves containing internal vessels made of gold, platinum, or a resistant organic polymer have found use in analytical practice substantially later than the dissolution methods in sealed ampules. Using an autoclave closed system, hydrofluoric acid can also be used for dissolution of analytical samples, which is of prime importance not only for analyses of quartz and resistant silicates, but also for decompositions of some oxides, nitrides, and other substances with a high purity.

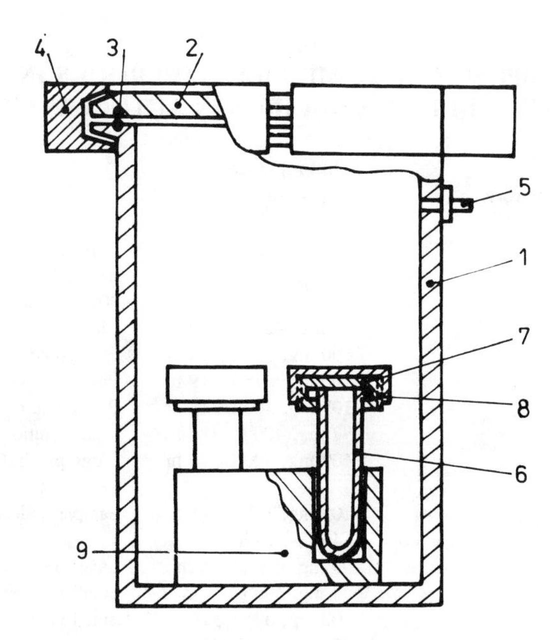

FIGURE 2. High pressure autoclave.[55-57] (1) pressure chamber, (2) lid of the pressure chamber, (3) O-ring, (4) ring retainer, (5) pressure gas inlet, (6) quartz vessel, (7) quartz lid, (8) steel screw cap, (9) heating block.

A. Decompositions in Autoclaves with Platinum or Gold Internal Vessels

Autoclaves with metallic internal vessels of various constructions have been proposed for a study of reactions in inorganic systems at high temperatures. May et al[58] employed experience thus obtained in their design of an autoclave with a platinum inlay. The vessel, with a volume of 3.5 mℓ and a wall thickness of 4.8 mm, is shaped to fit the cavity in a Nichrom alloy block and protected by a Hasteloy metallic mantle. It is tightly closed by a platinum disk pressed down by a copper plate fixed in place by a steel head. Samples of 2 to 100 mg of minerals were decomposed in the vessel by 1 to 2 mℓ of acids at temperatures of 375 to 425°C for 3.5 to 20 hr. The authors primarily studied the behavior of resistant phases that could not be decomposed by hydrofluoric acid in open systems.[59] Samples of zircon, staurolite, cyanite, chromite, chrysoberyl, and other minerals and rocks were dissolved in this acid or in its mixtures with sulfuric acid. The maximal reaction time and temperature were necessary for complete dissolution of 25 mg of separated zircon (Table 3).

The volume of the reaction vessel is too small for decompositions of larger amounts of resistant solids, and thus the authors[60] have constructed another type of autoclave with a gold internal vessel. The autoclave slowly rotates inside a box heated to 400 to 420°C. Only 5% of the total amount of finely pulverized (120 to 200 mesh) zircon remains undecomposed after a one-hour treatment with 40% HF at 400°C. On increasing the HF concentration to 71 to 75%, more than 99% of the mineral was dissolved within 15 min. These results provided a basis for decompositions in determination of traces of thorium and uranium in granitic rocks. The autoclave was modified and the volume of the internal gold vessel increased to 85 mℓ, to permit treatment of 10-g samples of the rock. The fluorides formed during the decomposition were displaced by evaporation of the solution with sulfuric acid and the residue, after dissolution of the salts in hydrochloric acid, was analyzed by the XRF method. Even under optimal conditions for total dissolution of zircon (400°C, 71 to 75% HF), Th, U, and Zr have been found, together with Al, Si, and Na, in the isolated insoluble

Table 3
DECOMPOSITION OF MINERALS OR ROCKS IN A PLATINUM-LINED CRUCIBLE IN AUTOCLAVE[59]

Mineral or rock	Decomposition procedure	Remarks
Garnet	20 mg, 425°C, 5 hr	Small precipitate[a]
Zircon	20 mg, 400°C, 3,5 hr	Incomplete decomposition[a]
	25 mg, 400°C, 4 hr	30% unattacked[a]
	25 mg, 425°C, 20 hr	Clear solution
Staurolite	100 mg, 420°C, 19 hr	Fine precipitate[a]
	100 mg, 375°C, 19 hr	Precipitate of AlF_3
Beryl	102 mg, 400°C, 18 hr	Precipitate of AlF_3
Phenacite	97 mg, 400°C, 18 hr	Clear solution
Chrysoberyl	100 mg, 400°C, 20 hr	Large precipitate, AlF_3 + $KAlF_4$ (?)
	100 mg, 1.5 mℓ HCl, 400°C, 19 hr	Sample undecomposed
Kyanite	99 mg, 400°C, 18 hr	Small precipitate
Sapphirine	28 mg, 400°C, 17 hr	Small precipitate
Diabase, W1	102 mg, 375°C, 20 hr	Small precipitate[b]
Granite, G1	27 mg, 400°C, 17 hr	Very small precipitate
Chrome Refractory—NBS-103a	50 mg, 425°C, 17 hr	Large green precipitate
	50 mg, 1.5 mℓ HCl, 425°C, 17 hr	Small gelatinous precipitate without Cr

Note: 1.5 mℓ 48% HF or 1 mℓ of the same acid[a] or 1 mℓ HF and 0.5 mℓ H_2SO_4 (1 + 1)[b] were used. Complete decomposition unless otherwise stated. The structure of the precipitate was identified by X-ray diffraction.

fraction. Complete decomposition of accessoric zircon in granitoids is, therefore, apparently more difficult than dissolution of the isolated mineral. Residues of silicon dioxide interfere and probably prevent total destruction of zircon. Part of the liberated uranium and thorium is also built into complex aluminum fluorides and cannot be liberated by evaporation with sulfuric acid or treatment with hydrofluoric acid. These difficulties have been overcome by sample pretreatment with perchloric, sulfuric, and hydrofluoric acids. Only the isolated residue, insoluble in hydrochloric acid, is subjected to pressure decomposition with 71 to 75% HF at 400°C for 30 min. The reaction mixture is removed from the autoclave, evaporated with sulfuric acid, and then dissolved in hydrochloric acid and aqua regia. The gold vessel resists the effect of fuming hydrofluoric acid better than platinum vessels.

B. Decompositions in Autoclaves with Internal Vessels Made of Organic Polymers

The reaction medium can be arbitrarily varied in dependence on the test material composition neither in glass ampules, nor in autoclaves with platinum or gold internal vessels. Strongly oxidizing acid solutions (aqua regia, acids containing chlorine or bromine, etc.) cannot be used in autoclaves with vessels made of precious metals. Hydrofluoric acid is excluded in work with quartz and glass tubes. Only after appearance of fluorinated hydrocarbon polymers in the analytical practice could efficient pressure reactors be designed, with which the selection of the reaction mixture is not limited by the danger of corrosion of the internal vessel. The behavior of polytetrafluoroethylene (PTFE, Teflon®) and its application to decompositions are discussed in Chapter 3. Here, only those properties of Teflon® that play a role in application of the material to pressure decompositions are pointed out, namely:

1. A sufficient thermal stability of the material, due to the strong C-F bonds in the polymer. Teflon® is decomposed at temperatures higher than 400°C.

2. The vessels are sufficiently clean. Most impurities are not introduced during the polymerization, but in the surface treatment of the material. The impurities can readily be leached from the surface, e.g., by pressure treatment with hydrochloric acid containing an oxidant.
3. The material resists aqueous solutions of virtually all reagents, including aqua regia, bromine, perchloric acid, hydrofluoric acid, etc., even at reaction mixture temperatures of 220 to 240°C, due to the compact structure and high bonding energy values. The surface is also inert toward redox processes.
4. Sorption of metal ions on the material from strongly acidic solutions is minimal.

The main drawback of Teflon® lined autoclaves is thermal expansion of the polymers. On a change in the temperature from 20 to 380°C, the reversible volume increase amounts to *circa*[10] 28%. Difficulties are also caused by an increased plasticity of Teflon® that is especially pronounced above 150°C and causes distortion of the vessels, with acid vapors escaping from the reaction space. It is thus recommended to use precompressed materials with lower thermal expansion. The latent porosity of the walls leads to permeability for acid vapors (e.g., for HF in prolonged decompositions). Nitrogen oxides, iodine vapors, and, to a limited extent, also mercury vapors are especially absorbed.[61] These properties of Teflon® must be considered when the apparatus is being designed and when test solids are decomposed.

The difficulties connected with liberation of arsenic from the test material and with the volatility of arsenic(III) fluoride were removed more than 30 years ago by Lounamaa,[62] who decomposed rocks by acids in a Teflon® crucible hermetically closed in a metallic mantle. The simple construction of the decomposition apparatus (''bomb'' or an autoclave with a teflon® lining) has been perfected by other authors, mainly Riley,[63] Ito,[64] Wahler,[65,66] Langmyhr et al.,[67-69] Bernas,[70] and Doležal et al.[71] The original design was simple; a teflon crucible with a lid of a suitable shape was inserted into the cavity of a metal cylinder of an autoclave. The whole system was hermetically sealed with a metallic lid with a thread or with fixing bolts. The successful designs have given rise to extensive commercial production. One of the most recent models of pressure vessel is depicted in Figure 3.

The extensive application and new possibilities connected with this decomposition technique have necessitated further modifications in the autoclave design, concerning:[72-90]

1. Construction and hermetic sealing of the internal vessel
2. Construction and the material of the autoclave metallic mantle
3. Safety in the operation of pressure vessels
4. The design of heating blocks for simultaneous decompositions of several samples
5. Problems of solution contamination, especially in analyses of high-purity substances
6. The modes of stirring the reaction mixture during the dissolution

Perfect sealing of the reaction vessel at temperatures of *circa* 150 to 240°C, even during long-term operation, is the principal condition for successful use of autoclaves, and the selection of the shapes of the Teflon® vessel and the Teflon® lid play an important role here. The crucible walls are usually thick, up to 5 mm.[88] The contact surfaces are either smooth or provided with fine, matching grooves. Sealing inserts (e.g., O-rings) are sometimes placed between the lid and the crucible.[10] The contact surface is usually horizontal, but in some designs, the surface is slanted to increase its length and improve the sealing.[75] The crucible is placed in a metallic mantle and sealed by a lid fixed with bolts. Another successful design employs a system of classical or disk metal springs in the metal head of the autoclave (e.g., apparatus made by Parr, Perkin Elmer, Berghof, etc.; see Figure 3). The pressed springs create a compensating counterpressure and prevent the volatile products from escaping during heating of the crucible contents and Teflon® contraction.

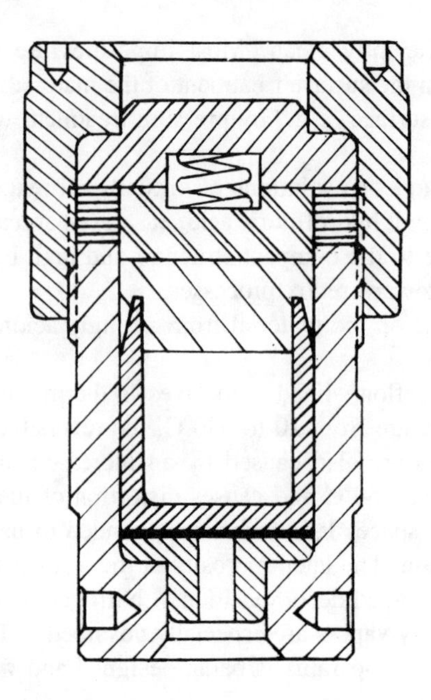

FIGURE 3. Parr's high pressure bomb. Teflon® vessel closed in the mantle made of stainless steel; for temperatures up to 285°C.

In designing the pressure vessel, the Teflon® plasticity at elevated temperatures and pressures, and its volume changes must be considered. The head of the autoclave metal mantle must provide for volume expansion of Teflon®, otherwise, the reaction vessel is irreversibly distorted and cannot be sealed properly.[10,89] An additional Teflon® plate is sometimes placed between the metallic lid and the closed vessel.[75,89] Leaks also develop due to distortion of the edges of the vessel and the lid. There are various opinions on whether the most common system (a Teflon® vessel-Teflon® lid-metal lid) is the optimal sealing method. If the shapes of the vessel and the lid are suitable, then the thermal expansion of the polymer should enhance the sealing of the reaction space.[88] However, according to Uhrberg,[86] the increased plasticity of Teflon® leads to leaks in the reaction vessel, and the author recommends the use of a thick-walled quartz test tube (1.7 to 7 mm) with a Teflon® collar. A suitably shaped Teflon® lid, placed over a quartz vessel which is inserted in a Teflon® case, is almost ideal, and its sealing effect at elevated pressures and temperatures is better than that obtained with the polymer alone. With glassy carbon internal vessels, a lid of the same material provided with a Teflon® foil and a Teflon® sealing ring is satisfactory.[91]

The Gordon equations and graphs,[14] together with Pitt's data,[92] permit an estimation of the pressures for some acids (e.g., hydrochloric and hydroidoic acids and aqua regia) at an increased temperature in a closed system. These caculations are practically important, not only for glass ampules, but also for autoclaves with Teflon® lining. The maximal permissible pressures for commercial devices are usually specified in the manual and mostly amount to units of MPa, except for the Parr type 4746 vessel (Figure 3), with which pressures higher than 30 MPa are permissible for short time intervals. However, practical data on operation at extremely high pressures are not yet available. A sudden increase in the pressure is the main cause of explosions of pressure vessels and may occur with fast chemical reactions connected with the formation of gaseous products, especially when the autoclave is uncontrollably overheated, and the vessel is overfilled with the solvent. The volume of the liquid is increased on heating, and the pressures formed may distort the vessel or even cause an explosion of the whole device. The appearance of unexpected, short-time pressure peaks,[93] caused by rapid interaction of the solvent and the test substance, is dangerous. These effects have been observed, e.g., in decompositions of biological materials and have not been accompanied by an increase in the temperature. In decompositions of inorganic substances, these pressure anomalies are less probable.[94] The behavior of perchloric acid in pressure decompositions of inorganic substances is still not unambiguously clarified. The operating instructions for commercial pressure vessels (e.g., from Perkin Elmer and Parr) explicitly prohibit the use of this acid. Our experience and other accessible data do not exclude the use of perchloric acid; no difficulty has been encountered when dissolving rocks and silicate materials in mixtures of perchloric and hydrofluoric acids with other acids. However, the well-known rules should be observed in work with perchloric acid; dilute acid should be used, e.g., in a mixture with nitric acid or other acids, under conditions when no rapid interaction may occur with reducing substances (carbonaceous compounds, sulfides, etc.).

The main purpose of the pressure springs in the autoclave head (Figure 3) is proper sealing of the system.[10] If the internal pressure increases to values that cannot be compensated for by the springs, the lid is loosened, and the volatile products escape, together with acid vapors, underneath it and may also escape out of the decomposition device through holes (or pressure valves) in the mantle. The internal vessel is then closed again. However, the pressure may increase rapidly, and the springs may not respond fast enough, so that an explosion occurs or the vessel is distorted. In some pressure vessels (Figure 3), the operational safety is improved by placing a thin metal foil under the Teflon® crucible. When high-pressure vapors escape, the foil breaks, and the gases expand through a hole in the stainless-steel mantle. At present, commercial pressure vessels permitting direct measurement of the pressure in the reaction space are also available.

The resistance of the "Teflon® bomb" proposed by Kotz et al.[74] (manufactured by the Berghof company) has been experimentally verified by Eustermann and Seifert,[95] who placed a defined amount of a mixture of lead(II) azide and tetranitropentaerythritol — a powerful explosive whose detonation effects are thoroughly known — in the Teflon® vessel contained in a metallic mantle. The results obtained cannot be transferred to the conditions of pressure decompositions without modifications, but they are valuable for improvements of the design of the reaction vessels and the autoclave metallic mantle from the point of view of the safety of operation. Decomposition of samples in a protective box proposed would substantially decrease the risk involved in the use of this progressive technique in routine analysis.[95]

The corrosive, acidic vapors diffusing through the vessel walls at increased temperature and pressure (e.g., with hydrofluoric acid during prolonged decompositions[75]) corrode the internal walls of the metallic mantle. The corrosion products penetrate to the outer surface of the vessel and cannot be leached with acids. The analyte is then often contaminated during transfer, e.g., by nickel, chromium, iron, and other metals. The sample contamination can be suppressed by a suitable modification of the decomposition device. The internal walls of the metal mantle, including the lid, are covered with a thin Teflon® layer to prevent contamination of the surface of the reaction vessel with the mantle material. For the same purpose, the vessel is inserted into a cylindrical Teflon® wrapping, whose upper and bottom parts are protected by flat Teflon® disks.[75] Reaction vessels made of borosilicate and quartz glass or of a resistant polymer are inserted into the internal Teflon® vessel.[78,84-86,88,90] A quartz test tube may directly be placed inside the autoclave and its contact with the metal autoclave wall prevented by a Teflon® ring around the test tube.[86] A quartz test tube with a quartz lid may also be placed in a heated metallic block located on the bottom of a cylindrical metallic mantle. The system is sealed by the effect of a counterpressure of an inert gas introduced through the side of the autoclave.[55-57]

Even the fluorinated hydrocarbon polymers (including Teflon®) are not completely inert toward the effects of oxidizing acids under demanding reaction conditions.[61,84] The organic substances leached complicate, e.g., differential pulse voltammetric determinations of submicrogram amounts of metals. It is then better to use glass or quartz test tubes inserted in Teflon® vessels.

Compared with Teflon,® glassy carbon vessels exhibit a low permeability for acid vapors,[91] and the material is not distorted on heating. The impurities can be leached with acids from the surface of the material. No interaction of the added radioisotopes, ^{74}As, ^{75}Se, and ^{203}Hg, with the vessel walls has been observed, even in acidic solutions. As the carbon crucible walls are not porous, a small overpressure remains in the vessel even on completion of the reaction and cooling of the liquid. The system can be sealed with a carbon lid with a Teflon® foil even at 220°C. These vessels have given good results in decompositions of biological and geological objects and have been successfully marketed.

Stainless steel is most commonly used as the material of the autoclave protective mantle, but aluminum and its alloys have also been employed.[65-69,72,73,79,89] Titanium is also suitable,[79,86,87,89] provided that hydrofluoric acid is not involved. The material of the autoclave mantle may be varied depending on the test elements. A pure copper mantle is, e.g., used for decomposition of glass sands in determination of iron.[78] On the other hand, a stainless-steel mantle suffices when aluminum and copper are to be determined in this material.

The quality and the construction of the metallic mantle also affect the possibilities of continuous stirring of the reaction mixture during decomposition, by a Teflon®-covered magnetic stirring bar, or by means of rocking of the whole autoclave.[71,76,77] The dissolution process is substantially accelerated by stirring, and the amount of sparingly soluble compounds formed at the solid-liquid interface, e.g., sparingly soluble fluorides, is decreased.

To decompose inorganic materials, aqueous solutions of inorganic acids are primarily used (especially HNO_3, HCl, HBr, HI, HF, $HClO_4$, H_2SO_4, H_3PO_4), exceptionally, also water, alkali hydroxides, and solutions of some complexing electrolytes. Among the above acids, only HF is a weak acid, but it exhibits pronounced complexing properties. Similar to open systems, the main problem in dissolution of a sample in hydrofluoric acid is rapid and complete dissolution of the reaction products[96] — insoluble complex fluorides (Chapter 4, Section IV). The rate of conversion of fluorides into soluble fluoroborates by the action of boric acid solutions depends on the composition of the solid phase formed. It is recommended to add boric acid immediately after opening the pressure vessel and to maintain the liquid volume minimal; otherwise, the results obtained for aluminum and other elements are subject to negative error.[97,98] In determination of lead in silicate rocks, the fluorides formed must be completely dissolved because the Pb^{2+} ions are built into the precipitate; similar difficulties have been encountered in analyses of ores and concentrates.[99] Rapid and complete dissolution of fluorides is attained in two-stage decomposition of the test substance. After the decomposition, the autoclave is opened, and boric acid is added. The pressure vessel is closed again and heated for 30 to 60 min at 150 to 160°C. This procedure has been successful in analyses of slags, cements, refractory bricks, and fireclays (with sample sizes from 10 to 200 mg)[100] and in determination of arsenic and tellurium in soils and rocks.[101,102] The rate of dissolution of Al^{3+} from a fluoride precipitate was studied with synthetic mixtures of silicon dioxide and aluminum oxide of known compositions.[100] In our experience, even this two-stage procedure is not a universal method for obtaining a clear solution of silicate materials with varying contents of the individual components.

Special technologies (especially the production of highly pure substances) have necessitated development of new instrumental techniques of solution analysis, leading to a great progress in the application of pressure decompositions in virtually all fields of inorganic analysis. As the commercial pressure vessels are of satisfactory quality and readily obtainable, the closed-system decompositions have become a commonly used laboratory technique.

Closed-system decomposition with hydrofluoric acid has been found exceptionally well suited for the preparation of solutions of silicon dioxide.[71,74,103-107] The starting material is pure synthetic or natural quartz, crushed to a grain size of *circa* 1 mm (Table 4). The solution

Table 4
METHODS FOR DECOMPOSITION OF SOME RESISTANT OXIDES

Oxide	Decomposition procedure	Notes	Ref.
α-Al_2O_3	0.5 g; 15 mℓ H_2SO_4 (2 + 1); 240°C; 3--4 hr, stirring	The dependencies on the temperature, H_2SO_4, and thermal pretreatment followed	71
	0.5 g; 30 mℓ HCl; 200°C; 24—30 hr	Determination of Na, K	109
γ-Al_2O_3	0.5 g; 30 mℓ HCl; 200°C; 8—12 hr		
Al_2O_3	1 g; 8 mℓ HCl; 250°C; 8 hr	Crystal modification unspecified	79
α-Al_2O_3	1 g; 10 mℓ 6 *M* HCl; 2—4 hr; 170—215°C	Gradual leaching of impurities; 30—40% of the Al_2O_3 dissolved	110
$Al(OH)_3$	1 g; 10 mℓ 2.5—7.5 *M* HCl; 205°C; 3 hr	Determination of admixtures	110
Corundum	0.3 g; 5 mℓ HF and 5 mℓ H_2SO_4 (1 + 1); 240°C; 4 hr		64
Rutile	Same as for corundum; reaction time, 16 hr		64
Synthetic rutile	0.5 g; 22 mℓ H_2SO_4; 240°C; 16 hr	Decomposition with stirring	71
Synthetic anatase	1 g; 25 mℓ H_2SO_4; 240°C; 12 hr		71
TiO_2	(a) 1 g; 2.5 mℓ 20 *M* HF; 3 mℓ HCl; 125°C; 8 hr (b) 1 g; 6 mℓ 20 *M* HF; 125°C; 4—10 hr	Technical TiO_2; determination of As; procedure (b) used after neutron irradiation; TiO_2 modification unspecified	111
ZrO_2	(a) 1 g; 15 mℓ HF; 190°C; 8 hr (b) 1 g; 10 mℓ HF, 10 mℓ H_2SO_4 (1 + 1); 240°C; 12 hr	For electrochemical stripping determination of Cu, Pb, Zn, and Cd, it is better to use HF alone	112
Baddeleyite	0.3 g; 5 mℓ HF and 5 mℓ H_2SO_4 (1 + 1); 240°C; 12 hr		64
Synthetic SnO_2	0.5—2.0 g; 240°C (a) 15 mℓ HCl; 7 hr (b) 15 mℓ HBr; 6 hr	Determination of impurities	71
Synthetic tetragonal SnO_2	0.3—0.5 g; 15 mℓ HCl (1 + 1); 235°C; 16 hr	Determination of impurities	108
Cassiterite	0.5 g; 15 mℓ HCl; 240°C; 7 hr	Determination of impurities	71
	0.1—0.3 g; 15 mℓ HCl (1 + 1); 225°C; 6—8 hr	Determination of Sn; dilute HCl(1 + 1) is more efficient than concentrated acid	113
CeO_2	(a) 1 g; 25 mℓ HNO_3; 240°C; 3 hr (b) 1—5 g; 25 mℓ HCl; 240°C; 3 hr	Determination of the purity of the preparation	71
CeO_2 ignited at 1000°C	0.5—1.0 g; 15 mℓ HCl; 160°C; 5 hr	Procedure suitable for the preparation of a stock solution of a cerous salt	114
Nb_2O_5, Ta_2O_5	1—4 g; 15 mℓ HF; 250°C; 2 hr	Preparation of a stock solution	79
	HF	Determination of Na, K; more detailed conditions not given	115
Nb_2O_5	Mixture of HF + HNO_3; 120°C; 16 hr	Stock solution preparation; sorption into SiO_2 gel	116
ThO_2 ignited at 1000°C	0.1 g; 10 mℓ HCl; 160°C; 5 hr	Preparation of a stock solution of a Th(IV) salt	114
Nuclear fuels	H_3PO_4	Details not given; a special construction of the pressure vessel	30
UO_2	HNO_3 or HCl		
(Pu, U)O_2	HNO_3 and HF or HCl and $HClO_4$,		
Ignited PuO_2	270—275°C		

obtained is used for determination of trace impurities (especially B, Fe, Cr, and Mn) that determine the suitability of the materials for further technological treatment, e.g., for production of quartz glass or pure silicon for electronics. On dissolution, silicon dioxide is converted into complex fluorosilicate. Direct analysis of such solutions prepared from large samples is difficult when using some spectrometric methods, and thus, the reaction mixture is evaporated to dryness with a mineral acid (usually $HClO_4$) in a stream of purified air or nitrogen. Preliminary removal of silicon fluorides from the reaction mixture is more advantageous for pressure decompositions of samples with impurities bound to resistant heterogeneous phases, e.g., some glass sands containing grains of tourmaline, chromite, staurolite, and rutile as the main carriers of impurities. Surprisingly, few procedures have been published[108] for dissolution of these raw materials. The best results have been obtained with a mixture of hydrofluoric acid with hydrochloric or nitric acid, aqua regia, or hydrobromic acid at 160 to 200°C, with a reaction time of 2 hr. To completely dissolve pure silicon, an oxidant must be added to a solution of hydrofluoric acid, usually hydrogen peroxide.[74]

The complexing effect of hydrofluoric acid is pronounced in pressure decompositions of resistant oxides of the elements that form stable complex compounds with fluorides. This decomposition method is used not only for the preparation of stock solutions with defined concentrations, but also in determinations of admixtures in the oxides of niobium, tantalum, and zirconium. The reaction time for dissolution of zirconium dioxide is shorter with HF alone than with its mixture with sulfuric acid. Langmyhr et al.[68] observed an analogous effect in open systems with some minerals.

The data published on decomposability of oxides are not always complete, and detailed data on the thermal history of the test substance and its crystal structure are often lacking. Both the factors have principal effects on decomposability of oxides. The corundum modification of aluminum oxide is more difficult to decompose than the other forms of the compound. Rutile is difficult to dissolve, even on prolonged reaction, whereas anatase is readily attacked by acids even in open systems. Oxides ignited at high temperatures (CeO_2, ThO_2, ZrO_2, PuO_2, Al_2O_3, etc.), especially in the form of single crystals, are highly resistant toward acids, even in pressure decompositions.

Some procedures for dissolution of aluminum oxide are listed in Table 4. In analyses of corundum ceramics, hydrochloric acid is preferred as the solvent. As chlorides do not form stable complexes with aluminum ions, a great excess of the acid is required for dissolution. According to Debizha et al., the water activity and the reaction mixture temperature play an important role in the decomposition.[106] Impurities in corundum can be quantitatively dissolved by pressure leaching with dilute hydrochloric acid, when only 30 to 40% of the initial substance are dissolved. The admixtures are not completely liberated[108] from corundum single crystals, thermally pretreated at 1700° C. Fused corundum single crystals resist hydrochloric acid,[15] even on prolonged heating at 250°C.

The conditions published for decomposition of rutile are sometimes difficult to compare.[31,71] Rutiles from some localities are not completely attacked. Synthetic compounds with the rutile structure are dissolved more easily. Cerium dioxide is reduced with hydrochloric acid, with formation of a soluble cerous salt.[114] A mixture ot nitric acid with hydrogen peroxide is less suitable for the purpose. Decomposition with HCl was useful in dissolution of resistant cerianite formed on thermal treatment of acid-leached, insoluble residues of bastnaesite-parisite ores.[114]

Minerals with the spinel structure are usually attacked with difficulties. Among these industrially important complex oxides, the greatest attention has been paid to decompositions of chromite. This mineral was dissolved, e.g., by mixtures of dilute sulfuric acid (1 + 1) with hydrofluoric acid or hydrogen peroxide at 180 to 240°C. The reaction time varied,[71,117] depending on the test material composition, from 4 to 24 hr. The mineral must be very finely pulverized (350 mesh) prior to decomposition. Analogous conditions also hold for dissolution of the spinel $MgAl_2O_4$.

Decompositions in autoclaves have now become an important part of analytical strategy for highly pure substances.[118-120] The closed reaction space and minimal volumes of pure acids required have led to a pronounced decrease in the blank values connected with dissolution of the test material. The analytes obtained after the pressure decompositions contain minimal amounts of salts and accompanying impurities, which permits full utilization of the potentialities of the present instrumental analysis.

Dissolution in a Teflon® vessel enclosed in a metallic mantle has found important application in elemental analysis of extraterrestrial materials, in isotopic geology, and especially in analytical procedures involved in geo- or cosmochronology. The methods used previously for determination of the age of geological objects could not dissolve the test solid phases without contamination from the reagents, vessels, and the laboratory atmosphere. Especially high contributions to contamination originated from fusion agents and the fusion crucible material. For example, in determining the age of zircon by the U/Pb method, fusion with borax in a platinum crucible was the main source of contamination by lead. As acid decompositions in an open system were ineffective, because of resistivity of the mineral, Krogh[75] has modified the original procedure proposed by Ito[64] for the purposes of geochronological analysis. He has improved the design of the reaction vessel and thus prevented the analyte contamination. He used 48 to 50% HF as the solvent and increased the reaction time up to 7 days. The clear solution obtained was evaporated in a laminar stream of purified air. The residue was spiked with ^{235}U and dissolved in a small volume of 3.2 *M* HCl by heating in an autoclave for 12 to 14 hr, to equilibrate the reaction mixture with the isotope added.

The rate of dissolution of zircon depends on the particle size, hydrofluoric acid concentration, and temperature. The radiation damage to the crystal structure accelerates the decomposition. It depends on the age of the mineral and the amount of isomorphous admixtures of the actinides, U(IV) and Th(IV). Metamict archaic zircons do not require fine pulverization and are largely dissolved after a few hours of heating; a clear solution is obtained after 7 days. Zircons of cretaceous age (with 50 ppm of U) are dissolved after this time, with a small amount of undecomposed residue. The fraction with the isolated mineral is divided into two parts prior to decomposition, one part being directly dissolved and solutions spiked with uranium and thorium isotopes being added to the other part. The reaction time may be extended to 21 days in exceptional cases.[121] Zircon grains may contain parts altered by external effects that are readily decomposed even in an open system. The altered parts of minerals must be removed in geochronological analyses; otherwise, the age determined is subject to error. The altered parts are leached for 1 hr with 50% hydrofluoric acid in an ultrasonic bath and cleaned and washed grains of the mineral are analyzed.[122]

The work of Krogh has successfully explained the differences observed in the decomposability of zircon.[75] It has formed the basis for numerous analytical procedures used in studies of isotopic composition of geological and cosmic objects. A closed-system decomposition has become an indispensable initial operation that affects the reliability of the analytical data in geo-and cosmochronology. To prevent contamination, this dissolution procedure is also used with less resistant minerals that could also be dissolved in open systems (e.g., eudialyte).[134,135]

Prior to the analysis, the reaction vessel is carefully cleaned, best by the same mixture of solvents as that used for decomposition of the solid phase. By repeated leaching with highly pure acids, the amounts of lead and uranium that are leached during a 4-day heating of the vessel could be decreased down to 40 to 50 and 10 to 20 pg, respectively.[133] At present, the individual elements are almost exclusively determined by mass spectrometry (MS) mostly employing the isotopic dilution method. The high sensitivity of the method makes it possible to successfully analyze even small amounts of the initial material or minerals with trace contents of the test components, maintaining a minimal contamination level. For example, individual grains of zircon weighing less than 1 mg have been analyzed. In the geochronological method based on the nuclear reaction,

$$^{176}_{71}\text{Lu} \rightarrow ^{176}_{72}\text{Hf} + e^- \qquad (4)$$

an amount of a substance containing 1 μg of Hf suffices for the analysis.[129]

In determining the age of meteorites and lunar rocks (e.g., using the $^{87}Rb/^{87}Sr$ or $^{40}K/^{40}Ar$ method, or from the ratio of the actinoids present and their decay products), lumps of the material with a weight of up to 1 g are mostly dissolved, to prevent contamination of the samples during crushing. During the decomposition and evaporation of the reaction mixture to dryness, sparingly soluble fluorides precipitate and are dissolved by repeated pressure leaching of the residues with 3 to 6 *N* HCl at 120 to 180°C for several hours.[75,125] The radioisotope checking has shown that uranium and lead are then quantitatively dissolved. If thorium is also to be determined in chondrites, the resultant solution must be perfectly clear, as any solid residue contains from 10 to 70% of the thorium present.[125]

With larger samples, hexafluorosilicic acid is removed from the solution to avoid difficulties in separating the components. Silicon is usually eliminated by evaporating the reaction mixture after dissolution of the solid phase. However, preliminary decomposition of the material, connected with displacement of silicon tetrafluoride, facilitates pressure decomposition of rocks containing resistant accessoric minerals. The major part of silicon dioxide volatilizes with excess hydrofluoric acid on heating of the mixture in an open pressure vessel. A new portion of acids added then attacks the resistant phases in the subsequent pressure decomposition. The acid is not consumed by reactions with the main rock-forming minerals. If silicon does not affect the subsequent analytical operations, it is not removed from the solution. This approach has been used not only for zircon, but also for other zirconium minerals, e.g., baddeleyite and eudialyte, for stone meteorites, and many rocks with increased contents of zirconium.[129,134] This decomposition procedure (Table 5) has been used, e.g., in isolation of hafnium using ion-exchange chromatography.

Monazite is not very much attacked by short-time heating with concentrated hydrochloric acid (180°C, 6 hr), even if it is very finely powdered.[114] However, small amounts of the isolated mineral (3 to 8 grains) are dissolved[133] on prolonged heating with 22% HCl. Monazite, intruded in granitic rocks together with zircon, is only attacked by hydrofluoric acid if the reaction time is extended to several days.[129] However, the precipitated fluorides of the rare earths and thorium must be dissolved by exaporation with perchloric acid. The rare earth elements are separated from the reaction mixture by ion-exchange chromatography. This procedure has been utilized in determining the age of granitic massifs by the $^{147}Sm/^{143}Nd$ method and in detailed study of the distribution of the neodymium isotopes in granite batholites.[140]

Autoclave dissolutions make it possible to use for dating even nuclear reactions involving elements that are generally present (e.g., the $^{40}K/^{40}Ca$ method). Any open-system decomposition would lead to an increase in the contamination level, thus preventing the use of the above method.[142]

Even pressure decompositions are not a universal dissolution method applicable without considering the phase composition of the test material. Among natural silicates, topaz is most resistant toward acids.[68] After 1 hr heating of the mineral (0.2 g) at 250°C in a mixture of HF and $HClO_4$, 65% of the solid remains undissolved. Minerals of the andalusite group are totally destroyed only after a very fine pulverization and a prolonged reaction time. They are best dissolved in a mixture of phosphoric and hydrochloric acids.[71] However, it should be emphasized that no solid has been so thoroughly studied from the point of view of the decomposition conditions as the minerals selected for dating, especially zircon. It is probable that a reaction in a suitably modified mixture for an extremely long time would lead to total dissolution of even the most resistant compounds, such as topaz and single crystals of some oxides.

Pressure decompositions have been successful in analytical practice for decomposition of

Table 5
THE USE OF PRESSURE DECOMPOSITIONS FOR ISOTOPIC ANALYSIS AND DETERMINATIONS OF ELEMENTS IN EXTRATERRESTRIAL MATERIALS

Material	Decomposition procedure	Notes	Ref.
Chondrites	Lump samples; 5 mℓ HF and 0.5 mℓ HNO_3 per gram sample; 150°C; 24—48 hr	Determination of Tl isotopes	123
Chondrites, rocks	Lump samples; 5 mℓ HF and 0.5 mℓ HNO_3 per gram sample; 180—200°C; 24—48 hr	U/Pb method of dating; checking of decomposition by ^{210}Pb; 98.5—99.2% decomposition	124
	Lump samples of up to 1 g; 5 mℓ HF and 0.5 mℓ HNO_3; 180—200°C; 24—48 hr; after evaporation, pressure decomposition again with 5 mℓ 6 *M* HCl at 120—180°C, up to 14 hr	U/Pb and Th/Pb dating methods; complete dissolution necessary mainly in determination of Th	125
Silicate slag after fusion of meteorites with Fe and Ni	HF, H_2SO_4, 12 hr	Distribution of elements between metal and silicate phases; no details given	126
Lunar and terrestrial rocks, lunar regolith	5—50 mg; 10 mℓ HF + $HClO_4$ (1 + 2); 85°C; 15 hr	Determination of trace elements; NAA after preconcentration	127
Lunar rocks from the Sea of Tranquility	Rock lumps of up to 0.15 mg; 3 mℓ HF and 1 mℓ $HClO_4$; 120—140°C	Dating by the Rb/Sr method	128
Meteorites, basalts, resistant minerals, monazite, zircon	0.5—1 g; 8 mℓ 29 *M* HF and 0.5 mℓ HNO_3; 160°C; 4 days	Dating of minerals and rocks by the $^{176}Lu/^{175}Hf$	129
Meteorites; Pt minerals	0.5 g; 15 mℓ HF and 5 mℓ H_2O_2; 170°C; 2 hr	Determination of the Pt metals	130
Lunar regolith	50 mg; 0.5 mℓ aqua regia and 5 mℓ HF; 110°C; 0.5—1.0 hr	Determination of major components, AAS	131
Meteorites, rocks	0.5—1.0 g; 17 g HF; 0.1 mℓ 12% Na_2SO_3; 180°C; 3 hr	Determination of iodine; addition of sulfite prevents sorption of I_2 on insoluble fluorides	132
Zircon	2.5—50 mg; 2—3 mℓ 48% HF and a few drops of HNO_3; 220°C; up to 7 days; HF diffusion through the teflon crucible observed	The basic work on dissolution of zircon without contamination for dating by the U/Pb method	75
	Separated fraction washed with 50% HF prior to analysis in ultrasound bath to dissolve altered and damaged parts of the mineral; 25°C; open system	Mineral pretreatment prior to pressure dissolution	122
Zircon	3—8 grains; 50% HF; 214—220°C	Dating by the U/Pb method	133
Monazite	Isolated grains; 22% HCl; 214—220°C; 4 days		
Zircon, eudialyte, baddeleyite	Dissolution according to Ref. 129	Dating by the $^{176}Lu/^{175}Hf$ method	134
Enriched fractions of perovskite and ilmenite from kimberlite	HF dissolution in a bomb	No details given	135
Rocks	0.2 g; 5 mℓ HF; 130°C; 14 hr	Pb isolation for isotopic measure; anex separation of the bromide complex	136
Rocks, minerals	0.15 g; HCl + HF + $HClO_4$ (10 + 3 + 1); 130°C; 5 hr	Determination of Rb and Sr for geochronology	137

Table 5 (continued)
THE USE OF PRESSURE DECOMPOSITIONS FOR ISOTOPIC ANALYSIS AND DETERMINATIONS OF ELEMENTS IN EXTRATERRESTRIAL MATERIALS

Material	Decomposition procedure	Notes	Ref.
Diabases, basalts, volcanic glasses	0.05—0.6 g; 10 mℓ 6 *N* HCl; 130°C; 40—50 hr	Determination of posteruptive alteration effects by the $^{87}Sr/^{86}Sr$ measurement	138
Tonalites, anorthosites, gabbro, pyroxenic granulites	0.1—0.4 g; HF with HNO_3; 150°C	$^{147}Sm/^{143}Nd$ geochronology; no details given	139
Granites with zircon and monazite	HF; 205°C; 3—7 days; residue evaporation with $HClO_4$	No details given; $^{147}Sm/^{143}Nd$ dating; isotopic characteristics of the massif for Sm, Nd, Rb, Sr	140
Amazonite, microcline	50 mg; 10 mℓ HF and 1 mℓ H_2SO_4; 180°C; several hours for dissolution	Reference analyses — determination of K for geochronological purposes	141
Amazonite, microcline	1 g; 16 mℓ HF and 4 mℓ $HClO_4$; 180°C; several hours	Determination of Ca for $^{40}K/^{40}Ca$ geochronology	142

silicate minerals that cannot be dissolved in acids (e.g., tourmaline, staurolite, some kinds of garnets, and amphiboles),[71] or which are attacked only slowly (e.g., silicotitanates and some Ni-Mg hydrosilicates), with the danger of contamination of the analyte during dissolution. These silicate minerals are attacked by hydrofluoric acid alone or, more often, in a mixture with sulfuric or perchloric acid.[64,68,71] The reaction temperature is usually maintained at 180 to 250°C for 1 to 7 hr. The crystal structure of most layered silicates is destroyed by hydrochloric, sulfuric, or phosphoric acid, or aqua regia, in an autoclave, and pure silicic acid separates. The solution obtained can, e.g., be used for determination of alkalis in industrially important raw materials, such as halloyisite, kaolinite, and talc.[143] Under certain conditions, silicic acid is quantitatively precipitated, which can be utilized for gravimetric determination even in rocks with high contents of SiO_2, such as diatomite.[144] Chrysotile asbestos is degraded by hydrochloric acid even at low temperatures. Structurally different amphibole asbestos can only be destroyed at elevated temperature and pressure.[145]

The carriers of platinum in ultrabasic rocks (containing chromite and Cu sulfides) are mostly very finely dispersed forms of natural alloys, sulfides, tellurides and arsenides. They can be dissolved[130] by repeated pressure decomposition of the rock with solutions of hydrofluoric acid containing hydrogen peroxide, followed by leaching with aqua regia (Table 5). The platinum, palladium, rhodium, and iridium contents are determined by AAS, after preconcentration on a selective ion exchanger. A mixture of hydrogen peroxide and hydrochloric acid (17 mℓ concentrated HCl, 2 mℓ H_2O, and 2 mℓ 30% H_2O_2) has yielded good results in leaching platinum deposited on aluminum oxide (reforming catalysts). A catalyst sample of up to 3.5 g is dissolved after 14 hr at 250°C.[146]

Dissolution of iron ores in autoclaves yields solutions suitable not only for the determination of the major components, but also to that of admixtures.[99,147-149] The solvent composition varies depending on the character of the raw material. The mixture always contains hydrochloric acid, usually with small amounts of hydrofluoric and nitric acids. The reaction time is from 0.5 to 2 hr at 110 to 200°C. Trace elements are liberated from deep-ocean manganese concretions by pressure heating with hydrofluoric acid and aqua regia.[150] Manganese ores with braunite, rhodochrosite, and pyrolusite are dissolved analogously.[151] If volatile components (e.g., mercury[152]) are not to be determined, then these manganese oxides are dissolved more simply in open systems.

Pressure decompositions of ores and concentrates find principal use in determinations of trace components, especially Hg, As, Si, Ge, Se, and Re. The losses of volatile compounds of these elements have been studied in detail using radioisotopes.[74,111,153,154] The sealing of the system and interactions of the ions with the vessel walls have been followed, using reaction mixtures containing hydrofluoric and nitric or nitric and perchloric acids. Of 22 studied elements, only ruthenium exhibited significant losses after a 12-hr decomposition at 140°C,[154] whereas no loss has been observed in Hg and Se, even if present in nanogram amounts. Only with iodine, the loss in the amount of the radiotracer ^{131}I added amounted to 4% and was caused by diffusion into the vessel walls.[74]

To liberate Ge and As from sulfidic ores, oxidizing mixtures are used with compositions analogous to those in open-system decompositions (HNO_3-H_2SO_4 or HNO_3-H_3PO_4). The reaction time does not exceed 1 hr at 200 to 220°C. Mercury bound in sulfides is dissolved by dilute sulfuric acid with solid potassium permanganate.[99,155] A mixture of nitric, sulfuric, and hydrofluoric acids is efficient for decomposition of copper ores and pyrites.[156,157] Losses in selenium are prevented by pressure leaching of sulfides by aqua regia.[76,158] A mixture of hydrochloric and nitric acids (180°C, 4 hr) was suitable in determinations of arsenic in tetrahedrite ores and the products of their treatment. Sulfuric or hydrofluoric acid mixed with nitric acid has been used[113,159] for the same purpose (natural antimony sulfides). Part of the silver contained in sulfidic ores may be present in a sparingly soluble form; a pressure decomposition accelerates the dissolution process.[160] Complete dissolution is also attained when using hydrofluoric and nitric acids[161] at 190°C. Rhenium contained in molybdenites can be dissolved without losses by nitric acid containing chlorate.[162] Aqua regia with hydrofluoric acid (140°C, 45 min) permits dissolution of quartz and silicates without losses of SiO_2, with simultaneous decomposition of chalcopyrite concentrates, pyrite, and galena.[163] Boric acid is added to the reaction mixture, and SiO_2 is determined by AAS. The reducing properties of hydroiodic acid make it possible to dissolve even very highly resistant minerals, such as barite and anglesite.[164] Sulfates are quantitatively reduced to sulfides within 2 to 3 hr at 210°C. By the action of concentrated hydrochloric acid on tungsten-tin concentrates, both cassiterite, wolframite, and scheelite are decomposed. The reaction mixture temperature is maintained at 220 to 240°C for 12 hr. Columbite ores and concentrates are dissolved in autoclaves by complexing action of hydrofluoric acid, to which sulfuric acid may be added.[115] The solution obtained contains fluoride complexes of niobium, tantalum, and accessoric metals. Apatite rocks with silicate admixtures from certain localities (e.g., Florida, U.S. and Kola, U.S.S.R.) are pressure-decomposed by aqua regia with formation of a residue consisting mainly of resistant silicates.[165] A mixture of hydrofluoric acid and aqua regia hastens the decomposition of these raw materials.[166] Rare earth fluorocarbonates (bastnaesite and parisite) are dissolved in open systems by heating with acids and liberate carbon dioxide and hydrofluoric acid. The decomposition is not, however, complete, even if the mixture is heated in an autoclave for 14 hr at 200°C. The insoluble residue, consisting mainly of barite, celestin, and quartz, always contains various amounts of rare earth elements (0.X to X%). It has been found using the XRF method that the Ce + La/Nd + Pr ratio is different in the isolated insoluble residue from that in the original ore.[114] The undecomposed fraction is enriched in cerium and lanthanum, whereas the praseodymium and neodymium compounds are leached to a greater extent. The dissolution process is complicated by the presence of barite and celestine that, after partial dissolution and reprecipitation, sorb a considerable part of trivalent ions of the rare earth elements. Pressure leaching of uranium ores leads to dissolution of uranium in the form of UO_2^{2+}. The most common reaction mixture contains aqua regia, hydrochloric acid, and hydrogen peroxide, or dilute sulfuric acid with potassium permanganate.[167,168]

Pressure decomposition has become an indispensable technique in analyses of substances with polyphase composition, e.g., river, lake, and sea sediments, waste sludges, coal ashes, slags, soils, and rocks. Analysts try to dissolve the material in a single operation without losses in the elements, without respect to the compositions and contents of the individual phases. Total decomposition of the material is mainly used in simultaneous determination of trace components when it is not known to which carriers the test elements are bound. The solvent composition and the reaction conditions must then be selected so that all the resistant phases present are dissolved.

In verifying new analytical procedures, the efficiency of various decomposition procedures is compared, with pressure decomposition as one of the methods tested.[149,169-173] In determining the level of contamination of soils by heavy metals, six different decomposition procedures were tested on seven different soil types.[171] Nitric acid was found to be a suitable solvent (150°C, 3 hr). The degree of leaching of arsenic from soils was followed under similar conditions for environmental protection pruposes.[101,172] The results obtained after a closed-system decomposition could readily be used, but exhibited a greater scatter than those obtained after the simpler open-system dissolution in a mixture of nitric and sulfuric acids. The efficiency of ten decomposition methods has been verified.[172] The completeness of leaching of an element from standard rocks (granite, syenite, slate) by aqua regia was compared for an open and closed system (190°C, 5 hr). The amount of arsenic dissolved after pressure leaching was substantially higher and approached the attested values. However, it has not been found whether the losses in the open system were due to incomplete decomposition or to volatilization of arsenic compounds during dissolution.[114] Pressure leaching of soils, rocks, and ashes with nitric acid or aqua regia led to only partial dissolution of the test substance. Quartz, some silicates, and resistant oxides, together with part of the organic substances present, remained in the undecomposed residue. The completeness of leaching of compounds and the elemental forms of certain elements (As, Sb, Se, Te, and Hg) with strong mineral acids from polyphase silicate matrices has not yet been unambiguously established. These elements virtually do not form mineral phases resistant toward acids, but may be enclosed as heterogeneous admixtures in grains of quartz or resistant silicates. The large ionic radius of Hg^{2+} prevents this element from entering silicate structures;[174] however, this assumption has not been unambiguously confirmed. In closed systems, the problems of volatility of the chlorides and fluorides formed during dissolution of materials in hydrofluoric and hydrochloric acid are avoided. The compounds of arsenic and selenium cannot volatilize, although they are formed in the reaction mixture during dissolution.

Lounamaa[62] has pointed out the necessity of destroying the silicate structures in determination of arsenic in eruptive rocks. Organic substances complicate leaching of arsenic and selenium by their reducing effects, even if arsenic has been quantitatively liberated from soils of certain types already after partial destruction of the material by nitric and sulfuric acid in an open system.[172] Most authors prefer pressure dissolution in aqua regia and hydrofluoric acid.[101,175] The temperature of the mixture of the acids is 100 or 160°C. The reaction time required becomes longer with increasing content of soil humus,[101] up to 8 hr. This procedure is suitable for the simultaneous determination of As, Sb, Se, Bi, Pb, Sn, and Te in coal ashes, using generation of the hydrides of the elements followed by an AAS determination.[175] With tellurium, the pressure decomposition of rocks by a mixture of aqua regia with hydrofluoric acid[102] can be replaced by an open-system decomposition.[176,177]

To liberate mercury from silicate rocks, a short-time heating (1 hr) to 140°C with a mixture of nitric and hydrochloric acids in an autoclave is sufficient.[174] The effect of temperature and the reaction time on the completeness of the liberation of nanogram amounts of mercury was studied in detail, using a mixture of nitric and sulfuric acids and standard rocks.[178] The reaction time required was 1.5 hr at 180°C. The presence of resistant cinnabar is apparently improbable in these materials, not even in submicrogram amounts. Pressure leaching with a solution of hydrofluoric acid and aqua regia (110°C, 2 hr) has given good results for sedimentary rocks and soils.[179] The analytical data obtained were compared with the values obtained after dissolution in an oxidizing mixture of nitric and hydrochloric acids or of nitric, sulfuric acid, potassium permanganate, and peroxodisulfate. The closed-system decompostion has yielded the best results, but their scatter has been greater than with the other methods tested. Teflon vessels are sometimes unsuitable for determination of mercury. As verified by testing with the ^{203}Hg isotope, mercury may penetrate into the walls.[61] The mercury vapors liberated on heating rocks in a distillation apparatus can be trapped on gold beads. Elemental mercury or its amalgam can be dissolved by heating in nitric acid at 130°C in an autoclave.[180]

A pressure decomposition has made it possible to determine silicon dioxide simultaneously with the major components in a single solution, e.g., in slags, cements, airborne ashes, and silicate rocks.[97,100,163,181-184] Most of the determinations employ spectrometry of silicomolybdate, after conversion of the fluorides into fluoroborates. A sample of 0.01 to 0.20 g of the material is dissolved in several milliliters of HF and aqua regia by a brief heating at 100 to 150°C (depending on the phase composition of the material). The scatter of the results is greater with pressure decomposition than with the analysis of the solution prepared by dissolving the melt with lithium metaborate.[170] The results obtained after fusion with sodium peroxide and after the pressure decomposition exhibit a similar precision. Aqua regia can also be replaced by nitric acid. Limestones are dissolved in a mixture of hydrochloric and hydrofluoric acids; this procedure is usually applied to the residue insoluble in hydrochloric acid. If the sample contains an organogenic admixture, it is first ignited at 950°C.

The closing of the system prevents volatile compounds of sulfur from escaping and enables their conversion into sulfates.[185,186] The test material is usually dissolved in a mixture of hydrochloric acid with aqua regia or nitric acid. Rock samples and ashes can be decomposed with acids in autoclaves without volatilization of chlorides,[187] bromides,[188] and iodides.[132] The blank value is then many times lower than with fusion or sintering of the substance, e.g., with a mixture of sodium carbonate and zinc oxide.[132] Hydrofluoric acid is used either alone, or mixed with sulfuric acid (180°C, 1 to 3 hr). In analyses of soils, ashes, cements, and rocks, either the original sample is dissolved, or the insoluble residue after acid decomposition in an open system.[169] The major and minor components, as well as some trace elements, can be determined in the solution obtained. Hydrofluoric acid mixed with hydrochloric and nitric acids is common; less often is hydrofluoric acid combined with nitric and perchloric acids.[170,175,182-184,189-194] The reaction temperatures required rarely exceed 160°C. An analogous procedure has been used in determinations of individual components, e.g., strontium or iron.[149,195,196]

In determinations of heavy metals (mainly Pb, Cu, Cd, and Zn), stripping voltammetry and differential pulse polarography are progressively more used,[197,198] in addition to AAS. The stock solution is prepared by decomposition of rock and soil samples by a mixture of hydrofluoric acid with nitric or perchloric acids (140 to 150°C, 12 hr). With ashes, the reaction temperature is increased[199] up to 230°C. Lead and cadmium contained in some kinds of soils are dissolved without complete destruction of the silicate structures by pressure leaching of the sample with nitric acid or aqua regia.[171,200]

Organic substances in soils and ashes are oxidized slowly and incompletely, even during pressure decomposition. However, soils containing up to 25% or organic matter can be completely oxidized and dissolved by a two-stage decomposition in an autoclave.[201] The organic compounds are first oxidized with 30% hydrogen peroxide with an addition of nitric acid (125°C, 4 to 8 hr). The hydrogen peroxide is evaporated off the reaction mixture, and the latter is dissolved in hydrofluoric and nitric acids at 160°C within 2 to 8 hr. A suitably modified procedure might be promising for sedimentary rocks with high contents of organic substances. It is impossible to liberate all the components bound to organic compounds in oil shales (e.g., phosphorus pentoxide), even during pressure decomposition with aqua regia and hydrofluoric acid, and the material must be ignited prior to the decomposition.[98] Certain difficulties have also been encountered in determination of chromium in coal ashes, and a pressure decomposition has yielded lower results than an analytical decomposition in an open system.[202] These difficulties have not been observed with rocks,[203] Other trace elements, e.g., Cu, Ni, Co, Ba, and V, can be determined simultaneously with chromium. A stock solution is obtained by dissolving the material in a mixture of hydrofluoric, nitric, and perchloric acids. Vanadium bound to silicate rocks is quantitatively dissolved in hydrofluoric acid;[204,205] the oxidizing medium is provided by the presence of nitric acid.[205] Coal ashes are dissolved in a mixture of hydrofluoric, sulfuric, and nitric acids. The decomposition process (150°C, 5 hr) and an extraction separation of vanadium were tested[206] using the ^{58}V isotope. A mixture of hydrofluoric and nitric acids has been successfully applied to a determination of tin in standard granite, basalt, and andesite.[207] The same solvent is capable of quantitatively liberating beryllium from a rock matrix[208] within 0.5 hr at 225°C. Perchloric acid[209] exhibits the same efficiency as nitric acid; the dissolution time is extended up to 24 hr at 110°C. A group separation of the trace elements (including the rare earths) is carried out with this material after irradiation of the sample with neutrons and dissolution. The reaction mixture contains hydrofluoric acid and fuming nitric acid to convert the radioisotopes to the maximal valence.[210]

For atmospheric pollution control, progressively greater attention is being paid to the contents of trace elements in coal. Not only principal components of ashes are determined, but also valuable trace elements (e.g., Ge, Ga, and U) and toxic elements (e.g., Hg, As, Se, Cd, and F). Most of these elements are converted into mobile forms during energy production in coal power stations and pass into the atmosphere in the form of gases or aerosols. The formation of ashes by direct combustion leads to losses in many elements that form volatile compounds (Chapter 10). To prevent losses, pressure decompositions of coal with acids have been tested. The mixture of acids must be highly oxidizing, to liberate even the elements bound to the organic matter. A mixture of hydrochloric, nitric, and hydrofluoric acids is most common.[211-214] Aqua regia has found an especially extensive use in multi-elemental analysis of coal, mostly employing the AAS and OES-ICP methods. A binary mixture of nitric and hydrofluoric acids is less effective, and coal is better oxidized with fuming nitric acid.[215] However, the metallic mantle of the autoclave may be corroded, and the analyte contaminated.[212] The dissolution is substantially accelerated on adding perchloric acid to the reaction mixture.

The oxidation of coal can be carried out in two stages.[216] The sample is first oxidized with excess nitric and perchloric acid. On completion of the oxidation of the organic matter, silicates are destroyed with hydrofluoric acid. Heavy metals and compounds of Fe, Ti, Mo, and As can sometimes be dissolved in the same oxidizing mixture with sulfuric acid added, and the elements determined by differential pulse polarography.[217] The reaction temperatures are usually from 110 to 150°C, with the reaction time from 2 to 10 hr. Various kinds of coal are decomposed at various rates. Only small samples (25 to 200 mg) of finely pulverized coal can usually be treated; samples of 0.5 to 1.0 g can be used only exceptionally.[213,216] Samples are often incompletely decomposed and contain unoxidized organic matter. Filtering off of the organic residues causes considerable losses in the test components.[212,218,219] The losses in mercury observed[220] were probably due to leaks in the system or diffusion of the mercury into the Teflon® vessel walls. In spite of partial successes, the pressure decomposition of coal with acids is still not a reliable method of obtaining an analyte containing all the elements present. Decomposition of coal ashes,[221] prepared under controlled conditions, is preferred at present (Chapter 10).

Accumulation of metals in various sediments and waste sludges has recently been systematically studied. The test environmental materials may contain the required elements, even in the form of volatile organometallic compounds. A closed-system decomposition enables oxidative destruction of organic compounds without losses in the test elements. The composition of solid phases suspended in sea water and the metal contents in atmospheric fallout are also determined in this way. The reaction conditions for decomposition of various materials are given in Table 6.

Table 6
PRESSURE DECOMPOSITION OF SEDIMENTS, ATMOSPHERIC FALLOUT, AND METALLURGICAL DUSTS

Material	Decomposition procedure	Notes	Ref.
River, sea, and estuary sediments, red mud with high SiO_2 contents	Up to 0.2 g; 4 mℓ HNO_3; 2 mℓ $HClO_4$ and 4 mℓ HF; 150°C; 5 hr	Increased amount of $HClO_4$ ensures oxidation of organic matter; determination of Pb and Cd	222
River, sea, and lake sediments	1 g; 4.5 mℓ HNO_3 and 1.5 mℓ HCl; 150°C; 3 hr	Determination of heavy metals, including Hg and Ag; addition of HCl ensures transfer of Ag to soluble Cl^- complexes; the same results using reaction mixture without HF	223
Sediments, wastes	0.5 g; 10 mℓ HF, 4 mℓ HNO_3, and 1 mℓ $HClO_4$; 140°C; 1 hr	Determination of heavy metals, including Hg; incomplete oxidation of organic matter	224
Various sediments and waste sludges	HNO_3 + HF; HNO_3 + $HClO_4$; HNO_3 + HF + $HClO_4$	No details given; determination of heavy metals	225
Sediments and soils	0.3 g; $HClO_4$ + HF + HNO_3 (2 + 1 + 1); 185°C; 4.5 hr	Determination of heavy metals; comparison with an open system	226
River sediments	Up to 2 g; 4 mℓ HNO_3; 4 hr at 140°C; no details given	NAA determination of Li	227
River sediments, coal ash	0.1 g; wetting with 0.5 mℓ 0.01 *M* HNO_3 and then addition of 1 mℓ conc HNO_3; 135°C; 12 hr	AAS determination of Sn	228
River and lake sediments	0.5 g; 3 g $K_2S_2O_8$ and 5 mℓ H_2SO_4; 135°C; 2 hr	Determination of total phosphorus; $S_2O_8^{2-}$ effectively oxidizes organic matter	229
Lake sediments	0.1 g; 4 mℓ HNO_3 and 1 mℓ $HClO_4$ and 6 mℓ HF; 145°C; 3.5 hr	Determination of 20 trace elements	230
Lake sediments, rocks	0.5 g; 6 mℓ HF and 1 mℓ aqua regia; 110°C; 2 hr	Determination of Hg; pressure decomposition yields highest results and higher standard deviation	179
Sediments	1 g; 5 mℓ HNO_3;140°C; 0.5 hr	Stripping voltam. determination of Hg	231
Sea, estuary, river, and lake sediments	0.25 g; 4 mℓ HNO_3; 2 mℓ $HClO_4$, and 4 mℓ HF; 130°C; 5 hr	Determination of trace elements	232
Waste water sediments	0.3 g; 1 mℓ HNO_3 and 0.4 mℓ 170°C; 3.5 hr	Decomposition controlled by the ^{7}Be, ^{75}Se, ^{131}I, ^{203}Hg	74, 153
Waste sludges, soils	0.5 g; possibly preignited at 450°C; leaching with 10 mℓ 6 *M* HCl; 120°C; 6 hr	Small part of test heavy metals remains in insoluble residue even after pressure decomposition	233
Waste sludges	0.2 g; 2.5 mℓ HNO_3; 105°C; 12 hr	Determination of Pb, Cd; Pb yield after pressure leaching is higher than in open system	234
Suspended particles from sea water on filter	50 mg; 0.3 mℓ HNO_3 + 0.3 mℓ $HClO_4$ + 0.1 mℓ HF; 150°C; 5 hr	AAS determination of metals	235
Suspended material from water	Membrane filter mineralized by 4 mℓ HNO_3 + 4 mℓ 30% H_2O_2; 150°C; 4 hr	Differential pulse polarography determination of traces of metals	236

Table 6 (continued)
PRESSURE DECOMPOSITION OF SEDIMENTS, ATMOSPHERIC FALLOUT, AND METALLURGICAL DUSTS

Material	Decomposition procedure	Notes	Ref.
Atmospheric aerosols	Membrane filter; mineralization by 5 mℓ HNO_3; 150°C; 1.5 hr	Determination of Hg	237
Fly metallurgical dust	0.5 g; 6 mℓ HF and 1 mℓ aqua regia; 110°C; 2 hr	Determination of Hg	238
	1 g; 15 mℓ HF and 16 mℓ HNO_3; 150°C; 4 hr	Determination of Co, Ni, Mn, Cr	239

Pressure decompositions have also given good results in the control of metallurgical treatment of Fe, Mn, Ti, Cr, and Cu ores. Both the initial materials and the products, such as chromomagnesites, various sintered materials, slags, and metallurgical dusts, are analyzed.[96,163,240-243] Dissolution of chromomagnesites in a mixture of nitric, hydrochloric, and hydrofluoric acids at 130°C or a mixture of hydrofluoric and sulfuric acids at 170°C takes 12 or 30 hr, respectively. Metallurgical slags are destroyed by hydrochloric acid or aqua regia, in the presence of hydrofluoric acid, within a reaction time from less than 1 to 12 hr, depending on the material composition.[160,240,242] The final method of determination is AAS or OES-ICP. Major components are mostly determined, including SiO_2, as well as some metals, depending on the kind of the ore treated,[241,242] e.g., Cr, Zn, V, and Sn.

The reaction conditions for dissolution of the platinum metals, some alloys, borides, nitrides, and phosphides are listed in Table 7. Powdered Ir and Rh and their alloys are dissolved analogously as described for glass ampule decompositions. Steels and alloys are first dissolved in an open vessel in acids, with formation of hydrogen. The decomposition is then completed in a pressure vessel.[244-248] The presence of hydrofluoric acid in the solvent prevents adsorption of aluminum and zirconium on the silicic acid precipitate.[244,247]

Decomposition of nitrides by acids should ensure that all the nitrogen is converted into an ammonium salt, and losses in the elemental form are prevented. Nitrides of boron, silicon, and aluminum belong among the materials of refractory, nonoxidic ceramics. The method of their thermal treatment affects their crystal structure, and, thus, also their resistance toward molten metals and inorganic acids. Various methods of thermal pretreatment are probably the main cause of the discrepancies in the data on decomposability of boron nitride, BN. Small samples of this material can be dissolved by heating with a mixture of hydrofluoric acid and hydrogen peroxide in an autoclave[251] (Table 7); AlN and Si_3N_4 behave analogously. The latter compound is highly resistant toward hydrofluoric acid alone, but even lumps of the material can be dissolved in a mixture of hydrofluoric and nitric acids.[252] The interaction of vanadium nitride, VN, with hydrofluoric acid is also slow and is accelerated, similar to boron nitride, by addition of hydrogen peroxide. The hydrated vanadium(III) fluoride formed is oxidized to the complex oxyfluoride

$$VF_3 \cdot 3\ H_2O + n\ HF \rightarrow (V^{IV}OF_{3+n})H_{n+1} + 2\ H_2O + 1/2\ H_2 \quad (5)$$

However, the conversion of vanadium nitride into an ammonium salt is not quantitative and part of the nitrogen escapes in the elemental form.[255]

Pressure decomposition of solids, in which polyvalent elements are dissolved without a change in the valence, is especially demanding as far as the maintenance of the reaction conditions is concerned. This approach is applied in analytical practice primarily to determinations of ferrous oxide in resistant, natural, and synthetic oxides with the spinel structure and in some silicates (e.g., tourmaline, staurolite, garnets, and amphiboles). Direct titration

Table 7
PRESSURE DECOMPOSITIONS OF METALS, ALLOYS, NITRIDES, BORIDES, AND PHOSPHIDES

Material	Decomposition procedure	Notes	Ref.
Steel, pure metals—Cu, Zr, Nb, Ta	8—350 mg; 120—140°C; 0.5—12 hr HCl + $HClO_4$ (10 + 1), HF + $HClO_4$ (10 + 1), HF + 30% H_2O_2 (10 + 1)	Determination of nitrogen in pure metals; Zr decomposition within 0.5 hr; total acid volume 0.5—4.0 mℓ	249
Ferrosilicon	0.2 g; 5 mℓ H_2O and 5 mℓ HF, then 1 mℓ conc HNO_3; 110°C; 0.5 hr	Determination of impurities	250
Ferrosilicon, ferromanganese, ferrotitanium	0.2—0.5 g; 3 mℓ aqua regia and 3 mℓ HF	Determination of Si, Al	163, 240
Boron nitride BN	30 mg; 4 mℓ 5 *M* H_2SO_4; 240°C; 6 hr	Determination of N	71
Nitrides BN, AlN, Si_3N_4	0.1 g BN, 0.04 g Si_3N_4, or 0.16 g AlN; 3—5 mℓ HF and 0.5 mℓ H_2O_2 or H_2SO_4; 120°C; 18—24 hr	Time dependence of the BN dissolution followed; determination of N	251
Si nitrides, including Si_3N_4	Lumps of up to 0.6 g; 3 mℓ HF and 1 mℓ HF; 150°C; 14 hr	Dissolution of lump samples	252
Si nitrides	60 mg; 3 mℓ HF + HCl (1 + 1); 150°C; 12 hr	Determination of major components and impurities	253
Separation of SiC, Si_3N_4, SiO_2, Si	Gradual decomposition by HF + HCl and HF + HNO_3	Gradual dissolution	254
Vanadium nitrides VN_{1-x}	0.15 g; 10 mℓ HF + 2 mℓ 30% H_2O_2; 150°C; 6—8 hr	Nitrides of various compositions cannot be used to determine N; incomplete formation of NH_4	255
Rh powder	1 g; 10 mℓ HCl + 1 mℓ $HClO_4$; 240°C; 10 hr		71
Rh and In powders, Pt-Rh and Pt-Ir alloys	0.25—0.5 g; 16.5 mℓ HCl + 1 mℓ HNO_3; 250°C; 24 hr	0.25-g sample for Ir	243
Steel	0.1—0.5 g; 20 mℓ aqua regia; 200°C; 12 hr	Determination of Zr; dissolution first in open system	244
	1 g; 2 mℓ HNO_3; 4 mℓ HCl; 2 mℓ HF + H_2O; 180°C; 1 hr	Determination of impurities, including Al, Si, B; DCP-AES	245
Special steels containing Pt, Pd, Ru, Cr	0.25 g; 4 mℓ HCl + 1 mℓ HNO_3; 150°C; 14 hr	XRF spectrometry in solution; 30 min standing after acid addition	246
Steel	1 g; 200°C; preliminary dissolution in HCl, HNO_3; subsequent decomposition (a) 3 mℓ HCl; 2.5 hr (b) 3 mℓ HCl + 4 mℓ HF; 2 hr	Decomposition (b) yields better results for determination of Al	247
Multicomposition alloys Fe-Si-Ba-Ca-Al	0.2 g; 1 mℓ HNO_3 + 5 mℓ HF; 1 hr; 140°C		248
Phosphides (arsenides) of various metals	2—60 mg; various mixtures of HNO_3; HNO_3 + HCl; HNO_3 + HCl + HF; HNO_3 + HF + H_3PO_4; acid volume 0.7—1.5 mℓ; 90—140°C; 2—16 hr	Decomposition of RuP_3 by HCl + HNO_3 takes up to 16 hr without losses in phosphan (arsane)	88

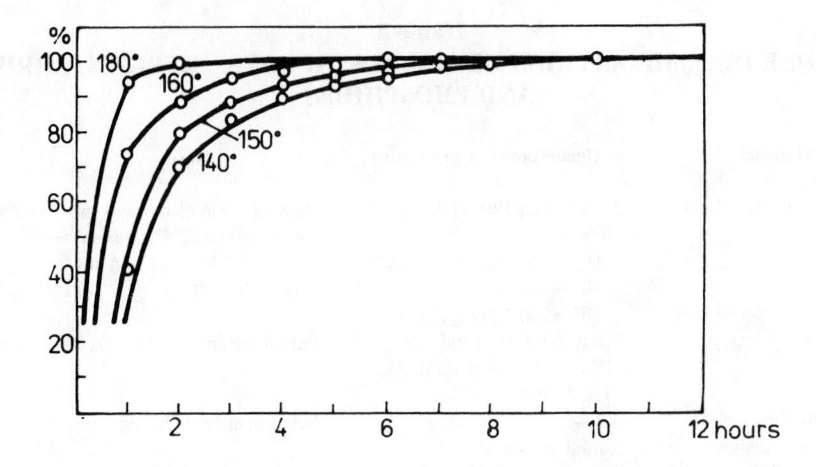

FIGURE 4. Time dependence of schorl decomposition for various temperatures.[258] The solution contains hydrofluoric and sulfuric acids with excess ammonium metavanadate.

of ferrous ions with potassium permanganate after the sample decomposition has given poor results. At high temperatures and long reaction times (250°C, 2 to 4 hr), the ferrous ions are almost completely oxidized in solutions of hydrofluoric and sulfuric acids. Negative errors arise from difficulties in the maintenance of an inert atmosphere during the decomposition and from interactions with other sample components in the presence of fluorides. This dissolution procedure has been successfully applied to analyses of ferrites and rocks on semimicroscale, but is unsuitable for the obtaining of reliable data on the FeO content in routine practice.[256,257] The results are subject to considerable negative error, especially after prolonged dissolution. More reliable results are obtained with decomposition in a closed system, in the presence of a known excess of an oxidant. As the oxidant must be stable at temperatures around 200°C, the use of ceric sulfate, potassium permanganate, and bichromate is excluded, because they are partially reduced on heating with hydrofluoric and sulfuric acid in a closed teflon crucible. The stability of the reagents decreases in the order, potassium bichromate — ceric sulfate — potassium permanganate.[114] The instability is probably caused by interaction with the surface of the Teflon® vessel. Under the given conditions (15 to 30% wt H_2SO_4, 200°C), the most stable oxidant is vanadate, similar to open systems. Less than 1% of ammonium vanadate added is reduced during a 6-hr heating.

The presence of hydrofluoric acid does not affect the blank magnitude and reproducibility. Addition of an oxidant to sulfuric and hydrofluoric acids accelerates destruction of the solid. The procedure is simple;[258] 0.25-g sample is wetted with a few milliliters of water, 10 mℓ 0.1 *M* NH_4VO_3 in 5.5 *M* H_2SO_4, 5 mℓ 38% HF are added, and the Teflon® crucible with the mixture is heated in an autoclave for 5 hr at 180 to 200°C. The reaction mixture is allowed to cool and is transferred to a beaker with an excess of a saturated solution of boric acid. The excess oxidant is back-titrated with a ferrous salt. An increase in the temperature to 200°C substantially decreases the reaction time for tourmaline and almandine. The reaction time must be extended to 16 to 20 hr for chromite. The dependence of the dissolution time on temperature is given in Figures 4 to 6 for various mineral phases. The vanadate stability decreases when the sulfuric acid concentration is increased above 30% wt.[114] Less resistant minerals and rocks can be dissolved even at 60°C after 18 hr.[259]

The atmospheric oxygen trapped in the pores of the Teflon® vessel is liberated during the pressure decomposition and may oxidize the components dissolved in the reaction mixture. This phenomenon has been observed during determination of vanadium(III) oxide in the vanadium mica roscoelite.[260] Therefore, the Teflon® vessel is heated for 3 days in an evacuated oven at 100°C prior to the analysis. The mineral is dissolved in hydrofluoric and

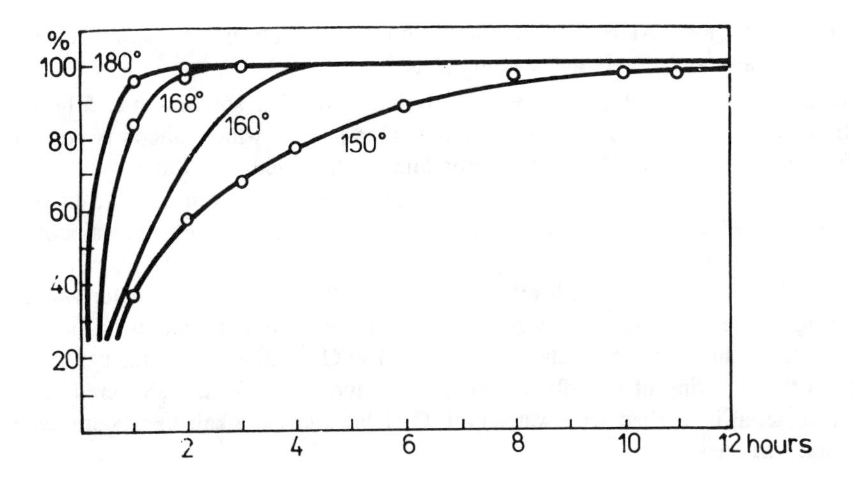

FIGURE 5. Time dependence of staurolite decomposition for various temperatures.[258] The solution contains hydrofluoric and sulfuric acids with excess ammonium metavanadate.

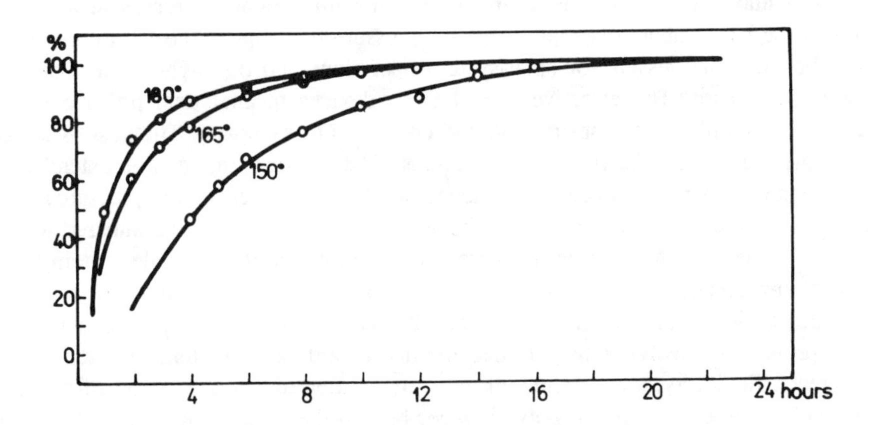

FIGURE 6. Time dependence of chromite decomposition for various temperatures.[258] The solution contains hydrofluoric and sulfuric acids with excess ammonium metavanadate.

sulfuric acids in an autoclave heated for 3 hr at the same temperature. The dissolved vanadium(III) ions are determined spectrophotometrically, using the reaction with thiocyanate. The formation of fluoride complexes prevents interaction between vanadium(III) and ferric compounds. All the operations must be performed in a nitrogen atmosphere. Hydrochloric acid cannot be used for this purpose, as a 5 M solution of the acid, mixed with hydrofluoric acid, reduces part of the vanadium(IV) compounds to trivalent vanadium salts. A similar reduction has been observed during a determination of impurities in tin dioxide; after sample dissolution (0.5 g; 15 mℓ HCl 1 + 1; 240°C; 16 hr), the solution contained 30 to 70% of stannous ions. The causes of this phenomenon have not yet been unambuguously explained; probably selective effects of Teflon® are responsible for it.[108]

Reagents other than acids are rarely employed in pressure decompositions. Interesting results have been obtained in pressure leaching of basic and ultrabasic rocks, minerals, and iron ores by lithium hydroxide. A sample of 2.5 to 3.0 g of a rock is mixed with 35 g of 4.5 M $LiOH{\cdot}H_2O$ and 4 mℓ of 1.6 M HNO_3. The reaction mixture is leached 60 to 70 min at 180°C, with continuous stirring. The Cs^+, Rb^+, K^+, and Na^+ ions are quantitatively transferred to the aqueous solution, whereas 30 to 50% of Li^+ remain in the solid phase in the form of sparingly soluble metasilicates and aluminates. The aqueous phase contains a mere 1% of the total Al content. Resistant minerals, such as tourmaline, cyanite, and chromite, are attacked slowly and incompletely.[261]

Silicon dioxide is dissolved in a 30% potassium hydroxide solution after 5 hr at 240°C. A defined solution of the alkali silicate is thus obtained, in which silicon is present in the reactive monomeric form. Synthetic corundum is dissolved under the same conditions. However, silicon is not completely leached from halloyisite, even on a 15-hr heating at 240°C with a 30% sodium hydroxide solution.[143] On dissolution of elemental silicon in 0.5% NaOH using an autoclave of a special design, the boron present is liberated from the sample.[262] Pressure leaching with dilute NaOH (pH 10.2) or with tap water was used for the testing of the resistance of the granite walls of underground containers for hot water; lumps of the material were leached 1 week at 150°C.[263] Similar testing has given good results in the checking of hydrolytic resistance of industrial glasses; shattered or coarsely crushed material is leached with water (110°C, 1 hr) and the alkali metals are determined in the aqueous phase.[264]

IV. DECOMPOSITIONS IN CLOSED VESSELS WITHOUT A METALLIC MANTLE

Vessels without a metalic mantle and provided with a tight lid represent a simplified version of the basic laboratory equipment for pressure decompositions. They are an intermediate between the vessels for open-system operations and the high-pressure autoclaves with a metallic case. The effective use of these devices in analytical practice is mainly caused by the simplicity of construction and consequent low prices. The vessels are easily handled and are suitable for routine applications. The reaction time can be extended (e.g., overnight), and no control is necessary. They retain the advantages of high-pressure autoclave dissolution to a considerable extent; the efficiency of acids is increased compared with open systems, and their amount can be decreased. Contamination of the whole system from the laboratory atmosphere is prevented, and losses of substances in the form of volatile compounds are eliminated or strongly suppressed. The vessels are made of plastics with suitable physical properties, involving the polymer stability toward increased temperatures and chemical reagents, flexibility, tensile strength, etc., most often of Teflon®, polysulfon, polycarbonate, polypropylene, or linear polyethylene, best in the transparent or translucent form, and exceptionally also of chemical or quartz glass. The vessels are shaped as bottles, conical flasks, test tubes, or crucibles and must be provided with a tight lid.

The sealing of the system is the principal problem of the construction of the whole system. The internal pressure fluctuates, depending on the reagents used, and the working temperature and usually amounts to tenths or, exceptionally, units of MPa. Plastic vessels are usually provided with a screw-on lid, and the tightness is improved by a Teflon® tape placed outside or by a special extension;[265] the vessel can also be pressed between two metal plates.[266] Glass vessels can be sealed with a silicone rubber lid coated with a Teflon® foil fixed in place by metal springs.[267-269] Quartz lids on quartz test tubes, fixed with these springs,[270] are tight at overpressures of up to 0.4 MPa. The overpressure may also be compensated when the plastic vessels are placed in a pressure cooker.[271,272] The maximal permissible temperature depends not only on the material quality and the lid tightness, but also on the technical level of the whole device. The reaction vessels are usually heated in a boiling water bath or are placed in an oven with controllable temperature. Closed polypropylene bottles (12 vessels with a volume of 250 mℓ) containing the samples and acids can be fixed between two metal plates of a holder and placed in a boiling water bath.[266] For geochemical prospecting analyses, an aluminum block with a controlled temperature has been found useful; up to 42 samples can be decomposed simultaneously. The upper part of the hot plate is thermally isolated and provided with a metallic extension containing cooling water inlet and outlet.[268,269] The lid is thus protected against overheating, and a temperature gradient is formed inside the test tubes, which leads to reflux circulation of the acids. Under these conditions, losses of volatile compounds of arsenic, antimony, and selenium during decomposition are prevented, and the temperature of the reaction mixture in the bottom part of the vessel may be increased up to 150°C. The simplest design is represented by a polyethylene vessel covered with a polyethylene foil fixed with a rubber ring;[273] the working temperature does not exceed 90°C. There is a certain confusion in the literature describing the decomposition devices. The term "Teflon®-bomb" is used both for metallic autoclaves with Teflon® crucibles inserted and for closed teflon vessels without a metallic mantle.

The closed plastic vessels are most often employed for decompositions of rocks and minerals. Both major components (including silicon dioxide) and trace elements are determined, usually combining various spectrometric methods.[274-276] Feldspar and olivine can be decomposed within several minutes[277] in a mixture of acids (0.1-g sample, 5 mℓ HF and 0.2 mℓ of aqua regia) in a polypropylene bottle with a screw-on cap, heated on a water bath. Quartz and biotite are also readily dissolved. Other rock-forming minterals, e.g., some garnets and pyroxenes, are dissolved within 1 hr. The fluorides formed are dissolved on addition of boric acid. Andalusite, silimanite, and tourmaline are decomposed very slowly (5 hr), even if very finely pulverized samples are used. Cyanite and staurolite cannot be decomposed even after heating for 30 hr. The amount of the undecomposed solid phase decreases linearly with the logarithm of the reaction time. The behavior of the individual minerals also determines the rate of dissolution of the corresponding rocks. The reaction time can be substantially decreased when the polypropylene bottles with the same mixture of acids are placed in a pressure cooker. A disadvantage of polypropylene and polycarbonate vessels is their poor stability toward strong oxidants. A mixture of hydrofluoric and sulfuric acid with hydrogen peroxide attacks polypropylene faster than the same mixture with hydrochloric acid replacing sulfuric acid. Aqua regia also exerts corrosive effects.

The decomposition of rocks for determination of tin[278] was checked using the ^{118m}Sn isotope. The causes of the losses in tin, observed during evaporation of the reaction mixture containing hydrofluoric and sulfuric acids, have not been satisfactorily explained and have been ascribed to volatile reaction products of unknown composition. Therefore, closed polypropylene vessels have been recommended for the decomposition. The test rocks have been decomposed by mixtures of hydrofluoric and hydrochloric acids. Carbonaceous substances from soils and sedimentary rocks have been oxidized by hydrogen peroxide. Cassiterite, which is commonly present in granitoids, remains unattacked.

Similar to autoclave decompositions, this dissolution procedure has been utilized for determining the oxidation states of iron in silicate rocks.[279-281] The Teflon® decomposition vessel may be provided with sealable inlet and outlet for an inert gas. After displacement of oxygen from the reaction space, the openings are closed, and the Teflon® vessel is heated in a water bath.[282] A mixture of hydrofluoric and sulfuric acid is mostly used as the solvent. After the decomposition, Fe(II), Fe(III), and other components can be determined spectrophotometrically in a single solution.

Direct dissolution of lumps of rocks in a mixture of nitric and hydrofluoric acid in a closed Teflon® crucible heated in a water bath can also be considered as a decomposition at a mildly elevated temperature. Destruction of silicates in closed Teflon® vessels at laboratory temperature (25°C) requires long reaction times, but the test substance is dissolved, even in minimal amounts of acids, without contamination of the reaction mixture. It is possible to determine the major components including silicon dioxide, as well as the heavy metals and other trace elements.[283-287] Hydrofluoric acid with aqua regia is mostly used for the purpose. The reaction time varies from 16 to 36 hr, but can be decreased to 15 min[288] when the sample weight is decreased to 50 mg. The recoveries of the dissolved compounds of chromium and zirconium amount to only 83 and 53% of the total content, respectively. The recovery of zirconium increases with increasing temperature, but all the phases containing

Table 8
DECOMPOSITIONS IN CLOSED VESSELS WITHOUT EXTERNAL METALLIC MANTLE

Material	Procedure	Vessel	Notes	Ref.
Coal ash, cement, refractory bricks	50 mg; 2.5 mℓ HF, possibly with HCl, HNO_3; 105—110°C; 30 min or longer	Thick-walled Teflon® 100 mℓ vessel	Results for Al lower than with reference methods	289
Glass sand, feldspar, nephelinic syenite	0.5 g; 5 mℓ H_2SO_4 (1 + 1) + 5 mℓ HCl + 10 mℓ HF; 250°C; 1 hr; evaporation to SO_3 vapors	Thick-walled 35 mℓ Teflon® vessel with screw-on cap	Determination of Fe; unsuitable for sands with resistant minerals; HCl necessary for magnetite dissolution	290
Glass sand	0.2 g; 5 mℓ HF; 110°C; 15 min	Polycarbonate conical flask	Determination of Fe, Ti, and other impurities, unsuitable for sands with resistance minerals	291
Fused quartz	Cleaned lumps up to 5 g; 30 mℓ HF + 0.5 mℓ H_2SO_4 + 1 mℓ HNO_3; 120°C; 48 hr	Teflon® 250-mℓ bottles with screw-on caps	Determination of impurities	292
Pure silicon, trichlorosilan	1 mℓ HF + 1 mℓ HNO_3 + 1 mℓ 0.1% mannitol; cooling in liquid N_2, addition of 0.1 g Si; heating at 100°C; after dissolution, evaporation to dryness	Closed Teflon® crucible	Determination of boron	293
Glass, silicates	0.3 g; 2 mℓ HF + 0.5 mℓ 30% H_2O_2; 48 hr	Polypropylene bottle	Determination of B	103
Glass	1 g; 40 mℓ HF + H_2SO_4 (20 + 1); 25°C; 30 min with stirring	Polypropylene vessel	Determination of B, for easily soluble glasses	294
Bauxite	0.5 g; 10 mℓ 13.5% NaOH; 120°C	Polypropylene bottles	Procedure for leachable Al_2O_3	295
Sea sediments	0.5 g; 3mℓ HNO_3 + 1 mℓ HClO4 + 3 mℓ HF in boiling water bath; 1—2 hr	Teflon® vessels according to Ref. 297	Resistant minerals are not decomposable; for Ti, the yield is maximum 80%	296
Sea sediments, silicate rocks	0.1—1 g; 1 mℓ aqua regia + 6 mℓ HF; boiling water bath; 1 hr	Specially constructed Teflon® vessel	Major components and trace elements	297
Substances suspended in sea water	2 mg material on membrane filter; 0.75 mℓ HCl, water bath, 30 min cooling, 0.25mℓ HNO_3 and 30 min again; cooling, 0.05 mℓ HF and 60 min	Teflon® vessel with lucite closing	Major and minor components, also Si	265

Table 8 (continued)
DECOMPOSITIONS IN CLOSED VESSELS WITHOUT EXTERNAL METALLIC MANTLE

Material	Procedure	Vessel	Notes	Ref.
Stone meteorites	1 g after neutron irradiation; 7 mℓ HNO_3 + 3 mℓ $HClO_4$; 250°C	Teflon® crucible with loaded lid	Determination of rare earths, control with ^{144}Ce	298
Geochemical samples	5 g; 10 mℓ $HClO_4$ overnight at 150°C; then 20 mℓ HCl + 2 mℓ HF; 85°C overnight	Teflon® crucible with screw-on closing	Geochemical prospecting; decomposition in separate stages	299
Soils, sediments, silicate rocks, minerals	0.25 g; 5 mℓ HCl; 150°C; 2 hr	Borosilicate glass test tubes closed with silicone rubber and Teflon® foil	Determination of As, Sb, Bi; organic matter worsens leaching of Bi; pyrite and marcazite are undecomposed	267
Geochemical samples, soils containing sulfides	0.25 g; 1 mℓ HNO_3 + HCl(4 + 1); closed only after gas liberation; 175°C in reaction space; 1 hr	Glass test tubes with closing according to Ref. 267; closing is cooled	Suitable for Hg, Se, Fe, Bi, Sb, and heavy metals	268
Soils, black slates, silicate rocks	0.25 g; 5 mℓ 6 *M* HCl + 0.1 mℓ Br_2; 2—3.5 hr	Vessels according to Ref. 267 and 268	Determination of Mo, leaching complete only in the presence of Br_2; for geochemical prospecting	269
Soils, sediments, atmospheric dust	(a) 0.6 g; 3 mℓ HNO_3 + 1 mℓ $HClO_4$; allowed to stand 2 hr, then 30 min in pressure cooker; cooling; 10 mℓ HF and the procedure repeated (b) 0.2 g; 2 mℓ HNO_3 + 0.5 mℓ $HClO_4$ + 4 mℓ HF for 1 hr; 1 hr heating in pressure cooker (c) 20 mg of particles; 2 mℓ HNO_3 + 0.5 mℓ $HClO_4$; 30 min heating in pressure cooker	Six Teflon® closed vessels in pressure cooker	Trace elements, heating in pressure cooker shortens the reaction time to one half	271
River sediments, volcanic ash, refractory bricks	1 g; 25 mℓ 6 *M* HCl; 80°C; 1 hr	Polyethylene bottles	ICP-AES + major and trace elements	300
Soils	20 g; 40 mℓ H_2O + 0.5 mℓ 10% $BaCl_2$; 7 min in boiling water bath	Closed plastic vessel	Leaching of B from soils	301
Steel slags, slags from Cu ore treatment	0.2 g; 5 mℓ HF + HCl(7 + 3) in pressure cooker	Polycarbonate vessels with polypropylene screw-on caps	Major components; Zn, Pb, As, Sb also determined	270

Table 8 (continued)
DECOMPOSITIONS IN CLOSED VESSELS WITHOUT EXTERNAL METALLIC MANTLE

Material	Procedure	Vessel	Notes	Ref.
Slags, matte, claystones, phosphate rocks	0.5 g; 5 mℓ HF + HCl(7 + 3) + 2 mℓ HNO_3; 15 min, in boiling water bath	Simultaneous decomposition of 12 samples in 250 mℓ polycarbonate bottles with polypropylene caps on a holder in water bath	Also suitable for elements forming volatile compounds; NO_x escapes through the walls	266
Cu treatment slags	0.2 g; 1 mℓ aqua regia; 110°C; 40 min	125 mℓ polypropylene vessels	Comparison with autoclave decomposition; easier dissolution of fluorides	96
Mn nodules	0.5 g; 5 mℓ HF + HCl(3 + 7) + 2 mℓ HNO_3; 30 min in water bath; 100 mℓ 1.5% H_3BO_3 added and heating repeated	Polycarbonate vessels	Determination of rare earths; dissolution of fluoride precipitates in closed system	302
Nb, Ta minerals	0.5 g; 25 mℓ HCl + 20 mℓ HF on a boiling water bath	Polyethylene beaker closed with polyethylene foil	Ion-exchange separation	273
Orthoclase, biotite, amphibole	0.5 g; 10 mℓ HF + 1.5 mℓ HNO_3; 3 hr in a boiling water bath	Polyethylene vessel	Refer to analysis for K_2O; dating by the $^{40}K/^{40}Ar$ method	303
Fluorite raw material, bauxite, barite	0.1 g; 3 g HF; 80°C to decompose; possibly mixture of HF and $HClO_4$	Polyethylene bottle	Spectrophot. determination of SiO_2	304
Glass, sand	0.1 g; HF + H_2SO_4; 65°C	Polyethylene tube	Determination of SiO_2 after K_2SiF_6 precipitation	305
Glass	1 g; 12 mℓ HF + 5 mℓ HCl + HNO_3 (1 + 1); 70°C; 30 min	Polyethylene vessel	Determination of Se; fluorides dissolved by heating with H_3BO_3	306
Silicate atmospheric fallout, soils, river and lake sediments	1 g; 1 mℓ HCl + HNO_3 + 10 mℓ HF; 90°C; 4 hr	Polypropylene vessel	AAS determination of Cd, Cr, Cu, Pb; Cr is contained in the carbonaceous residue and undissolved	307

influencing the reliability of the results. Autoclave decompositions with acids rarely permit determinations of elements at concentrations lower than 10^{-4}%. The amounts of impurities contined in the acids used or leached from the surface of the pressure vessels often exceed many times those in the test materials. Vapors of acids in a closed system introduce minimal amounts of impurities into the reaction mixture, as they always contain less impurities than the corresponding solutions. The temperature and vapor pressure may be considerably increased in a closed system. The vapors become strongly corrosive and readily attack the structure of solids. The decomposition apparatus mostly consists of a cylindrical Teflon® vessel closed with a lid. The acid is placed on the bottom of the vessel which is heated; the

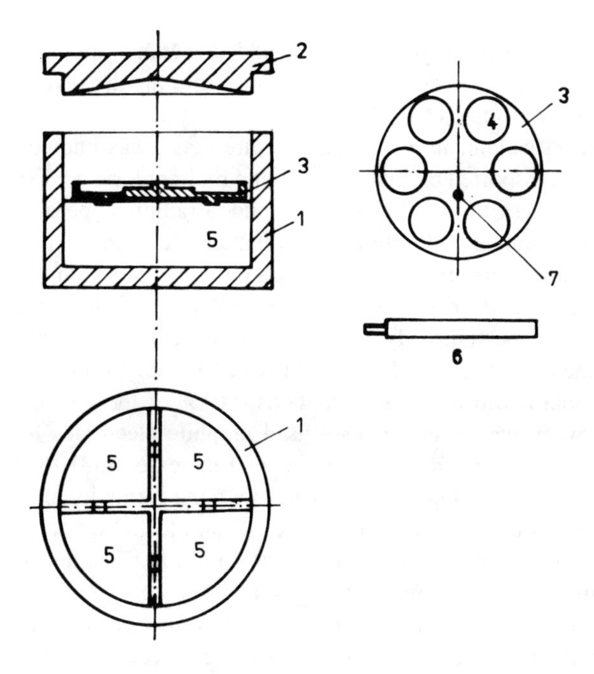

FIGURE 7. A vessel for decomposition in acid vapors.[309] (1) Teflon vessel, (2) lid, (3) sample holder, (4) sample dishes, (5) space for storage of acids, (6) threaded rod placed in the hole, (7) used for removal of the sample holder. All the parts made of Teflon® are bolted inside an aluminium mantle.

this element are not dissolved even at 130°C.[286,287] The decomposition methods for technical products, some sediments, soils, slags, glass, and minerals are summarized in Table 8.

V. DECOMPOSITIONS WITH ACID VAPORS

Decomposition with acid vapors meets even the demanding conditions of determinations of trace impurities in highly pure substances. The amounts of the impurities are often in the nanogram range, where the magnitude and reproducibility of the blank is the decisive factor vapors formed dissolve the samples placed in Teflon® dishes located above the acid level in a suitable Teflon® holder. A vessel closed with a tight screw-on cap suffices for temperatures of up to 110 to 120°C. At higher temperatures the whole apparatus is closed in a metallic mantle. Vessels suitable for this temperature interval have been designed by Wooley,[308] and one of the modified designs is depicted in Figure 7.[309]

Vapors of volatile inorganic acids are employed most often and involve those of hydrofluoric, hydrochloric, hydrobromic, and nitric acids. The decomposition conditions must be selected so that the compounds of the test trace element, contained as impurities in the acids used, do not volatilize. Volatile compounds (e.g., $AsCl_3$ from HCl) are transported with the acid vapors and may contaminate the samples. Volatile compounds may also be formed during interaction of the acid vapors with the sample and lead to losses of the test elements, as well as to cross-contamination among the samples that are dissolved simultaneously. The contamination of the test substances has been verified experimentally in determination of traces of arsenic in coal ashes and silicate rocks.[310] During the dissolution, the samples were contaminated by volatile As(III) compounds — probably arsenic trifluoride from the acid mixture used. This compound is only slowly oxidized in solutions of hydrofluoric and nitric acids and solid potassium permanganate must be added to attain rapid oxidation to nonvolatile As(V) form. No loss of arsenic due to volatilization from the dissolved coal ashes has been observed under these conditions.

The migration of trace elements within the pressure vessel was studied using radioisotopes. The fraction of ^{59}Fe transported by hydrofluoric and hydrochloric acid vapors at 190°C from a test glass sand was less than 0.003% of the added radioisotope amount.[309] Even if the hydrofluoric acid was spiked with a ferric salt to attain a content of 10^{-2}%, the condensate collected was more than four orders of magnitude purer.[106] In decompositions of silicon and quartz, the blank can be decreased down to 3 to 8 ng/g of the sample.[311] If solids are decomposed by acid vapors directly in the craters of graphite electrodes,[312-315] then the increase in the contents of Fe, Al, Mg, Mn, Ni, Co, Cu, and Cr amounts to less than 10^{-10} g, even if the original hydrofluoric acid contains 0.01% of these elements. No transfer of trace elements between the simultaneously used graphite electrodes has been observed at 220°C, even when the impurity contents in the acid increased to 0.1%.[315]

In an ideal case, only the sample matrix reacts with the formation of volatile compounds, whereas the residue on the dish or in the crater of an electrode contains only nonvolatile compounds formed by dissolution of the admixtures present. This decomposition procedure leads not only to an effective preconcentration of the test components, but also to elimination of interferences caused by the matrix in the final determination. Under suitable conditions, the residue of the matrix does not exceed $n \cdot 10^{-4}$ g. Preconcentration of trace elements after decomposition by acid vapors has been analytically utilized for determination of 10^{-4} to 10^{-7}% of admixtures in materials for electronic industry, e.g., in silicon, quartz, quartz optical fibers,[106,311,312,315-318] germanium metal, and its oxide.[313,314] The volatile reaction products (SiF_4, H_2SiF_6) are transported into a container with the acids. Due to a temperature gradient inside the pressure vessel, the hydrofluoric acid condenses on the walls and flows down to the bottom of the reactor.[106] The vapors of germanium tetrachloride are dissolved in stock hydrochloric acid. In the presence of water vapors the tetrachloride hydrolyzes with formation of the oxide or oxychloride.[313,314] To decrease the sample contamination through leaching of the walls of the vessels or through a solution transport, the coarsely granulated material is placed directly in the crater of a graphite electrode of a suitable shape. After volatilization of silicon tetrafluoride or germanium tetrachloride, the remaining impurities, concentrated in the bottom part of the graphite electrode, are determined by an OES method.

The reaction product remaining after decomposition in a dish of an electrode is formed by dry or wet salts, mixtures of solutions and crystals, or a clear solution. The state of aggregation depends on the reaction conditions, mainly the kind of acids used and of the test material.[309,319] A condensate was formed in a system of hydrofluoric and hydrochloric acid only when a small amount of perchloric, sulfuric, or phosphoric acid was added to the glass sand analyzed.[309]

The impurities remaining after volatilization of the matrix may be in the form of readily soluble nitrates.[308] However, they are mostly dissolved in small amounts of highly pure acids. In decompositions of glass sands by hydrochloric and hydrofluoric acid vapors,[309] the fluoride content in the residue fluctuates from 0.02 to 0.2 mmol. On formation of a condensate, the fluoride concentration increases to $n \cdot 10^{-3}$ to 10^{-2} mol·ℓ^{-1}. The fluorides are displaced from the reaction mixture by evaporation with a minimal volume of pure sulfuric or perchloric acid.[311] In dissolution of the oxides of tantalum, niobium, zirconium, and titanium in hydrofluoric acid vapors, complex compounds of the H_2MeF_6 and H_2MeF_7 types were identified in the residue using X-ray diffraction. The volatile fluorides TiF_4 and ZrF_4 were not formed in a closed system, due to interaction of hydrofluoric acid vapors with water.[319]

Destruction by acid vapors is suitable for decomposition of substances with a known phase composition; however, some resistant substances are decomposed only partially. Whereas

Table 9
DECOMPOSITION BY ACID VAPORS

Material	Procedure	Notes	Ref.
GeO_2	0.5 g; HCl; 200°C; 3 hr	Samples in graphite electrode	313
Ge	1 g; aqua regia; 220°C; 6 hr	Samples in graphite electrode	314
SiO_2	0.5 g; HF; 220°C; 2 hr	Samples in Teflon® dish with 20 mg of graphite	106
Quartz glass	0.5 g; HF; 220°C; 3 hr	Samples in graphite electrode	312
Quartz	0.5 g; HF; 220°C; 3 hr	Samples in graphite electrode	315
SiO_2	2.4 g; HF vapors; conditions according to Ref. 316	SiO_2 doped with trace element isotopes	317
Si, SiO_2, quartz glass, glass	10 g; HF; 110—160°C	Exposure of 0.5—24 hr according to grain size	316
Si, quartz, quartz optical fibers	1 g; HF and HNO_3 or HF; 110°C; 14 hr	Liquid residue evaporated with H_2SO_4	311
Glass sands, minerals	1 g; HF and HCl or HBr; 180—195°C; 6 hr	Determination of Fe and Ti, and/or Na, K	309
Glasses, Na, and Ca silicate	0.2 g; HF and HNO_3; 110°C; 14 hr	Nitrates obtained dissolved in H_2O	308
Coal ash, slag, silicate rocks	2 g; HF and HNO_3 (1 + 1) and 50 mg $KMnO_4$; 70°C; 16 hr	Decomposition in wide-neck Teflon® bottle with a screw cap	310
Atmospheric particulates	Filters; HF + HNO_3 mixture; 100°C; 14 hr	Decomposition in a polypropylene bottle	321
Ti, Zr, Nb, Ta oxides	1 g; HF	Reaction time 4 hr at 200—220°C; α-Al_2O_3 undecomposed	319
α-Al_2O_3, γ-Al_2O_3	1 g; HCl		
Ba metaphosphates	HCl, HNO_3, and HF		

α-Al_2O_3 is highly resistant toward hydrochloric acid vapors, the γ-form of the oxide can be quantitatively dissolved.[319] Tourmaline reacts with hydrochloric and hydrofluoric acid vapors (220°C, 6 hr) with formation of a gel-like precipitate of complex aluminum fluorides. Iron can be leached from the precipitate by hydrochloric acid. Synthetic and natural tin dioxide is attacked at 190 to 200°C by hydrochloric acid vapors, as well as by hydrobromic acid vapors with bromine. The tin volatilizes as the appropriate halide and is trapped in a container with acids. Rocks with higher contents of aluminum oxide react with hydrofluoric acid vapors with formation of sparingly soluble fluorides.[309] Milligram amounts of zircon are dissolved in hydrofluoric acid vapors on prolonged exposure.[320] The blank value for uranium and lead is then lower than in decomposition of the sample by acids in an autoclave.[75]

Samples of silicon, quartz, and germanium metal are dissolved in the form of tiny lumps, whose surface is rinsed with dilute hydrochloric, hydrofluoric, or nitric acid or with hydrogen peroxide prior to decomposition.[311] Some important practical applications of this method of dissolution and the appropriate reaction conditions are summarized in Table 9.

An interesting variant of the decomposition vessel design has been proposed for dissolution of quartz and quartz technical glass.[322] A Teflon® crucible has holes in the upper part of the wall, and matching grooves are made in a turnable Teflon® lid. The samples are placed in Teflon® vessels and fixed in a holder above hydrofluoric acid level. To dissolve 5 to 20 g of coarsely crushed quartz, 2 mℓ of 60% perchloric acid are added to the sample, and the bottom of the vessel is covered with 40% HF. The whole system is closed by turning the lid and heated to a temperature of 110°C in the space of the sample holder. After 12 to 72 hr of heating (depending on the sample weight and SiO_2 particle size), the temperature is increased to 150°C. The lid is turned, and the reaction space is rinsed with a stream of nitrogen introduced into the vessel through a closable capillary. The rapid displacement of the gaseous products decreases the amount of fluorides in the dry evaporation residue. The dish with the residue is allowed to cool, taken out of the vessel, and the residue is wetted with 1 mℓ $HClO_4$. The contents of the dish are evaporated to obtain a damp residue, which is dissolved, and the solution is used for determination of the alkali metals, alkaline earths, iron, cobalt, nickel, and copper using the AAS method. The SiO_2 remaining in the residue could not be detected even by OES. However, in decomposition of boron glasses, the boron trifluoride did not completely volatilize in all cases. The temperatures of the gaseous and liquid phases are controlled during the whole dissolution process, using a thermocouple protected by a corundum case. The dissolution of a 0.1-g sample of a technical glass takes 6 hr.

REFERENCES

1. **Carius, G. L.,** *Ann. Chem.*, 136, 1, 1860; *Ber. Dtsch. Chem. Ges.*, 3, 697, 1870.
2. **Mitscherlich, A.,** *J. Pract. Chem.*, 81, 116, 1860.
3. **Jannasch, P.,** *Ber. Dtsch. Chem. Ges.*, 24, 273, 1891.
4. **Jannasch, P. and Vogtherr, H.,** *Ber. Dtsch. Chem. Ges.*, 24, 3206, 1891.
5. **Jannasch, P.,** *Z. Anorg. Chem.*, 6, 72, 1894.
6. **Wichers, E., Schlecht, W. G., and Gordon, C. L.,** *J. Res. Natl. Bur. Stand.*, 33, 363, 451, 1944.
7. **Novák, J.,** *Chem Listy*, 61, 581, 1967.
8. **Novák, J.,** *Acta Geol. Geogr. Univ. Comenianae*, No. 15, 35, 1968.
9. **Šulcek, Z., Povondra, P., and Doležal, J.,** *CRC Crit. Rev. Anal. Chem.*, 6, 255, 1977.
10. **Jackwerth, E. and Gomiššek, S.,** *Pure Appl. Chem.*, 56, 479, 1984.
11. **Des Marais, D. J. and Hayes, M.,** *Anal. Chem.*, 48, 1651, 1976.
12. **Foner, H. A.,** *Anal. Chem.*, 56, 856, 1984.
13. **Stevenson, F. J.,** *Anal. Chem.*, 32, 1704, 1960.
14. **Gordon, C. L., Schlecht, W. G., and Wichers, E.,** *J. Res. Natl. Bur. Stand.*, 33, 457, 1944.
15. **Foner, H. A.,** *Analyst*, 109, 1469, 1984.
16. **Crossley, D.,** *Proc. Anal. Chim. Chem. Soc.*, 16, 149, 1979.
17. **Mallett, R. C., Breckenridge, R., Palmer, I., and Dixon, K.,** *Natl. Inst. Metall. Repub. S. Afr. Rep.*, No. 1401, 1972.
18. **Mallett, R. C., Breckenridge, R., Palmer, I., Dixon, K., and Wall, G.,** *Natl. Inst. Metall. Repub. S. Afr. Rep.*, No. 1852, 1976.
19. **Mallett, R. C., Wall, G. J., Jones, E. A., and Royal, S. Y.,** *Natl. Inst. Metall. Repub. S. Afr. Rep.*, No. 1864, 1977.
20. **Pearton, D. C. G., Breckenridge, R. L., Dubois, M., and Steele, T. W.,** *J. S. Afr. Chem. Inst.*, 25, 244, 1972.
21. **Ashy, A. M. and Headridge, J. B.,** *Analyst*, 99, 285, 1944.
22. **Ginzburg, S. I., Ezerskaya, N. A., Prokofeva, I. V., Fedorenko, N. V., Shlenskaya, V. I., and Belskii, N. K.,** *Analiticheskaya Khimiya Platinovykh Metallov*, Nauka, Moskva, 1972, 24.
23. **Beamish, F. E. and Van Loon, J. C.,** *Analysis of Noble Metals. Overview and Selected Methods*, Academic Press, New York, 1977, 69.
24. **Watanabe, K.,** *Talanta*, 31, 311, 1984.
25. **Vogel, C. H. and Etten, N.,** *Fresenius Z. Anal. Chem.*, 275, 349, 1975.
26. **Jackson, H.,** *Analyst*, 75, 414, 1950.
27. **Fleet, B., Liberty, K. V., and West, T. S.,** *Analyst*, 93, 701, 1968.
28. **Young, P. N. W.,** *Analyst*, 99, 588, 1974.
29. **Novák, J. and Coufová, P.,** *Collect. Czech. Chem. Commun.*, 32, 2644, 1967.
30. **Krtil, J.,** *Jad. Energ.*, 29, 440, 1983; *Jad. Energ.*, 30, 91, 1984.
31. **Neirinx, R., Adams, F., and Hoste, J.,** *Anal. Chim. Acta*, 48, 1, 1969.

32. **Pucci, J. R. and Maffei, F. J.,** *An. Assoc. Quim. Brasil,* 3, 61, 1944; *Chem. Abstr.,* 39, 1816, 1945.
33. **Mulder, B. J.,** *Anal. Chim. Acta,* 72, 220, 1974.
34. **Hillebrand, W. F., Lundell, G. E. F., Bright, H. A., and Hoffman, J. I.,** *Applied Inorganic Analysis,* 2nd ed., John Wiley & Sons, New York, 1953, 840.
35. **Novák, J., Funke, A., and Kleinert, P.,** *Fresenius Z. Anal. Chem.,* 237, 339, 1968.
36. **Novák, J. and Pollert, E.,** *Fresenius Z. Anal. Chem.,* 283, 363, 1977.
37. **Donaldson, E. M.,** *Methods for the Analysis of Ores, Rocks and Related Materials,* Energy, Mines and Resources, Canada Mines Branch Monogr. 881, Ottawa, 1974, 305.
38. **Funke, A. and Kleinert, P.,** *Fresenius Z. Anal. Chem.,* 246, 362, 1969.
39. **Van Oosterhaut, G. W. and Wisser, J.,** *Anal. Chim. Acta,* 33, 330, 1965.
40. **Novák, J. and Arendt, H.,** *Silikáty,* 9, 59, 1965.
41. **Funke, A.,** *Fresenius Z. Anal. Chem.,* 244, 105, 1969.
42. **Dippel, P.,** *Silikattechnik,* 13, 51, 1962.
43. **Budyak, N. F. and Gryaznova, I. S.,** *Zh. Prikl. Chem.,* 44, 669, 1971.
44. **Malinowska, A.,** *Chem. Anal. (Warshawa),* 13, 1081, 1968.
45. **Kuznetsov, V. I., Malofeeva, I. V., and Nikolskaya, I. V.,** *Zavod. Lab.,* 24, 1178, 1958.
46. **Sung, J. F. C., Nevissi, A. E., and Dewalle, F. B.,** *J. Environ. Sci. Health,* A19(8), 959, 1984.
47. **Jeffery, P. G. and Hutchison, D.,** *Chemical Methods of Rock Analysis,* 3rd ed., Pergamon Press, Oxford, 1981, 265.
48. **Nakamura, S., Azuma, N., and Arai, Z.,** *Kogyo Kagaku Zasshi,* 63, 903, 1960; *Chem. Abstr.,* 56, 9693, 1960.
49. **Cosgrove, J. D. and Shears, E. C.,** *Analyst,* 85, 448, 1960.
50. **Brožek, V., Novák, J., and Dufek, J.,** *Silikáty,* in press.
51. **Hejduk, J. and Novák, J.,** *Fresenius Z. Anal. Chem.,* 234, 327, 1968.
52. **Long, G. and Foster, L. M.,** *J. Am. Ceram. Soc.,* 42, 53, 1959; *J. Am. Ceram. Soc.,* 44, 255, 1961.
53. **Hájek, B., Kohout, V., Novák, J., and Brožek, V.,** *Silikáty,* 26, 169, 1982.
54. **Hulmston, P.,** *Anal. Chim. Acta,* 155, 247, 1983.
55. **Knapp, G.,** *Fresenius Z. Anal. Chem.,* 317, 213, 1984.
56. **Knapp, G.,** *ICP Inf. Newsl.,* 10, 91, 1984.
57. **Knapp, G.,** *Trends Anal. Chem.,* 3, 182, 1984.
58. **May, I., Rowe, J. J., and Letner, R.,** *U.S. Geol. Surv. Prof. Pap.* 525B, B165, 1965.
59. **May, I. and Rowe, J. J.,** *Anal. Chim. Acta,* 33, 648, 1965.
60. **Borchert, W. and Donderer, E.,** *N. Jb. Miner. Abh.,* 110, 142, 1969.
61. **Kaiser, G., Götz, D., Tölg, G., Knapp, G., Maichin, B., and Spitzy, H.,** *Fresenius Z. Anal. Chem.,* 291, 278, 1978.
62. **Lounamaa, K.,** *Fresenius Z. Anal. Chem.,* 146, 122, 1955.
63. **Riley, J. P.,** *Anal. Chim. Acta,* 19, 413, 1958.
64. **Ito, J.,** *Bull. Soc. Chem. Jpn.,* 35, 225, 1962.
65. **Wahler, W.,** *Aluminium,* 39, 323, 1963.
66. **Wahler, W.,** *N. Jb. Miner. Abh.,* 101, 109, 1964.
67. **Langmyhr, F. J. and Graff, P. R.,** *A Contribution to the Analytical Chemistry of Silicate Rocks,* No. 230, Norges Geologiske Undersökelse, Oslo, 1965, 44.
68. **Langmyhr, F. J. and Sveen, S.,** *Anal. Chim. Acta,* 32, 1, 1965.
69. **Langmyhr, F. J.,** *Acta Geol. Geogr. Univ. Comenianae,* No. 15, 23, 1968.
70. **Bernas, B.,** *Anal. Chem.,* 40, 1682, 1968.
71. **Doležal, J., Lenz, J., and Šulcek, Z.,** *Anal. Chim. Acta,* 47, 517, 1969.
72. **Langmyhr, F. J. and Paus, P. E.,** *Anal. Chim. Acta,* 49, 358, 1970.
73. **Paus, P. E.,** *At. Absorpt. Newsl.,* 10, 44, 1971.
74. **Kotz, L., Kaiser, G., Tschöpel, P., and Tölg, G.,** *Fresenius Z. Anal. Chem.,* 260, 207, 1972.
75. **Krogh, T. E.,** *Geochim. Cosmochim, Acta,* 37, 485, 1973.
76. **Tereshchenko, B. S.,** *Zavod. Lab.,* 40, 1452, 1974.
77. **Tereschenko, B. S.,** *Razved. Okhr. Nedr,* No. 3, 49, 1976; *Chem. Abstr.,* 86, 25462 r, 1977.
78. **Knoop, P.,** *Anal Chem.,* 46, 965, 1974.
79. **Vasnev, A. N., Kreongold, S. V., Zherebovich, A. S., and Chupakhin, M. S.,** *Zavod. Lab.,* 42, 657, 1976.
80. **Debizha, E. V., Sukhanovskaya, A. I., Chupakhin, M. S., Voroiskii, F. S., and Tatevosyan, R. A.,** *Zavod. Lab.,* 45, 117, 1979.
81. **Stoeppler, M. and Backhaus, F.,** *Fresenius Z. Anal. Chem.,* 291, 116, 1978.
82. **Rombach, N., Apel, R., and Tschochner, F.,** *GIT Fachz. Lab.,* 24, 1165, 1980.
83. **Kasper, H. U.,** *Fresenius Z. Anal. Chem.,* 320, 55, 1985.
84. **Oehme, M.,** *Talanta,* 26, 913, 1979.

85. **Okamote, K. and Fuwa, K.**, *Anal. Chem.*, 56, 1758, 1984.
86. **Uhrberg, R.**, *Anal. Chem.*, 54, 1906, 1982.
87. **Döscher, B. and Knöchel, A.**, *Fresenius Z. Anal. Chem.*, 322, 776, 1985.
88. **Buresch, O. and Schnering, H. G.**, *Fresenius Z. Anal. Chem.*, 319, 418, 1984.
89. **Krasilshchik, V. Z.**, *Zavod. Lab.*, 51(8), 19, 1985.
90. **Schramel, P. and Quiang, X. L.**, *Anal. Chem.*, 54, 1333, 1982.
91. **Kotz, L., Henze, G., Kaiser, G., Pahlke, S., Veber, M., and Tölg, G.**, *Talanta*, 26, 681, 1979.
92. **Pitt, M. J.**, *Chem. Ind. (London)*, No. 20, 804, 1982.
93. **Stoeppler, M., Maüller, K. P., and Backhaus, F.**, *Fresenius Z. Anal. Chem.*, 297, 107, 1979.
94. **Philips, C. V. and Westall, S.**, *Lab. Pract.*, 30, 598, 1981.
95. **Eustermann, K. and Seifert, D.**, *Fresenius Z. Anal. Chem.*, 285, 253, 1977.
96. **Bailey, N. T. and Wood, S. J.**, *Anal. Chim. Acta*, 69, 19, 1974.
97. **Gill, R. C. O. and Kronberg, B. I.**, *At. Absorpt. Newsl.*, 14, 157, 1975.
98. **Nadkarni, R. A.**, Geochemistry and Chemistry of Oil Shales, in *ACS Symp. Ser.*, American Chemical Society, Seattle, WA, 230, 477, 1983.
99. **Brandvold, L. A.**, *Circular 142*, New Mexico Bureau of Mines and Mineral Resources, Socorro, N.M., 1974.
100. **Price, W. J. and Whiteside, P. J.**, *Analyst*, 102, 664, 1977.
101. **Ilgen, G., Buduan, P. V., and Fiedler, H. J.**, *Chem. Erde*, 42, 309, 1983.
102. **Hughes, T. C.**, *J. Radioanal. Chem.*, 59, 7, 1980.
103. **Grallath, E., Tschöpel, P., Kölblin, G., Stix, U., and Tölg, G.**, *Fresenius Z. Anal. Chem.*, 302, 40, 1980.
104. **Jütte, B. A. H. G., Heikam, A., and Agterdenbos, J.**, *Anal. Chim. Acta*, 110, 345, 1979.
105. **Lange, J.**, *Silikattechnik*, 31, 44, 1980.
106. **Debizha, E. V., Krasilshchik, V. Z., Sokolskaya, N. N., and Chupakhin, M. S.**, Zh. Anal. Khim., 36, 1939, 1981.
107. **Chruscinska, T. J.**, *Chem. Anal. (Warsaw)*, 29, 301, 1984.
108. **Šulcek, Z. and Mužík, L.**, unpublished results, 1981.
109. **Seger, J.**, unpublished results, 1982.
110. **Sukhanovskaya, A. I., Shuginina, A. V., and Chupakhin, M. S.**, *Zh. Anal. Khim.*, 38, 1018, 1983.
111. **Erdtmann, G. and Aboulwafa, O.**, *Fresenius Z. Anal. Chem.*, 272, 105, 1974.
112. **Štulík, K., Beran, P., Doležal, J., and Opekar, F.**, *Talanta*, 25, 363, 1978.
113. **Balaž, V.**, unpublished results, 1979.
114. **Šulcek, Z.**, unpublished results, 1984.
115. **Malyutina, T. M., Namvrina, E. G., Shiryaeva, A. O.**, *Zavod. Lab.*, 47(9), 8, 1981.
116. **Date, A. R.**, *Analyst*, 103, 84, 1978.
117. **Adam, J.**, *Chem. Listy*, in press.
118. **Tölg, G.**, *Pure Appl. Chem.*, 50, 1075, 1978.
119. **Tölg, G.**, *Pure Appl. Chem.*, 55, 1980, 1983.
120. **Tschöpel, P.**, *Pure Appl. Chem.*, 54, 913, 1982.
121. **Fisher, L. B.**, *Anal. Chem.*, 58, 261, 1986.
122. **Krogh, T. E. and Davis, G. L.**, *Carnegie Inst. Wash. Yearb.*, 73, 560, 1973/74; *Carnegie Inst. Wash. Yearb.*, 74, 619, 1974/75.
123. **Arden, J. W.**, *Anal. Chim. Acta*, 148, 211, 1983.
124. **Arden, J. W. and Gale, N. H.**, *Anal. Chem.*, 46, 2, 1974.
125. **Arden, J. W. and Gale, N. H.**, *Anal. Chem.*, 46, 687, 1974.
126. **Rammensee, W. and Palme, H.**, *J. Radioanal. Chem.*, 71, 401, 1982.
127. **Krähenbühl, U. and Wegmüller, F.**, *Radiochem. Radioanal. Lett.*, 36, 31, 1978.
128. **Hurley, P. M. and Pinson, W. H.**, *Science*, 167, 473, 1970.
129. **Patchett, P. J. and Tatsumoto, M.**, *Contrib. Mineral. Petrol.*, 75, 263, 1980.
130. **Kritsatakis, K. and Tobschall, H. J.**, *Fresenius Z. Anal. Chem.*, 320, 15, 1985.
131. **Schnetzler, C. C. and Nava, D. F.**, *Earth Planet. Sci. Lett.*, 11, 345, 1971.
132. **Heumann, K. G. and Weiss, H.**, *Fresenius Z. Anal. Chem.*, 323, 852, 1986.
133. **Schärer, U. and Allegre, C. J.**, *Earth Planet. Sci. Lett.*, 63, 423, 1983.
134. **Patchett, P. J., Kouvo, A., Hedge, C. E., and Tatsumoto, M.**, *Contrib. Mineral. Petrol.*, 78, 279, 1981.
135. **Kramers, J. D. and Smith, C. B.**, *Chem. Geol.*, 41, 23, 1983.
136. **Taylor, P. J.**, unpublished results, 1984.
137. **Katz, A.**, *Chem. Geol.*, 16, 15, 1975.
138. **Verma, S. P.**, *Chem. Geol.*, 41, 339, 1983.
139. **Cliff, R. A., Gray, C. M., and Huhma, H.**, *Contrib. Mineral. Petrol.*, 82, 91, 1983.

140. **McCulloch, M. T. and Chappel, B. W.,** *Earth Planet. Sci. Lett.,* 58, 51, 1982.
141. **Kubassek, E. and Heumann, K. G.,** *Fresenius Z. Anal. Chem.,* 289, 41, 1978.
142. **Heumann, K. G., Kubassek, E., Schwabenbauer, W., and Stadler,I.,** *Fresenius Z. Anal. Chem.,* 287, 121, 1977.
143. **Mužík, L. and Lenc, J.,** *Sb. Ustred. Ustavu Geol.,* 12, 161, 1972.
144. **Manoliu, C., Bulac, E., and Zugravescu, P.,** *Rev. Chim.,* 35, 940, 1984; *Anal. Abstr.,* 47, 11B106, 1985.
145. **Chen, Jo Yun, T.,** *J. Assoc. Off. Anal. Chem.,* 60, 1266, 1977.
146. **Sýkora, V. and Dubský, F.,** *Sb. Vysoke Skoly Chem. Technol. Praze,* H 12, 177, 1977.
147. **Langmyhr, J. F. and Paus, P. E.,** *Anal. Chim. Acta,* 45, 157, 1969.
148. **Tomljanovic, M. and Grobenski, Z.,** *At. Absorpt. Newsl.,* 14, 52, 1975.
149. **Sixta, V. and Šulcek, Z.,** *Sklar Keram.,* 28, 364, 1978.
150. **De Carlo, E. H., Zeitlin, H., and Fernando, Q.,** *Anal. Chem.,* 54, 898, 1982.
151. **Marabini, A., Barbaro, M., and Passariello, B.,** *At. Spectrosc.,* 3, 140, 1982.
152. **Toth, J. R. and Ingle, J. D.,** *Anal. Chim. Acta,* 92, 409, 1977.
153. **Kaiser, G., Grallath, E., Tschöpel, P., and Tölg, G.,** *Fresenius Z. Anal. Chem.,* 259, 257, 1972.
154. **Van Eenbergen, E. and Bruninx, E.,** *Anal. Chim. Acta,* 98, 405, 1978.
155. **Brandvold, L. A. and Marson, S. J.,** *At. Absorpt. Newsl.,* 13, 125, 1974.
156. **Omang, S. H. and Paus, P. E.,** *Anal. Chim. Acta,* 56, 393, 1971.
157. **Trybula, Z., Krzyzanowska, M., and Krzyzanowski, P.,** *Chem. Anal. (Warsaw),* 30, 137, 1985.
158. **Sighinolfi, G. P. and Gordoni, C.,** *Talanta,* 28, 169, 1972.
159. **Martinez, C. and Castillo, J. R.,** *At. Spectrosc.,* 4, 63, 1983.
160. **Solozhenkin, N. M., Pupkov, V. S., Usova, S. V., Saidova, M. B., and Yunusov, M. M.,** *Zh. Anal. Khim.,* 39, 2165, 1984.
161. **Skorko-Trybulowa, Z., Boguszewska, Z., and Rozanska, B.,** *Micr. Acta,* I B151, 1979.
162. **Elliott, E. V., Stever, K. R., and Heady, H. H.,** *At. Absorpt. Newsl.,* 13, 113, 1974.
163. **Guest, R. J. and Macpherson, D. R.,** *Anal. Chim. Acta,* 71, 233, 1974.
164. **Takano, B. and Watanuki, K.,** *Jpn. Analyst,* 10, 1376, 1972.
165. **Langmyhr, F. J., Solberg, R., and Thomassen, Y.,** *Anal. Chim. Acta,* 92, 105, 1977.
166. **Hendel, Y., Ehrental, A., and Bernas, B.,** *At. Absorpt. Newsl.,* 12, 130, 1973.
167. **Pakalns, P.,** *Anal. Chim. Acta,* 69, 211, 1974.
168. **Mainka, E., Coerds, W., and Koenig, H.,** *Kernforschungszentrum Karlsruhe Ber.,* KFK 2458, 1977; *Anal. Abstr.,* 35, 3B78, 1978.
169. **Fogl, J., Urner, Z., and Šucha, L.,** *Sb. Vysoké, Skoly Chem. Technol. Praze,* H16, 169, 1981.
170. **Funk, D. K., Dubois, J. P., and Kubler, B.,** *Analusis,* 11, 291, 1983.
171. **Cottenie, A. and Verloo, M.,** *Fresenius Z. Anal. Chem.,* 317, 389, 1984.
172. **Van der Veen, N. G., Keukens, H. J., and Vos, G.,** *Anal. Chim. Acta,* 171, 285, 1985.
173. **Cresser, M. S., Lebdon, L. C., McLeod, C. W., and Burridge, J. C.,** *J. Anal. Spectrom.,* 1, 1R, 1986.
174. **Bartha, A. and Skrenyi, K.,** *Anal. Chim. Acta,* 139, 329, 1982.
175. **Nadkarni, R. A.,** *Anal. Chim. Acta,* 135, 363, 1982.
176. **Beaty, R. D.,** *At. Absorpt. Newsl.,* 13, 38, 1974.
177. **Beaty, R. D. and Manuel, O, K.,** *Chem. Geol.,* 12, 155, 1973.
178. **Sighinolfi, G. P., Gorgoni, C., and Santos, A. M.,** *Geostand. Newsl.,* 4, 233, 1980.
179. **Agemian, H., Aspila, K. I., and Chau, A. S. Y.,** *Analyst,* 100, 253, 1975.
180. **Heinrichs, H.,** *Fresenius Z. Anal. Chem.,* 273, 197, 1975.
181. **Langmyhr, F. J. and Paus, P. E.,** *At. Absorpt. Newsl.,* 7, 103, 1968; *Anal. Chim. Acta,* 43, 397, 1968.
182. **Buckley, D. E. and Cranston, R. E.,** *Chem. Geol.,* 7, 273, 1971.
183. **Domingues, A.,** *Ceramica,* 27, 1, 1981.
184. **Gomez Bueno, C. O., Rempel, G. L., and Spink, D. R.,** *Talanta,* 29, 461, 1982.
185. **Li, M. and Filby, R. H.,** *Anal. Chem.,* 55, 2236, 1983.
186. **Botto, R. I., White, B. H., and Karchmer, J. H.,** *Fuel,* 59, 157, 1980.
187. **Heumann, K. G., Beer, F., and Kifmann, R.,** *Talanta,* 27, 567, 1980.
188. **Heumann, K. G., Schrödl, W., and Weiss, H.,** *Fresenius Z. Anal. Chem.,* 315, 213, 1983.
189. **Wickert, H.,** *Zem. Kalk Gips,* 27, 375, 1974.
190. **Betinelli, M., Pastorelli, N., and Baroni, U.,** *At. Spectrosc.,* 7, 45, 1980.
191. **Green, J. B. and Manahan, S. E.,** *Anal. Chem.,* 50, 1975, 1978.
192. **Shan, X.-Q., Ni, Z.-M. and Yuan, Z.-N.,** *Anal. Chim. Acta,* 171, 269, 1985.
193. **Liese, T.,** *Fresenius Z. Anal. Chem.,* 321, 37, 1985.
194. **Vigler, M. S., Varnes, A. W., and Strecker, H. A.,** *Int. Lab.,* 51, October 1980.
195. **Harju, L. and Hulden, S. G.,** *Talanta,* 27, 811, 1980.

196. **Carter, D., Regan, J. G. T., and Warren, J.,** *Analyst,* 100, 721, 1975.
197. **Cammann, K.,** *Fresenius Z. Anal. Chem.,* 293, 97, 1978.
198. **Lee, A. F.,** *Natl. Inst. Metall. Repub. S. Afr. Rep.,* No. 2124, 1981; No. 2083, 1980; No. 2061, 1980.
199. **Block, C.,** *Anal. Chim. Acta,* 80, 369, 1975.
200. **Weitz, A., Fuchs, G., and Bächmann, K.,** *Fresenius Z. Anal. Chem.,* 313, 38, 1982.
201. **Brückner, H. P., Drews, G., Kritsotakis, K., and Tobschall, H. J.,** *Fresenius Z. Anal. Chem.,* 322, 778, 1985.
202. **Kamata, E., Nakashima, R., and Shibata, S.,** *Bunseki Kagaku,* 33, 173, 1984.
203. **Warren, J. and Carter, D.,** *Can. J. Spectrosc.,* 20, 11, 1975.
204. **Vačkova, M., Streško, V., and Žemberyová, M.,** *Chem. Listy,* 72, 408, 1975.
205. **Ohta, K. and Suzuki, M.,** *Anal. Chim. Acta,* 108, 69, 1979.
206. **Suzuki, N., Takahashi, M., and Imura, H.,** *Anal. Chim. Acta,* 60, 79, 1984.
207. **Ohta, K. and Suzuki, M.,** *Anal. Chim. Acta,* 107, 245, 1979.
208. **Campbell, E. Y. and Simon, F. O.,** *Talanta,* 25, 251, 1978.
209. **Korkisch, J. and Sorio, A.,** *Anal. Chim. Acta,* 82, 311, 1976.
210. **Smet, T., Hertogen, J., Gijbels, R., and Hoste, J.,** *Anal. Chim. Acta,* 101, 45, 1978.
211. **Ward, A. F. and Marciello, L.,** *Jarrel Ash Plasma News,* 1, 10, 1978.
212. **Nadkarni, R.,** *Anal. Chem.,* 52, 929, 1980.
213. **Farreira, G. G. and Pinheiro, L. B. M.,** *Metal ABM,* 36, 485, 1980; *Anal. Abstr.,* 41, 4B199, 1981.
214. **Rigin, V. I.,** *Zh. Anal. Khim.,* 41, 46, 1986.
215. **Hartstein, A. M., Freedman, R. W., and Platter, D. W.,** *Anal. Chem.,* 45, 611, 1973.
216. **Nakashima, R., Kamta, E., Goto, K., and Shibata, S.,** *Bunseki Kagaku,* 32, T92, 1983.
217. **Somer, G., Cakir, O., and Solak, A. O.,** *Analyst,* 109, 135, 1984.
218. **Kellerman, S., Haines, J., and Robert, R. V. D.,** *MINTEK Rep.,* M121, 1983.
219. **Mahante, H. S. and Barnes, R. M.,** *Anal. Chim. Acta,* 149, 395, 1983.
220. **Doolan, K. J.,** *Anal. Chim. Acta,* 140, 187, 1982.
221. **Davidson, R. L., Natusch, D. F. S., Wallace, J. R., and Evans, C. A.,** *Environ. Sci. Technol.,* 8, 1107, 1974.
222. **Hsu, C. G. and Locke, D. F.,** *Anal. Chim. Acta,* 153, 313, 1984.
223. **Breder, R.,** *Fresenius Z. Anal. Chem.,* 313, 395, 1982.
224. **Sakata, M. and Shidoma, O.,** *Bunseki Kagaku,* 31, T81, 1982; *Water Res.,* 16, 231, 1982.
225. **Griepink, B., Muntau, H., and Colinet, E.,** *Fresenius Z. Anal. Chem.,* 318, 490, 1984.
226. **Henrion, G., Bode, K., and Pelzer, J.,** *Z. Chem.,* 23, 424, 1983.
227. **Yang, J. Y., Tseng, C. L., Lo, J. M., and Yang, M. H.,** *Fresenius Z. Anal. Chem.,* 321, 141, 1985.
228. **Jin, L.-Z.,** *At. Spectrosc.,* 5, 91, 1984.
229. **Aspila, K. I., Agemian, H., and Chau, A. S. Y.,** *Analyst,* 101, 187, 1976.
230. **Agemian, H. and Chau, A. S. Y.,** *Anal. Chim. Acta,* 80, 61, 1975.
231. **Jagner, D. and Aren, K.,** *Anal. Chim. Acta,* 141, 157, 1982.
232. **Bettinelli, M., Pastorelli, N., and Baroni, U.,** *Anal. Chim. Acta,* 185, 109, 1986.
233. **Berrow, M. L. and Stein, W. M.,** *Analyst,* 108, 277, 1983.
234. **Ritter, Ch. J.,** *Int. Lab.,* p. 30, October 1982.
235. **Noriki, S., Nakanishi, K., Fukawa, T., Uematsu, M., Uchida, T., and Tsunogai, S.,** *Hokkaido Daigaku Suisangakubu Kenkyu Iho,* 31, 354, 1980; *Anal. Abstr.,* 42, 2H82, 1982.
236. **Gillain, G.,** *Talanta,* 29, 651, 1982.
237. **Trujillo, P. E. and Campbell, E. E.,** *Anal. Chem.,* 47, 1629, 1975.
238. **Chiu, C. H. and Hilborn, J. C.,** *Analyst,* 104, 1159, 1979.
239. **Lachowitz, E. and Kaliczuk, A.,** *Fresenius Z. Anal. Chem.,* 323, 54, 1986.
240. **Guest, R. J. and McPherson, D. R.,** *Anal. Chim. Acta,* 78, 299, 1975.
241. **Fritsche, H., Wegschneider, W., Knapp, H., and Oertner, H. M.,** *Talanta,* 26, 219, 1979.
242. **Buresch, O. and von Schnering, H. G.,** *Fresenius Z. Anal. Chem.,* 321, 681, 1985.
243. **Sýkora, V. and Dubský, F.,** *Sb. Vysoke Skoly Chem. Technol. Praze,* H12, 167, 1977.
244. **Ashton, A., Fogg, A. G., and Burns, D. T.,** *Analyst,* 99, 108, 1974.
245. **Fernando, L. A.,** *Anal. Chem.,* 56, 1970, 1984.
246. **Eddy, B. T.,** *MINTEK Rep.,* No. M269, 1986.
247. **Headridge, J. B. and Sowerbutts, A.,** *Analyst,* 98, 57, 1973.
248. **Jurczyk, J., Glenc, T., Sheybal, I., and Swiderska, K.,** *Chem. Anal. (Warshawa),* 29, 541, 1984.
249. **Werner, W. and Tölg, G.,** *Fresenius Z. Anal. Chem.,* 276, 103, 1975.
250. **Langmyhr, F. J. and Paus, P. E.,** *Anal. Chim. Acta,* 45, 173, 1969.
251. **Novák, J., Flašarová, M., Andrštová, E., Dufek, V., and Brožek, V.,** *Silikáty,* 30, 365, 1986.
252. **Davis, W. F. and Merkle, E. J.,** *Anal. Chem.,* 53, 1139, 1981.
253. **Parker, A. and Healy, C.,** *Analyst,* 95, 204, 1970.

254. **Julietti, R. J.,** *Trans. J. Br. Ceram. Soc.,* 80, 175, 1981.
255. **Lengauer, W.,** *Fresenius Z. Anal. Chem.,* 322, 23, 1985.
256. **Kiss, E.,** *Anal. Chim. Acta,* 161, 231, 1984; *Anal. Chim. Acta,* 89, 303, 1977; *Anal. Chim. Acta,* 72, 127, 1974.
257. **Riley, J. P. and Williams, H. P.,** *Micr. Acta,* p. 516, 1959.
258. **Povondra, P.,** unpublished results, 1977.
259. **Whitehead, D. and Malik, S. A.,** *Anal. Chem.,* 47, 554, 1975.
260. **Wanty, R. B. and Goldhaber, M. B.,** *Talanta,* 32, 295, 1985.
261. **Goguel, R.,** *Anal. Chim. Acta,* 169, 179, 1985.
262. **Luke, C. L. and Flaschen, S. S.,** *Anal. Chem.,* 30, 1406, 1958.
263. **Ronge, B. S. H. and Bernardsson, H.,** *At. Absorpt. Newsl.,* 13, 70, 1974.
264. **Paleček, M.,** unpublished results, 1985.
265. **Eggimann, D. W. and Betzer, P. R.,** *Anal. Chem.,* 48, 886, 1976.
266. **Farrell, R. F., Matthes, S. A., and Mackie, A. J.,** *U.S. Bur. Mines Rep. Invest.,* No. 8480, 1980.
267. **Pahlavanpour, B., Thompson, M., and Thorne, L.,** *Analyst,* 105, 756, 1980.
268. **Hale, M., Thompson, M., and Lovell, J.,** *Analyst,* 110, 225, 1985.
269. **Thompson, M. and Zao, L.,** *Analyst,* 110, 229, 1985.
270. **May, K. and Stoeppler, M.,** *Fresenius Z. Anal. Chem.,* 317, 248, 1984.
271. **Farrell, R. F., Mackie, A. J., and Lessick, W. R.,** *U.S. Bur. Mines Rep. Invest.,* No. 8336, 1979.
272. **Van Loon, J. C.,** *Selected Methods of Trace Analysis,* John Wiley & Sons, New York, 1985, 259.
273. **Kallmann, S., Oberthin, H., and Liu, R.,** *Anal. Chem.,* 34, 609, 1962.
274. **Langmyhr, F. J. and Paus, P. E.,** *Anal. Chim. Acta,* 43, 397, 1968; *Anal. Chim. Acta,* 45, 176, 1969; *Anal. Chim. Acta,* 47, 371, 1969.
275. **Rantala, R. T. T. and Loring, D. H.,** *At. Spectrosc.,* 1, 163, 1980.
276. **Kiss, E.,** *Anal. Chim. Acta,* 140, 197, 1982.
277. **French, W. J. and Adams, S. J.,** *Anal. Chim. Acta,* 62, 324, 1973.
278. **Smith, J. D.,** *Anal. Chim. Acta,* 57, 371, 1971.
279. **Rice, T. D.,** *Analyst,* 107, 47, 1982.
280. **Ayranci, B.,** *Schweiz, Mineral. Petrogr. Mitt.,* 56, 513, 1976; *Schweiz. Mineral. Petrogr. Mitt.,* 57, 299, 1977.
281. **Hey, M. H.,** *Mineral. Mag.,* 46, 111, 1982.
282. **Antweiler, J. C.,** *U.S. Geol. Surv. Prof. Pap.,* 424B, B322, 1963.
283. **Uchida, T., Nagase, M., Kojima, I., and Iida, C.,** *Anal. Chim. Acta,* 94, 275, 1977.
284. **Kojima, I., Uchida, T., Nanbu, M., and Iida, C.,** *Anal. Chim. Acta,* 93, 69, 1977.
285. **Uchida, T., Nagase, M., and Iida, C.,** *Anal. Lett.,* 8, 825, 1975.
286. **Uchida, H., Uchida, T., and Iida, C.,** *Anal. Chim. Acta,* 116, 433, 1980.
287. **Uchida, H., Iwasaki, K., and Tanaka, K.,** *Anal. Chim. Acta,* 134, 375, 1982.
288. **Uchida, H., Uchida, T., and Iida, C.,** *Anal. Chim. Acta,* 108, 87, 1979.
289. **Pearce, W. C., Thornewill, D., and Marston, J. H.,** *Analyst,* 110, 625, 1985.
290. **Szeto, M. S., Opdebeeck, J. H. L., Van Loon, J. C., and Kreins, J.,** *At. Spectrosc.,* 5, 186, 1984.
291. **Langmyhr, F. J. and Paus, P. E.,** *Anal. Chim. Acta,* 43, 506, 1968.
292. **Sugawara, K. F. and Su, Y.-S.,** *Anal. Chim. Acta,* 80, 143, 1975.
293. **Liu, C. Y., Chen, P. Y., Lin, H. M., and Yang, M. H.,** *Fresenius Z. Anal. Chem.,* 320, 22, 1985.
294. **Lange, J.,** *Silikattechnik,* 33, 214, 1982.
295. **Coulson, R. E.,** *Trans. Inst. Min. Metall. Sect. C,* 85(September), C164, 1976; *Chem. Abstr.,* 86, 32204m, 1977.
296. **McLaren, J. W., Berman, S. S., Boyko, V. J., and Russell, D. S.,** *Anal. Chem.,* 53, 1802, 1982.
297. **Rantale, R. T. T. and Loring, D. H.,** *At. Absorpt. Newsl.,* 14, 117, 1975; *At. Absorpt. Newsl.,* 12, 97, 1973.
298. **Becker, R., Buchtela, K., Grass, F., Kittl, R., and Müller, G.,** *Fresenius Z. Anal. Chem.,* 274, 1, 1975.
299. **Boucetta, M. and Fritsche, J.,** unpublished results, 1979.
300. **McCarthy, J. P., Caruso, J. A., Wolnik, K. A., and Fricke, F. L.,** *Anal. Chim. Acta,* 147, 163, 1983.
301. **Mahler, R. L., Naylor, V. D., and Fridrickson, M. K.,** *Commun. Soil Sci. Plant Anal.,* 15, 479, 1984; *Anal. Abstr.,* 47, 2G12, 1985.
302. **Fries, T., Lomothe, P. J., and Pesek, J. J.,** *Anal. Chim. Acta,* 159, 329, 1984.
303. **Rice, T. D.,** *Anal. Chim. Acta,* 91, 221, 1977.
304. **Fresenius, W. and Schneider, W.,** *Fresenius Z. Anal. Chem.,* 214, 341, 1965.
305. **Maxwell, J. M. R. and Budd, S. M.,** *J. Soc. Glas. Technol.,* 40, 509, 1956.
306. **Hermann, R.,** *At. Absorpt. Newsl.,* 16, 44, 1977.
307. **Farmer, J. G. and Gibson, M. J.,** *At. Spectrosc.,* 2, 176, 1977.

308. **Woolley, J. F.,** *Analyst,* 100, 896, 1975.
309. **Šulcek, Z. and Mužik, L.,** *Silikáty,* 28, 67, 1984.
310. **Feldman, C.,** *Anal. Chem.,* 49, 825, 1977.
311. **Phelan, V. J. and Powell, R. J. W.,** *Analyst,* 109, 1269, 1984.
312. **Pimenov, V. G., Gaivoronskii, P. E., Shishov, V. N., and Maksimov, G. A.,** *Zh. Anal. Khim.,* 39, 1072, 1984.
313. **Pimenov, V. G., Timonin, D. A., and Shishov, V. N.,** *Zh. Anal. Khim.,* 41, 1173, 1986.
314. **Pimenov, V. G., Pronchatov, A. N., Maksimov, G. A., Shishov, V. N., Shcheplyagin, E. M., and Krasnova, S. G.,** *Zh. Anal. Khim.,* 39, 1636, 1984.
315. **Pimenov, V. G., Gaivoronskii, P. E., and Shishov, V. N.,** *Zavod. Lab.,* 50(2), 36, 1984.
316. **Mitchell, J. W. and Nash, D. L.,** *Anal. Chem.,* 46, 326, 1974.
317. **Mitchell, J. W.,** *Int. Lab.,* January/February, 12, 1982.
318. **Zief, M. and Mitchell, J. W.,** *Contamination Control in Trace Element Analysis,* John Wiley & Sons, New York, 1976, 166.
319. **Krasilshchik, V. Z., Zhiteleva, O. G., Sokolskaya, N. N., and Chupakhin, M. S.,** *Zh. Anal. Khim.,* 41, 586, 1986.
320. **Krogh, T.,** *Geol. Surv. Open File Rep.,* 78, 1978; *Chem. Abstr.,* 90, 9011, 1978.
321. **Stolzenburg, T. R. and Andren, A. A.,** *Anal. Chim. Acta,* 118, 377, 1980.
322. **Křesťan, V., Paleček, M., and Habrman, Z.,** unpublished results, 1986.

Chapter 6

MICROWAVE ACID DIGESTION SYSTEM

Decomposition with acids heated in a microwave oven has certain peculiarities, not only in the mode of transport of the thermal energy to the reaction medium, but also in the decomposition technique itself. The dissolution process is considerably accelerated and takes only a few minutes. Microwave radiation has so far been used in analytical chemistry primarily for drying of materials (e.g., sulfidic concentrates[1]) and for determination of moisture in inorganic and organic substances.[2,3]

The classical technique of heating suspensions of solid in acids is based on transport of thermal energy from a heating body (electrical heating at a frequency of 50 to 300 Hz or heating by a burning gas) through the vessel material to the reaction medium. Direct heating by infrared radiation is also used. Losses in heat occur primarily through radiation into the surroundings of the heating body and through heating of the vessels. Thus, only a fraction of the thermal energy produced is absorbed in the solution.

Compared with the classical heating, microwave heating is many times more efficient. Of the microwave spectrum (300 MHz to 300 GHz), a frequency of 2450 MHz has been used for heating of aqueous solutions. This frequency is employed in domestic microwave (MW) ovens that are still the basic instruments for MW acid digestion systems. This type of oven has recently been adjusted for dissolution of solids and is successfully produced commercially.[4]

The source of MW radiation is magnetron, and its power can be regulated, usually from 0 to 650 W (100% power). The amount of energy transported into the solution also depends on the losses in the waveguides and on the arrangement of the oven cavity. The magnetron must be protected against corrosive acid vapors and against unconsumed or reflected MW energy;[5-7] therefore, a beaker with water is often placed in the oven.[8] If samples to be decomposed are in vessels placed in a caroussel, bottles with water must be placed in the empty dishes on the caroussel, in order to absorb the excess MW energy. When using vessels transparent for microwaves, the radiating energy is only absorbed in the acid solution, due to polarization effects of water moleule dipoles. For absorption, the dielectric constant of the solvent must be high. The value of the dielectric constant of water decreases with increasing temperature. Aqueous solutions of acids absorb microwaves less than water alone. On dilution of the acid, the amount of the absorbed energy increases and approaches the value obtained for water. It has been shown experimentally that nitric acid absorbs the greatest amount of energy.[9] The efficiency of absorption of MW radiation decreases in the series: hydrofluoric, sulfuric, and hydrochloric acid. Acid vapors do not absorb MW radiation. A relationship has been verified[9] for the rate of heating of nitric acid in the interval from 25 to 180°C, which permits prediction of the reaction time required.[9]

The selection of the material and shape of the decomposition vessel has a decisive importance in microwave dissolution techniques. Glassy carbon and platinum vessels cannot be used, and chemical or quartz glass and plastics are suitable, as they are transparent for microwaves. Aqueous solutions of acids in glass or plastic vessels are heated within several minutes to a required temperature in a MW field. The acid vapors thus formed escape from the vessels and corrode the interior of the oven (the magnetron and waveguides); therefore, they must be aspirated into a fume cupboard through holes in the oven walls. Continuous rinsing of the reaction space by pressurized air or carbon dioxide during the whole dissolution process is also effective.[5-7] The walls of the vessels, heated from inside by hot acids, are cooled by the gas stream, their lifetime is thus prolonged, and the tightness improved.

Basic experiments have mostly been carried out in conical glass flasks,[10,11] in open plastic vessels enclosed in a plastic shield or container, or in a partially evacuated glass dessicator.[6,8,12,13] This technique is used progressively less at present, and MW decompositions in closed systems are preferred, employing mainly 60 to 250 mℓ polycarbonate bottles with polypropylene screw-on caps or Teflon® vessels with screw-on caps (Table 1).

The best practical results have been obtained with closed vessels made of translucent Teflon PFA. The screw-on caps are provided with Teflon® tape to improve the tightness. The lifetime of polycarbonate bottles depends on the conditions in the MW oven and the composition of the solvent used. They last through up to ten decompositions with a mixture of hydrochloric, nitric, and hydrofluoric acids.[5] Teflon vessels are substantially more durable. The screw-on caps made of Teflon PFA break when a pressure of *circa* 1.1 MPa is exceeded.[9] Direct measurement of the pressure and temperature inside the vessel has shown that this limit is exceeded within a temperature interval from 200 to 260°C (decomposition of biological materials with nitric acid).

The MW energy is nonuniformly distributed within the oven cavity, and there exist hot spots.[5,9,14] Therefore, the samples are usually placed on a polypropylene or polyethylene holder located on a caroussel. The holder containing 12 samples rotates at a rate of *circa* 3 rpm, which ensures a uniform distribution of the MW energy over each vessel. The holder space can be closed to trap vapors and microdroplets of acids escaping along the caps of the vessels.[7,15] Recently, a decomposition apparatus has been designed that is very similar to the autoclaves for pressure decompositions, but has no metallic parts.[16] The internal Teflon® vessel with an O-ring seal is placed in a polymeric resin mantle. The temperature in the internal Teflon® vessel should not exceed 250°C, and the pressure should be lower than 8.3 MPa. Under these conditions the temperature of the external mantle is below 50°C. Safety of operation is ensured by a disk of an organic polymer, placed at the bottom of the mantle. If the pressure inside the vessel exceeds 10.3 MPa, the disk is pressed down and the upper seal is open. The acid vapors expand into the space of the internal mantle, so that the whole apparatus is usually not damaged (Figure 1).

The first papers[10,12,13] describing dissolution of biological materials in a MW acid digestion system were published in 1974 and 1978. The technique found use in inorganic analysis substantially later: the basic paper[5] on dissolution of a slag, a Ni-Cu alloy, and potassium feldspar was published as late as 1983. The number of papers substantially increased in 1986, primarily in the field of analysis of geological materials, soils, and sediments. Nitric, hydrofluoric, and perchloric acids have so far been used as decomposition agents. A brief survey and characteristic of the individual procedures are given in Table 1.

The efficiency of a mixture of hydrochloric, nitric, and hydrofluoric acids has been tested on decomposition of 56 SRM samples, including minerals, rocks, soils, and various ores (involving bauxite and fluorite) and sediments.[7] ICP-OES analyses of the solutions after decomposition have demonstrated a high efficiency of the MW digestion. The results obtained are very promising, although the procedure is not yet universally applicable. Minerals such as chromite, rutile, corundum (probably also other Al oxides), cassiterite, and zircon have been found resistant, which could have been expected on the basis of experience with closed systems. Spodumen was dissolved with difficulties. The resistance toward mixtures of hydrochloric, nitric, and hydrofluoric acids has been manifested in low values obtained for Al, Ti, Zr, Cr, Sn, and the rare earth elements in analyses of rocks and soils containing the above minerals (SRM such as G-2, PCC 1-1, DTS-1, MRG-1, etc.). On the other hand, basalt BCR 1, volcanic glass RGM 1, and diabase W1 (except for Cr) have been dissolved without difficulties. The losses in Cr(III) and Pb, observed during heating of synthetic solutions in an open system,[8] have not occurred in a closed Teflon® vessel.[7] The resistance of quartz[7,17,18] is surprising and cannot be satisfactorily explained (possibly a short reaction time and an insufficient excess of hydrofluoric acid?). Of a 0.1-g sample, 4.5% of the rock or mineral remained undecomposed (the average of 33 samples of various types). The

Table 1
DISSOLUTION OF MATERIALS IN ACIDS IN A MICROWAVE OVEN

Material	Test elements and method of determination	Dissolution conditions	Notes	Ref.
Bones from archaeological findings	Sr, AAS	0.2 g + 3 mℓ HNO_3 and heat 5 min; add 3 mℓ 60% $HClO_4$ and heat 5 min; dilute with H_2O to 10 mℓ	Bones cleaned in ultrasonic bath; dissolution in open 25-mℓ conical flasks (glass)	10
Synthetic or natural volcanic glass, K-feldspar, zircon, SRM-rocks	U, Pb; determination of ratio of Pb isotopes; MS, isotopic dilution	0.1—0.5 g + 1 mℓ HNO_3, 3 mℓ HF, 0.5 mℓ 50% $HClO_4$ + 1—2 mℓ isotope additional; in vessel with cap closed and holes open, heat 5 and 15 min at 15 and 23% oven output, respectively. Cool, add 4 mℓ HF, close the holes; heat 1 hr (feldspar) or 5 hr (rocks) at 23% output; evaporate almost to dryness in open vessel. Add 6 mℓ HNO_3, heat to disappearance of $HClO_4$ vapors; dissolve in 15 mℓ 5% HCl	60-mℓ Teflon® PFA vessels with screw-on cap and 2 holes in the cap that can be closed by Teflon® seals; carousel with a closed polyethylene sample holder	6
Basic blast-furnace slag	Fe, Mn, Mg, Ca, Si	0.5 g + 5 mℓ HCl + HF (7 + 3), then 2 mℓ HNO_3; heat at full output 3 min	250-mℓ polycarbonate bottles with polypropyl screw-on caps; polypropylene turntable for 12 samples	5
K-feldspar	Na, K, Ca, Al, Si	Cool with CO_2 stream, open, add 93 mℓ 1.5% H_3BO_3, and heat 2 or 8 min		
Ni-Cu alloy	Cu, Ni, Mn, Fe, Al, Si, AAS			
Sea and river sediments, rocks, minerals. Cr, Mn, Fe, Mo, Si ores; sulfidic ores and concentrates. Soils. Total of 56 SRM samples	Si, K, Al, Li; sulfidic ores, also Ag, As, Bi, Cd, Cu, Pb, Sn, Zn; ICP-OES	0.1 g + 2 mℓ HNO_3 + 5 mℓ HCl + HF (7 + 3); heat at full output 2.5 min. Cool with air stream and in ice water; open, add 93 mℓ 1.5% H_3BO_3, and heat 10 min	The same vessels as Ref. 5. Cap sealed with Teflon® tape	7
Coal, coal fly ash, rocks, sediments, biological materials	25 elements, including As, Si, Cr, Ba, Ti; ICP-OES	0.2 g + 5 mℓ aqua regia + 2 mℓ HF. Heat 3 min at full output; add 1 g H_3BO_3 and heat 10 min on water bath. Filter if necessary, dilute to 100 mℓ	Open polycarbonate or Teflon® vessel	8
Cu-Ni ores, concentrates, and wastes	Cu, Ni; AAS	0.5 or 1 g (wastes), add 1.5 g $KClO_3$, 10 mℓ HNO_3, and 5 mℓ HF; heat 3 min at output of 477 W; cool the vessels in ice water and dilute with H_2O to 1000 mℓ	150 mℓ Teflon® PFA vessel with screw-on cap; MW oven with output of 750 W	17

Table 1 (continued)
DISSOLUTION OF MATERIALS IN ACIDS IN A MICROWAVE OVEN

Material	Test elements and method of determination	Dissolution conditions	Notes	Ref.
Buddingtonite, $NH_4AlSi_3O_8 \cdot {}^1/_2H_2O$	Na, K, NH_4^+ HPLC	0.1 g add 1 mℓ 3 *M* HCl and 4 mℓ HF; close vessels, heat 30 sec at full output of 650 W; cool, open, close again, and heat 2 min	Feed with pressurized air; 250-mℓ polycarbonate bottles with polypropylene caps sealed with Teflon® tape	15
Ashes of active charcoal or resin	Au, Ag, Al, Ca, Co, Cu, Fe, Mg, Ni, Pb, Si, Zn; ICP-OES	Ash of 1 g of material + 7 mℓ HCl + HF (4 + 3) + 2 mℓ HNO_3; digest 5 min at full oven output; cool, add 80 mℓ 6.2% HCl + 2.5% H_3BO_3 + 1 mℓ diethylenetriamine; mix, close vessel, and heat 10 min	250-mℓ Teflon® bottles with screw-on caps; cool-in with air stream	18
SRM steels	Al, Mn, P, Cu, Ni, Cr, V, Mo, Sn, Si, Ti, possibly also As, W; ICP-OES	1 g + 3 mℓ HNO_3 + 3 mℓ HCl + 2 mℓ HF in open vessel; after reaction, close vessels and heat 80 sec at full oven output; cool and dilute to 100 mℓ	60-mℓ teflon PFA vessels with screw-on caps	14
SRM of river and estuary sediments	Ca, Fe, Mn, Cr, Pb, Zn; AAS	Leaching sequence: 1 *M* $MgCl_2$; acet-buffer, pH 5; 0.04 *M* hydroxylammonium chloride in 25% acetic acid; aqua regia, possibly with HF	Metal speciation	19

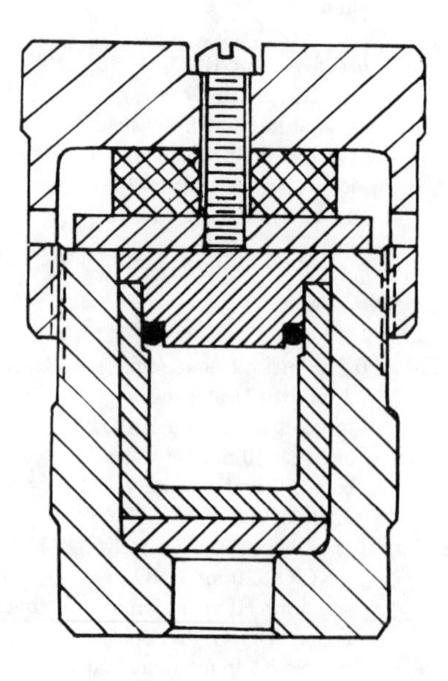

FIGURE 1. Parr's microwave acid digestion bomb.

efficiency of the decomposition may be improved by changing the composition of the acid mixture. A single solvent cannot act as an ideal, universal decomposition agent for such an extensive set of materials with variable elemental and phase composition.

The considerable shortening of the reaction time and minimal contamination introduced during dissolution have made this technique also prospective for isotopic geology and geochronological U/Pb methods.[6] The rock samples are decomposed by a mixture of nitric, hydrofluoric, and perchloric acids, and the uranium and lead contents are determined mass spectrometrically in the solution, after adding the isotopes. Special attention has been devoted to the behavior of zircon, especially to leaching of uranium and lead from the mineral. The leaching was quantitative within 20 to 120 min, using MW radiation with radiation-damaged (metamict) zircons with high uranium contents. The amounts of dissolved metals depends on the structure of the mineral grains, especially on the shape of the growth zones and the existence of microcracks along which the solvent penetrates, and is rapidly heated by microwaves inside the solid phase.

Sulfidic Ni-Cu ores and the products of their treatment are dissolved within 3 min in a mixture of nitric and hydrofluoric acids and potassium chlorate.[17] Closed Teflon® vessels are more suitable for the purpose than polycarbonate bottles that turn brown and become opaque by the influence of the solvent. Wastes after ore treatment are dissolved with a white residue with a high silicon dioxide content, but sulfidic minerals are quantitatively extracted from these products.

In dissolution of steel, atmospheric oxygen is removed from the reaction vessel, as it forms an explosive mixture with the hydrogen liberated by the reaction of the metals with the acids.[14] Preliminary dissolution is an open system is safer: After degassing, the decomposition is completed in a MW oven. Traces of aluminum are dissolved from steels within a mere 80 sec. The above difficulties have not been encountered in dissolution of Ni-Cu alloys, and the metallic material has been directly dissolved in a closed polycarbonate bottle, without a preliminary decomposition.[5]

Coal samples (with a grain size below 150 mesh) appear unchanged after treatment with a mixture of aqua regia and hydrofluoric acid,[8] but the inorganic admixtures are extracted from the sample. Parallel determinations, carried out after coal decomposition or dissolution of ashes that were prepared, gave the same contents, close to the recommended ones, for major and some trace elements (except for Ba, Ti, and Cr). The sulfur contained in coal in the form of inorganic sulfides or sulfates is also dissolved (except for $BaSO_4$). Organically bound sulfur is minimally attacked in this way. This dissolution procedure makes it possible to differentiate among the various forms of sulfur present in coal.

The reaction time required for sequential leaching of metals from sediments by various solvents can be shortened from 24 to 4 hr by application of the MW technique, with the results comparable with those obtained by conventional dissolution.[19] The possibilities of this progressive decomposition technique are summarized in a study under preparation.[20]

REFERENCES

1. **Steger, H. F., Mark, E., and Desjardins, L. E.,** *Talanta,* 25, 181, 1978.
2. **Pyper, J. W.,** *Anal. Chim. Acta,* 170, 159, 1985.
3. **Kraszewski, A.,** *J. Microwave Power,* 15, 209, 1980.
4. CEM Corporation, Prospectus, Indian Trail, N.C., 1985.
5. **Matthes, S. A., Farrell, R. F., and Mackie, A. J.,** *U.S. Bur. Mines Tech. Prog. Rep.,* No. 120, 1983.
6. **Fischer, L. B.,** *Anal. Chem.,* 58, 261, 1986.
7. **Lamothe, P. J., Fries, T. L., and Consul, J. J.,** *Anal. Chem.,* 58, 1881, 1986.

8. **Nadkarni, R. A.,** *Anal. Chem.,* 56, 2233, 1984.
9. **Kingston, H. M. and Jassie, L. B.,** *Anal. Chem.,* 58, 2534, 1986.
10. **Brown, A. B. and Keyzer, H.,** *Contrib. Geol.* (*Univ. Wyo.*), 16, 85, 1978.
11. **De Boeur, J. L. M. and Maessen, F. J. M. J.,** *Spectrochim. Acta,* 38B, 739, 1983.
12. **Abu Samra, A., Morris, J. S., and Koirtyohann, S. R.,** *Anal. Chem.,* 47, 1475, 1975.
13. **Barrett, P., Dawidowski, L. J., Penars, K. W., and Copeland, T. R.,** *Anal. Chem.,* 50, 1021, 1978.
14. **Fernando, L. A., Heavner, W. D., and Gabrielli, C. G.,** *Anal. Chem.,* 58, 511, 1986.
15. **Klock, P. R. and Lamothe, P. J.,** *Talanta,* 33, 495, 1986.
16. Parr, Microwave Acid-Digestion Bombs, Prospectus, Moline, Ill., 1986.
17. **Smith, F., Cousins, B., Bozic, J., and Flora, W.,** *Anal. Chim. Acta,* 177, 243, 1985.
18. **Russell, G. M.,** *MINTEK Rep.,* No. M289, 1986.
19. **Mahan, K. I., Foderaro, T. A., Garza, T. L., Martinez, R. M., Maroney, G. A., Trivisonno, M. R., and Willging, E. M.,** *Anal. Chem.,* 59, 938, 1987.
20. **Šulcek, Z., Novák, J., and Vyskočil, J.,** *Chem. Listy,* in press.

Chapter 7

DECOMPOSITION BY ACIDS IN ULTRASONIC BATH

Ultrasound has found use in inorganic analysis primarily in special methods of gas analysis, in electrochemistry for acceleration of electrolytic deposition of metal ions in solution, and in degassing of solutions.[1] Its dispersing effects have been utilized not only for nebulization of solutions in plasma spectrometry, but also in dissolution of solids in acids. An ultrasonic source converts the line voltage into a high-frequency electric energy (20,000 kHz) that is fed to a converter in which it is changed into mechanical vibrations of the same frequency. The vibrations are transferred to the solution by means of a horn. The vibrations of the horn tip create longitudinal sound waves that produce in the solution pressure waves connected with positive and negative pressure areas. An enormous number of microscopic cavities are formed, and the cavities either expand or collapse, depending on the local pressure conditions. The cavitation produces miniature shock waves. The solvent molecules are intensely stirred and activated, and the dissolution process is accelerated. Ultrasound also substantially accelerates leaching of melts by acids and water, as the melt disintegrates rapidly, and the increased surface area is more easily accessible to the solvent. This procedure has been used for disintegration of the melt of a ceramic material with a mixture of sodium carbonate and zinc oxide.[2] Boron has been determined by ICP-OES in the aqueous extract of the melt. Analogously, disintegration of glassy melts of alkali borates and their dissolution in mineral acids are accelerated.[3]

Samples are dispersed by ultrasound to form fine suspensions that are readily attacked by acids, e.g., with dissolution of viscous reaction products that are formed in decompositions of iron ores by condensed phosphoric acids.[4]

In mixed solvents with hydrofluoric acid, ultrasound suppresses the formation of insoluble surface films at the interface and thus facilitates penetration of the acid into the solid. This procedure has been successful, e.g., in determination of K_2O and FeO in rocks and minerals[5,6] and CaO and MgO in glass fragments.[7] Ultrasound can accelerate the interaction between sparingly soluble complex fluorides and boric acid.[8] After dissolution of 1 g of a fly ash in 10 mℓ of 48% HF, 80 mℓ of a saturated solution of boric acid is added to the reaction mixture. The solution is agitated for 4 hr on a mechanical shaker and exposed for 2 hr to ultrasound. An analogous procedure has been used in the leaching of coal ashes with hydrochloric acid.[9]

The rate of dissolution of natural antimony(III) oxide in 1.5 *M* tartaric acid is increased in an ultrasound field, and the reaction time is decreased from 14 to 2 hr, while the amount of the oxide dissolved increases. Antimonite and auripigment are minimally attacked under these conditions.[10] An ultrasound bath has also been employed in determination of Pu in soils,[11] using 8 *M* HNO_3 for acid extraction, with a yield of 23% of the element. Hydrogen ions and some other cations sorbed on soil materials are liberated more easily by the effect of ultrasound than by agitation on a mechanical shaker.[12,13] For extraction, 0.1 *M* HNO_3, 1 *M* KCl, and H_2O are used.

Special attention has been paid to rapid decomposition of various types of natural calcium phosphates and products of their chemical treatment.[14,15] The rate of liberation of phosphorus pentoxide from magmatic and sedimentary apatites, superphosphates, and other fertilizers has been followed. The solid was dissolved in an ultrasound bath with intense mechanical agitation, in solutions of aqua regia, 20% HCl, 0.01 *M* EDTA, 2% citric acid, and in water, at temperatures of 25 and 75°C, and at the boiling point. The shape of the extraction curves indicates that the dissolution rate is limited in this system by interface processes. The authors[14] report a substantial shortening of the reaction time. However, the efficiency of ultrasound leaching has not been confirmed in the later detailed work.[15] Similar results can be attained for most analyzed objects in the same time when vigorously stirring the solution (up to 700 rpm). An increase in the reaction rate has been objectively demonstrated for dried, powdered plant samples.[16] A 4-min leaching of the material in 1.5 *M* HCl in an ultrasound bath yielded higher results for Cu, Mn, Ca, and Mg than the same leaching without ultrasound.

Selective leaching of zircon grains by 50% hydrofluoric acid in an ultrasound bath permits removal of the altered parts of the mineral.[17] At present, increased attention is being paid to the effects of ultrasound, in order to accelerate decomposition of rocks and ores by acids, and to automation of the whole apparatus.[18]

Ultrasound also liberates solid particles mechanically trapped on the surfaces. This phenomenon has been utilized for cleaning of plastic vessels in determinations of submicrogram amounts of elements in various materials.[19,20] Teflon vessels, closed in polyethylene, are leached with 4 *M* HNO_3 for 12 hr in an ultrasonic bath. An analogous principle has been applied to analyses of atmospheric particulates trapped on glass fiber, cellulose filters, or plastic membranes. Hydrochloric or nitric acid and its mixtures are mostly employed as the solvents.[21-26] The filter is placed in a Teflon® beaker containing the solvent, and the beaker is placed in an ultrasonic bath for 10 to 50 min. The fine particles, liberated from the filter, are completely or partially dissolved in the acids. Quartz grains are only dissolved when the solvent contains hydrofluoric acid.[25] Some resistant phases with high contents of titanium and chromium are dissolved incompletely.[24] The metal ions are then determined in the solution obtained, using the ICP-OES or electrothermal AAS method.

REFERENCES

1. **Zolotov, Yu. A.,** *Zh. Anal. Khim.,* 13, 408, 1958.
2. **Debras-Guedon, J.,** *Bull Soc. Fr. Ceram.,* 123, 29, 1979.
3. **Feldman, C.,** *Anal. Chem.,* 55, 2451, 1983.
4. **Mizoguchi, T. and Yshii, H.,** *Talanta,* 25, 311, 1978.
5. **Rice, T. D.,** *Talanta,* 23, 359, 1976.
6. **Kiss, E.,** *Anal. Chim. Acta,* 161, 231, 1984.
7. **Catterick, T. and Wall, C. D.,** *Talanta,* 25, 573, 1978.
8. **Silbermann, D. and Fisher, G. L.,** *Anal. Chim. Acta,* 106, 299, 1979.
9. **Ichikumi, M. and Tsurumi, M.,** *Bunseki Kagaku,* 34, 268, 1985.
10. **Vasilev, V. V., Kovenya, A. V., and Shutova, Yu. M.,** *Vestn. Leningr. Gos. Univ. Ser. Fiz. Khim.,* 3(16), 152, 1966; *Anal. Abstr.,* 14, 6775, 1967.
11. **Veselsky, J. C.,** *Anal. Chim. Acta,* 90, 1, 1977.
12. **Simeonov, V., Asenov, I., and Diadov, V.,** *Fresenius Z. Anal. Chem.,* 285, 252, 1977.
13. **Tamari, Y., Ynoue, Y., and Tsuji, H.,** *Benseki Kagaku,* 31, E409, 1982.
14. **Pleskach, L. I. and Zaitseva, N. I.,** *Zh. Anal. Khim.,* 29, 1433, 1974.
15. **Belyakova, N. I., Pleskach, N. I., and Zaitsev, P. M.,** *Zh. Anal. Khim.,* 40, 648, 1985.
16. **Kumina, D. M., Karyatin, A. V., and Gribovskaya, I. F.,** *Zh. Anal. Khim.,* 40, 1184, 1985.
17. **Krogh, T. E. and Davis, G. L.,** *Carnegie Inst. Wash. Yearb.,* 73, 560, 1973/1974.
18. **Sixta, V.,** private communication, 1987.
19. **Kinsella, B. and Willix, R. L.,** *Anal. Chem.,* 54, 2614, 1982.
20. **Gretzinger, K., Kotz, L., Tschöpel, P., and Tölg, G.,** *Talanta,* 29, 1011, 1982.
21. **Sneddon, J.,** *Talanta,* 30, 631, 1983.
22. **Cresser, M. S., Ebdon, L. C., McLeod, C. W., and Burridge, J. C.,** *J. Anal. At. Spectrosc.,* 1, 1R, 1986.
23. **Van Loon, J. C.,** *Selected Methods of Trace Analysis,* John Wiley & Sons, New York, 1985, 270.
24. **Harper, S. L., Walling, J. F., Holland, D. M., and Prager, L. J.,** *Anal. Chem.,* 55, 1553, 1983.
25. **Begnoche, B. C. and Risby, T. H.,** *Anal. Chem.,* 47, 1401, 1975.
26. **Janssens, M. and Dams, R.,** *Anal. Chim. Acta,* 65, 41, 1973; *Anal. Chem.,* 70, 25, 1974.

Chapter 8

DECOMPOSITION BY FUSION

I. INTRODUCTION

Fusion, taking place at high temperatures, leads to deep changes in the structure of the test material. The mechanism of the reactions occurring during fusion is very similar to that of dissolution of solids in liquids. The rate of decomposition primarily depends on the surface area of the substances to be dissolved. A limiting factor is a high viscosity of melts compared with solutions, a low rate of diffusion of the ions liberated, and the formation of insoluble products at the interface. The process is also affected by the chemical character of the solvent and the properties of the substances to be decomposed, e.g., the bonding energy of the atoms or ions in the structure.

In general, fusion is employed for decomposition of substances that are insoluble in acids. Readily soluble products need not necessarily be obtained. More often, several new phases are obtained during fusion that are more easily soluble in acids than the original phase. The formation of compounds with various solubilities is then used to advantage for group separation of the substances.

The decomposing effect of fusion agents is mainly given by high fusion temperatures. The supply of a sufficient amount of energy causes the heterogeneous reactions to proceed relatively rapidly, and the original, stable substance is completely decomposed, provided that a sufficient excess of the fusion agent is present.

The heterogeneous reactions occurring in the melt can be divided into two groups, namely, (1) acid-base and (2) redox reactions:

1. (a) Alkaline fusion (carbonates, borates, and hydroxides)
   (b) Acid fusion (disulfates, fluorides, and boron oxide)
2. (a) Oxidizing fusion (alkaline fusion agents + oxidants, peroxides)
   (b) Reducing fusion (alkaline fusion agents + reductants, fusion with sulfur and alkali)

Some kinds of fusion, e.g., with borates and reducing fusion, are treated in separate chapters, in view of new findings in the field and extensive application in analytical practice.

II. FUSION WITH ALKALI CARBONATES

The fusion agents involve anhydrous sodium and potassium carbonates, less often also the lithium and cesium salts. Mixtures of these substances, mostly with eutectic compositions, are also employed. The melting points of some alkali fusion agents are listed in Table 1. The most common agent is sodium carbonate. The commercial preparations are sufficiently pure and suitable for determinations of major and minor components. However, the agent is exceptionally used for determinations of trace elements, as the system may be contaminated by impurities introduced with the fusion agent and by corrosion of the fusion vessels. For purification of fusion agents see Chapter 3. Mitchell[1] has described a separation of impurities by multistep extraction and ion-exchange chromatography in the preparation of pure Na_2CO_3 from $NaNO_3$. The sampling and testing of industrial Na_2CO_3 are described in the British Standard[2] and an AAS-ETA determination of trace elements in Na_2CO_3, after volatilization of the matrix.[3] It is advantageous to use $NaHCO_3$ as the fusion agent, as it is available highly pure and is converted into normal carbonate at 300°C. When using this substance for decomposition, a very fine dispersion of the fusion agent around the sample grains is attained.[4]

Table 1
MELTING POINTS OF SOME FUSION AGENTS

Compound	Symbol	Melting point (°C)
Lithium carbonate	Li_2CO_3	720
Sodium carbonate	Na_2CO_3	851
Potassium carbonate	K_2CO_3	891
Cesium carbonate	Cs_2CO_3	610
Sodium potassium carbonate	$NaKCO_3$	500
Sodium hydroxide	NaOH	314
Potassium hydroxide	KOH	360
Sodium peroxide	Na_2O_2	675
Potassium superoxide	KO_2	380
Ammonium hydrogen sulfate	NH_4HSO_4	147
Sodium hydrogen sulfate	$NaHSO_4$	185
Potassium hydrogen sulfate	$KHSO_4$	214
Sodium pyrosulfate	$Na_2S_2O_7$	401
Potassium pyrosulfate	$K_2S_2O_7$	414
Ammonium hydrogen fluoride	NH_4HF_2	125
Potassium hydrogen fluoride	KHF_2	239
Potassium fluoride	KF	856
Sodium nitrate	$NaNO_3$	306
Potassium nitrate	KNO_3	339

During fusion with carbonates, some compounds volatilize — As and Se partially, Tl and Hg completely. Bock and Jacob[5] state that 60% of $^{75}Se(IV)$ is lost during fusion with 2 g $NaKCO_3$.

The fusion vessels are usually made of platinum and some alloys (see Chapter 3, Section II.A). The losses in the platinum through corrosion by the melt depend on the texture of the metal (the crucible age), the fusion temperature, and the composition of the material decomposed. At temperatures around 900°C the fusion agent partially thermally dissociates to carbon dioxide and sodium oxide; the latter causes serious damage to the platinum. Especially great damage to platinum is caused by lithium oxide that is formed at temperatures as low as *circa* 700°C. The Fe^{2+} and Fe^{3+} ions also exhibit very unfavorable effects, as they are readily reduced to iron metal, especially when the fusion is carried out in a gas flame, and the iron forms an intermetallic alloy with the platinum which is poorly soluble in mineral acids. Analogous behavior is exhibited by Sn^{4+}, Pb^{2+}, and compounds of Sb and As, in the presence of which the fusion crucibles may be seriously damaged (see Chapter 3). To prevent these phenomena leading to contamination of melts and losses in some ions, lower melting mixtures are recommended (e.g., $NaKCO_3$) or fusion in an electrical crucible furnace in the atmosphere of carbon dioxide or an inert gas.[6] Crucibles made of a Pt-Zr alloy (0.1%) have an increased resistance toward fusion agents.

Fusion with carbonates is mainly used to decompose silicate materials, such as rocks and glasses. During the fusion, polysilicate and aluminosilicate bonds dissociate with formation of simple alkali silicates, soluble in water or in mineral acids. Sulfates, molybdates, tungstates, polycomponent structures of phosphates, and halides are decomposed analogously. In the alkaline Na_2CO_3 melt, sulfides, manganese(II), and chromium(III) ions are oxidized by atmospheric oxygen to sulfates, manganates, and chromates that are soluble in water. Amphoteric elements, e.g., aluminum, are converted, depending on their basicity, to the anion of the corresponding acid or form a basic carbonate in the melt. Base-forming cations then yield sparingly soluble carbonates. (For the carbonate complexes of Fe, Sc, Cu, Co, Be, and U in the melt see Reference 7.) Some of these complexes are stable even in an aqueous extract of the melt (U); some are hydrolyzed with formation of hydroxides or basic carbonates. The Ce^{4+} ions form complexes even in a melt of alkali carbonates; the complexes are not formed in an aqueous extract of the melt, and all the cerium is contained in the precipitate of the hydroxides. The aqueous extracts of the melts can be employed for group separation of the oxidic anions from the insoluble hydroxides and carbonates (S, Cr, W, V, Mo).

For thorough fusion of the sample, it is necessary that the substance to be composed be very finely pulverized, usually to less than 200 mesh, and intimately mixed with the fusion agent. The amount of the fusion agent depends on the content of bases in the sample. A three- to fivefold excess of the fusion agent suffices for acidic substances with *circa* 70% or more of SiO_2, while up to a 15-fold excess is required for more basic substances with 40% or less of SiO_2.

The fusion procedure is as follows. The weighed sample is thoroughly mixed with a weighed amount of the fusion agent in a platinum crucible, covered with a thin layer of the latter, and the crucible is closed by a tightly fitting lid. The crucible is placed in a cold furnace and heated, with a break at 300°C, to expel the adsorbed water from the fusion agent and the sample. The temperature is slowly increased until the mixture begins to melt slowly; the temperature is then maintained constant, with occasional stirring, to obtain a homogeneous matter. On completion of the decomposition, the melt is either spread over the crucible walls and allowed to cool slowly, or is cooled rapidly so that it cracks and can be mechanically removed from the crucible. The melt can also be mechanically disintegrated, using a platinum wire that is placed into the melt above the liquidus temperature and pulled out with the melt cake after solidification.

The melt can be decomposed by prolonged leaching with hot water, or by dissolution in acids directly in the crucible. Manganates must be reduced with methanol during their dissolution, otherwise the amount of platinum dissolved by the action of the chlorine formed is too large. The work[8] deals with decomposition of carbonate and hydroxide melts with hydrochloric acid vapors in a closed space under a reduced pressure.

The alkali carbonates are, as pointed out above, suitable fusion agents for decomposition of silicates. Framework silicates, e.g., feldspars, the nepheline group, and zeolites, are readily dissolved in the melt and so are sheet silicates, such as micas, the series of chlorites, and clay minerals. With micas containing high amounts of fluorine, silicon losses may occur during the decomposition through volatilization as SiF_4. A fusion agent with a lower melting point must be selected in this case and fused at a low temperature. However, it seems that fluorides are preferentially bound in the complexes with aluminum in the mildly alkaline medium of carbonate melt, and thus, there is no danger of losses in silicon and fluoride. Sorosilicates-minerals containing the Si_2O_7 group, chain silicates with an n-fold SiO_3 chain (amphiboles and pyroxenes), and ring silicates with a ring arrangement of the SiO_4 tetrahedra are readily decomposed by fusion with carbonates. The latter group includes tourmalines, cordierites, and beryl, minerals that are relatively resistant toward HF. Beryl causes difficulties during decomposition. The excess of the fusion agent must not be high (two- to fourfold), while the temperature must be high; the use of a blower is most suitable. The alpha beryllium hydroxide formed in the melt is also difficult to dissolve. Simple orthosilicates, e.g., the $Al_2[SiO_4/O,(OH,EF)_2]$ mineral group, titanite, staurolite, and zircon, belong among the minerals that are difficult to decompose in the soda melt.

Mg-rich olivines and garnets are easily decomposed by fusion, but the melt is difficult to dissolve; on acidification of the extract with HCl, poorly soluble $MgSiO_3$ is formed. In the presence of excess magnesium salts, it is difficult to dehydrate silicic acid. An optimal procedure, also for other magnesium-rich phases, e.g., serpentinite, involves[9] dissolution of the melt in dilute $HClO_4$. Titanite behaves analogously, and sparingly soluble modification of hydrated TiO_2 separates during the decomposition of the melt. Its film covers the crucible walls, and is difficult to remove mechanically. The crystals formed are so fine that they cannot be separated by paper filters. Aluminum-rich minerals (andalusite, sillimanite, cyanite, topaz, and staurolite) are decomposed very slowly and must be very finely pulverized, and the fusion temperature must be high (blower). Topaz is negligibly decomposed by fusion with Cs_2CO_3, and boron oxide must be present.[10] Zircon, $ZrSiO_4$, is gradually decomposed by fusion with sodium carbonate, but its behavior is problematic when it is present as an accessory in rocks. Uchida et al.[11] have compared various kinds of fusion and pressure decomposition for an ICP-OES determination of traces of zirconium in standard rocks, and have found that direct fusion of the rock with an eightfold excess of $NaKCO_3$ for 30 min at a higher temperature is sufficient. Fusion with this substance under the same conditions has been recommended[12] for decomposition of coal ashes, rocks, ores, and slags in an AAS determination of 14 components. The master solution is prepared by a classical procedure, after precipitation and removal of silicic acid.

Synthetic oxides and the corresponding minerals are very poorly decomposed in carbonate melts, and the use of various mixtures of sodium carbonate with borax is more suitable. Cassiterite, SnO_2, is especially resistant toward carbonate fusion, even if the procedure is repeated. Other oxygen-containing compounds, such as phosphates and sulfates, are readily decomposed by fusion with carbonates. The procedure has been used for determination of the REE in rocks[13] and ores containing monazite and xenotime.[14] However, the phosphates cannot be quantitatively separated from the hydroxides of thorium and the rare earths in an aqueous extract of the melt.

The final method of the determination is ICP-OES, either direct, or following a separation of given elements from excess alkali salts by extraction or chromatography. Donaldson[15] has recommended carbonate fusion for decomposition of resistant phosphates of Zr, Ti, and Th in determination of phosphorus in ores and rocks. Fusion is either used to decompose the acid-insoluble residue, or, at high contents of the metals ($>$15 mg), the material is fused directly.

Barium sulfate and the mineral barite can be completely decomposed by a 30-min treatment with a tenfold excess of sodium carbonate. The sulfate ions pass quantitatively into the aqueous extract of the melt, whereas Ba, Sr, and Ca are converted into sparingly soluble carbonates. The procedure has also been used[16] in determination of Ba and Sr in a mixture of natural niobates; the melt is dissolved in HCl + H_2O_2, and the solution obtained is used for AAS.

Carbides are also readily decomposed in melts of alkali carbonates;[17] the procedure has, e.g., been used for determination of impurities in silicon carbide.[18] Nonmetallic borides have been decomposed by fusion with $NaKCO_3$ in a nickel or glassy carbon crucible.[19]

Another example of the use of carbonate fusion in determinations of elements in ores and rocks is an NAA determination of tin[20] and spectrophotometric determinations of tantalum,[21] vanadium,[22] and selenium.[23,24] A determination of tungsten (wolframite and scheelite concentrates) has also been described, employing an aqueous extract of carbonate melt.[25] In determining higher contents of Fe in silicates and iron ores, fusion with Cs_2CO_3 or Na_2CO_3 and an oxidant has given good results.[26]

Carbonate decomposition has often been applied to determinations of halides, whose anions quantitatively pass into an aqueous extract of the melt and can be determined, e.g., photometrically, potentiometrically with ISEs or by HPLC. Some examples of these applications are given in Table 2.

The anion of boric acid also passes into an aqueous extract of the melt, which has been utilized in an ICP-OES determination of traces of boron in rocks.[41] If the potassium salt is used for the fusion, then the excess fusion agent cations can be removed from the extract in the form of the poorly soluble perchlorate.[42] From a carbonate melt of borides (10 to 100 mg + 1.5 g of soda at 1000°C for 20 min), boron can be separated by distillation of its

Table 2
DETERMINATION OF HALIDES AFTER DECOMPOSITION OF THE MATERIAL BY CARBONATE FUSION

Material	Fusion agent	Temp (°C) and time	Methods of separation and determination	Ref.
Silicate rocks	Na_2CO_3 1.2g sample 200mg	900/30 min	Precipitation with NH_4 carb. photometry of F, Cl	27
Sn-W ores, slags, silicate rocks	$NaKCO_3$ 5g sample 0.1-0.5g	750/20 min	Precipitation with NH_4 carb. photometry of F	28
CaF_2, K_2 SiF_6, Na_3AlF_6	Na_2CO_3 10—20-fold excess	750/20 min	Distillation from $HClO_4$, or H_2SO_4, F by ISE	29
Rocks	$NaKCO_3$ 4 g	950/30 min	Solution in HCl + citrate + EDTA, F by ISE	30
Mineralogical material	K_2CO_3 1 g	900/20 min	Water extract and photometry of F	31
Rock, raw material	$NaKCO_3$ 1 g	1050/15 min	F-distillation from H_2SO_4 ISE - determ.	32
Mineralogical material	$NaKCO_3$ 3 g sample 50-250 mg	900/25 min	Hydrolysis with Fe^{3+}, F-ISE	33
Carbonate rocks, soil, sediments	Na_2CO_3		Diffusion separation after Conway; distillation, ISE	34
Coal	Na_2CO_3 5 + 2 g	475/24 hr combustion 1000/15 min fusion	Water extraction, ISE	35
Rock	Na_2CO_3 0.2 g sample 0.1 g	1000/15 min	Water extract, HPLC-separation, F, Cl	36,37
Soils	K_2CO_3 0.6 g sample 0.5-1 g	460/30 min	Water extract, photometric catalytic determination of I	38
Rocks	Na_2CO_3 4 g sample 0.5 g	1000/30 min	Separation by ion exchange, ISE	39, 40

methyl ester, after acidification.[43] The same fusion agent is suitable for decomposition of elemental boron,[44] and the procedure has been applied to a mass-spectrometric determination of ^{16}B in enriched boron.

The decomposing effects of alkali carbonates can be substantially enhanced by an addition of oxidants to them. Many cations are oxidized to higher valence states merely by the action of atmospheric oxygen in the mildly alkaline medium of the melt, e.g., Mn^{2+} and Cr^{3+} up to manganate and chromate. As these ions yield readily soluble anionic compounds in the melt, the reaction equilibrium is shifted toward the more soluble compound, and thus, it is possible to decompose even stable substances containing the two elements in the lower valence state. Highly resistant chromites can thus be decomposed using a tenfold excess of sodium carbonate in a stream of oxygen.[45]

The most common oxidant is an alkali nitrate and is extensively used in a 1 + 10 mixture with sodium carbonate for decomposition of silicates. The reduction of ferric ions to the metal and the formation of intermetallic alloys with the platinum of the fusion crucible are thus prevented. The losses in the iron from the sample are then decreased, but the amount of the platinum liberated increases. This mixture has also found application in determination of sulfidic sulfur in pyrite-rich rocks.[46] A sample containing not more than 100 mg of sulfur is fused with 5 g of the fusion agent until most of the nitrate is decomposed; the melt is leached with water, and the insoluble residue is filtered off. The filtrate is acidified, and the remaining nitrate is reduced to elemental nitrogen by hydroxylamine chlorohydrate.

A mixture of Na_2CO_3-KNO_3 (1 + 1) oxidizes, on gradual increase in the temperature, all arsenic and antimony to the pentavalency.[15] The oxidation of selenium and tellurium to Se(IV) and Te(IV) is quantitative, using this mixture in a porcelain crucible at temperatures of up to 700°C. In iron and nickel crucibles, significant losses of Te have occurred,[47] due to interaction with the crucible material and sorption on the hydroxide precipitate.

A fivefold excess of a 1 + 1 mixture of KNO_3 and K_2CO_3 is an efficient fusion agent for tantalum pentoxide. The fusion temperature should not exceed 800°C, as the melt is otherwise insoluble in acids.[48] Boron nitride can be easily decomposed[49] by a mixture of Na_2CO_3 and KNO_3.

Another oxidant used in mixture with carbonate is sodium peroxide. The mixture is most often employed for decomposition of chromites and has also been proposed[15] for determination of chromium in silicates. The fusion can be carried out to advantage in crucibles made of glassy carbon[50] or zirconium metal.[51] Chromites can also be decomposed by a mixture of Na_2CO_3 and $KClO_3$; soda containing $KMnO_4$ (40 + 1) has been used to decompose copper ores.[52] An uncommon fusion agent has been used[53] to decompose binary fluorides of the II group of the periodic system. A 50-mg sample is fused with 1 g $KNaCO_3$ and 0.5 g $K_2S_2O_8$; the melt obtained is dissolved in a phosphate buffer of pH 8.2 containing EDTA, and the fluoride concentration is determined by ISE potentiometry.

A 3 + 1 mixture of K_2CO_3 and KCN is advantageous for decomposition of cassiterite and tin-containing silicates.[54] A 0.5-g ore sample is heated with an eightfold excess of the fusion agent in a nickel crucible until a clear melt is obtained, and the heating is then continued for another 30 min. The aqueous extract with the precipitate is acidified with sulfuric acid (fume cupboard!) and evaporated to dryness with HF, and the residue is dissolved in HCl. The AAS determination of Sn is carried out after extraction of the phenylhydroxamates. The procedure has an advantage in that it is not necessary to use platinum vessels, and the corrosion of nickel is negligible.

The decomposing effect of carbonates is substantially broadened in the presence of boron compounds. Borates act as solvents of oxides, and thus, various mixtures of alkali carbonates with borates are used as outstanding fusion agents for resistant compounds, such as simple and binary oxides of Al, Ti, Zr, Cr, etc. Some decompositions of this type are discussed in Chapter 8, (Section VIII. C and D), especially when the solutions of the melts are to be used as master solutions for AAS and ICP-OES.

Russel et al.[55] studied in detail the corrosion of platinum vessels in fusion with a 1 + 1 soda-borax mixture using a gas burner. The 10 + 1 mixture is commonly used for cleaning of platinum and for dissolution of oxidic residues after volatilization of "raw" SiO_2 in silicate analysis. The 3 + 1 mixture has been recommended[56] for decomposition of pure Al_2O_3 in preparation of Al^{3+} standard solutions. Similar mixtures can be used in analyses of rubies, sapphires,[57] and ignited bauxites.[58] Fusion agents with a composition of 2 + 1 to 1.5 are suitable for decomposition of natural oxides and TiO_2-based synthetic pigments,[59] titanates, e.g., ilmenite[60] (H_3BO_3 + $NaKCO_3$, 3 + 4), and titanates of strontium[61] and cobalt.[62] A determination of titanium in Guinea bauxites involving the same decomposition has been described.[63] The mixture has also found use in analyses of zircon[64] and ZrO_2-based refractory ceramics.[65]

The above mixture of fusion agents at various ratios is further suitable for decomposition of chromites (Fe,Mg)Cr_2O_4 and refractory materials derived from them.[66-68] A determination of chromium[69] and vanadium[70] in this matrix has been described, employing this decomposition procedure. This procedure has also solved the problem of decomposition of cassiterite, either pure,[71] or finely dispersed in poor ores and silicate rocks.[72,73] Fusion with a 3 + 1 mixture of K_2CO_3 and borax has been recommended[74] in an uncovered platinum crucible at 1150°C, using a burner with a blower. At lower temperatures, Sn^{4+} is readily reduced to the metal and losses in the metal occur through alloying with the crucible material.

Fusion with various mixtures of soda with borax has been employed in silicate analysis; dissolution of the melt in an acidified solution of ammonium molybdate prevents polymerization of the monosilicate.[75] The effect of the amount of B_2O_3 in the fusion agent on the quantitativeness of the SiO_2 separation on dehydration with acids has been reexamined.[76] In dissolution of the melts of synthetic Mn-silicates (after fusion with soda and borax) in dilute nitric acid, manganese precipitates as MnO_2 and can be separated by filtration from compounds of Si, Al, Ca, and Mg that pass into the extract.[77] Fusion has also been used in determinations of scandium,[78] rhenium, tungsten, molybdenum,[79] and fluoride in rocks[80] and coal,[81] often in the water extract of the melt.

The presence of SiO_2 is necessary for decomposition of materials with high contents of fluoride, such as fluorite and cryolite, by fusion with carbonates. Various mixtures of this fusion agent type have been proposed for these minerals.[82] For determination of fluorine in apatites, the following ratio of the mineral and the reagents has been recommended:[83] 1 part sample + 5 parts SiO_2 + 10 parts $NaKCO_3$, with fusion at 900°C. It has also been found[84] that in decomposition of Al-rich tourmalines by fusion with alkali carbonates, more than 20% of B_2O_3 remain in the residue after leaching the melt, if a twofold excess of SiO_2 over Al is not present.

Various mixtures of alkali carbonates and fluorides are suitable for direct fluorimetric determinations of uranium in soils and rocks. The efficiency of the method has been discussed,[85] the compositions of fusion agents specified,[86] and newer applications of the method to geochemical prospecting described.[87] A 3 + 2 soda-NaF mixture has been found effective[88] for decomposition of boron carbide; even borides of nonmetals have been reliably decomposed[19] by an analogous mixture (5 parts of soda and 2 parts of NaF).

III. FUSION WITH ALKALI HYDROXIDES

The melting points of the alkali hydroxides (Table 1) are substantially lower than those of the alkali carbonates, and thus fusion also proceeds at lower temperatures. Moreover, the temperatures given hold for the pure substances and are decreased in the presence of water and carbon dioxide in the fusion agents. The alkali hydroxides readily dissociate at high temperatures, yielding the corresponding oxides that are the cause of the high reactivity of the fusion agents with metals. The alkali hydroxides (KOH, NaOH, and exceptionally, also LiOH) are efficient decomposing agents. During fusion, the cationic components of the phases decomposed are converted into hydroxides exhibiting varying solubilities in the fusion agent and in water, whereas acid-forming elements yield anionic compounds that are readily soluble both in the melt and in water. Leaching of the melt with water has thus often been used as the first separation step after the decomposition.

For selection of alkali hydroxides as decomposition agents, the applicability of various fusion crucibles is of decisive importance. Fused hydroxides attack various metals to various extents, especially in the presence of atmospheric oxygen. Platinum, palladium, and their alloys are relatively strongly corroded; the more so, the higher the temperature. The corrosion is economically acceptable for temperatures below 500°C. Gold and a 80 + 20 gold-palladium alloy are rather resistant toward these fusion agents. Silver and some of its alloys with gold and nickel are most often used as the crucible material for these decompositions. The amount of Ag^+ passing into the solution is of the order of tens of mg, but these ions either do not interfere in the subsequent analysis or can be readily separated by precipitation, ion exchange, reduction, etc. Zirconium metal and glassy carbon are highly resistant toward fused hydroxides; teflon vessels can be used at temperatures below 300°C. In determinations of individual components of the materials decomposed, fusion in nickel, corundum, and ceramic crucibles is advantageous. The material selection depends on the ion to be determined and on the effect of the corroded vessel on the determination procedure.[6,89] The resistance of various materials is discussed in more detail in Chapter 3.

Fusion with alkali hydroxides is less universal than fusion with alkali carbonates, owing to considerable contamination of the analyte by the vessel material and to a lower temperature of the melt. Nevertheless, the use of high-purity glassy carbon permits utilization of the advantages of alkali hydroxides as efficient fusion agents for rapid decomposition of many resistant phases, applicable even to determinations of minor and trace components.

A 10- to 15-fold excess of the fusion agent over the sample is common. The weighed fusion agent is prefused in the crucible and the melt is spread over the walls by spinning the crucible and is allowed to cool. A weighed sample is placed over the solidified melt and the crucible is cautiously heated to attain liquefaction of the mixture and fusion at a lowest possible temperature. On completion of the decomposition, the melt is again spread over the crucible walls and leached with wate issolved in a mineral acids. In silicate analyses, the melt can be rapidly poured into a platinum dish and dissolved in it. The solution formed is then prepared for precipitation of SiO_2. A better wettability of solid particles by the melt can be attained by dampening of the material to be decomposed with an alcoholic solution of an alkali hydroxide prior to fusion.

Some elements volatilize considerably during fusion, e.g., Hg, Se, Te and Re. Significant losses also occur during leaching of the melt with water, mainly through adsorption and occlusion of ions by the hydrates separated; e.g., TeO_3^{2-} is sorbed on ferric or nickel(II) hydroxide when the fusion is carried out in iron or nickel crucibles.[47]

The alkali hydroxides are excellent agents for decomposition of silicates. All the polymerized and chained forms of SiO_4 are decomposed in the melt and converted into simple alkali monosilicates. Iron compounds are not reduced to the metal during fusion, interelemental compounds are not formed, and all the iron is dissolved. Provided that the fusion takes place in an electric furnace in a pure nitrogen atmosphere, the decomposed sample can be used for both determination of major components and for trace element analysis.[90] Determination of 50 elements in samples from prospecting has been described,[91] using ICP-OES after fusion decomposition with a tenfold excess of NaOH in a graphite crucible. Silicates that are more difficult to fuse and require a prolonged treatment involve simple aluminum silicates $Al_2[SiO_4(O,F_2,OH_2)]$, e.g., cyanite, andalusite, sillimanite, and topaz. Zircon is also difficult to decompose, probably due to the formation of a sparingly soluble alkali zirconate that coats the undecomposed grains of the mineral and thus prevents further reaction.

Oxides, including cassiterite and chromite group minerals, are difficult to fuse with alkali hydroxides alone, but the fusion efficiency can be improved on addition of reductants or oxidants. However, the fusion procedure is efficient for natural and synthetic phosphates, arsenates, vanadates, sulfates, molybdates, tungstates, and halides. The decomposition is usually combined with subsequent group separation of anions and cations, by leaching the melt with water or a solution of an alkali carbonate.

Many recent works have been devoted to applications of alkali hydroxide fusion in determinations of halides. Fabbri and Donati[92] followed distribution of fluoride between the solution, undecomposed fraction, and a crystalline suspension formed on addition of a complexing buffer to the melt extract. They recommend fusion with a tenfold excess of the agent in a nickel crucible placed in a furnace at 490 to 600°C for 20 min, extraction of the melt with water, and an ISE-potentiometric measurement of the F^- activity in an aliquot part of the solution. Very similar procedures have been developed for determination of fluorine in oil shales,[93] coal and coal ashes,[92] ores and ore concentrates,[28,94] and various geological materials,[95] including the fluorine-rich minerals, fluorite and cryolite.[82] The same decomposition procedure is also useful for determination of chlorine in rocks.[96,97]

The decomposition procedure can further be used for dissolution of arsenic-containing compounds, and it has been applied, without losses in the arsenic content,[98,99] to geological materials, soils, and sea sediments.[100-102] Arsenic is mostly separated in the form of the hydride from the extract of the melt. Phosphorus behaves analogously during fusion: its content in the extract was tested using standard rocks,[103] and the procedure has been applied to analyses of natural phosphorites.[104] In our experience, not all of the phosphates passed into the aqueous extract of the melt in analyses of an apatite containing 1% REE. The losses are probably caused by the formation of sparingly soluble REE phosphates. Fusion with alkali hydroxides readily decomposes niobates and tantalates; a melt extract with dilute hydrogen peroxide can be used for a direct determination of niobium in the presence of tantalum.[105] Natural tungstates can be easily decomposed by fusion with KOH, and the procedure has found application in analyses of tungsten minerals, ores,[25] and concentrates.[106,107] Cemented carbides with complicated compositions,[17] some technologically important metals (e.g., V, U, Ti), and their alloys are thus also readily decomposed. If the latter are decomposed in a stream of damp hydrogen, all the nitrogen is released from the nitrides present in the form of ammonia and can be determined, e.g., photometrically.[108]

Other applications of alkali hydroxides involve determination of traces of tin in rocks,[109] decomposition of silicates containing zirconium,[110] and determination of zirconium in ceramic frits.[111] The formation of an alkali aluminate during fusion of alumosilicates has been utilized for determination of aluminum in iron ores. The losses in aluminum through sorption on ferric hydroxide are much smaller than during precipitation with an alkali hydroxide in solution.[112]

Lithium hydroxide, $LiOH{\cdot}H_2O$, has been used[113] for liberation of alkalis from ultrabasic rocks. The sample is fused with a two- to threefold excess of the agent for 30 min in a gold crucible, and the melt is extracted with water. (However, the lithium silicates formed are poorly soluble and cause losses in the alkali metals. The recovery has improved with pressure leaching of the melt with water for 6 hr at 150°C.)

The efficiency of fusion with alkali hydroxides can be improved by adding oxidants, such as KNO_3 and Na_2O_2. Various mixtures of KOH with KNO_3, with the nitrate predominating, are suitable for fusion of the platinum metals, especially iridium.[114] Decomposition with this mixture has been used in determination of osmium in ores, fusing for 30 min at 650°C in a corundum crucible.[115] Various mixtures of NaOH with KNO_3 (10 to 50 + 1) have found use as efficient fusion agents for the nitrides and carbides of boron and silicon, titanium boride, and chromium carbide. The fusion takes place in a glassy carbon crucible, with gradual heating of the mixture to 650°C, followed by a 10-min heating at the latter temperature. The crucible is minimally corroded by the melt and thus can be used for up to 70 decompositions.[49] Fusion with a 3 + 1 mixture of KOH and KNO_3 permits decomposition of various kinds of soils and rocks in determination of bromine[116] and iodine[117] that pass into the melt extract.

The decomposing effect is enhanced when adding sodium peroxide to alkali hydroxides, because of the alkaline and oxidizing character of the former. A 4 + 1 mixture of NaOH and Na_2O_2 decomposes even cassiterite.[28] A similar mixture (10 + 1) has been found suitable for decomposition of natural and precipitated fluorites and cryolites for determination of fluorine.[82] Sulfides and selenides are quantitatively oxidized to the hexavalency during decomposition without losses. In this way, natural sulfides (galena, pyrite, arsenopyrite, cobaltite, etc.) have been analyzed for selenium, which is distilled off from an aqueous extract of the melt and determined, e.g., photometrically.[118]

A combined fusion agent of NaOH and $Na_2B_4O_7$ has been successfully utilized to decompose chromites and zircon concentrates.

IV. FUSION WITH SODIUM PEROXIDE

Sodium peroxide, Na_2O_2 (mp 675°C), is an efficient, very strongly alkaline fusion agent that simultaneously acts as a strong oxidant. The substance alone melts without decomposition and danger of explosion. However, it dissociates at red heat, with liberation of oxygen. It is dangerous when its melt is in contact with substances that are easy to oxidize, mainly various forms of carbon, elemental sulfur, and aluminum or other metal powders. Because of its strong decomposing effects, Na_2O_2 would be an ideal agent for degradation of all resistant substances, but its use is limited for two reasons: (1) insufficient purity of the chemical, and (2) the fact that its melt corrodes to various extents all the materials of which fusion crucibles are made.

Most substances used for manufacture of fusion crucibles, e.g., porcelain, silver, iron, nickel, platinum, sillimanite, and corundum, are considerably corroded by sodium peroxide at temperatures above 400°C. The losses amount to tens of milligrams per gram of the fusion agent, depending on the fusion time. Platinum is relatively stable toward the reagent up to 500°C, and this fact has been utilized, especially in sintering decomposition. The crucible can be protected against the effect of the melt at higher temperature by a film of an alkali carbonate or sulfate, and the fusion should be performed at the lowest possible temperature. Nickel resists the melt up to 600°C, but then its solubility in the melt sharply increases. The loss of sintered corundum is up to 20 mg during a 15-min fusion at 600°C.

Zirconium resists molten Na_2O_2.[119] It is also somewhat dissolved in the melt, but it seems that the losses depend on the oxidation to ZrO_2 by atmospheric oxygen at higher temperatures (5 to 25 mg during a single fusion operation). It is thus advantageous to use a reducing flame or an oxygen-free atmosphere of an electric furnace (Ar is preferable to N_2).

Another substance that resists Na_2O_2 and other fusion agents is glassy carbon. It is suitable for fusion involving sodium peroxide alone or in a mixture with a carbonate at temperatures of up to 700°C. At higher temperatures the crucible is rapidly oxidized; as the material is sufficiently pure (the sum of 19 elements in the Soviet material SU-2000 equals $1 \times 10^{-3}\%$) and the products are gases, the sample is not contaminated by the crucible material. Moreover, glassy carbon is sufficiently mechanically strong to resist the strain during rapid cooling of melts. The fusion is carried out at 650 ± 50°C, best in two muffle furnaces at various temperatures;[120-122] in the first furnace the mixture is predried and liquefied, and in the other its fusion is completed within the above temperature range.

In common procedures a weighed sample is thoroughly mixed with a 5- to 15-fold excess of the fusion agent in a crucible selected according to the purpose of the analysis and the mixture is covered with a layer of 1 to 2 g of the fusion agent. The crucible is covered with a lid and placed in a cold muffle furnace, and the temperature is slowly increased to attain the above limit. The liquid is then stirred by swirling to permit decomposition of any aggregated particles, and the heating is continued for another 5 min. The crucible is allowed to cool, and the contents are extracted with water or a dilute mineral acid.

To suppress corrosion of the crucible as much as possible, explosive decomposition by burning peroxide has been recommended. The sample is first mixed with the same amount of carbon that is prepared by carbonization of sucrose and then with an excess of the fusion agent. The mixture is properly pressed together, and the crucible is covered with a lid so that there is a gap of *circa* 5 mm and placed in water with about two thirds of it submerged. The mixture is lit using a cotton fiber, and the decomposition then takes only a few seconds. The melt is allowed to cool and treated in the same way as in fusion.

Explosive fusion with Na_2O_2 is also carried out in a closed system in a special bomb of a simple construction. The decomposition cup is placed in a case which is closed by a head; a nickel-rich alloy or pure nickel is the bomb material. The sample is mixed with the peroxide in the cup (the amounts depend on the working volume of the bomb and are recommended by the manufacturer), the bomb is closed and placed on a triangle in a steel housing, and the peroxide is ignited by cautious heating of the bottom with a gas torch. This operation requires certain skill and caution to prevent overheating and explosion of the bomb. The present type of sodium peroxide bomb is provided with an electrical ignition, with which the bomb can be cooled by immersion in water in a thick-walled container, and the operator can remain at a safe distance from the apparatus. The decomposition itself takes only a few seconds. It is used to advantage for substances that are difficult to burn in an oxygen stream, or, on the other hand, that react explosively with Na_2O_2 at laboratory temperature during mixing etc. It is suitable for determination of halides, As, P, S, Se, and B in analyses of solid caustobiolites, such as coal, coke, bitumens, various carbonaceous rocks, and oils.[123] Organic carbon is oxidized up to carbon dioxide during the decomposition. To ensure its complete burning when it is dispersed in an inorganic material, aluminum or magnesium powder must be added to the fusion agent, using its heat of combustion for acceleration of the oxidation.[124]

Fusion with Na_2O_2 ensures efficient decomposition of resistant oxygen-containing and oxygen-free compounds — borates, silicates, phosphates, sulfates, borides, carbides, nitrides, and halides. The procedure and its efficiency can be suitably modified by additions of alkali carbonates, hydroxides, and borates. The fusion agent alone or with modifying admixtures is often used to decompose refractory materials, e.g., chromite and zirconium oxide ceramics and resistant mineral materials, such as bauxites, chromites, zircon, tantalite-columbites, beryl, ilmenite sand, fired clays, etc. Chromites are decomposed by fusion with a 10- to 15-fold excess of the peroxide in an iron, nickel, or zirconium crucible, for 30 min at the temperature of liquefaction of the mixture. An aqueous extract of the melt or its acidified solution can be used for a determination of chromium, e.g., by titration, spectrophotometry, AAS, or OES-ICP.[125-127] These procedures have been applied to determination of chromium in ultrabasic rocks, chromium ores, alloys and slags. An explosive decomposition with a sodium peroxide-carbon mixture has also been used for chromites.[15]

These fusion agents have found wide use in decompositions of the PGM and are indispensable, especially with some alloys. The PGM are converted into oxygen-containing anions of higher valence states that are readily soluble in water (Os, Ru) or in mineral acids, e.g., HCl, HBr, and aqua regia. The decomposition procedure is usually not used for platinum and palladium, but to advantage for samples containing all the PGM.[114] Natural iridosmium (up to 90% Os + Ir plus other PGM) is also readily decomposed by fusion with sodium peroxide.[128] Novák[129] has pointed out poor fusibility of iridium powder (even on repeated fusion). In arbitration analyses of precious metal sweeps, a 2 to 2.5 + 1 mixture of Na_2O_2 and N_2CO_3 has been used for decomposition.[130] The PGM pass into an aqueous extract of the melt and are precipitated, after degradation of the peroxide by boiling, by reduction with an alkali formate. An alternative procedure involves fusion of the sample with a sixfold excess of the peroxide alone: the extract is then acidified, and the solution is used for a DCP spectrometric determination of the PGM. Various optical and electrochemical methods have been applied to determination of iridium[131,132] and ruthenium[133,134] in industrial concentrates and catalysts following this decomposition procedure.

In analyses of poor platinum ores the alkaline-oxidizing fusion is most often combined with separation of ballast material. Sulfidic Cu-Ni-Fe ores with chalcopyrite, cubanite, and pentlandite are preignited at 700 to 800°C and the residue is dissolved in a mixture of HCl and H_2O_2. Silicate admixtures, if present, are degraded by repeated evaporation of the residue with a mixture of HCl and HF, provided that their contents do not exceed 6%. The insoluble residue after acid leaching is filtered off, fused, usually with a fourfold excess of sodium peroxide, and the melt is dissolved in the original filtrate.[135] An analogous decomposition procedure has also been used in a spectrochemical determination of the PGM in mineral raw materials.[136] The alkaline-oxidizing fusion is the most common decomposition procedure in determination of osmium and ruthenium, especially when distillation separation of their tetroxides follows. Powerful oxidizing fusion produces oxygen-containing anions of the hexavalent ions that are further easily oxidized to octavalency by treatment with BaO_2 in an alkaline solution.[137] The perosmiates are quantitatively formed during the fusion.[115] Some authors[137,138] have pointed out losses in osmium during fusion and dissolution of the melt, due to unsuitable fusion conditions, e.g., a too-high temperature, a too-long fusion time, or a too-large excess of the fusion agent. Various physicochemical methods have been described for determination of osmium, ruthenium, and iridium in mineral raw materials,[139-141] Cu-Ni-concentrates,[142] matte,[143] and anodic nickel,[144] after an alkaline-oxidizing decomposition. Silver,[145] gold,[146] and platinum[147] have been analogously determined in mineral raw materials.

Alkaline-oxidizing fusion is often employed for isolation of the REE. All REE minerals are easily decomposed, and sparingly soluble REE hydrates are obtained in an aqueous extract that can be filtered off from soluble anions. Coprecipitation of cerium(IV) need not be quantitative, as the ion forms a soluble carbonate complex. For perfect precipitation of the REE hydroxides, salts of Fe(III) or Mg should be added to the solution as collectors.[148,149] The REE can be separated from the other elements of the ammonia group by leaching the peroxide melt with solutions of triethanolamine or mixed complexing buffers containing triethanolamine and EDTA. The REE hydroxides are quantitatively precipitated in this medium, whereas the ions of Fe, Mn, Ti, Zr, and other yield anionic complexes that are separated by filtration or centrifugation. This principle has been utilized in isolation of trace amounts of REE from ores,[150] tin concentrates,[151] and rocks.[152]

Fusion of rocks with sodium peroxide can also be used for the preparation of master solutions in OES-ICP determinations of the REE. A 0.2- to 0.5-g sample is fused with 2 to 3 g Na_2O_2 in a zirconium or graphite crucible at a dark red heat of Meker burner. The melt is cooled and dissolved in 15% HNO_3 with gradual addition of dilute HCl. The clear solution is diluted to 200 mℓ and used for the OES-ICP determinations.[153]

This decomposition procedure is very often employed in determinations of various elements in a great variety of materials. An example of the elements from group IV of the periodic system involves an AAS determination of silicon[154] and photometric determinations of germanium in zinc ores and concentrates[155] and of traces of thorium in rocks.[156] As cassiterite is reliably decomposed by peroxide, this decomposition is often utilized in determination of tin in ores, prospecting samples, and rocks. An aqueous extract of the melt contains a soluble alkali stannate. The fusion must not be carried out in nickel crucibles, as higher hydrated oxides of nickel(III) sorb not only stannate, but also part of tungstate and molybdate that may be present. The melt is removed from the crucible with a minimal amount of water and acidified to complete dissolution with hydrochloric acid.[157] Ascorbic acid is added, and stannic ions are extracted into methylisobutyl ketone for an AAS determination. Zircon is also readily decomposed by peroxide fusion, and thus the procedure has been employed for determination of zirconium in many materials. An aqueous extract of the melt can be used for a separation of anions and excess of salts, with a further treatment of the hydroxide precipitate,[158] or can be acidified and used whole for further procedures. These principles are also suitable for semimicroanalysis of zircons.[158-160]

The alkaline-oxidizing decomposition has also often been used for determinations of the elements of group V of the periodic system, especially phosphorus, vanadium, and niobium. Donaldson[161] studied in detail the quantitative oxidation of various bonding forms of antimony to Sb(V). In the melt of the peroxide alone, mixed oxides Sb_4O_8 are apparently also formed, and complete oxidation is only attained on an addition of KNO_3 to the fusing agent. The same author[162] has recommended the fusion agent alone for decomposition of polymetallic ores (Pb, Zn, Cu, S) and concentrates in determination of bismuth. The melt of 0.5 g sample and 3 g Na_2O_2 in a zirconium crucible is leached with water and acidified with H_2SO_4. Silicic acid is removed by evaporation to SO_3 vapors in the presence of HF and bismuth is preconcentrated by extraction into diethyldithio carbazon in chloroform and determined by AAS. The same procedure of fusion and melt treatment has been used[163] in determination of As, Sb, Se, and Te in silicate rocks and ores, using hydride generation followed by AAS.

Among the many other applications of the peroxide fusion, the master solution preparation for analyses of ore concentrates,[164,165] determination of tungsten and molybdenum in ores by the XRF method,[166] and an NAA determination of rhenium in mineral raw materials with a complex matrix[167] should at least be mentioned. Further uses involve analyses of raw materials and products in metallurgy, especially in decompositions of ferroalloys,[168,169] hard metals and cemented carbides,[17] iron ores,[170] and agglomerates.[171]

Decompositions with mixed fusion agents of peroxide-carbonate, hydroxide, or boron oxide are used for decompositions of the same materials for which sodium peroxide alone is used, in order to suppress the corrosive effect of the peroxide alone on the crucible material and to decrease the fusion temperature. Glassy carbon crucibles can be used to advantage, as they lose only 0.8% of their weight on fusion with a 3 + 1 mixture of the peroxide and an alkali carbonate (an amount of 2 g) at 500°C.[120] Kuteinikov et al.[122] recommend peroxide-carbonate mixtures of 4 + 1 to 1 + 1 in a 10- to 20-fold excess over the sample. The mixture is allowed to react with the sample at 300°C, and then the fusion is completed at 650 ± 50°C for 3 to 10 min. A 3 + 1 mixture has given good results for fusion of the residues of chromium ores and titanium concentrates that are insoluble in acids (HF + H_2SO_4 or $HClO_4$) or for direct fusion of ferrochromium.[121] The application to determination of sulfur in fly ash has been described,[172] with analysis of an aqueous extract of the melt. A 2 + 1 mixture readily decomposes iron ores,[173-175] and the iron is determined in the separated hydroxide precipitate. A small part of the iron is contained in the aqueous extract of the melt, apparently in the form of a colloid solution of the hydroxides or ferrates. On fusion of rocks, molybdenite, and Cu-Bi concentrates with a 4 + 1 mixture (500°C for 90 min), alkali perrhenates pass into an aqueous extract, in addition to other anions,[176] and can be extracted and then determined by ICP-OES. To decompose zirconium sands and concentrates, a sevenfold excess of a 2.5 + 1 mixture is used.[177,178] Titanium and chromium ores that are difficult to decompose by sodium carbonate alone are readily attacked by an eightfold excess of a 3 + 1 Na_2O_2-$NaKCO_3$ mixture; an aqueous extract of the melt is acidified and heated until a clear solution is obtained that can be used for determination of many elements, not only in the above materials, but also in refractory materials, coal, rocks, and slags,[12] best using AAS. Cassiterite is easily decomposed by a fourfold excess of a 1 + 1 mixture of soda and sodium peroxide.[179]

Mixtures of an alkali peroxide and hydroxide are used when the melting point of the mixture is to be lowered while maintaining a high alkalinity of the fusion agent, especially for a transfer of anions into an aqueous extract. Examples involve separation of rhenium from other elements in ore concentrates[180] and combustible slates,[181] or determination of tin in ores.[182] Gallorini et al.[183] have described fusion with this mixture in a closed system for decomposition of materials from city incinerators. The material is irradiated and fused on a nickel boat in a quartz tube connected with a trap cooled by liquid nitrogen; the fusion temperature is 850°C. Volatile elements are trapped and extracted by nitric acid.

Fusion with potassium superoxide, KO_2, is suitable for decomposition of some mineral materials containing REE that require a strong alkaline, oxidizing fusion agent. This agent has especially been used when an excess of other ions (e.g., Na^+) interferes in the subsequent analysis. The sample is mixed with a threefold excess of the fusion agent alone or in a 1 + 3 mixture with potassium hydroxide and fused in a zirconium crucible at red heat. The melt is allowed to cool and dissolved in HCl. The agent strongly corrodes most common metals, but zirconium is resistant.[184] Decomposition with KO_2 alone has been used in a determination of thorium in arcose sandstone (SRM Canmet) containing uraninite, brannerite, and monazite.[185] Rigin[186] has used it for decomposition of coal; the carbon is oxidized by a large excess (50- to 70-fold) of the agent by fusion in a furnace for 20 min at 900 K.

V. DECOMPOSITION WITH HYDROGENSULFATES AND DISULFATES

These fusion agents are weakly oxidizing and strongly acidic. On heating, hydrogensulfates first release water and are converted into disulfates which, as can be seen from Table 1, melt at relatively low temperatures. On heating above the melting point, the melt is decomposed to yield sulfate and sulfur trioxide, the latter being the actual effective component of the agent. As the decomposition usually takes a rather long time, during which the fusion agent is decomposed to produce an alkali sulfate, the fusion must occasionally be interrupted and concentrated sulfuric acid be added to produce more hydrogensulfate; the procedure is then continued as described above. Potassium and sodium salts are mostly used for fusion, and their selection depends on the solubility of the sulfates produced; e.g., K-Cr(III), K-Ta(V), and other binary sulfates are poorly soluble. If excess cations present interfere in subsequent determination, e.g., in numerous electrochemical and optical methods, many materials can then be decomposed to advantage by ammonium hydrogensulfate, which melts at a very low temperature (mp 147°C) and decomposes most substances as equally well as the alkali disulfates, and its corrosive effects on platinum and other materials are negligible. After completion of the decomposition, the agent can be almost completely volatilized at a temperature greater than 200°C. Other special agents of this group are cesium hydrogensulfate and potassium peroxodisulfate.

Quartz and platinum crucibles are most commonly used for fusion, but, in view of the low melting points of the agents and their small reactivity, borosilicate glass and silica-rich Vycor glass vessels can also be used. Platinum is perceptibly corroded by the melt (a loss of a few mg for a single fusion operation), and thus quartz crucibles are more suitable for prolonged fusion. Corrosion of quartz is almost negligible, the melt foams less than in platinum vessels, and the fusion procedure can be monitored visually. Glassy carbon is highly resistant toward the fusion agents; a 10-min action of 2 g of the fusion agent at 500°C causes a loss in the crucible material of only *circa* 0.02% of the original weight.[120] Zirconium is poorly stable toward these agents; it is considerably corroded at temperatures above 500°C in an oxidizing atmosphere and thus is not used for this type of decomposition (see Chapter 3).

The procedure first involves fusion of a selected amount of the agent; the disulfate produced is allowed to cool. The weighed sample is then introduced and the crucible is covered and heated at progressively increasing temperature to complete liquefaction of its contents. The temperature is then maintained with occasional stirring, until the liquid is perfectly clear. If the melt is becoming viscous, heating is interrupted, a required amount of sulfuric acid is added as described above, and the procedure is repeated. On completion of decomposition, the solidified melt is wetted with a few drops of sulfuric acid, and the contents of the crucible are dissolved in 10% sulfuric or hydrochloric acid. It is advantageous to combine these solvents with complexing agents, e.g., hydrogen peroxide, oxalic, tartaric, and hydrofluoric acids, which accelerate the dissolution of the melt and stabilize some ions as complexes in solution (e.g., Ti, Zr, Nb, Ta, etc.), and can also be used for group separations, e.g., of Nb and Ta from the REEs, whose oxalates or fluorides are precipitated.

Disulfates are mainly employed as fusion agents for simple and complex oxides of Al, Fe, Mn, Co, Cr, Ti, Th, Nb, Ta, Mo, W, U, etc. Sulfides are also readily decomposed, as are fluorides that are poorly soluble in acids (Ca, Al) and whose decomposition is connected with quantitative removal of the fluoride ions. In metallurgical laboratories disulfates have been applied to analyses of ferroalloys, basic metal oxides, and slags. They have been widely used to decompose actinoids. Silicates are rather difficult to decompose by these agents, which have been mainly used to decompose aluminum monosilicates $Al_2[SiO_4(O,OH,F)_2]$ that are difficult to fuse otherwise, such as cyanite, andalusite, topaz, and others. Disulfates are often used in geochemical prospecting of soils, as decomposition is possible using a gas burner and vessels made of laboratory glass under the field conditions.[187] In silicate analysis, disulfate is primarily used for final fusion of insoluble residues after decomposition by a mixture of hydrofluoric and sulfuric acids and in determination of individual elements, mainly titanium, zirconium, niobium, and tantalum in silicate matrices.

Titanium dioxide and titanates are very often decomposed by fusion with disulfates, e.g., in analyses of rutile and other modifications of TiO_2 — anatase and brookite, plus ilmenite, perovskite, and more rare titanates of manganese, magnesium, and barium. This decomposition procedure has been applied to determination of titanium[188] and vanadium[189] in titanium ores and pigments[190] and used for the preparation of titanium standard solutions. The PGM are often concentrated in titanomagnetites, natural solid solutions of magnetite, and ulvite. As these phases are exceptionally difficult to decompose by reductive fusion, disulfate fusion is used. Relatively large samples of up to 10 g can be decomposed by fusion with a sevenfold excess of the fusion agent at a temperature not exceeding 440°C. Under these conditions, $TiOSO_4$ is produced in the melt, which is readily soluble in sulfuric acid. Titanium oxides, formed at higher temperatures, hydrolyze during dissolution.[135] Fusion with a threefold excess of the fusion agent has been applied to decomposition of metamict titanoniobates (betafite), and the solution obtained has been used for separation of many elements by ion exchange chromatography.[191]

Disulfate fusion is very efficient for decomposition of niobates and tantalates, such as columbite and tantalite. The material is fused with a large excess of the fusion agent (up to 50-fold) for as long as 60 min. The melt is dissolved in hydrochloric and tartaric acids, and the solution is used for an AAS and flame emission determination of lead,[192] manganese,[193] titanium, niobium and, tantalum.[194,195] Accessoric amounts of niobates in rocks can be decomposed in the same way, after acid decomposition.[15] Many niobium-tantalates of the rare earths with uranium from the group of euxenite and aeschinite can be decomposed to advantage by this procedure, with simultaneous separation of the REE on dissolution of the melt in a complexing agent. The REE precipitate is filtered off, and the filtrate is used for chromatographic separation of Nb, Ta, Zr, and Ti on an anion exchanger, for selective hydrolysis of Nb and Ta and for other determinations. Selective effect of the disulfate melt on rhodium metal is utilized for quantitative dissolution of rhodium and its separation from iridium and platinum.[114]

Disulfate fusion is advantageous for decomposition of a great variety of materials in determinations of individual elements, as well as in complete analyses. Decomposition is easy for isolated REE oxides, various natural and synthetic spinels, zirconium ores containing baddeleyite (ZrO_2), manganese nodules from the sea bed, concentrates of nonferrous metal sulfidic ores, and others. In our experience, decomposition of chromites is often incomplete, and the undecomposed residue must be refused. The mineral zircon is difficult to decompose, but refractory ceramics[196] containing up to 90% ZrO_2 or containing corundum[197] are perfectly decomposed. Disulfate fusion retains its importance in geochemical prospecting of soils[198] for nonferrous metals, e.g., Cu, Pb, Zn, Bi, Ag, and Mo. As the decomposition does not cause losses in arsenic and antimony, these elements can be utilized for indirect prospecting for gold ores. The melt of the soil sample is dissolved in hydrochloric acid, and the trace amounts of the elements of interest are preconcentrated and determined by AAS.

Among substances not containing oxygen, carbides and borides, resisting many fusion agents and acids, are decomposed by this fusion agent. Disulfate fusion has been applied to determination of aluminum in iron and steels,[199] titanium in its carbide,[200] and to decomposition of tungsten carbide.[17] For decomposition of boron carbide and titanium boride, a mixture of disulfate with peroxodisulfate or the latter alone have been found more suitable.[17,49,201,202]

Disulfates have been used for rapid and total destruction of actinoid compounds; e.g., high-temperature ignited plutonium dioxide can be decomposed by fusion with a 4 + 1 mixture of sodium and potassium disulfate or with the sodium salt alone.[203,205] A great excess of the fusion agent is required. Mixed oxides of uranium and thorium can be decomposed in the same way, by fusing for 30 min at 400°C in a glass vessel. The potassium salt alone yields with plutonium poorly soluble binary sulfate that precipitates on prolonged standing, even from a solution of 5 *M* HNO_3; the sodium salt is sufficiently stable in this medium.[204] Not only oxides, but also phosphates of thorium and the REE, oxidic ores, and technical dusts containing uranium oxides can be decomposed analogously, and this principle has been used in determinations of a number of natural actinoids. In the latter case, the material is first decomposed by hydrofluoric and sulfuric acids in a teflon vessel until sulfur trioxide fumes appear, and the sulfate solution is transferred to a flask containing anhydrous sodium sulfate. After dehydration of the system, the insoluble residue is decomposed by fusion with the disulfate formed, with appearance of a transparent melt. The melt is then quantitatively dissolved in 3 *M* HCl and the solution is used for subsequent determinations. Direct fusion of materials containing radionuclides with disulfate in platinum crucibles leads to an up to 20% loss in the ^{212}Pb isotope through its alloying with platinum[206,207] and thus, cannot be used.

The importance of ammonium hydrogensulfate as a fusion agent is mentioned above. The agent has been used with good results for decomposition of oxides of titanium, zirconium,[208] and niobium[209] and in determination of these elements in rocks, river sediments, and other geological objects. The agent also decomposes plutonium dioxide, including the ceramics supporting nuclear fuels based on uranium and plutonium.[210,211] The time of fusion of 200 mg of finely pulverized material is up to 4 hr at 400°C. The decomposition has also been used for rapid extraction of plutonium from soils.[212]

An interesting modification of sulfate fusion agents is the use of ammonium sulfate. The melting point of this compound (350°C) enables work at higher temperatures. During thermal decomposition, sulfuric acid is liberated in statu nascendi, and thus, its action is more powerful than that of its aqueous solutions; the ammonium ion liberated simultaneously behaves as an ampholyte, dissociating a proton and enhancing the acidic character of the ammonium salts. The decomposition is performed with a five- to tenfold excess of the reagent at 400 to 450°C and lasts 1 to 2 hr. It is possible to decompose oxides of aluminum, titanium, niobium, tantalum, zirconium, and tin (cassiterite). The melts are leached with 4 *M* H_2SO_4 (Nb and Ta in the presence of complexing agents) with formation of clear solutions.[213,214]

VI. FUSION WITH FLUORIDES

Potassium and ammonium hydrogenfluorides belong among acidic fusion agents and are efficient agents for decomposition of highly resistant minerals, such as zircon, beryl, and a number of niobotantalates. During fusion, binary complex fluorides are mostly formed (Be, Al, Fe, Ti, Zr, Nb, Ta, U, and others), but also insoluble fluorides of calcium, thorium, and the rare earths. An extract of the melt in 5% HF is then used for group separation of soluble and insoluble fluorides. Elements yielding fluorides with low boiling points volatilize during fusion with KHF_2. Thus, this fusion procedure is advantageous for decomposition of silicates with simultaneous removal of most of the silicon. On the other hand, the fact can be utilized that on rapid fusion with excess fusion agent silicon yields binary silicon-potassium hexafluoride that can be separated from soluble fluorides by leaching of the melt with dilute nitric acid with an excess of potassium nitrate; the isolated salt can be used for a titration determination of silicon. This decomposition procedure has been used for silicates to a limited extent, as complete removal of fluoride ions from the solution is difficult. Evaporation with nonvolatile acids can be replaced by a combination of fluoride and disulfate fusion.

Platinum vessels are almost exclusively used for this decomposition. It has been found experimentally[55] that 150 μg of platinum is liberated from the crucible during a single fusion with 2 g KHF_2. The fusion has two stages. First, it is necessary to fuse at a very low temperature, with the agent thermally decomposing to yield HF in statu nascendi that strongly reacts with the phases present and liberates volatile fluorides. The end of this stage is heralded by solidification of the melt. The temperature is then increased to 700 to 800°C (best in a furnace), and the undecomposed fractions are fused. This stage has the character of alkaline fusion, and the mixture is maintained at the above temperature until a clear melt is obtained. The melt is allowed to cool, and 5 to 10 mℓ concentrated sulfuric acid is added. The acid is evaporated almost to dryness, and the residue is fused with the potassium hydrogensulfate formed at 500°C. The cooled melt is leached with 10% HCl and boiled to dissolution of the sulfates.

Alternative leaching with nitric or hydrofluoric acids is discussed above. The application of this fusion procedure in analytical chemistry and in determination of trace elements has been reviewed.[215,216] Oxidic substances without a silicate matrix can be readily decomposed by direct fusion with normal alkali fluorides at temperatures of up to 900°C.

Potassium hydrogenfluoride is one of the few agents capable of decomposing beryl. The above procedure is used, including the final displacement of fluorides by sulfuric acid.[217,218] For extraction-fluorimetric determination of low contents of beryllium in rocks and alloys, it is recommended[219] to use acid decomposition prior to the fusion. A reagent special for beryl is sodium silicon hexafluoride.[220] Among RO_2-type oxides, cassiterite,[221] rutile, and baddeleyite can be readily decomposed. In addition to oxides, some silicates of zirconium, especially zircon, and certain rare minerals containing zirconium, thorium, and uranium can be decomposed. Fluoride fusion has also found use in fluorescence determinations of traces of uranium in rocks. Cesium hydrogenfluoride has rarely been used for decomposition of tourmalines and determination of the alkali metals in it. The fusion agent is prepared by the action of hydrofluoric acid on a mixture of the sample with cesium carbonate, and the fusion itself takes only a few minutes. The fluorides are removed by the following fusion with boric acid, in which cesium tetraborate is formed.[222] Potassium hydrogenfluoride is often employed for decomposition of heavy concentrates with columbite and tantalite predominating, as well as for the separated minerals.

Potassium hydrogenfluoride serves relatively often for isolation of the REE and thorium from silicate materials and their minerals, especially monazite. The silicate matrix is first degraded by evaporation with hydrofluoric acid, and the residue is fused with a tenfold excess of the agent, followed by leaching of the melt with slightly acidified water. The fluorides separated on digestion (REE + Th + Ca + U) are filtered off, dissolved, and used for chromatographic separation or an ICP-OES determination.[223-227]

The alkali fluorides have been applied analogously as the alkali hydrogenfluorides, e.g., for decomposition of beryl, monazite, columbite, cassiterite, zircon, and other stable compounds. A 3 + 1 mixture of potassium fluoride and nitrate is an efficient fusion agent for silicon carbide-carborundum.[228] The alkali fluorides have attained considerable importance in the control of the state of nuclear fuels and in determination of actinoids, in combination with a subsequent disulfate fusion.[229,230] The procedure for determination of the actinoids from radium to californium in silicate matrices[231] first involves preliminary decomposition of a weighed sample with nitric and hydrofluoric acids in a platinum vessel. The salts formed are fused with a threefold excess of potassium fluoride at a high temperature. The melt is allowed to cool, and it is repeatedly evaporated with concentrated sulfuric acid to form hydrogensulfate, with which the mixture is then fused. The melt is dissolved in 2 *M* HCl, and the solution is used for separation of the actinoids. Disulfate has principal importance for the solubility of protactinium salts, as it prevents the formation of polymeric hydrolytic compounds that are produced during dissolution in acids. The actinoids are usually separated by coprecipitation on barium sulfate. Various modifications of the procedure have been used in determinations of isotopes of uranium, thorium, protactinium, and transuranium elements in ores,[232,233] in soils, silicates, and environmental samples.[228,233-235] Determination of strontium isotopes in soils has been described,[236] after an analogous decomposition and separation. Actinoids in soils can also be isolated by coprecipitation on neodymium fluoride,[237] after the same decomposition procedure.

Ammonium fluoride and hydrofluoride are only rarely used for decomposition of inorganic materials. They can be employed for the same oxidic phases as the other fluorides; among silicates, however, cyanite, andalusite, topaz, zircon, and resistant ceramic materials react with difficulty. The agents have found use for decomposition of ultrabasic rocks containing olivine and enstatite in determination of the PGM dispersed in titanomagnetites of these rocks.[135] A sample of up to 30 g is decomposed in a large nickel or glassy carbon dish with 50 g NH_4F at 150°C, and the temperature is increased to 500°C after solidification of the matter, to ensure complete degradation of the titanium minerals and volatilization of SiF_4. The melt is then wetted with 10 mℓ of concentrated hydrofluoric acid and fused again at the same temperature. Excess fluoride is then removed by repeated evaporation with concentrated sulfuric acid, and the residue in the crucible is fused with the ammonium disulfate formed. The cooled melt is leached with 10% sulfuric acid, and the solution is used for determination of the PGM.

Ignited plutonium dioxide can also be decomposed by fusion with ammonium hydrogenfluoride in an argon stream.[238]

The use of ammonium hydrogenfluoride in inorganic analysis has been dealt with by Biskupski.[239]

VII. SPECIAL FUSION PROCEDURES

Alkali nitrates are oxidizing fusion agents that reliably and safely decompose organic matter. Their melting points are relatively low (Table 1), and they decompose to oxygen and nitrogen, with formation of nitrites, at temperatures around 400°C. The oxidation of carbon leads up to the formation of carbon dioxide.[240] An equimolar mixture of sodium and potassium nitrates has been used in determination of total sulfur and phosphorus in soils. The sample is mixed with a twofold excess of the fusion agent, covered with a small amount of the agent, and heated to attainment of 450°C, which is then maintained for 1 hr. The melt is dissolved in boiling mineral acids (HCl and HNO_3), and the solution is made to a suitable volume. Sulfate is then determined turbidimetrically and, phosphate spectrophotometrically in aliquot parts of the solution.[241]

Sodium hexametaphosphate is sometimes used to prepare beads for XRF. Its advantage over the classical borate glasses is a low melting point and solubility of the bead in water.[242] The substance has found use as an efficient solvent for chromium(III) compounds and has been applied to analyses of chromium ores and chromium-containing refractory materials.[243] It can also be employed for decomposition of natural sulfates, e.g., barites and anhydrite, that liberate sulfur trioxide on fusion at 850°C. The latter is reduced to sulfur dioxide and then can be used for isotopic analysis of sulfur.[244] Fusion with KH_2PO_4 has been used for decomposition of silicates in an ICP-OES determination of boron. The procedure is performed in a glassy carbon crucible for 1 hr at 800°C, in an electrical furnace rinsed with nitrogen. The melt is then fused with potassium hydroxide, and an aqueous extract of the melt is further treated.[245]

Selenium dioxide is a very efficient fusion agent for oxides of Nb, Ta, Fe, In, Ga, and others. The oxides are refused under a decreased pressure with a tenfold excess of the agent at 420°C in a quartz ampule[246] (see Chapter 5).

Cupric sulfide has very rarely been used as the fusion agent in determination of oxygen in zinc selenide.[247] Among special techniques, levitation melting should be pointed out, as it is suitable for analyses of extremely pure metals and simple alloys. The material is fused without reagents, in an alternating electromagnetic hf field.[248,249]

VIII. FUSION WITH COMPOUNDS OF BORON

Boron oxide and many alkali metal and alkaline earth borates are efficient nonoxidizing decomposition agents for resistant minerals, their synthetic analogues, heat-resistant materials, carbides, and various refractory materials. The group of these fusion agents involves substances with widely varying chemical character, from acidic boric anhydride to the most alkaline alkali metal tetraborates. These substances are mostly used in the anhydrous form, after dehydration of the compounds, which is best attained by fusion at a sufficiently high temperature. They are used for decompositions either alone, or in mixtures with substances modifying the decomposition procedure; oxidants, compounds altering the acid-base character of the fusion agent, or admixtures that affect the following analytical procedures, such as internal standards, substances absorbing X-rays, etc., are often added to the fusion agent. The fusion is carried out in platinum crucibles that are sufficiently resistant toward the effects of the fusion agents. However, a disadvantage lies in the fact that platinum is readily wetted with the melt, so that the borate glass formed adheres strongly to the crucible surface. Glassy carbon crucibles have been found much more convenient, and the melt can readily be poured out of them or the solidified bead can be removed mechanically. The same properties are exhibited by the binary alloy, Pt-Au (95 + 5), or the ternary alloy, Pt-Au-Rh (92 + 5 + 3). Crucibles made of these materials are not wetted at all by borate melts and behave analogously as those made of glassy carbon.

The melts obtained with borate fusion agents are generally highly viscous, and the viscosity increases with increasing acidity of the fusion agent; this fact then affects the fusion temperature. High temperatures are mostly used (a Meker burner with a blower, electric furnaces, hf heating). The high temperatures during the fusion lead to thermal dissociation of alkali borates, especially the lithium salts. The free alkali metal oxide then considerably corrodes both pure platinum and its alloys. At high temperatures, some ions in the melt are also reduced to the elemental form, especially iron(III), cobalt(II), lead(II), and tin(IV) ions. The metals formed then react with platinum with the production of intermetallic alloys, causing changes in the composition of the test material and permanent contamination of the crucibles. The temperatures of fusion of some fusion agents of the borate group are listed in Table 3.

A. Boron Oxide

This compound is the weakest fusion agent in the borate group and exhibits virtually no oxidizing effect. It has been introduced in the analytical practice by Davy,[250] and its application has mainly been developed by Jannasch.[251,252] The agent has found use primarily in decompositions of resistant minerals and in the determination of SiO_2 in materials with high fluorine contents. The fusion agent is either prepared by gradual dehydration of boric acid followed by fusion of the product, or is obtained commercially. (For a discussion of the purity of the fusion agents see Reference 253.) Anhydrous boron oxide is highly hygroscopic and must be stored in the form of large lumps in a tightly closed vessel. For a decomposition, the required amount of the agent is sufficiently finely pulverized in an agate mortar. An advantage of this agent is the fact that alkalies are not introduced into the sample, and that the agent can be completely removed. As the B_2O_3 melt is highly viscous, the fusion must be performed at a high temperature; consequently, alkalies may partially volatilize during the fusion. Fluorine is completely volatilized during the fusion, but no loss of silicic acid has been reported. However, some authors[254] have demonstrated that this method of separation of fluorine from silicon is not quantitative. The contents in the crucible must be often

Table 3
MELTING POINTS OF SOME BORATE FUSION AGENTS

Base components	%[a] (w/w)	Melting point (°C)
Boron trioxide		
Sodium metaborate		
Sodium tetraborate		450
Sodium tetraborate	80 }	966
Sodium fluoride	20 }	741
Lithium metaborate		839
Lithium tetraborate		845
Lithium tetraborate	66.5 }	930
Lithium metaborate	33.5 }	875
Lithium tetraborate	80 }	
Lithium fluoride	20 }	780
Lithium metaborate	80 }	
Lithium tetraborate	20 }	830
Calcium metaborate		1154

[a] The composition of the mixed fluxes refer to composition prior to fusion.

stirred during the decomposition, using a platinum spatula or a thick wire. Excess boric acid, after dissolution of the borate glass, is removed by volatilization as the methyl ester, evaporation with hydrofluoric acid, distillation with water vapor, etc. in most cases (e.g., in the gravimetric determination of SiO_2). During fusion, B_2O_3 is partially volatilized in the form of a white aerosol irritating the respiration tract. Solubility of the air in fused B_2O_3 has been pointed out.[256] It is interesting[257] that fusion of silicates with this agent does not cause complete degradation of the tetrahedral SiO_4^{4-} groups on dissolution of the melt in an acid, and some Si-O coordination groupings, leading to the formation of various polymers, are preserved, even in solution.

The fusion agent has not found wide use in the silicate analysis because of the above disadvantages.[258] It is used more often for decomposition of resistant oxides, such as corundum, zircon, ceramics based on these minerals, and compounds with high contents of fluorine.[255] In addition to the classical determination of SiO_2 in fluorite and cryolite, a decomposition of a synthetic, fluorine-rich phlogopite has also been described;[259] the authors consider losses in SiO_2 to be negligible. The same decomposition procedure has been used for a rapid determination of SiO_2 in glass-making sands[260] containing more than 90% SiO_2. The cooled melt is treated with hydrofluoric acid at an elevated temperature, thus removing the fusion agent, together with silicon dioxide.

Boron oxide has also been used for a determination of alkalies and calcium[261,262] in materials with a high Al_2O_3 content. A 0.5-g sample is decomposed by fusion with 0.85 g B_2O_3 in a platinum crucible, first mild and then, after disappearance of bubbles (*circa* 5 min), vigorous, using full flame of a gas burner. The cooled melt is dissolved in warm HCl, 0.5% v/v, and the solution obtained is analyzed.

Boron oxide fusion has also been applied to a determination of total uranium in lean ores[263] and rocks. The residue from the acid decomposition is fused with boron oxide, the fusion agent is removed by heating with HF and $HClO_4$, and the solution obtained that does not contain excess salts is reduced in volume to 1 mℓ. The whole operation can be carried out in a small platinum crucible. The use of the eutectic of the binary mixture, B_2O_3-PbO, to liberate elemental nitrogen from slags[264] is also interesting.

B. Alkali Borates

These fusion agents, derived from meta- and tetraboric acids, are important reagents in analytical chemistry. Anhydrous sodium tetraborate, either alone or in mixture with an alkali carbonate, has long been used as an efficient fusion agent. However, its application has been limited to decompositions of selected resistant materials and refractory ceramics. Only the efforts to speed up the classical procedures of the silicate analysis and their replacement with other rapid methods have led to the renaissance in the use and development of new fusion agents based on boric acid. These fusion agents have found use not only in wet chemical analysis, but their efficiency and simple techniques of the preparation of solutions have been found useful in many instrumental analytical methods, especially XRF and its instrumental modifications that require rapid and quantitative transfer of the test substance into stable solid solution. Borate melts meet these conditions.

In direct-reading methods, such as optical emission spectroscopy, borate melts are important in two ways. First, the sample is perfectly homogenized, and the borate glass obtained by the fusion process has a uniform composition; this process is called "iso-formation". Further, the borate fusion agent functions as a spectral buffer affecting the thermal equilibrium in the electric arc, owing to low ionization potentials of the fusion agent components. The dilution of the sample with the fusion agent also minimizes the effect of the changing matrix on the quanta of the emitted radiation due to the test elements. Alkali borates are especially suitable because they yield simple spectra interfering with only a few analytical lines of the test elements. The interelemental effect can be minimized by using internal standards (Co, Sr) that can be contained in the fusion agent, the fusion process ensuring their perfect homogenization.

The greatest importance of borate melts lies in application to XRF. In analyses of pulverized materials alone, the effect of macroheterogeneity of the material and uneven distribution of the grain-size fractions in the sample on the intensity of the excited X-rays is prominent. To eliminate or suppress these effects, pulverization of the sample to a very fine grain size has been recommended, but the required effect has not been attained, because particles of various hardness are pulverized at various rates; the effect of a nonuniform granulometry is thus not removed. Prolonged milling is also connected with a danger of segregation of heavier phases and increased contamination caused by abrasion. Borate fusion completely removes the individual phases, and the sample is converted into a homogeneous glass. During the fusion, an internal standard can be introduced to compensate for the interelemental effects, and a suitable dilution ratio can be selected.

The development of instrumental techniques has also affected the techniques of the preparation of the test samples. Many borate fusion agents of a high purity have appeared on the market, and considerable progress has been made since the advantages of the lithium salts for most instrumental analytical methods were recognized. Simultaneously with the development of the fusion agents, the problem of adhesion of borate glasses to platinum was solved, as platinum is most often used for the manufacture of fusion crucibles. Platinum has been replaced by glassy carbon or by nonwettable alloys. The fusion process has been substantially accelerated and unified by introducing automated fusion apparatus with hf heating of the system.

The composition of alkali borate melts and their temperatures of fusion can be readily found in the binary concentration vs. temperature diagrams for the Na_2O-B_2O_3 and Li_2O-B_2O_3 systems,[265] depicted in Figures 1 and 2. The completely different behavior of the sodium and lithium compounds is quite conspicuous. It can be seen from the diagram what the melting points are of the two compounds that can be considered, $R_2O \cdot B_2O_3$ and $R_2O \cdot 2B_2O_3$, as well as the eutectic temperatures and the composition of the phases below the liquidus temperature. On the basis of this diagram, two basic fusion agents were chosen for analytical purposes, as well as some of their mixtures, as given in Table 3. These are, e.g., products

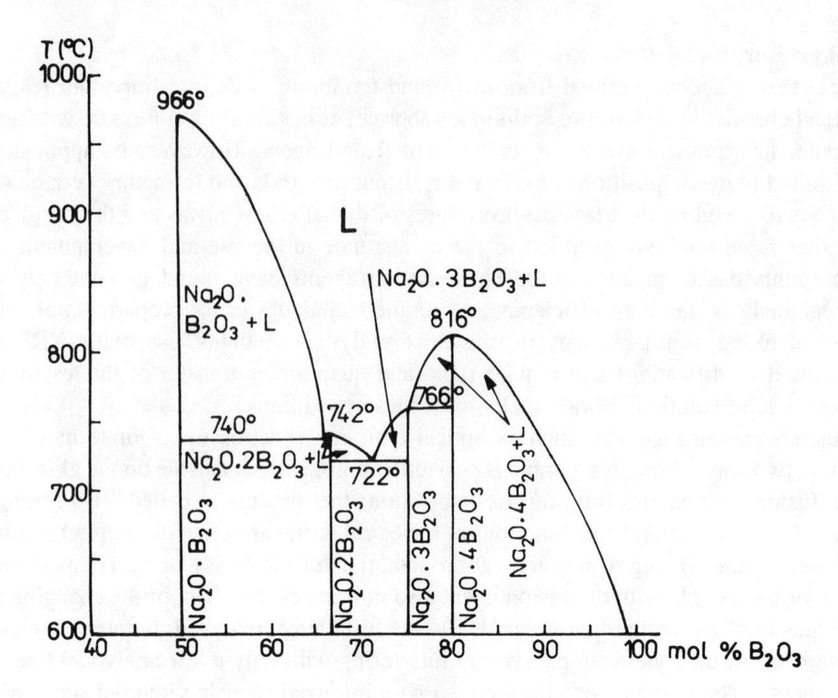

FIGURE 1. Isobaric TX section through part of the Na_2O-B_2O_3 system.[265]

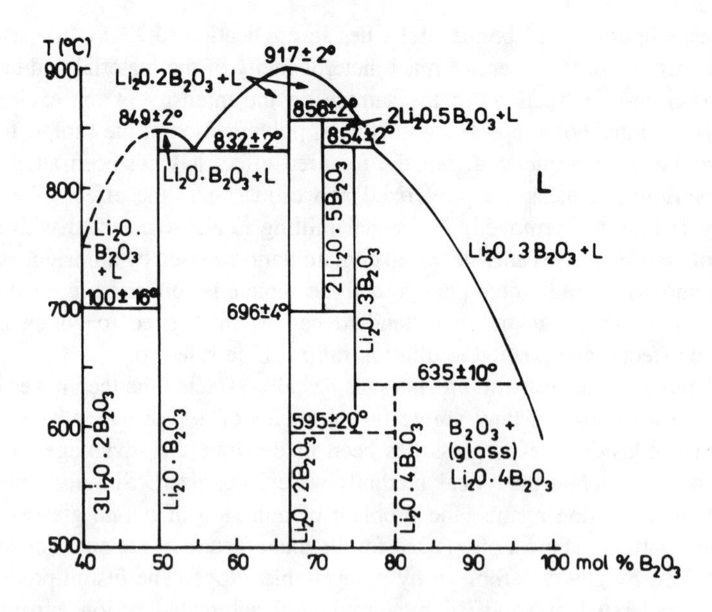

FIGURE 2. Isobaric TX section through part of the Li_2O-B_2O_3 system.[265]

of the Johnson Matthey Chemicals, London, G. B., with the trade name Spectroflux.® The 80 + 20 mixture of lithium meta- and tetraborate corresponds to the eutectic with a melting point of 830°C; the 33.5 + 66.5 mixture has the composition $2Li_2O{\cdot}B_2O_3$, containing 60% mol B_2O_3 and mp 875°C. An analogous series of fusion agents is supplied by E. Merck (Darmstadt, FRG) under the trade name, Spectromelt®, consisting of sodium and lithium tetraborates, lithium metaborate, and a mixture of the lithium salts containing 60.5% mol B_2O_3 with a liquidus temperature of 890°C.

A theory of the decomposing effect of borates has been developed by Bennett and Oliver[266] and Oliver.[267] The authors observed the efficiency of fusion of lithium tetra- and metaborate alone and of their 1 + 4 mixture and studied the degree of decomposition of various types of materials used in the ceramic industry. In the ceramic technology and petrology, silica is considered as an acidic oxide, alkalies and the alkaline earth oxides as strongly alkaline, and alumina as an amphoteric oxide with a tendency toward increasing the melt basicity. Analogously, with alkali borates, the alkali oxide is the basic component, and boric oxide the acidic component. In fusion of an acidic oxide with lithium metaborate, the alkaline component participates in the reaction, and the liberated boron oxide produces tetraborate in the melt. On the other hand, tetraborate is primarily considered as the source of boron oxide that reacts with basic oxides, with formation of the metaborates and an equivalent amount of an alkali metaborate. Amphoteric oxides, such as Al_2O_3, react more readily with B_2O_3 than with R_2O.

These concepts are in a good agreement with the experimental results, where materials with high SiO_2 contents react more readily with metaborates and those with high Al_2O_3 contents with tetraborates. It is also important that only tetraborate of the three above fusion agents yields a suitable glass bead in the absence of a sample. This fact suggests that the fraction of tetraborate in the bead must exceed the value, 1:4 with $LiBO_2$, except when other elements participate in the glass formation. Thus, it can be expected that in fusion of samples with high contents of basic oxides, e.g., those of the alkali metals and the alkaline earths, well-castable glasses will only be formed with tetraborate. The selection of the ratio of tetraborate to a basic oxide is discussed in the section dealing with the preparation of glasses for XRF.

C. Sodium Tetraborate, Borax, and Its Use in Wet Chemical Analysis

Borax is a strong, nonoxidizing fusion agent used to decompose substances that are difficult to fuse, such as cassiterite, zircon, oxidic minerals of chromium, titanium, and niobium, spinels, and a refractory ceramics. The disadvantages of the agent involve relatively poor fusibility of the glasses formed, a high viscosity of the melt, and the necessity to remove a great amount of boron, introduced with the fusion agent, prior to the next analytical step. High temperatures during the fusion cause losses in some components during fusion; e.g., volatilization of the alkali metal oxides, especially during prolonged heating. Losses in selenium were found experimentally,[268] using the ^{75}Se isotope; a 5-min fusion of a standard with 1 g of anhydrous borax led to a loss of up to 20%.

Fusion is carried out in a platinum crucible in which the required amount of the fusion agent is first fused, and the melt is poured over the crucible walls. The sample is introduced after cooling, the crucible is covered with a lid, and the contents are heated at gradually increasing temperatures. The melt is stirred occasionally with a platinum spatula; the process takes 30 to 120 min. On termination of the fusion, the melt is poured over the crucible wall and allowed to cool under the lid; during cooling the melt cracks. The cooled melt is transferred to an evaporation vessel and covered with 50 mℓ of anhydrous methanol saturated with gaseous hydrogen chloride. The mixture is heated on a water bath under a watch glass. The watch glass is removed after dissolution of the melt, and the solution is evaporated to dryness. The residue of the melt in the crucible is leached with the same reagent, and the removal of boric acid methyl ester is repeated several times after adding methanol.

The borax fusion has been rarely used during the last two decades in complete silicate analysis. Neuberger et al.[269] used this decomposition for the determination of the major and some minor elements in iron and manganese ores, steel-mill slags, and other metallurgical materials. The 1 + 10 melt is dissolved in dilute nitric acid (1 + 9). The solution obtained is used as the stock solution for photometric determinations of most of the elements in suitable portions of the solution. An analogous preparation of the stock solution for an ICP-

OES determination of some elements in coal and airborne ashes has been described by Bennett[255] and Botto;[270] the latter author used an automated fusion device. In about 95% of analyses, the sum of the oxides determined after a single fusion step lies within an interval from 97.5 to 101.5%. The borax fusion has further been applied to a determination of silica in iron ores[271] and fluorites.[272] The former author[271] points out difficult degradation of magnetite in a borax melt. Broekaert et al.[273] used this fusion procedure for degradation of insoluble residues isolated from silicate rocks after dissolution in a mixture of hydrofluoric and sulfuric acids. The melt is then dissolved in 2 *M* hydrochloric acid (1 g sample in 100 mℓ), and the stock solution is used to determine the rare earth elements by the ICP-OES method. An analogous procedure was adopted[274] in determinations of trace elements in the GA, GH, and BR international rocks standards.

Silicates were decomposed by fusion with borax for an AAS determination of lithium.[297] The sample is thoroughly mixed with the fusion agent (1 + 5) and fused in a graphite crucible for 15 min at 1000°C in a muffle furnace. The melt obtained is transferred to 3 *M* nitric acid and dissolved with continuous stirring. The solution is used for the determination of lithium without any other treatment.

Determination of oxidic substances is discussed extensively in an older paper by Jeffery,[275] devoted to decomposition of titanium (ilmenite, rutile perovskite) and niobium-tantalum (niobite, tantalite) minerals, possibly containing rare earths (pyrochlor, microlite, euxenite, samarskite, etc.) in a borate melt. The decomposition takes up to 2 hr at about 1200°C.

It has long been known that borax is the most powerful and universal fusion agent for zirconium minerals — zircon and baddeleyite. Among special procedures, separation of zirconium from minimal samples of its minerals (0.1 to 1 mg) has been described,[276] using fusion with a 20-fold excess of anhydrous borax. The melt is dissolved in a minimal amount of dilute hydrochloric acid, and zirconium is extracted with a β-diketone and determined mass spectrometrically. The authors used an analogous method[277] to determine zirconium in geological and lunar samples on macroscale. The fusion with borax is carried out at 1200°C, and zirconium is determined in the solution of the melt using isotope dilution analysis. The same decomposition has also been used for isolation of lead produced by the radioactive decay of uranium and thorium isotopes present in isomorphously in zircon.[278] The melt is dissolved in 2 *M* hydrochloric acid and lead is separated in the form of its chlorocomplex on an anion exchanger and determined mass spectrometrically in the aqueous eluate. (This decomposition has recently been replaced by dissolution of zircon in hydrofluoric acid in an autoclave.)

Mixtures of borax with alkali carbonates are used in wet analysis more often than borax alone. These mixtures are more basic than borax alone, as metaborate is formed during fusion, whose eutectic with tetraborate has a relatively low melting point of 722°C. Various mixtures of these substances are useful for decomposition of oxides and some resistant materials.

A soda-borax mixture (3 + 1) is often employed to fuse the insoluble residue after decomposition of silicate rocks with hydrofluoric acid. The residue contains small amounts of accessory minerals resistant toward mineral acids and melts of alkali carbonates. The same mixture is used to fuse the mixture of oxides remaining from the "raw" silicon dioxide after volatilization of SiF_4 with hydrofluoric acid.

A 1 + 1 mixture of the two reagents is suitable for decomposition of cassiterite;[279] however, the test material must be very finely pulverized. Weiss[280] used a 3 + 1 mixture of borax and potassium carbonate in a determination of traces of tin in rocks to decompose the insoluble residue after volatilization of SiO_2 with hydrofluoric acid. The fusion is performed in a platinum crucible placed in an electric furnace at 1150°C, and the solidified melt is dissolved in a mixture of 1 *M* sulfuric acid and 0.25 *M* oxalic acid. Losses of tin have been observed when gas burner is used for the fusion, as the gases diffusing through the crucible reduce Sn^{4+} to the metal that forms alloys with platinum.

Among oxygen-containing minerals, aluminum oxide-corundum and its natural and synthetic analogues can be decomposed by a soda-borax mixture (3 + 2).[281] The procedure has been applied to a determination of chromium in ruby. A similar mixture has been used to decompose titanates,[282] the synthetic silicotitanate-fresnoite, and the raw materials for its synthesis.[283] Determinations have been described of major components in magnesium-aluminum spinels[285] and a fluorescence determination of uranium in these minerals,[284] after decomposing the material with a 1 + 1 soda-borax mixture. The latter mixture is suitable for decomposition of spinels of the type, $(Fe,Mg){\cdot}Cr_2O_4$, and refractory materials based on these spinels. After 50-min fusion at 1000°C the cooled melt is dissolved in 1 *M* sulfuric acid, thus preparing the stock solution for the determination.[286] The decomposition of chromites by this method is more universal than fusion with hydroxides and alkali peroxides used previously, and thus, is has been selected as the obligatory procedure by the British Ceramic Research Association Committee.[287] A fusion decomposition in the determination of calcium in chromites is described in Reference 288.

To decompose zirconium compounds, its oxides, silicates, and ores, a borax-richer mixture (2 + 1) is used; the fusion takes place at 1000 to 1200°C.[289] The decomposition has also been used to determine scandium in zirconium dioxide.[290]

In order to improve fusibility of zircon and its concentrates, up to 25-fold excess of a 1 + 1 soda-borax mixture was used,[291] with a small amount of a calcium salt added. Knyazeva et al.[292] have developed a complete spectrophotometric analysis of zircon using relatively small samples of 20 to 100 mg; they have used a tenfold excess of anhydrous borax with sodium nitrate added.

Mixtures richer in borax (3 + 1) have been successfully applied to decompose resistant silicates of the Al_2SiO_5 type in determinations of silica[293] and traces of arsenic[294] in silicates. Jeffery[295] gives another procedure for decomposition of cyanite, sillimanite, corundum, and other resistant minerals. A 1-g sample is mixed in a platinum crucible with 6 g sodium carbonate and 0.5 g borax glass. The mixture is heated for 30 min at 650°C, thus sintering the materials with liberation of carbon dioxide. The temperature is then increased to 1000°C, and the melt is maintained at this temperature for 10 to 15 min.

Cobb and Harrison[296] used a soda-borax mixture (2 + 1) for decomposition of iron ores, slags, and refractory materials. The melt was dissolved in 20% v/v nitric acid with an addition of hydrogen peroxide at a maximum temperature of 80°C. After adding a lanthanum(III) salt, the solution can be used for the AAS determination of aluminum.

D. Fusion with Alkali Metaborates

Sodium metaborate is only occasionally employed for decompositions. The commercial product, $NaBO_2{\cdot}H_2O$, is insufficiently pure and must be dehydrated prior to use. The high melting point of this compound (966°C) is also disadvantageous. On the other hand, the glasses obtained during fusion are readily soluble in mineral acids. The fusion mixture is usually prepared by mixing equimolar amounts of anhydrous sodium carbonate and boric oxide. Usually the components are directly mixed with the material to be decomposed in a platinum or graphite crucible. The fusion is carried out on a burner, first at a low temperature with the crucible covered with a lid (liberation of carbon dioxide), and then the temperature is increased to *circa* 900°C. The contents of the crucible are periodically stirred; the melt formed is readily mobile. On completion of the decomposition, the melt is allowed to cool slowly and is spread over the crucible walls. The cold melt is dissolved in a suitable mineral acid. By varying the ratio of the two components, the basicity of the fusion agent can be varied to obtain values suitable for decomposition of materials with various chemical compositions.

Hazel[298] used the fusion agent to decompose silicates and considers it a universal agent, capable of decomposing even relatively coarse materials (<25 mesh) within 20 min. For the following analysis, the boron can be removed as boric acid methyl ester, but the author states that contamination of the separated silicic acid with boron can also be removed by washing the precipitate with a suitable solution.

A mixture of sodium potassium carbonate and boric acid has been used[299] to decompose silicate materials with coagulation of the silicic acid sol without dehydration. For materials rich in silicon, a threefold excess of the 8 + 1 mixture is suitable, whereas a twofold excess of the 5 + 1 mixture suffices for materials rich in aluminum (up to 85% Al_2O_3). The fusion process takes 10 to 30 min at a high temperature of 1200°C. The melt is dissolved in warm 8 *M* hydrochloric acid and yields a clear solution used for further determinations. Fusion with an equimolar mixture of sodium carbonate and boric oxide has been applied to a gravimetric determination of silica in rocks. The melt is dissolved in dilute hydrochloric acid, and the silicic acid is coagulated with polyethylene oxide, followed by the gravimetric determination.[300,301]

The effect of the amount of the fusion agent on the separation of the silicic acid gel in the gravimetric determination of silica has also been studied.[302] An eight- to tenfold excess of a 3 + 2 mixture of sodium carbonate and boric acid has been found to be optimal. Silicate decomposition by fusion with an up to 20-fold excess of a 7 + 1 mixture was used in an AAS determination of beryllium and vanadium.[303]

Julietti and Williams[304] have proposed a universal method for decomposition of a great variety of silicate and refractory materials and oxides, yielding clear solutions for AAS or ICP-OES determinations of silicon. A finely pulverized sample is fused in a platinum crucible with a 2.8-fold excess of a 6 + 1 mixture at 1200°C for 25 min. After cooling, the surface of the melt is covered with a twofold weight excess (over the sample weight) of potassium carbonate, and the mixture is fused for 5 min at the same temperature. After rapid cooling of the crucible in water, the melt is dissolved in 60 mℓ of warm 6 *M* hydrochloric acid with occasional stirring. The solutions obtained are perfectly clear, and all the silicon is retained in solution.

The materials that have been decomposed in this way involve aluminosilicates, such as clays, graphite, and coal ashes, sillimanite, cyanite, blast-furnace slags and silicides, up to 95% SiO_2, aluminum-rich materials, up to 95% Al_2O_3, barium titanate, titanium dioxide, etc.

Various mixtures of sodium carbonate and boric oxide have also been found to be excellent fusion agents for refractory oxides. For example, zircon ($ZrO_2 \cdot SiO_2$)[305] has been decomposed using a 12-fold excess of the 2 + 1 mixture; the melt is dissolved in HCl and yields a clear solution suitable for instrumental determinations of various elements. The good fusibility of aluminum oxide mentioned above has been used for determinations of some trace elements in bauxites.[306] Mixtures richer in boric oxide can even be applied to decomposition of cassiterite,[307] but a large (40- to 200-fold) excess of the fusion agent is necessary.

The advantages of the mixed fusion agent containing sodium potassium carbonate (e.g., a lower melting point and easy dissolution of the melt), as pointed out by the authors,[304] were utilized in decomposition of chromites[308] and titanium oxides.[309] Chromites are decomposed in a mixture of $Na_2CO_3 + K_2CO_3 + H_3BO_3 = 1 + 1 + 1$, using a 12-fold excess over the sample weight. The glass formed is dissolved in sulfuric acid. Titanium oxides (edisonite, rutile, and ilmenite) are decomposed in a 4 + 3 mixture of $NaKCO_3$ and H_3BO_3 used in a 3.5-fold excess. The mixture is first sintered and then fused at 1000°C. On completion of the decomposition, the melt is poured on a ceramic tile and dissolved separately from the residues in the crucible in warm (70°C) 8 *M* hydrochloric acid.

E. Fusion with Lithium Metaborate

A great turn occurred in the use of alkali borates as decomposition agents when fusing effects of the lithium salts were discovered. The behavior of the two substances that are used most often can be found from the diagram for the Li_2O-B_2O_3 binary system (Figure

2). Compared with the analogous Na_2O-B_2O_3 system (Figure 1), the differences in the melting points of lithium metaborate and tetraborate (849 and 917°C, respectively) are not as large as those for their sodium counterparts. The mixture of the two salts produces an eutectic with a relatively high melting point of 832°C. The composition of the eutectic is approximately tetraborate + metaborate = 1 + 4. The mixture has been selected as a universal fusion agent for materials with silicon and aluminum oxides as major components.[266] The two compounds differ in basicity. Metaborate is a better fusion agent for acidic rocks and materials, whereas tetraborate is more suitable for decomposition of basic materials.

Lithium metaborate was first used as a fusion agent by Keith[310] in XRF and emission spectrometric analyses. It has been found to be primarily an efficient fusion agent for silicates and other rock-forming minerals. It yields suitable glasses that are mechanically strong, exhibit a low hygroscopicity, and are relatively easily dissolved; their solutions can be used in instrumental methods employing liquid phases, such as absorption and emission spectrometry, AAS, ICP-OES, etc. Ingamells[311-313] and co-workers[314,315] applied lithium carbonate for these purposes.

Lithium metaborate is supplied commercially with suitable purity and grain size, so that it can directly be used in chemical analysis. It can also be prepared[314] by fusing equimolar amounts of lithium carbonate and boric acid powder (1 mol Li_2CO_3 + 2 mol H_3BO_3). The mixture is slowly heated on a platinum dish and is maintained at 250 to 300°C. The cake formed is crushed, pulverized, and heated again at a temperature below 625°C; a yield of the product of *circa* 95% is obtained after sufficiently long heating, and the product can usually be directly used as a fusion agent. If necessary, the substance can be purified by recrystallization from an aqueous solution and saturated with the substance at 90°C over several days at laboratory temperature. The separated octahydrate of the salt is gradually converted into the dihydrate by slow, several-day drying at 40°C in an aired oven, followed by heating to 200°C. The dihydrate is further dehydrated by heating at 620°C. The anhydrous product is pulverized and sieved to completely pass through a 16-mesh sieve. The purity of the preparation is tested by a 10-min ignition at 950°C in a platinum crucible; the loss in the weight must not exceed 1%. The salt is converted into lithium sulfate by treatment with HF and H_2SO_4, and the lithium content must correspond to the theoretical value within 1 mg.

For fusion with lithium metaborate, platinum, gold, Au-Pt and Pd-Au alloy, glassy carbon, or boron carbide vessels are employed. The fusion in platinum or platinum alloy vessels must always be carried out in the oxidizing flame of a Meker burner. A simple method for the testing of the flame quality has been recommended; a nickel crucible is ignited in the same position and under the same conditions. At a correct oxygen-to-gas ratio, the crucible is covered with an oxide layer; the formation of bright spots on the surface indicates a reducing flame, in which iron, copper, lead, cobalt, manganese, etc. are readily reduced in melts and form intermetallic alloys with platinum. Such a reaction not only depletes the test sample, but also creates the danger of slow oxidation of these alloys and contamination of the melts in future analyses.

Ideal materials for crucibles involve alloys of gold and platinum and of rhodium and palladium that are not wetted by the melt, from which the melts can be poured out quantitatively. Glassy carbon crucibles are used for fusion in furnaces with a controlled atmosphere. If the air circulation in the furnace is too rapid, the crucible is rapidly burned. On the other hand, complete prevention of the air access causes the formation of volatile forms of metals and carbonyls (e.g., alkalies, Ca, Zn, Co). The optimal conditions of the furnace operation, including the temperature and the time of fusion, must be verified experimentally. Molybdenum crucibles can be used for fusion in vacuo, for extraction of gases from minerals etc.

The fusion agent ratio to the material depends on the sample composition and on the purpose for which the glass is prepared. In silicate analyses involving the preparation of a master solution, the fusion is usually performed in a platinum crucible at fusion agent-to-sample ratios from 5 + 2 to 7 + 1. For carbon crucibles, a 3.5- to 5-fold excess of the fusion agent is recommended at a somewhat lower temperature and with a shorter fusion time (900°C/15 min). In both cases, successful decomposition depends on good mixing of the components; vibrational mixers are often recommended.

Similar to other borate agents,[257] the decomposition of some silicate structures by fusion with lithium metaborate is incomplete, so that polymerized Si-O bonds are found in solid glasses and their solutions and exert negative effects on quantitative determinations of silica, both in wet methods and in instrumental ones. The formation of these bonds can probably be prevented by fusion at a high temperature and rapid cooling of the melt, e.g., by pouring it into a solvent.

The glass formed can be dissolved in three ways: (1) after cooling, without crushing, (2) after cooling, followed by crushing of the bead, and (3) by pouring the molten material into a dilute acid. The latter method is most effective. As the solvents, dilute mineral acids, but also solutions of citric acid, EDTA, etc., are used. The bead has also been dissolved in hydrofluoric and fluoroboric acids.[316] To hasten decomposition of some hydrated oxides (Mn), several drops of hydrogen peroxide or sulfurous acid must be added to the solvent.

If the sample contains larger amounts of sulfur or ferrous iron, the two components must be removed, best by preignition at 550°C for 12 hr. An addition of SiO_2 to the fusion agents has led to good results in analyses of iron ores and iron-containing silicates. The following procedure has been given[312] for decomposition of silicates. An ignited, 0.2-g sample (<200 mesh) is thoroughly mixed with 1 g $LiBO_2$ in a platinum crucible, and the crucible is placed on a clean silica triangle. It is covered with a lid and heated by a strongly oxidizing flame. The crucible contents are periodically stirred, and the decomposed sample is heated for another 5 min (or longer, if refractory minerals are present). The viscous melt is then spread over the walls and the hot crucible is immersed into 25 mℓ dilute (1 + 24) HNO_3. After cooling, another 50 mℓ of the acid is added, so that the crucible is submerged in it. A small stirring bar covered with PTFE is placed in the crucible, and the contents are stirred until all the glass is dissolved. The solution is then quantitatively transferred to a 100-mℓ volumetric flask, the beaker and the crucible are rinsed, and the flask is filled to the mark. The solution is then prepared for determination of Si, P, Al, Fe, Ti, Mn, Ni, Co, Cr, and others.

The efficiency of silicate decomposition with lithium metaborate has most often been verified using various standard rocks. Cremer and Schlocker[317] have critically evaluated decomposability of mineral materials by this procedure, studying 26 N.B.S. standards and 47 separated minerals of various chemical compositions. The decomposition was carried out by the standard procedure, a 15-min fusion with $LiBO_2$ at a 7 + 1 ratio in a graphite crucible at 950°C. The bead formed was dissolved in 100 mℓ 4% HNO_3, the solution filtered, and the insoluble residue was investigated optically and by X-ray diffraction. The results have shown that the silicate phase is completely fused; certain accessory minerals, however, remain undecomposed and involve zircon, transition metal oxides, REE phosphates and fluorides, and numerous sulfides. Examples of poorly decomposable phases are (with percent of the insoluble residue in parentheses): bastnaesite (20), chalcocite (92), chalcopyrite (40), chromite (47), fluocerite (6), galena (35), ilmenite (16), monazite (12), pyrite (40), xenotime (5), and zircon (21). Copper minerals "freeze" in the melt.

An analogous critical study of the decomposition of accessory minerals has been carried out by Feldman.[318] The author not only studied the completeness of the decomposition, but also dealt with (1) the formation of insoluble carbides by the reaction of the melt with the graphite of the crucible (e.g., Ti, Cr, W, Nb, Ta, Zr, Hf), (2) the formation of hydrolytic products of some ions during the dissolution of the bead in 3% HNO_3 (e.g., Sn, Ti, W, Nb, Ta), and (3) the reactions among some components in the solution. The test minerals

— chromite, ilmenite, rutile, cassiterite, wolframite, magnetite, zircon, columbite, beryl, and corundum — were decomposed (up to 100 mg) with a fivefold excess of the fusion agent at 950°C within 15 to 20 min. However, hydrolytic products are formed in most dissolved glasses, but can be removed when complexing solvents are used (a solution of 3% HNO_3, 2.5% tartaric acid, and *circa* 1% H_2O_2). Only in decomposition of chromites is chromous oxide formed in the melt that is difficult to dissolve in acids. Its formation can be prevented by adding an oxidant, best an alkali perborate, to the fusion agent. Monazite, a REE phosphate, is readily decomposed by fusion with $LiBO_2$ alone, but insoluble phosphate is reprecipitated during dissolution, even in 9 *M* HNO_3. A clear solution is obtained on melt dissolution in 6 *M* HCl and can be further diluted, down to a concentration of 0.3 *M* HCl, without the danger of hydrolysis. The apparently contradictory results of the authors of the works[317,318] can readily be explained by the different conditions of the melt dissolution.

The only tested mineral that is not decomposed in the above way is molybdenite, MoS_2. It seems that its passivity toward the reagent is caused by hydrophobic surface of this phase. A similar phenomenon has been observed in decompositions of some molybdenites in mineral acids.[319]

Metaborate fusion seems to be an ideal and rapid method for the preparation of a stock solution for photometric determinations of the individual components. Procedures for rapid analyses of silicates are based on this principle.[312] Some authors[314] add cobalt(II) nitrate to the leaching solutions as an internal standard for the subsequent procedures. Shapiro[320] has simplified the original, two-solution method (fusion with NaOH and acid decomposition with HF + H_2SO_4) and introduced a single-solution one. A 100-mg sample is fused with a sixfold excess of lithium metaborate, and the bead obtained is dissolved in 1 ℓ of 1.5% hydrochloric acid. Part of the solution is used for the photometric determination of Si, Al, Ti, Fe, and P; the other cations are determined by AAS, after adding a lanthanum salt. The scheme described by Abbey[321] and co-workers[322] is based on the decomposition of silicates by fusion with a fivefold excess of lithium metaborate in a graphite crucible at 950 to 1000°C for 15 min in an electric furnace. The hot melt is poured in water placed in a transparent, polypropylene vessel with a screw-on cap. The residues of the solidified melt in the crucible are transferred mechanically to the vessel. To its content is added 25 mℓ of cold dilute hydrofluoric acid (6 + 19), with the cap on. The contents is stirred by an electromagnetic stirrer, until all of the melt is dissolved. The solution is allowed to cool in a refrigerator, 5 g boric acid is added dissolved in 100 mℓ of water, the vessel is closed again, and the solution is stirred until it becomes clear. Then it is transferred to a 200-mℓ volumetric flask. The master solution prepared in this way is used for a photometric determination of Si and P. Further portions of the solution are used, after adding suitable buffers, for an AAS determination of 12 major and minor silicate components. A comprehensive review on the application of metaborate fusion in rock analysis has been published by Abbey et al.[323]

The greatest difficulties encountered in application of these procedures concern the poor stability of silicic acid in solution. In many schemes the solutions obtained are considered sufficiently stable, but many authors have pointed out a disagreement between the results of the gravimetric and instrumental determination of SiO_2.[255,324-329] These fluctuations may have two causes; glass may be partially devitrified during cooling of the borate melt, and thus the Si-O bonds may be reconstituted. The main cause of losses in silica in solution is its polymerization. The degree of hydrolysis of Si^{4+} and, therefore, also the solution stability can be disturbed by changes in the temperature, vigorous stirring, the time of standing, the silicate concentration, as well as the concentrations of acids and the other components of the solution. The solution stability can be improved by application of depolymerization agents. Burdo and Wise[325] have proposed molybdate as a stabilizer. Glasses and silicates are decomposed by fusion with a soda-borax mixture, and the melt is dissolved in 100 mℓ of a 20% solution of ammonium molybdate in dilute (0.2 *M*) nitric acid. The resultant pH, after diluting the solution to 500 mℓ with 0.126 *M* nitric acid, is 1.05 to 1.1. The solution of silicomolybdenic acid obtained is stable for *circa* 3 weeks. The rate of polymerization and the effect of inhibitors were also followed[326] during dissolution of a metaborate melt in 5% nitric acid alone and in the presence of EDTA, or EDTA plus BF_4^- ion. Polymerization in acid alone starts, at a suitable dilution, after 8 hr and attains an equilibrium within 20 hr. The above complexing agents perfectly stabilize the silicate anion and also have a favorable effect on AAS determinations of other cations. However, the same authors[327] later found that these solutions cause mechanical faults during the measurements, such as clogging of the capillaries in flame spectrometers by lithium fluoride, and potassium fluoroborate separated.

Fluorides were used as stabilizers in dissolution of borate glasses in analyses of ceramic raw materials and zirconium oxide and chromite ceramics.[329] The lithium-fluoroborate scheme[321,322] mentioned above also utilizes the stabilizing effects of fluoroboric acid. Some authors[324,328] do not consider lithium metaborate fusion of silicates as suitable for the subsequent determination of silica and prefer direct dissolution of the materials in mixtures of hydrofluoric acid with other acids in a closed system.

Other spectrophotometric determinations connected with complete sample decomposition without removal of silicon involve, e.g., the determination of phosphorus in silicate rocks by the molybdenum blue method,[330] the determination of uranium in bituminous slates with rhodamine B,[331] the determination of tungsten with dithiol,[332] etc.

Borate fusion is very well suited to rapid and reliable AAS determination of most major and minor elements in silicate rocks, ashes, slags, ceramic products, cements, and glasses. The procedure requires no chemical separation, and the presence of lithium and borate ions does not interfere in most determinations. Suhr and Ingamells[314] used this procedure for the analysis of silicate rocks as early as 1966. The procedure has later been applied in many schemes of rapid silicate analyses, some of which are given in Table 4.

The relatively low temperatures of fusion of silicates have been selected as a compromise, to decompose refractory accessories in some rocks (chromites, zircon), and to minimize losses in the alkali metals through volatilization. Owens and Gladney[343] have pointed out that in the decomposition procedure that they used, the losses in selenium and arsenic amount to as much as 80 and 15 to 50% of the total amount, respectively, so that the elements cannot be determined. The results of analyses of silicate rocks using this procedure have been compared with those obtained with the classical decomposition with hydrofluoric and perchloric acids. The efficiencies of the two procedures have mostly been found to be the same.[345]

International rock standards have usually been used as materials for the comparison and often also for the preparation of control and calibration solutions under the conditions identical with those for the analyzed solutions. In this way the differences in the solution viscosity are decreased, and the matrices of the test rocks are modeled. Especially reliable results are obtained in measurements when the test sample is bracketed with standards that do not differ much from the preliminary values of the test component contents.

In addition to silicate analyses, AAS methods have also been applied to determinations of some elements in resistant minerals, after decomposition by fusion with lithium metaborate. For example, Rodrígez and Pacz[346,347] dealt with decomposition of chromites and bauxites. For chromites they recommend fusion at a 1 + 10 ratio and a temperature of 1100°C for 30 to 60 min. Lower fusion times or lower temperatures lead to incomplete decomposition of these oxides. Addition of an oxidant is useful when chromospinellides with higher contents of divalent iron are to be decomposed. Hydrated aluminum oxides, bauxites, are also readily decomposed, at a ratio of 1 + 5 and 1000°C; still, a preliminary acid decomposition has been recommended.[347] Zircons are also readily decomposed by fusion; the procedure has been applied to an OES determination of zirconium and hafnium.[348]

Table 4
FUSION OF INORGANIC MATERIALS WITH LITHIUM METABORATE FOR THE PREPARATION OF THE STOCK SOLUTIONS FOR AAS

Material	Fusion-agent to sample ratio	Crucible	Temp(°C)/time (min)	Solvent	Ref.
Rocks	5:1	Graphite	1000/60	0.3% HNO_3	294
	6:1	Graphite	1000/60	1.5% HCl	320
	5:1	Graphite	1000/30	1 + 24 dil HNO_3	333
	7:1	Graphite	900/15	1 + 24 dil HNO_3	334
Rocks, clays, limestones	6 + 1	Platinum	980/15	Hot 1 + 4 HCl	335
Silicates	5 + 1	Graphite	1000/15	3% HNO_3	336, 337
	5 + 1	Platinum	1000/15	3% HNO_3	338
Silicates, coal ash	4 + 1 + VO_3^-	Platinum	900/to clear melt	5.5% HNO_3	339
Silicates	4 + 1	Graphite	900/15	6 + 19 HF	321, 322
Silicate rocks	5 + 1	Gold	950/30	3.8% HBF_4	316
Silicates, carbonates, bauxites	5 + 1	Platinum	950/2 to clear melt	0.9 *M* HCl	340
Silicates	5 + 1	Platinum	950/15 + 5	0.5% HNO_3 + EDTA + HF	326
Fly ashes	10 + 1 + VO_3	Platinum	900/clear melt	0.5% HNO_3 + EDTA	341
Rocks, fly ashes	7 + 1	Platinum	950/10	4 *M* HNO_3	342, 343
Rocks, soils	5 +1	Platinum	950/15	10% HNO_3	344
Slags from Cu-melting	6 + 1	Graphite	950/10	1 *M* HNO_3	345
Rocks	2.5-5 +1	Platinum	1000/3 + 3	1.5 *M* HCl	376
Ceramic material	6 + 1	Platinum	1000/15	Dil. HCl	377
Rocks	5 + 1	Platinum	950/15	Dil. HNO_3	326, 377
Slags, ashes cement	—	Graphite	—	Dil. HCl	378, 379

[a] A mixture Li_2CO_3 + B_2O_3.

AAS has been applied to analyses of magnetite concentrates, rutile, and titanomagnetites. These oxides are poorly decomposed when using the fusion agent alone; considerable improvement is attained when the same amount of silica is added to the oxide to be decomposed.[349] Metaborate fusion has further been applied to determinations of many trace elements in fluorite materials.[350] In some works,[346-350] hydrochloric acid is used to dissolve the borate glass.

The presence of Li^+ in the system is used to advantage in the determination of the alkali metals in silicates that are difficult to decompose,[351] e.g., in tourmalines[352] and synthetic Al_2O_3 single crystals,[353] using AAS. Highly precise results have been obtained in the determination of potassium in rocks and minerals;[354] if the conditions of the metaborate fusion are controlled well, there is no danger of volatilization of potassium. The completeness of the decomposition has been compared with the results of pressure decomposition with HF and of dissolution of the material in an open system of HF-H_2SO_4-HNO_3.[355] AAS and emission spectroscopic determinations of barium and strontium in silicates have also been described.[315] Fusion with lithium metaborate has also been used to decompose an Al,Y-garnet; the sample melt (1 + 3, at 950°C, 30 min) was dissolved in 4% HNO_3 and the solution used for an AAS determination of neodymium.[356] Lithium metaborate is also an efficient fusion agent for cassiterite and other tin oxides; therefore, the method has been used in determinations of small amounts of tin in rocks and lean ores, using AAS.[357-362] Tin is introduced into the flame in the form of the hydride[358-360] or is separated by extraction (TOPO-MIBK), and the extract is atomized electrothermally in a graphite cuvette, impregnated by tungstate ions.[361] Weiss[362] has pointed out easy reducibility of tin(IV) ions in the melt and the formation of an intermetallic alloy of tin with platinum. Metaborate decomposition is further applicable to the AAS determination of indium,[363] tungsten,[365] and lead[366] in geological materials. The borate melt is dissolved in HBr.[363] Prior to the determination of lead, silica should be depolymerized using dilute hydrofluoric acid. The solution of the melt in nitric acid can also be used for a determination of fluoride with an ion-selective electrode, after addition of a suitable buffer.[364]

Fusion with lithium metaborate is very often replaced by fusion with a mixture of lithium carbonate with boric acid (boric oxide), which is considered to be more efficient for decomposition. The mixture has been applied to virtually the same materials as metaborate alone. Various combinations of solvents then lead to the preparation of master solutions for instrumental analytical methods.

Govindaraju and co-workers of Centre National des Recherches Scientifiques (Nancy, France) have proposed various fusion mixtures based on Li_2CO_3-H_3BO_3 for silicate analysis. Initially, cobaltous-cobaltic oxide was added to the fusion agent, as an internal standard for a subsequent spectrochemical determination. The glass obtained by fusion was mechanically crushed, part was used for OES, and another part was dissolved in dilute nitric acid for a determination of the alkali metals.[367] Later the fusion agent contained also a strontium salt as a spectral buffer for the determination of the alkali metals (20 parts H_3BO_3 + 6 parts Li_2CO_3 + 3 parts $SrCO_3$ + 1 part Co_3O_4); the glass was dissolved in citric acid.[368,369] However, in application of these solutions to AAS determinations, it has been found that some determinations are adversely affected by the presence of silica and large amounts of boron in the solution. Therefore, the melt was dissolved in an aqueous suspension of a strongly acidic cation exchanger in a column in which the cation exchanger and the melt form a fluid bed, maintained by a stream of air passing through the column. The anions remaining in the solution are washed from the column with water; the sorbed cations are eluted from the ion exchanger with 2 *M* HCl, and the eluate is used for AAS determinations of the individual elements.[370] The same procedure was used to determine silicon in the effluent.[371]

A new scheme of silicate analysis, involving the determination of 16 major, minor, and trace elements and based on metaborate fusion, has been developed[372] on the basis of analyses of several tens of thousand rock samples. A 650-mg sample is fused in a graphite crucible with 650 mg Li_2CO_3 and 1300 H_3BO_3 at 1050°C in an automated, programmed fusion device (*circa* 50 min). The melt obtained is poured into a milling device in which a foil 80 μm thick is produced, which is readily manually pulverized in an agate mortar. Part of the melt is decomposed by the above effect of a cation exchanger; the decomposition takes 3 hr, with agitation of the glass with the cation exchanger suspension by an electromagnetic stirrer, in 200 mℓ 0.01 *M* nitric acid. The cation exchanger is separated from the anions in the solution by repeated decanting with water; then it is dried at 100°C and is prepared for the following direct-reading OES determination. Another aliquot of the melt is dissolved in dilute nitric acid, and the solution is used for some AAS and other flame photometric determinations. An analogous fusion procedure, including the desintegration of the melt, has also been used for XRF and OES with MW plasma excitation.[373-375]

Some applications of decomposition of silicate materials with an equivalent mixture of lithium carbonate and boric acid to AAS determinations are summarized in Table 4. The decomposition has been used in analyses of aluminum oxide and other aluminum-rich compounds, especially in determinations of the alkali metals and some trace elements.[380-382]

Zirconium concentrates are efficiently decomposed with this fusion agent.[326] The mixed fusion agent has found application in photometric determinations of titanium,[383] calcium,[384] phosphorus,[385,386] and of silicon by flow injection analysis (FIA) in silicates.[387] The authors fused the sample with a sixfold excess of the 1 + 1 fusion agent for 15 min at 1000°C. The glass formed was then dissolved in 1 *M* hydrochloric acid. Analyses of synthetic molecular sieves employed this fusion agent to which lead carbonate was added.[388]

A fusion agent with an acidic component predominating has been proposed by Ohlweiler and co-workers.[389] It consists of a 10 + 1 mixture of B_2O_3 and Li_2CO_3 that is mixed with the sample in a 10 + 1 ratio. The fusion takes place in a platinum crucible in the flame of a Meker burner (1000°C), for 15 to 30 min. The melt is rapidly cooled by immersing part of the crucible in icy water. The glass formed is dissolved in dilute (0.015 to 0.1 *M*) hydrochloric acid, employing a 1-min action of ultrasound waves followed by *circa* a 1-hr stirring with an electromagnetic stirrer. A volume of 1 to 2 mℓ of 30% hydrogen peroxide is added to suppress hydrolysis of titanium, zirconium, and other elements. For complete decomposition, the material must be finely pulverized (200 mesh) and thoroughly mixed with the fusion agent. Lithium carbonate improves the fusion ability of boric oxide, increases the solubility of the glasses, and suppresses polymerization of silica in solution.

The above decomposition procedure has been verified on the following resistant materials: andalusite, sillimanite, cyanite, tourmaline, garnet, staurolite, minerals of the spinelide group, corundum, and quartz. Microamounts (25 to 50 mg) of zircon and rutile have further been completely decomposed; at higher sample amounts opaque glasses are formed, and the decomposition is incomplete. The behavior of fluoride and sulfate ions during fusion has also been studied. In an open crucible, up to 95% of the fluorine volatilizes, whereas up to 75% of the fluorine remains in the melt, if the crucible is covered with a lid. Losses in sulfate amount to *circa* 8% under the same conditions. The above fusion has been applied to the gravimetric determination of silicon in rocks containing phosphorus, zirconium, and titanium, using precipitation of the oxinate of silicomolybdic acid. The elements interfering in the precipitation are separated from the melt solution by batch treatment with strongly acidic cation exchanger and strongly basic anion exchanger.[390-391] The fusion procedure has also been used in an AAS determination of potassium in silicate rocks.[392]

F. Fusion with Lithium Tetraborate

This fusion agent, available in a highly pure form, has found main use in OES and XRF; there are relatively few applications to wet chemical analysis. Lithium tetraborate has a relatively high melting point of 917°C, the melt is highly viscous, and the glasses formed are difficult to dissolve in mineral acids. Similar to the other borate melts, this glass also strongly adheres to platinum, so that it is advantageous to use crucibles made of unwettable Pt-Au alloys. Lithium tetraborate is a weakly acidic, nonoxidizing fusion agent, suitable mainly for decomposition of basic materials. In fusion at high temperatures, there are losses of some elements through volatilization or a reaction with the fusion vessel material. For example, in fusion of biological materials at 1100°C, labeling with radionuclides indicated that the losses in As due to volatilization amount to 40%, whereas Ba, Na, and Zn are lost completely. The retention loss in As is 60% of the total content.[393] During fusion in graphite crucibles (1 + 6 to 1 + 9, 1200°C, 15 min), Co^{2+} and Fe^{3+} are partially reduced to the metals, and the losses are proportional to the fusion time.[394]

Bettinelli[395,396] used lithium tetraborate fusion for the preparation of master solutions in analyses of silicate standards, airborne coal ashes, blast-furnace dust, solid fallout, river sediments, and coal. The sample (0.3 g for silicates and 3 g for dried coal) is fused with 1.5 g $Li_2B_4O_7$ in a platinum crucible at 1000°C for 1 hr. The melt is dissolved in two portions of 50 mℓ of 5% HCl at 50 to 60°C and with stirring. The clear solution is diluted to 250 mℓ and used for flame or ETA-AAS determination of the major and trace elements. It has

Table 5
MODIFICATION OF THE MIXED BORATE FLUX
$4\ LiBO_2 + Li_2B_2O_7$[406]

Material type	Sample mass (g)	Fusion mixture (g)	Boric acid (g)	Sodium carbonate (g)
Aluminosilicate				
High silica	1	3	0.4	—
Aluminous	1	1.5	0.2	—
Zircon bearing	1	5	2	—
Frits and glazes				
Main	1	3	—	—
B_2O_3 determination	0.5	—	—	10
Lead bisilicate	1	—	0.6	5
White opacified glaze	1	—	—	7
Chrome-bearing				
Main analysis	1	8	4	—
SiO_2 determination	0.1	4	2	—
Cr_2O_3 determination	0.25	4	2	—

been found that volatilization of arsenic from the melt is suppressed by adding a small amount of nickel(II) nitrate to the fusion agent.[395]

Analogous fusion procedures have been applied to determinations of the alkaline earths and refractory oxides in geochemical standards using AAS,[397] or of alumina in feldspars by differential spectrophotometry.[398]

Brown et al.[399] have shown that lithium tetraborate is a more efficient solvent for aluminosilicates than the metaborate. The bead formed by fusion of a sample (1 + 10) in a graphite crucible at 1200°C for 60 min, was quite homogeneous; it was optically isotropic and inert in X-ray diffraction measurements. The solutions of the beads in acids were stable for up to 4 months; after 11 months, the silica content increased due to evaporation of the solvent by 1%.

Rechenberg[400] has found that fluorine losses do not occur during a limited time of fusion with lithium tetraborate (15 min at 1000°C, 10 min at 1100°C, and 5 min at 1200°C) and applied this finding in analyses of cement and clinker and dust emissions,[401] as well as to a photometric determination of silica in analogous materials.[402]

The preparation of stock solutions for multielemental analysis is considerably accelerated when automated fusion devices are employed.[270,403] The material is fused with a tenfold excess of the fusion agent at 950°C for 15 min, and the melt is dissolved in 15% HCl. The procedure has been applied to analyses of coal, oil-bearing shales, ashes, airborne dust,[404] and of geological materials,[405] using the AAS and ICP-OES methods.

G. Use of Mixed Borate Fusion Agents

Mixed fusion agents, based on alkali borates, are primarily used in the XRF analysis. An addition of basic or acidic components modifies the properties of the glass formed that should be perfectly homogeneous and should not undergo devitrification. As shown above, the most suitable fusion agent consists of a 4 + 1 mixture of lithium metaborate and tetraborate, corresponding to the eutectic with a melting point of 832°C.[266] For wet chemical analyses of ceramic raw materials and products, this basic mixture was modified with B_2O_3 or alkalies, to attain rapid fusion and good solubility of the melt.[406]

The amount of B_2O_3 in the mixture should generally be proportional to the silica content in the test material. Examples of the compositions and applications of these fusion agents[267] are summarized in Table 5.

The above eutectic mixture has also been used for the preparation of master solutions in silicate analysis.[407,408] The sample is fused with a fivefold excess of the mixture, and the melt is dissolved in 6 *M* nitric acid. An analogously obtained solution in 4% HNO_3 was used, after adding a buffer, to determine fluorine with an ion selective electrode; the procedure was applied to biotite, phlogopite, tourmaline, granite, syenite, etc.[409] Welsch[410] verified the efficiency of the decomposition of geological materials (rocks, soils, stream sediments) by fusion with $LiBO_2$, a mixture of Li_2CO_3 + H_3BO_3, and the above mixture on a determination of tin and molybdenum. The highest results were obtained after fusion with the mixture of borates, but the difference among the other procedures was insignificant.

Shapiro[411] proposed a mixed fusion agent (2 parts of lithium tetraborate and 1 part of lithium metaborate) for the preparation of the master solution in rapid silicate analysis. A 0.2-g sample is fused with a sixfold excess of the mixture in a graphite crucible at 1000°C for 1 hr. The melt is dissolved in dilute nitric acid, and the solution (200 mℓ) is used for spectrophotometric and AAS determinations of the major components. Fluoride is used as a depolymerization agent in the spectrophotometric determination of silicon after this fusion.[412] The same decomposition procedure was applied to the determination of phosphorus in silicates by the molybdenum blue method.[413]

Analogous fusion mixtures, close in the composition to the tetraborate or to the above eutectic, have been prepared by mixing metaborate with boric oxide, or tetraborate with lithium carbonate. The former mixture has been used to analyze ceramic materials[414] and to determine by AAS yttrium in synthetic zirconium dioxide.[415] The latter mixture has been applied to decompositions of synthetic yttrium garnets[416] and silicate rocks.[417] The melt solution was used for an automated spectrophotometric determination of phosphorus. To increase the basicity of the fusion agent, lithium carbonate was replaced by the hydroxide.[418] This mixture has given good results in decomposition of ZrO_2 ceramics.

The decompositing effect of borate fusion agents can be increased by adding oxidants; lithium perborate has been found very advantageous and decomposes in the melt into metaborate and free oxygen. It was used to advantage for decomposition of chromium spinels, with simultaneous oxidation of Cr^{3+} to Cr^{6+}, thus eliminating the formation of chromium(III) oxide that is insoluble in the metaborate melt.[318] Organic carbon is quantitatively oxidized to carbon dioxide by fusion with a mixture of 2 parts of perborate, 2 parts of metaborate, and 1 part of lithium tetraborate at 1100°C for 25 min.[419] The procedure was applied to analyses of slags and solid atmospheric fallouts and verified on an NBS-SRM 1648 standard of urban particulate matter.[420]

The strong oxidizing effect of the oxygen liberated by thermal dissociation of the fusion agents has been utilized to separate volatile oxides, such as OsO_4, from the test material. In the determination of OsO_4 in rocks, sulfidic minerals, nickel concentrates, and alloys, a mixture of 15% $LiBO_2$, 45% $Li_2B_4O_7$, 15% BaO_2, and 25% $KBiO_3$ has been used. The fusion is carried out in a quartz tube, and the volatile oxides are transported into a heated chromatographic column, where they are separated. The final determination (down to 2×10^{-7}%) is carried out by AFS.[421] A mixture of lithium metaborate with sodium peroxide has been found to be an efficient fusion agent for resistant substances (refractory bricks, carbides, nitrides, and some oxides).[422]

H. Decomposition with Alkaline Earth Metaborates

Metaborates of calcium, strontium, and barium are effective decomposition agents for silicate materials, rocks, soils, and resistant minerals, similar to the alkali borates. The reagents are prepared by mixing the appropriate carbonate with an equivalent amount of boric oxide or boric acid, or the commercial salts is used. The preparation of strontium metaborate has been described by Jeanroy;[423] the preparation separates as the insoluble salt from saturated solutions of lithium metaborate and strontium nitrate. The product is dehydrated by drying at 200°C and ignited at 1000°C. The preparation is contaminated with lithium. The fusion agent can also be prepared by the thermal reaction of equivalent amounts of strontium carbonate and boric acid at 1000°C.[424]

Decomposition with the alkaline earth metaborates are carried out analogously as with lithium metaborate; the melts formed are dissolved in dilute mineral acids and diluted to a suitable volume. For decomposition of most silicates, a tenfold excess of the fusion agent over the sample is used; a fivefold excess suffices for silica-rich samples (>90%), whereas excesses of up to 20-fold must be employed for aluminum-rich phases.[423] In view of the high melting points of these substances (Table 1), fusion must be performed at temperatures around 1100°C, e.g., in an electric furnace or in a flame enriched in oxygen. The solutions obtained are used as master solutions in AAS determinations of major and minor components of silicate rocks,[423-425] soils,[425-427] cement,[425] and resistant materials.[428] The latter authors employed the barium salt and pointed out a high efficiency of fusion and rapid dissolution of the melt. The strontium present in the solution functions as a releasing element in the AAS determination of calcium, aluminum, etc.

To decompose α-corundum and determine the alkali metals in it, calcium metaborate has been recommended as the fusion agent and compared with other metaborates,[429] as well as with other decomposition methods.[430] The author considers a pressure decomposition with mineral acids or a direct instrumental neutron activation analysis (INAA) determination of sodium as the best approach.

A mixed fusion agent of 2 parts of SrB_4O_7 + 1 part of Na_2CO_3 yields homogeneous glasses with a number of nonmetals, which are universally applicable to XRF.[431]

I. Use of Borate Fusion in OES and XRF

As mentioned elsewhere, the purpose of fusion is to homogenize the sample, i.e., to remove the matrix effect, the heterogeneity in the mineral composition, and the effect of varying granulometry. This is also the most effective method for homogenization of an internal standard. Borate glasses and their solutions are used in OES, and they have brought about a principal change in the XRF analyses of complex materials.

1. OES

In most applied methods the sample is mixed with a six- to tenfold excess of the fusion agent (lithium meta- or tetraborate or a suitable mixture of them) that may further contain an internal standard (V_2O_5, SrO, Co_3O_4, BeO) and is thoroughly fused, best in a graphite crucible. After cooling, the bead is mechanically cleaned and pulverized in a swing-mill grinder. The powder obtained is thoroughly mixed with carbon powder, placed in suitable electrodes, and excited by a DC arc or an AC spark. Analogously, the powdered glass can be mixed with pelletable graphite powder, a briquette pressed and used as the counterelectrode against a graphite rod.[432,434]

Roubault et al.[367] have used a tape machine, developed by Danielsson et al.,[433] to transport a mixture of a borate glass and graphite to the spectrograph spark gap. The sample is uniformly deposited by a mechanical device on an adhesive tape that passes through the spark gap of two graphite electrodes; new material is continuously fed to the source, and thus, the change in the intensity of the emitted radiation is minimized. The procedure was later modified, dissolving the borate melt with the use of cation exchangers and feeding the dried resin by means of the tape machine.[372]

The above procedures have been applied to the determination of major and minor components of silicate rocks,[367,372,434-436] as well as to analyses of slags, lean iron ores, and limestones,[436] a determination of beryllium in minerals,[437] etc.

The borate fusion has also found use in solution OES. The solution obtained by dissolving the melt in dilute nitric acid is analyzed using a rotating graphite electrode and AC spark excitation.[315,438,439] The procedure was used to determine barium and strontium in silicates.[315] Both the groups of methods, powder and solution, are greatly accelerated and refined by the use of direct-reading spectrographs.

It can be seen from the literature survey that borate fusion has not been very often applied to classical OES. However, the method has gained a new impetus with the appearance of new sources of radiation, denoted as Plasma Excitation Torch and involving the argon jet, MW-induced plasma cavities, and inductively coupled plasma torches, ICP. As all these sources have been designed primarily for solutions (and gases), it is evident that the well-developed techniques of the solution preparation, such as decompositions of solids by fusion with alkali borates and dissolution of the melts in various solvents, will find an important use here. The introduction of automated fusion apparatus[440] is a great contribution to the preparation of master solutions for ICP, AAS, and other chemical methods.

The outlined principles have primarily been used to determine major and minor components of silicate rocks,[441-445] phosphates,[444] and sediments.[446,447] Among further applications, a determination of strontium and barium in sediments[448] and diatomite earth ashes[449] has been described. Rare earths and yttrium have also been determined in geological materials after decomposition by fusion with lithium metaborate. The individual elements present at higher concentrations can be determined directly after dissolution of the borate melt in nitric acid,[450] but group separation is mostly needed, using ion exchangers, followed by ICP-OES analysis.[450-452] Tungsten can also be determined in silicate rocks and ores using this method after borate decomposition.[453]

2. The Fusion Agents Used in XRF and Their Properties

The most common method of sample preparation for XRF analysis is borate fusion. In this way problems connected with the use of free powders and pressed briquettes, e.g., the sample heterogeneity, the effect of the mineral composition, the material granulometry, and the matrix effects, can be overcome. The preparation of synthetic standards and homogeneous dispersion of heavy elements absorbing X-rays has been made easier. During the preparation of the borate glass, problems stemming from the chemical composition of the test material can also be eliminated, e.g., the presence of volatile components and the occurrence of elements in various valence states.

Borate fusion has been introduced into the XRF analytical practice by Claisse.[454,455] He used sodium tetraborate as the fusion agent and applied the procedure to ores, minerals, ceramics, and other chemical compounds. The original procedure involved casting of the molten glass into rings placed on a smooth metallic support. However, the preparation of disks in this way was difficult, as internal tension caused by nonuniform cooling of the glass led to uncontrollable cracking of the disks. Materials releasing gases during the decomposition developed numerous bubbles that decreased the precision of the subsequent determination. Therefore, two modified techniques have been introduced: (1) the use of Pt-Au (5%) crucibles and (2) the use of automated fusion devices.[456] Borax has been recommended for determination of elements at higher concentrations, while lithium tetraborate is suitable for determination of minority contents and the elements with the atomic numbers less than 22. Barium peroxide or lanthanum(III) oxide have been recommended as additives for increasing the absorption of radiation by the glass. A technique of casting of disks into molds of a special shape with perfectly smooth surface has further been introduced. To decrease the amount of melt adhering to the crucible, releasing agents are added to the fusion mixtures, such as potassium, lithium, and cesium iodide, or sodium bromide in an amount of 2% in the mixture. The fusion agent ratio to the sample may exceed 100 + 1. The 20 + 1 ratio is suitable for refractory ores, 5 + 1 for most rocks and 2 + 1 for cements. At a high dilution, the matrix effect is zero, and the dependence of the X-ray intensities on the element concentration in the sample is linear from 0 to 100%. The fusion method has finally been modified[457] for XRF determinations at unknown fusion agent-to-sample ratios.

Application of borate melts to XRF can be classified into three groups: (1) pulverization of the melt and pressing of briquettes (plus a binder); (2) casting of the melt into rings placed on a heated plate, possibly combined with the pressing of the disks, and (3) casting of disks into suitable molds.

Method 1 found use in the initial stage of analytical applications and was only justified in the time when the technique of the disk casting to obtain mirror-like surface was not completely developed. In this way, the difficulties with devitrification of glasses could be overcome. The separated crystalline phases are homogenized by pulverization, and a surface of a sufficiently good quality can be obtained on pressing. The necessity of adopting this procedure has been verified by a study of the fusibility of slags, fireclays, and chrommagnesites with borax. When fused in a muffle furnace (1100°C), the latter materials were decomposed to 77%, when using hf heating (1600°C), to 92%. A reference analysis of a fireclay with a high aluminum content, using cast disks that were pulverized and pressed, yielded higher results for Si and Al in the latter procedure.[458] An older comparison study[459] of XRF silicate analyses, carried out with (1) free powders, (2) pressed briquettes, (3) pulverized and pressed melt, and (4) cast disks, has shown that the best results are obtained using the third analysis type. The results obtained after fusion with lithium tetraborate were compatible for dilutions 1 + 1 to 1 + 10.

Group 2 methods, many published, employ disks produced by casting the borate melt into brass, nickel-coated, aluminum, or graphite rings. The ring is usually placed on a graphite support heated to a suitable temperature (*circa* 220°C). The borate melt is rapidly poured from the fusion crucible into the ring and slightly pressed with an aluminum piston heated to the same temperature. Thus, the melt is cooled rapidly and solidifies. The bead formed is placed between two asbestos plates heated to 200°C, where it is tempered and later cooled. The decisive effect on the disk quality is exerted by the temperature and the velocity of pouring of the melt. If the melt is cooled during casting, microsegregations may occur, as well as La_2O_3 separation etc.[460] The procedure has been improved[461] by using a press producing a constant pressure, thus maintaining identical conditions for all the disks. The casting tools (the block and the piston) are made of polished graphite and preheated to 280°C. The pressed disks must be thermally stabilized for 1 hr at 250°C. Borate glasses must not be cooled too rapidly; however, upon too slow cooling, devitrification occurs and crystalline phases are formed.

Method 3, the most common method of the pretreatment of samples for XRF, is casting of the melt into suitable molds. The method has been introduced by Claisse[456] and has been made possible by the existence of unwettable Pt-Au alloys. If the bead was prepared by solidification of the glass directly in the fusion crucible, the product contained many bubbles and had an unsuitable surface for the measurement. As above, in this procedure the course of the temperature during cooling of the disk is again decisive. Because the properties of the glass are determined both by the composition of the initial material and that of the fusion agent, as well as their mutual ratio, each laboratory develops a special procedure. The cast disks can be used for further measurement without any treatment; some authors, however, polish the disk surface or use the convex side of the disk which is of better quality. To improve the surface, variously shaped molds and their surface pretreatment, e.g., polishing, are used.

Lithium tetraborate is suitable for XRF determinations of light elements. Lithium exhibits a low absorption of X-rays. Its glasses are mechanically strong and not very hygroscopic. However, on cooling, they have an undesirable tendency to devitrification.

Lithium metaborate is a suitable solvent for a great variety of substances, but its glasses recrystallize readily. Therefore, it is rarely used alone for XRF, and mixtures with tetraborate are preferred.

Table 6
THE COMPOSITION OF MIXED BORATE FUSION AGENTS FOR THE PREPARATION OF GLASSES SUITABLE FOR XRF[266]

Material	Part of flux to 1 part of sample (ignited)	
	$LiBO_2$	$Li_2B_4O_7$
Silica-alumina range	4	1
Steatite	4	1
Bone china	4	1
Bone ash (apatite)	0	3
Zircon	4	1
Zirconia, titania	0	12
Limestone, dolomite	0	5
Magnesite	0	10
Witherite	0	12
Barium titanate	0	12
Borax frit	4	1
Glaze, <20% PbO	4	1
Glaze, 20—60% PbO	0	12
Lead bisilicate, enamel	0	12
Iron(III) oxide	0	12
Cr-bearing refractory	10	12.5
Silicon carbide	6	1.5
Boron carbide	3.52	0.88

Borax is a universal fusion agent for many refractory compounds; drawbacks are a high viscosity of its melts and hygroscopicity of the solidified glasses. The disks must be stored in dessicators with a suitable dessicant. The borax glasses are marked by a high stability during cooling.

Bennett and Oliver[266] have dealt with empirical theory of the decomposition with borate fusion agents and the preparation of a universal fusion agent for decomposition of ceramic materials. The optimal properties have been exhibited by an eutectic mixture containing 72.5% mol B_2O_3 and 27.5% mol Li_2O, obtained from a mixture of 1 part of lithium tetraborate + 4 parts of lithium metaborate. The mixture yields homogeneous, clear glasses in fusion with materials greatly varying in the contents of alumina and silica. However, the composition of the fusion agent must be modified, if the test material contains basic oxides (Na, Ca, Mg) and oxides of iron, titanium, chromium, and zirconium. If the material is too basic, it is fused well, but the glass is strongly devitrified during cooling. Most of these substances require more acidic fusion agents and sometimes also an addition of silica. The fusion agent compositions and the ratios to various types of ceramic materials are given in Table 6.

Many fusion agents can be found in the literature that are based on borates and contain substances modifying their acid-base character, primarily carbonates, hydroxides, and boric oxide. Other substances that may be added involve oxidants, mostly nitrates, heavy cations as substances absorbing radiation, and substances decreasing the wettability of Pt-Au alloys. Some special fusion agents are used in analyses of metallurgical products.

1. Many elements (Pb, Zn, Sn, Bi, Fe, Cu and P) react with platinum and must be oxidized prior to or during fusion. This is especially important with ferroalloys. The sample is mixed with the fusion agent ($Li_2B_4O_7$ + $NaNO_3$ + $NaKCO_3$ + $NaIO_3$ +

$Sr[NO_3]_2$ mixed at various ratios for various alloys); in the first stage, heating to 500 to 700°C, the alloy is oxidized. A glass envelope is formed at the interface and separates the metal from the platinum. After the oxidation, more tetraborate is added and the oxides are fused to form a suitable glass.

2. Airborne dust and dusts containing Pb, Zn, C, S, and organic substances are decomposed by an analogous fusion agent without carbonate. Strontium plays an important role in this mixture, as it absorbs radiation and decreases the hygroscopicity of the borate glasses.
3. Substances with high melting points, e.g., carbides, are not decomposed by fusion with borates. Thus, a copper salt is added, and the corresponding binary alloy of the metal with copper is formed by reduction at a lower temperature, which is then oxidized and fused as described in (1).[462]

The quality of borate fusion also depends on the intensity of stirring during the decomposition. Comparison studies of manual and continuous stirring have shown that the time required for the decomposition is three times as long with manual stirring than with mechanical stirring.[463] The heating temperatures of melts also affect the disk quality. The variances of the results of determination of iron, cobalt, etc., are apparently increased by the reduction of the metals at high temperatures.[464] (For the techniques of borate fusion see References 465 and 466.)

3. Conditions for Glass Formation in Melts

The transition of a liquid phase into a solid is a thermodynamically complicated process that may lead to the formation of more stable solid phases or of a metastable glass. The latter phenomenon is called supercooling of liquids. The decisive temperature for the formation of a glass is the transformation temperature, T_g, not the liquidus temperature. The physical properties of the supercooled liquid, e.g., dilatation, strongly change around temperature T_g. Between the transformation temperature and the liquidus temperature lies the region of nucleation and that of the crystal growth. It has been shown experimentally that upon sufficiently rapid, monotonic decrease in the temperature, glass is formed within the whole bulk of the sample. Local fluctuations in the temperature in the critical regions of nucleation and crystal growth lead to rapid crystallization of the glass. This may happen, e.g., due to cooling of the melt surface on contact with a cold metal. Very rapid cooling of the melt down to laboratory temperature is also unsuitable for the preparation of glasses. Rapid changes in the region of T_g cause internal tension that is manifested in considerable fragility of the glass. Thus, it is necessary to decrease the sample temperature around T_g as slowly as possible, e.g., by placing the disk between two heated plates and prolonged tempering; the internal tension is then partially compensated.[467]

In R_2O-B_2O_3 binary systems, apparently clear glasses are formed from melts. However, detailed studies have shown that these glasses consist of two phases. One liquid is formed by glass enriched in alkalies and is dispersed in the basic liquid depleted in alkalies. It has been found that this tendency toward immiscibility is connected with the ionic potential of the alkali metals. Therefore, sodium salts are better suited for the preparation of homogeneous glasses than lithium ones, as sodium has a lower ionic potential and a greater ionic radius.[468] The formation of two immiscible phases in a ternary Li_2O-B_2O_3-SiO_2 system has been described by Sastry and Hummel.[469] The diagram given in Figure 3 exhibits a small field of the existence of a homogeneous glass that is located close to the composition of $Li_2B_4O_7$. In the glass-forming region, the boundary separating two liquids, and the limit of devitrification are denoted. Exnar[467] gives the maximum contents of impurities in alkali borate glasses that do not lead to the formation of solid phases under common experimental conditions. The values for some borate fusion agents are given in Table 7.

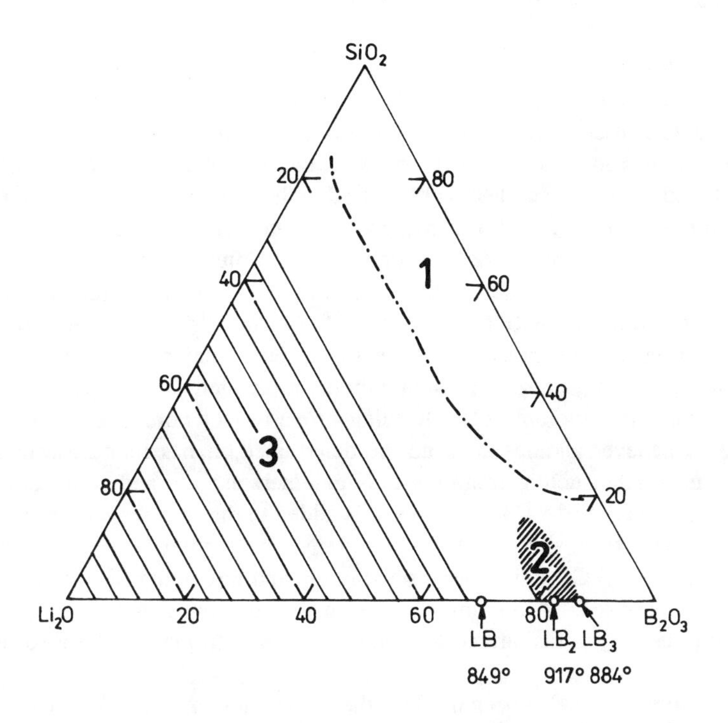

FIGURE 3. Ternary isobaric diagram of the Li_2O-B_2O_3-SiO_2 system.[469] LB = $LiBO_2$, LB_2 = $Li_2B_4O_7$, LB_3 = $Li_2B_6O_{10}$. (1) Two liquids, (2) region of homogeneous glass formation, (3) devitrification field.

Table 7
STABILITY LIMITS OF GLASSES IN BINARY SYSTEMS OF AN ALKALI BORATE WITH AN OXIDE[467]

	Maximum content in % wt		
Component	$Li_2B_4O_7$	66% $Li_2B_4O_7$ + $LiBO_2$	$Na_2B_4O_7$
Li_2O	—	—	5
Na_2O	25	24	—
K_2O	33	30	12
BeO	21	20	20
MgO	20	19	20
CaO	33	30	13
SrO	52	52	19
BaO	43	37	18
MnO	58	57	43
ZnO	41	28	56
CdO	55	52	32
PbO	86	84	50
Al_2O_3	39	38	42
La_2O_3	46	43	21
GeO_2	+[a]	+	—
TiO_2	14	16	19
ZrO_2	22	25	28
SnO_2	17	13	24
P_2O_5	50	56	51

[a] (+), Not limited.

4. Application of XRF

The method of disk casting has found the most extensive use in practice; the first of the above procedures is used only exceptionally. Briquetting of pulverized borate glasses was used in analyses of iron ores,[470] silicates,[471] and tungsten carbide.[472] An universal method of analysis, based on this principle, has been used in the Mintek laboratory:[473] poorly fusible silicates, coal dust, slags, ferroalloys, manganese, iron-titanium-vanadium ores, zirconium sands, quartzite, and alumina are fused with a mixture lithium metaborate and nitrate, boric oxide, and lanthanum oxide. The bead formed is milled, thoroughly mixed with a binder, and a disk is pressed for the XRF determination. Pressed melts of a sodium carbonate-borax mixtures have been recommended for the determination of major rock elements;[474] the melt is cast into a copper ring, pressed with an aluminum plunger, and tempered at 250°C.

Inmold casting, the favorable glass-forming properties of borax, is often utilized, and borax is used whenever alkalies need not be determined. It has found use in analyses of metallurgical materials, such as iron ores and agglomerates[476,477] and nonmetallic materials (fireclay, bauxite, etc.).[475,476] Effective oxidants added to the fusion agent involve the alkali nitrates, barium peroxide, and vanadium pentoxide. These mixtures are used for oxidative fusion of ferroalloys.[478] Ceramics based on alumina and zirconium dioxide have also been decomposed with borax.[479] A mixture of borax with a perborate makes it possible to burn organic substances during fusion and to analyze the glass formed by XRF for cations and some anions.[480]

Lithium metaborate has also been used for the preparation of cast disks for XRF, although it is not an ideal glass-forming agent. It yielded good results in fusion of silicate rocks of various compositions. With a small sample dilution (2 + 1), major and trace elements can be determined in a single disk.[481,482] A low-dilution method has been developed and applied to determination of nickel and gallium (down to 10 and 5 μ, respectively).[483] For a general treatment of decomposition of geological samples with a mixture of lithium tetraborate, carbonate and nitrate see Reference 484. Metaborate alone has been applied to analyses of soils.[485]

To prepare cast samples, lithium tetraborate alone or in mixtures with oxidants and substances absorbing radiation is at present most often used. Smales et al.[486] report the use of the Spectroflux® 105 mixture in XRF analysis. Crucibles made from unwettable gold-platinum alloys are almost exclusively used for fusion, and the same material is employed for the molds. Still, there are some works reporting the use of graphite crucibles for the fusion and disk-shaping.[487] The authors recommend polishing of the bottom side of the disk that was in contact with the crucible bottom, using fine grinding pastes. The properties of graphite crucibles and molds and the application of borate melts in XRF analysis are discussed in Reference 488. A fusion mixture of 1 part of $Li_2B_4O_7$ + 4 parts of $LiBO_2$ + 1 part of SiO_2 has been applied to decomposition of baddeleyite, loparite, and zircon in the determination of Y, Zr, Nb, and Ta. A graphite mold and a three-step polishing have been recommended.[491] Lithium tetraborate alone or its mixture with lithium nitrate and bromide has been used for a determination of major components of silicate rocks,[489-492] trace elements in them,[493,494] and especially for determination of sulfur in silicates.[495] The same procedure was employed in the determination of 16 major and microelements in river sediments.[496] After addition of Al_2O_3 and Li_2CO_3, the glass can be used in analyses of uranium concentrates.[497] Fusion has also been used in ED-XRF analyses of environmental samples.[498] Among industrial products, refractory materials,[491] and cement[499] have been analyzed after fusion with lithium tetraborate. Further applications are to analyses of iron ores, agglomerates, blast-furnace slags, and refractory bricks.[500] In analyses of ferroalloys, attention has been paid to preoxidation of the metals to the oxides, e.g., by mixing the alloy with V_2O_5[501] or by atmospheric oxygen at high temperatures.[502,503]

A characteristic feature of the last three procedures is the fact that the fusion agent is not mixed, but sandwiched with the sample. First, a protective layer is formed from part of the fusion agent or the admixtures (MgO, La_2O_3, $CaCO_3$), and the sample is spread over its surface and covered with the remaining fusion agent. Platinum is thus protected against direct contact with the metals. The melt must be vigorously stirred during fusion. Disks with high contents of chromium (ferrochrome) must be slowly cooled; otherwise the glass cracks.

5. Automated Fusion Devices

A great progress in automation of analytical operations has been achieved upon introduction of fusion devices. One of the first devices was proposed by Claisse,[456] whose Claisse Fluxer Model VI serves both for the preparation of solutions and for casting glass disks for XRF. The instrument is equipped with six crucibles and enables simultaneous preparation of six disks. Fusion is carried out using Fisher burners operated by liquid propane or butane. The fusion procedure, temperature, time, and stirring of the melt are programmed, with many combinations of the above parameters. The melt can be cast in molds to prepare beads or in beakers with a suitable solvent. On pouring the melt, intense rocking of the glass occurs, and the dissolution is hastened by programmed heating and stirring. For details on the use of the instrument see Botto.[270]

Other instruments employ induction heating. The temperatures used, around 1400°C, are just below the melting point of the Pt-Au alloy of the fusion crucibles. Due to eddy currents, the melt spins vertically and is thus vigorously stirred. An advantage of the induction heating is the fact that the fusion process starts at the maximum temperature, and its duration is much shorter than that of classical fusion in a furnace. The heating of the crucible is uniform. Fusion devices permitting sequential fusion of up to 12 samples have been made for both manual and automated casting; in the latter case, the mold is placed in another induction furnace permitting preheating of the mold to a required temperature and programmed cooling of the bead formed. Many fusion devices have been manufactured for special purposes of the preparation of disks for XRF and solutions for AAS, ICP-OES, and wet chemistry.

IX. FUSION WITH ALKALI FLUOROBORATES

As fusion agents, reagents containing boron and fluorine at a great variety of ratios are used, sodium and lithium being most often used among the alkali metals. The fusion comprises the advantages of the alkali borates and fluorides as fusion agents; poorly decomposable oxides and silicates are intensely attacked. During the decomposition, silicon partially volatilizes in the form of SiF_4. The alkali metafluoroborates have been introduced in analytical practice by Rowledge,[504] and various modifications of the procedure are termed the Rowledge fusion. Mixtures of boric acid with the alkali fluorides at various ratios are used, or the metafluoroborate, $Na_2B_2F_2O_3$, is prepared from these components in a molar ratio of 2NaF + B_2O_3. The prepared mixture is cautiously dehydrated in a large platinum dish, sintered, and fused at 900°C, with formation of a clear melt. This is rapidly cooled, pulverized, and stored in a tightly closed bottle for protection against the atmospheric effects. The decomposition is then carried out in a platinum crucible during 12 to 15 min, using a tenfold excess of the fusion agent. Many resistant materials are decomposed belonging to the groups of oxides (rutile, spinels, cassiterite, chromites, zircon) and silicates (tourmaline, topaz, beryl, sillimanite, cyanite, etc.).[295,505] Excess fluorine can be removed from the system by heating with sulfuric acid; in this way, excess boron and silicon are also removed. A 2 + 3 mixture of boric acid with lithium fluoride was used by Biskupski[506] for rapid decomposition of rocks and resistant minerals; a tenfold excess of the fusion agent is used, and the fusion takes 15 min at 800 to 850°C in a platinum crucible. The melt is dissolved in dilute nitric acid, and the solution is used for subsequent determinations. Russel et al.[507] have found that the fusion agent is rather aggressive and 6 to 10 mg of platinum are dissolved under these conditions during a single fusion.

As fluoroborates energetically decompose the above resistant minerals, they are used to fuse insoluble residues of rocks after decomposition by hydrofluoric plus sulfuric acids. The procedure has mainly been used in a determination of tin in geological materials,[508-510] zirconium,[511] gallium,[512] and barium.[513] In the latter case, it is interesting that higher concentrations of salts in the solution prevent reprecipitation of barium sulfate that is decomposed by this procedure. To decompose uranium minerals in rocks, ores, and technological samples, sodium fluoroborate, $NaBF_4$, has been used for fusion of acid-insoluble residues.[514,515] Metafluoroborate has been used for direct decomposition of uranium ores.[516] The Rowledge mixture has been found useful as a fusion agent for iron ores of various compositions.[517]

The completely inert behavior of the fluoroborate melt has been utilized in the determination of ferrous iron in rocks and, especially, resistant minerals. The test material is fused in a suitable apparatus (a tube or crucible furnace), permitting work in a highly inert atmosphere of CO_2, N_2, or Ar. After cooling, the melt is dissolved in dilute sulfuric acid in the presence of boric acid, and ferrous ions are determined oxidimetrically. This procedure has been proposed by Rowledge[504] and modified by many authors.[518,519] The latter authors have pointed out that the accuracy of the results depends on the initial Fe^{2+}/Fe^{3+} ratio in the test material. The especially difficult and tedious dissolution of the melt in sulfuric acid has been successfully replaced by dissolution in a mixture of cold hydrochloric acid and iodine monochloride.[520] Another modification involves dissolution of the melt in a solution containing a known excess of bichromate ions. After the melt dissolution, the unreacted bichromate is determined by reductometric titration with the Mohr salt. The procedure has been adapted for work on semimicroscale.[521]

Another possibility of an indirect determination of ferrous ions is a determination of the excess of an oxidant, after the reaction with the oxidant directly in the melt. The sample is fused with metafluoroborate containing a known amount of vanadium pentoxide in an inert atmosphere. After dissolution of the melt, either the unreacted vanadate, or the amount of vanadyl ions equivalent to the amount of the ferrous ions, is determined. The procedure has been applied to a determination of the FeO in resistant minerals,[522] especially in chromites.[523] An analogous principle has been utilized to differentiate tri- and pentavalency of vanadium in vanadium-containing muscovites and tourmalines.[524] Ferric oxide was added as the oxidant. The amount of ferrous ions, determined oxidimetrically, is equivalent to the content of trivalent vanadium.

X. REDUCTIVE FUSION

Reductive fusion belongs among special methods of decomposition of certain inorganic materials and utilizes the fact that the reduction products are more susceptible to the dissolving action of acids than the original substances. The ions liberated during the decomposition can sometimes be reduced as far as to the metals. The metallic phase thus formed can then be separated mechanically. For example, highly stable cassiterite can be reduced to tin metal, which is separated in the form of the regulus and weighed. The metal is readily soluble in hydrochloric acid, and the solution is suitable for an oxidimetric titration determination of Sn^{2+} formed during the dissolution.

The decomposition has two stages: (1) the material to be decomposed is fused with a suitable agent (carbonates, hydroxides, borates, etc.) to obtain a homogeneous melt, and (2) reducing agents are added during the fusion to obtain compounds with elements in lower valence forms. Reductive fusion is most often used in the fire assaying of precious metals. This procedure, whose character is technical rather than analytical, serves for concentration and separation of certain elements prior to an analytical determination. The procedures of reductive fusion have been less often applied to the concentration and separation of common metals. Alkaline fusion in the presence of elemental sulfur, called the Freiberg decomposition, can also be included in the methods of reductive fusion.

A. Fire Assaying

This is the method that has widely been employed for separation of silver, gold, and the platinum metals from ballast materials prior to their analytical determination. It has been used in the evaluation of ores and other raw materials containing these elements, as well as in analyses of technological concentrates, industrial intermediates, and wastes, up to the final products — the pure metals and their alloys. The method has been known for an exceptionally long time (about 4000 years), and its principle is virtually unchanged. Classical fire assaying is based on the fusion of the test material in a pot made of a refractory material with a suitable fusion agent, at a sufficiently high temperature. The fusion agent contains slag-forming components, such as soda, borax, silica, fluorite, etc., and substances acting as collectors of the precious metals, e.g., common metals obtained on the reduction, their alloys, and sulfidic phases that give rise to a metallic button and possibly matte. An important component of the fusion mixture are reductants; the sulfidic sulfur contained in the test raw materials and concentrates exhibits reducing effect; more often, a reductant must be added, e.g., carbon (coke, flour, starch, carbonized wood, alkali cyanides). During the fusion in the reducing medium, a metallic phase-collector is formed that sweeps precious metals from the melt through the formation of solid solutions (alloys) of the precious metal and the collector or through the formation of defined intermetallic compounds. With some platinoids, larger crystals grow that are concentrated at the pot bottom, below the silicate melt, owing to their high density. The reduced collector also forms larger drops in the silicate melt, and these drops react with the precious metals originally present or secondarily formed by the reduction in the melt. Owing to a high density of the collector, the metal drops slowly pass through the viscous melt, extract the precious metals, and are concentrated in the bottom part of the pot or in a conical iron mold.

On cooling, the collector solidifies and yields a regulus button, theoretically containing the whole amount of precious metals in the test material. The button is then treated thermally or chemically. A thermal treatment is usually preceded by mechanical cleaning of the surface and shaping, or by chemical dissolution of the slag residue. Lead button is then cleaned from base metals by oxidizing fusion (scorification) with slag-forming reagents, followed by oxidizing ignition, cupellation, during which the lead is rapidly oxidized to an oxide. The oxide is almost completely soaked in the cupel made of a porous material. The precious metals remain as a prill at the bottom of the cupel. This Ag-Au prill is freed of the silver by leaching with dilute HNO_3 (1 + 9) (parting process). However, dissolution only occurs when the Ag/Au ratio is greater than 5. If this condition is not met, the prill must be inquarted, e.g., alloyed with a known amount of silver.

Sometimes it is advantageous to treat the metallic or sulfidic buttons chemically by dissolution in mineral acids in open or closed systems. The precious metals, except for silver, are accumulated in the insoluble residue that is further chemically treated. Some platinum metals, e.g., osmium, volatilize during the button dissolution and must be trapped in suitable absorbents.

It is thus evident that the mixture of the chemicals and of the test ore during the pot fusion is a very complex system. The chemistry of the fusion process is virtually unknown. Using binary or ternary diagrams of the systems SiO_2-metal oxide or SiO_3-B_2O_3-metal oxide, the reaction taking place can roughly be explained, but a generalization of the fusion process is very complicated, even with simple initial raw materials. The technique of fire assaying of precious metals is thus empirical, and the prescriptions for the selection of the fusion agents and the procedure of fusion contain individual experience of the assayists. In general, such a mixture must be obtained which contains at least two immiscible melts, between which the precious metals are distributed. In this respect, fire assay resembles extraction separation methods. However, quantitative description of the distribution is complicated by difficulties in expressing the bonding of the precious metals in the two immiscible phases.

In our opinion, this chapter should not substitute specialized monographs dealing with the analytical chemistry of precious metals, including the problems of fire assaying. For the theory and practice of this method, the reader is referred to the books by Beamish,[525] Beamish and Van Loon,[526,527] Ginsburg et al.,[528] the manual by Haffty et al.,[529] and many reviews.[530-541] The chapter on fire assay in the book of Johnson and Maxwell[542] is also useful.

In the above works, the reader finds information on all the stages of fire assay including:

1. Sampling of raw materials and the products of their treatment; the sampling method and the size of a representative sample considering random distribution of precious metal minerals
2. Pretreatment, milling, homogenization, and selection of the sample weight for the assaying
3. Determination of the redox potential of the sample by trial fusion
4. Sample pretreatment by oxidative roasting or leaching with mineral acids
5. Calculation of the fusion agent composition and its ratio with respect to the sample, in view of the chemistry of the test material
6. Description of the apparatus — furnaces, pots, tools, and protective devices
7. Temperature program for the fusion, scorification, and cupellation, fusion time, and the redox properties of the atmosphere
8. Wet treatment of the metallic buttons
9. Treatment of the precious metal prills — purification, weighing, inquart, and parting
10. Practical procedures for various materials

It should be pointed out that in North America, it is customary to select the sample weights in terms of the assay ton weight (= 29,166 g) or integral fractions thereof. The weight of the prill obtained in mg then corresponds to the number of troy ounces per ton. In the countries using the metric system the results are given in ppm or in grams per tonne (metric).

As the precious metal collector, lead metal is most often used, less frequently silver, tin, Cu-Fe-Ni alloys, and Cu and Ni sulfides. The further description of the fire assay methods is, therefore, classified according to the properties of the collectors.

B. Lead Button

Fire assaying with the lead metal collector is a classical procedure that is used primarily for the collection of silver, gold, and some platinum metals, with an advantage that higher sample weights can be used for the fusion. The method is mainly employed for Ag, Au, Pt, and Pd; the collection of Rh, Ru, Os, and Ir by lead, as well as by silver, creates problems. The latter metals yield at high temperatures stable compounds that are difficult to decompose, mainly oxides, of which RuO_4 and OsO_4 are volatile. Nevertheless, Ru, Rh, and Ir are relatively readily alloyed with Pd and Pt, which makes their fusion easier. Lead is added to the test raw material usually in the form of the oxide (litharge) in an amount that is sufficient for the formation of Pb silicates and a sufficiently large regulus. For testing of wastes rich in precious metals, the fusion is performed with elemental lead. The precious metals are separated from the lead and other common metals obtained on reduction (Cu, Ni) by scorification and cupellation, obtaining a prill that is usually composed of an alloy of silver with other precious metals. Suitable composition of the fusion mixtures for fire assay of minerals, ores, and rocks are given in the manual by Haffty et al.[529] and in Table 12.5 in Reference 542.

So far, there exist no quantitative data that would enable estimation of an optimal button weight. This weight should be proportional to the sample weight and should increase from certain unspecified limits. The value is not critical for the collection of silver, but platinoids require a certain minimal amount for quantitative collection. It has been shown experimentally[525] that the optimal amount of lead lies within 25 and 35 g.

A little is known about the bonding of the precious metals in the lead button. Binary systems of Pb-Pt and Pb-Pd have mostly been studied. It has been found that the formation of solid solutions is very limited. A study of microstructures of binary alloys of the individual platinoids with lead has shown that two phases are formed (at Pt and Pd contents of up to 2% at, and a Rh content of up to 1.5% [atomic]): (1) alloys with lead and (2) intermetallic compounds, e.g., $PtPb_4$ and $PdPb_2$. At higher platinoid contents, an eutectic melt is formed. Iridium and ruthenium do not react at all with lead and remain at the button surface as impurities, which may cause losses of these elements during fusion in a pot. This phenomenon is especially pronounced with ruthenium, whose density is somewhat smaller than that of lead.[528]

1. Separation of Gold and Silver

A detailed description of fire assaying employing the lead button can be found in the manual of the U.S. Geological Survey (U.S.G.S.), published by Haffty et al.[529] More recently, the method was critically appraised by VanWyk and Dixon.[543] The authors mainly studied the redox properties of the atmosphere during the fusion and the effect of the regulus size on the completeness of the collection. It has been found that it is useful to use coke furnaces for the fusion, in which the CO/CO_2 ratio can be relatively easily regulated. It is further recommended to cover the mixture in the pot by a charcoal layer. The amount of gold and the platinoids obtained depends on the ratio of the litharge added with respect to the reductant, which should be from 9 to 12:1, as well as on the composition of the furnace atmosphere. The maximum yield of the sum of the precious metals is dependent on the maximum amount of the reduced lead (97% of the PbO added). Copper does not affect the collection, up to its amount of 1 g, but even 0.6 g of nickel prevents the formation of a prill of precious metals; to separate nickel, a great excess of lead must be used. To determine precious metals in mineral raw materials, a standard fusion mixture is used, consisting of 39.1 parts of soda, 21.5 parts of litharge, 24.5 parts of anhydrous borax, 12.75 parts of silica, and 2.1 parts of curcuma flour; the mixture has a redox ratio of 10:3.1. The sample weight is selected depending on the precious metal content, from 1 to 150 g, samples of up to 20 g being mixed with 30 to 50 g of silica. The maximum sample weight for chromites is 25 g. Ores containing carbon or sulfur are preignited with a small amount of silica. An amount of 360 g of the fusion agent is added to 100 g of the sample. The mixture is fused in a pot for 1 to 1.5 hr at 1000 or 1100°C (Au or Pt ores, respectively); the temperature is then increased to 1200°C, and after 30 min, the melt is cast into an iron mold. The regulus is separated from the slag after cooling. The slag is milled, borax, litharge, and flour are added, and the mixture is fused again. The joint buttons are cupelled in an air stream at 960°C, until all the lead oxide is soaked into the cupel or volatilizes. The prill is then transferred into a new cupel and ignited at 1350°C, thus removing the remaining lead and all the silver. The resultant prill then contains gold, platinum, palladium, rhodium, and a small part of iridium.

The classical fire assay was used for the determination of silver, gold, and the platinum metals in geological materials using the OES method.[544] The prill is treated in concentrated nitric acid, in which silver and palladium are dissolved; the other insoluble platinoids are decomposed by aqua regia. In analogous analyses for contents of gold, platinum, and palladium of 0.005 ppm and that of rhodium of 0.01 ppm, errors of ± 20 and 40% have been reported, respectively.[545] Fire assay can be employed to preconcentrate gold and silver from sulfidic concentrates containing 13 to 18% sulfur and large amounts of antimony, if the reductive fusion is carried out in the presence of a calculated amount of KNO_3.[546,547] It has been found that upon removal of lead to the attainment of a weight of 200 to 400 mg, a negligible amount of other metallic admixtures is left in the button. On completion of the cupellation, the prill obtained is weighed, quartered with silver, and parted using HNO_3.

For the determination of silver, gold, and palladium in technical concentrates, the metals were preconcentrated into a Pb button, the prill was dissolved, and the metals determined by AAS. To concentrate traces of precious metals, silver is added to trap traces of gold and palladium, or gold to trap traces of silver.[541]

Cocollectors have been used in an OES determination of gold and the platinum metals in geological materials, after fire assay with a Pb button.[548,549] For the collection, silver (for Au, Pt, Pd) or platinum (for Au, Pd, Rh) is added in an amount of 4 mg. Standards are prepared by evaporating the appropriate amounts of standard solutions with hydroxylamine in lead dishes, followed by cupellation. The prills obtained are directly excited in the cavities of undercut electrodes.

The finding that the greatest losses of precious metals occur during the complete removal of lead during the cupellation has led many authors to a chemical treatment of a small lead button.[550-554] The lead button obtained by reductive fusion is cupelled at lower temperatures of 800 to 900°C, to obtain a residue with a weight of about 0.5 to 4 g. The residue is then decomposed in dilute nitric acid, in which lead and silver are dissolved, as well as certain platinum metals. The insoluble residue is then treated with aqua regia, and the precious metals in the solution are determined by AAS. It has been found[551] that after the lead removal from a 50-g button, the 100-mg lead residue contained more than 95% of silver, gold, platinum, and palladium. The dissolution of lead in HNO_3 is catalyzed by an excess of mercuric ions added.

The especially large losses of silver during cupellation (on average, 5 to 10% and with Ag contents of less than 2 mg up to 30%) have led Diamantatos[555] to adopt a chemical treatment of the whole button. The author used a mixture of 50 g PbO, 20 g of borax, 60 g of anhydrous soda, 20 g of silica, and 5 to 6 g of flour for the reductive fusion. These substances are mixed with a sample weight of up to 50 g, which is, if necessary, freed of sulfur by ignition, and the mixture is fused for 1 hr at 1200°C. The melt is cast into a mold and cooled, and the regulus is freed of the slag residues by boiling in 30% NaOH. The parting is carried out in a large beaker using 300 mℓ 70% $HClO_4$ and 30 mℓ anhydrous acetic acid at a temperature of 180 ± 5°C, until all the lead is dissolved. The solution is cooled to 100°C and is cautiously diluted with 250 mℓ of water. The precious metals are reduced from the boiling solution with hydroquinone and formic acid, filtered off, dissolved in aqua regia, converted into the chlorocomplexes, and gold and silver are determined by AAS. Analyses of standard ores have shown that the method of partial diminishing of the button[550,553] is unacceptable for accurate determinations of gold and silver.[556]

To determine traces of precious metals, fire assay has been combined with NAA. Two approaches can be taken in the determination of gold: (1) the test material is irradiated first, and then the precious metals are concentrated into a lead button, after adding gold as the carrier. The lead is removed by cupellation, and the activity of the prill obtained is measured;[557] (2) after fire assay of the mineral material, the button is irradiated, and its activity is measured either directly,[558] or the prill activity is measured after cupellation.[559]

2. Losses of Gold during Fire Assay

Many authors, e.g., Reference 556, do not recommend classical fire assay with the formation of a lead button followed by cupellation for accurate determination of gold and silver in ores and concentrates. They point out that the greatest losses occur during the cupellation and the following parting of the prill. Coxon et al.[560] have confirmed in their radiochemical study that the cupellation temperature exerts a principal influence on the precision of the results. On an increase by 100°C the loss of gold (6 mg prill) increases by 0.5%, owing to absorption into the cupel. Sinclair[561] dealt with the distribution of the gold losses during the whole process and has also shown that silver is not quantitatively separated by the decomposition of the prill with an Ag/Au ratio of 4 with dilute nitric acid; however, on the other hand, a small amount of the gold is dissolved. In analyses of various alloys with precious metal contents of 333 to 999/1000 using cupellation in a lead foil with an addition of silver, the losses of gold were from 0.15 to 0.19% and were compensated for, as a surplus, by imperfect separation of the silver during the parting of the prill. Large losses, up to 20%, have been observed when using bone flour cupels.[562]

The completeness of the lead removal during the cupellation is affected by common metals dissolved in the regulus and formed by the reduction of their compounds, mainly sulfides. Yaguchi and Kaneko[563] have shown that copper in amounts above 3 g and nickel in amounts greater than 0.03 g in the initial sample make cupellation of the button completely impossible. A negative effect of tellurium is also without doubt; tellurium is transferred into the button during the fusion, and its presence increases losses of gold during the cupellation, apparently through the formation of gold telluride that is absorbed in the cupel.[530] In small amounts (up to 0.2 g in the sample) Te is transferred into the prill, but amounts greater than 1 g in the sample prevent the prill formation. Selenium behaves analogously as tellurium and in amounts greater than 5 g in the sample cause ''fritting'' of the button.

In determinations of gold by the ''iron nail'' method in ores containing chalcopyrite and other sulfides, copper and arsenic exert a negative effect. MAs-type compounds (''speiss'') are formed, and silver and gold penetrate into them, which causes great losses of them.[563]

Wall and Chow[564] studied the losses of precious metals during fire assay in great detail, using a simulated button made of a lead foil, and weighed amounts of gold and silver. It was found that the losses of gold during cupellation are unaffected by the button weight (6 to 40 g), the silver-to-gold ratio (10 to 30:1), and the cupel size. The losses amounted to 0.59 to 0.61% ($s = 0.08\%$ for $n = 12$). It has been concluded from the results of the parting of the prills obtained with nitric acid that the completeness of the leaching of silver depends on the thickness of the flattened bead and on the acid concentration: the losses of gold decrease with decreasing thickness of the prill and increasing acid concentration. Maximal losses have been encountered when the prill was freed of the cupel material mechanically. Further, losses of precious metals during the whole fire assay were followed.[565] Under ideal thermal conditions, these losses are about 1%. The process has been improved by mixing the sample with the fusion agent in a polyethylene bag preventing mechanical losses. The gold distribution during the procedure was followed using a tracer; the greatest losses occur through trapping in the fusion pot.

C. Platinum Group Metals (PGM)

The classical extraction of the platinum metals into a lead button is complicated by differences in the behavior of the individual members of this group toward lead. It is not exactly known what solutions or intermetallic compounds are formed in the Pb-PGM system; it seems unambiguous that elemental iridium, iridosmium, and ruthenium are present. It has also been shown that palladium need not be collected by the sinking reduced lead, but that it diffuses spontaneously to the bottom of the fusing vessel. The capability of lead of collecting PGM is also negatively affected by the presence of larger amounts of copper and nickel; these metals affect the button properties and cause losses of the precious metals into the slag and the fusion pot walls. It seems that the completeness of the collection also depends on the mineral form of PGM occurrence in the test material. The decisive effect on the collection completeness is undoubtedly exerted by the fusion agent composition, which should provide a neutral reaction of the fused mixture, i.e., the same contents of oxygen from bases and acidic components. Unsuitable compositions of the fusing mixtures have caused especially high losses of osmium; these losses further depend on the heating rate and the melt viscosity. Surprisingly, osmium can be quantitatively extracted during the oxidative "nitre assay" (fire assaying in the presence of KNO_3), in which the high volatility of osmium tetroxide is not encountered. A simple fusion is insufficient for quantitative extraction of most PGM into lead; the slag must be reassayed two to three times and treat the joint buttons.

Mechanical losses may occur during loosening and cleaning of the button, owing to segregation of some PGM on the button surface. Therefore, casting of the melts from pots into iron molds and pounding of the button into a cubic shape are not recommended. Especially great losses, mainly with ruthenium, osmium, and iridium, occur during scorification and cupellation.[525,528]

The methods of isolation and separation of the PGM, including fire assay, have been critically reviewed by Beamish.[540] The method of the PGM collection by lead has been evaluated by Zolotov.[566] The author considers the procedure tedious, energetically demanding, and dirty, connected also with the use and formation of toxic substances. He has also pointed out poor solubility of ruthenium, osmium, iridium, and rhodium in molten lead. He sees the only advantage in that it is possible to use large samples. In South Africa, more than 90% of the mined ores are analyzed by this procedure.

Although other metallic collectors have been introduced, the classical fire assay into lead regulus followed by preconcentration of platinum, palladium, and rhodium by cupellation into a silver prill cannot be replaced by other procedures, in view of the simplicity and rapidity of the operations, the recovery precision, and the general accuracy of the results.[526]

Guest et al.[567] statistically evaluated the precision of the determination of gold and the PGM by the fire assay method with Pb and NiS collectors during the preparation of two reference samples. The scatter of the results is large, but the results in a single laboratory vary only negligibly, as do the results of parallel determinations carried out by a single worker. The use of reference samples is useful in order to prevent differences among the results from various laboratories and among the day-to-day results from a single laboratory.

Anisimov et al.[568,583] dealt in detail with losses of the PGM during fire assay. The authors followed all the stages of the procedure, from the sample pretreatment to the cupellation, to a silver prill. The fusion lasted 1.5 hr at a temperature of 1000 to 1050°C, the 40-g button was slagged to a weight of 18 g, and the lead button was then cupelled at 900 to 950°C. The recovery of the PGM (except for Ru and Os) was from 98.3 to 100.0%. The decisive factor in the cupellation is a sufficient excess of silver; its ratio with respect to the PGM should be 15 to 20:1 and with high iridium contents up to 40:1. It is necessary to refuse the original slags and the crushed cupels. The procedure markedly contrasts with the observation[572] that the volatilization of ruthenium and iridium especially increases in the presence of larger amounts of silver. It has been confirmed[569] that high losses of rhodium and ruthenium occur during cupellation. The two metals are present in the lead film at the interface of the regulus and the cupel. The silver prill may contain up to 2% iridium.

The losses of the PGM during fire assay were followed using the radioisotopes formed by irradiation of the original ores. The distribution of ^{172}Ir, isolated from the ore of the Merensky Reef (0.5 g of the irradiated sample) + 8 g Na_2CO_3 + 4 g of borax + 3 g SiO_2 + 20 g PbO + 1 g of corn flour, fused at 1240°C for 1 hr, reassaying of the slag with 15 g PbO, and cupellation of the joint buttons) has shown that the losses of the metal in the two slags are minimal (up to 1%). The precious metal prill contained 53 to 77% iridium. The remaining iridium is concentrated in the zone of contact of the prill with the cupel. The walls of the cupelling furnace were also weakly radioactive, owing to contamination with iridium carried with the lead vapors.[570,571] Activated regulus has been used[572] to study losses of platinum, ruthenium, and iridium. The radiography of the button has shown that ruthenium and iridium are accumulated in the bottom part of the button. The losses of the two metals are minimal up to a certain weight of the lead residue and increase in the presence of silver. This phenomenon has been explained by lead forming a protective layer preventing oxidation of the two metals by the external atmosphere. On the other hand, molten silver dissolves a considerable amount of atmospheric oxygen which oxidizes the metals with formation of volatilize oxides, from a certain Ag/PGM ratio onwards. If lead is partially removed, the residue amounting to 100 mg and containing a maximum of 20 mg silver, the losses of microamounts of Ru and Ir do not exceed 8%. On removal of all the lead with obtaining a silver prill, the losses increase to 43 and 60%, respectively. The most reliable method of determining the PGM is the INAA of the lead button, diminished to a weight of 200 to 700 mg.[573]

An analogous radiographic technique was employed[574] to follow the distribution of all the PGM between the silicate melt and the lead button. In contrast to Reference 572, the distribution of iridium within the button has been found to be homogenous, and a nonmetallic character of the collection of osmium and ruthenium by lead has been verified. Virtually all iridium and a great part of ruthenium pass into the slag during scorification, and all the osmium is volatilized. In a study of cupellation, the conclusions of the above authors[570-572] were confirmed.

1. Cupellation to Obtain a Platinum Metal Prill

Classical fire assay followed by complete removal of lead with formation of a precious metal prill is often used for extraction of platinum, palladium, and rhodium. The laboratory method of National Institute for Metallurgy, S. Africa (N.I.M.)[575] is based on low-temperature cupellation of the lead button at 900°C, during which losses of these PGM do not occur. The following procedure has been recommended for ores and concentrates: a 1- to 116-g sample (sulfidic materials must be preignited) is mixed with 100 to 350 g of the fusion agent consisting of 40 parts of soda, 33 parts of litharge, 25 parts of borax, 13 parts of silica, and 2 parts of wheat flour. Fusion is started at 1000°C, and the temperature is increased to attain 1240°C within 1 to 1.5 hr; the melt is allowed to quiet and is cast into an iron mold. The slag is refused with 50 g of litharge, 20 g of borax, and 3 g of flour. The joint buttons are cupellated at 960°C in a preheated cupel to complete removal of lead. To assay chromites, a special fusion agent is employed. Coombes et al.[576] and Tello and Sepulveda[577] used a similar thermal program of fusion and cupellation in determination of platinum in ores. The former authors use 1- to 2-g samples of concentrates and 3- to 7-g samples of the original ores. Flotation concentrates are freed of sulfur by 1-hr roasting at 700°C. The latter authors[577] employ 10-g samples. The prill obtained is always dissolved in aqua regia, and the platinum content is determined by flame or electrothermal atomization AAS. Fire assay with extraction of platinum into a lead button and AAS have been applied to sulfidic ores and Fe-Cu-Ni sulfide concentrates.[578] Using the sample weight of 1 assay ton, the method is also suitable for determination of palladium and gold; the determination limit is estimated by the sensitivity of AAS.

Isolation of rhodium from chromites followed by its AAS determination was studied.[579] The optimal results are obtained with a 3-g sample fused with a mixture of 35 g of anhydrous soda, 11 g of silica, 19 g of borax, 50 g of litharge, and 4.2 g of flour, at 850 and 925°C for 40 and 10 min, respectively. Cupellation is carried out at 950°C, until a precious metal prill is formed. The prill is dissolved in aqua regia. The authors have pointed out that no existing flux permits complete decomposition of chromite. Optimal decomposition was not even attained using a series of fusion agents simulating various acidities of slags with model composition from $2Na_2O{\cdot}3SiO_2$ to $2Na_2O{\cdot}SiO_2$. The above optimal composition of the fusion agent permitted decomposition of a 3-g sample to 88%, but a 15-g sample was only decomposed to 55%, with a simultaneous decrease in the rhodium amount determined from 2.4 to 1.7 ppm.

Some platinoid is often added to the sample as the collector, especially in extraction of trace amounts. Greenland et al.[580] added platinum metal and rhodium to samples of rocks and meteorites for determination of traces of iridium. A 0.25- to 1-g sample is first activated by neutrons and then subjected to fire assay, using the standard procedure according to Beamish.[525] A 50-mg Pt-Rh wire is added to the sample as the collector, with 2 to 5 mg of NH_4IrCl_4. The button formed is closed in a container and the ^{172}Ir activity is measured gamma-spectrometrically.

To extract microgram amounts of iridium, ruthenium, and osmium, Broadhead et al.[581] used 10 mg of platinum and the collector. Two kinds of fusion agent were used: weakly acidic for quartz-containing ores and basic for sulfidic copper ores (65 and 125 g PbO). The regulus is cupelled at a very low temperature, with a gradient from 930 to 830°C. The prill formed is excited cathodically-anodically in a DC arc at 20 Å for 40 sec. Platinum is used as the internal standard. The U.S.G.S. also employs platinum as the collector for iridium and rhodium in geological materials.[582] The fusion agents and the analysis procedure are the same as in Reference 529, and the determination of the metals is described in Reference 581.

2. Cupellation with Formation of a Silver Prill

Silver is an efficient collector of the PGM.[525] When used in excess, the losses of ruthenium and rhodium during cupellation are minimized. Simultaneously, the parting of the prill of the precious metals by inquart with dilute nitric acid is facilitated.

Silver was added to separate small amounts of platinum, palladium, rhodium, ruthenium, and gold followed by an OES determination,[584,585] or determinations by pulse polarography,[586] XRF,[587] AAS,[588] and NAA.[589] For fire assay preconcentration, standard fusion procedures are usually employed. The authors[585] cupel the button in an air stream at 771 to 816°C. In a determination of platinum and palladium in geological materials[589] the silver prill formed is activated by neutrons and dissolved in aqua regia, and the radioisotopes are separated by chemical procedures. The NAA standards are prepared from pure quartz doped by standard solutions of platinum and palladium salts, using identical procedures. Silver is added as the element[583,584] or in the form of a Pb-Ag alloy[589] called the Herman inquart, silver oxide,[585] or a solution of silver nitrate.[588,590]

A standard procedure for the determination of platinum, palladium, and rhodium (including gold) has been given by Van Wyk.[590] The fusion agent consists of 39.1 parts of anhydrous soda, 21.5 parts of litharge, 24.5 parts of borax, 12.7 parts of silica, and 2.1 parts of corn flour. An amount of 360 g of the fusion agent is usually employed for samples from 1 to 150 g. At the upper limit, given by PGM contents of less than 1 ppm, 71 g of soda and 44 g of borax are further added. In decompositions of chromites, the sample weight does not exceed 25 g, and 30 g of silica are added to 360 g of the fusion agent. Materials rich in sulfur are first mixed with 20 g of silica and roasted for 1 hr at 650°C. A solution containing 0.5 mg Ag^+ is added to the mixture in the fusion pot; for gold ores the gold-to-silver ratio must be less than 3. The fusion is performed in a furnace within a temperature range of 1100 to 1240°C (ores containing platinum) or 1000 to 1150°C (gold ores) for 1 to 1.5 hr. The button is cast and separated, and the slag is refused with another addition of litharge, borax, and flour. The joint buttons are cupelled for *circa* 1 hr at 960°C. The gravimetric and AAS determination of the precious metals is carried out after the prill parting.

In determination of platinum and palladium in mineral materials[591,592] the melt was additionally washed with another portion of the fusion agent. After a 45-min fusion at 1200°C, when all the lead has diffused through the silicate melt, a mixture of 10 g of litharge, 5 g of soda, and 1 g of starch is added, and the fusion is continued for another 15 min. This procedure replaces the reassaying of the slag. The button is cupelled at 1100°C in an MgO dish.

Kallmann and Maul[593] used a silver collector in referee analyses of precious metal sweeps and similar materials. The fusion mixture consists of the common components whose ratios are determined by the matrix composition. The sample weight is selected so that the total content of the PGM does not exceed 50 mg, and the collector amount does not exceed 5 g. After fusion at 1150°C the melt is cast, the slag is reassayed, and the joint buttons are slagged. The diminished button is cupelled at 920°C, and the cupel used is rinsed after removing the precious metal prill by evaporation of 100 mg of silver wrapped in a lead foil. The silver beads are dissolved in warm dilute nitric acid, and the platinoids are determined in the aqua regia solution by AAS and OES-ICP.

Gold is also an excellent collector for the PGM. In one method,[594] a gold wire of 0.8 to 1.2 mg wrapped in a lead foil is added to the mixture. The fusion mixture contains 1 to 3 g of calcium fluoride, that considerably improves the decomposing effect of the standard fusion agent. Cupellation can be carried out at a higher temperature, provided that silver is not to be determined. The precious metals are determined by solution OES. The method was used to determine the PGM in six standard U.S.G.S. rocks. Grimaldi and Schnepfe[595] have found that the collection of iridium by gold involves no loss, in contrast to silver, with which the losses during cupellation were of up to 5%. However, the collection efficiency depends on the amount of the gold added: in repeated assay of 20 g of dunite, 95% of the iridium present were collected by 30 mg of gold. If the addition was decreased to 10 mg, the loss increased to 13%. Therefore, addition of 50 mg of gold is recommended in the procedure. The same collector for the PGM was used in fire assay of chromites.[596] The lead is removed at 960°C, and the temperature is then increased to 1300°C, in order that the silver quantitatively volatilize. The sum of the precious metals is weighed, dissolved in aqua regia, and the precious metals are determined (except for Os and Ir) by AAS.

3. Method of a Minimal Lead Button

Many authors have utilized the fact that the extraction of all the PGM into the lead regulus is quantitative, but significant losses occur during the following thermal treatment, especially through oxidation and loss of volatilize oxides, or soaking of some platinoids into the cupel, together with the lead(II) oxide formed. If, however, the lead is not completely removed, then it can be assumed that the PGM remain accumulated quantitatively in the residue. This principle was used by Turkstra et al.[597,598] in determination of the PGM and silver by NAA in ores and matte. The material is decomposed by current fire assay, and the regulus formed is diminished to obtain a 150- to 1500-mg bead. Part of the bead is irradiated with neutrons, and the spectrum of the gamma emmiters formed is measured using a Ga(Li) detector. Kolosova et al.[599,600] studied in detail the method of incomplete cupellation and have confirmed, using radioisotopes, that 70% of ruthenium and almost 100% of iridium and osmium are lost in complete cupellation to a silver bead. After an incomplete removal of lead, more than 90% of the PGM remains in the bead. Radiography of diminished buttons was employed to follow the distribution of the PGM within the lead. It has been found in the parting of the lead buttons that platinum, palladium, rhodium, and gold are readily dissolved in aqua regia and in a mixture of hydrochloric acid with hydrogen peroxide, whereas iridium and ruthenium remain undissolved. The assumption of Georgiev[603] that a lead mantle protects ruthenium and osmium against oxidation during cupellation has been confirmed.

Various fusion mixtures suitable for fire assay of platinum-bearing ores containing copper and nickel sulfides from the Norilsk region (U.S.S.R.) are described in Reference 601. The buttons formed are cupelled to a weight of 100 mg, and the platinum metals are determined after dissolution by OES or AAS. In practical analyses, the recoveries are 90 to 95% for platinum, palladium, gold, and silver and 80 to 90% for rhodium, ruthenium, and iridium. A universal method has simultaneously been developed for dissolution of the diminished button. A 100-mg amount of lead is treated with 15 mℓ dilute nitric acid (1 + 5), and then

15 mℓ of aqua regia are gradually added. After the decomposition, the solution is evaporated to a wet residue which is dissolved in hot 1 *M* nitric acid. The insoluble residue containing ruthenium and iridium is filtered off and decomposed by fusion with 0.2 g of sodium peroxide and 0.2 g of sodium hydroxide; the melt is dissolved in hydrochloric acid. This procedure of fire assay and shortened cupellation has been applied to primary rock materials, flotation concentrates, matte, and wastes.

Rakovskii et al.[602] also used incomplete cupellation for determination of the PGM. They add 0.15 g of silver nitrate as the collector to the fusion mixture. The buttons are cupelled to attain a weight of 0.5 to 2 g and dissolved, and the precious metals are determined by AAS, NAA, and extraction photometry. Georgiev[603] has pointed out a possibility of decreasing the losses of osmium and other precious metals in incomplete cupellation. His reasoning is based on the comparison of the free energies of the formation of lead and PGM oxides.

4. Method of Complete Dissolution of a Lead Button

The uncertainties concerning the behavior of the precious metals during cupellation have led Diamantatos[555,604-608] to chemical treatment of the whole button after classical fire assay. His procedure of the chemical purification of the button from the slag using an alkali hydroxide and parting[555] is described in Section III. To fuse 2- to 60-g samples, he uses 150 to 180 g of a fusion agent containing 50 parts of litharge, 20 parts of borax, 60 parts of anhydrous soda, 20 parts of silica, and 5 to 6 parts of flour. The mixture is fused one hour at 1200°C; the slag must be refused. Combined buttons are cleaned and leached with perchloric acid containing 10% (vol) of acetic acid. Iridium and ruthenium are not dissolved in this reagent,[604] whereas platinum, palladium, rhodium, and gold are completely dissolved. Osmium is only partially dissolved and part of it volatilizes.[605] Therefore, the dissolution is performed in a distillation flask, and the volatile osmium tetroxide is absorbed in an alkali hydroxide solution. On the basis of these findings, an integrated scheme was developed for the recovery of the platinum metals and gold.[606,607] The procedure has been applied to various types of platinum ores, nickel-copper matte, black sands, and depleted platinum catalysts, and has been compared with other fire assay methods, with complete lead cupellation and collection on tin or nickel(II) sulfide. In all the cases the method described[83] yields highest results for all the precious metals.

The same procedure has been recommended by Georgiev[603] for determination of osmium. Grinzaid et al.[609] employ an analogous procedure for determination of the PGM in rocks containing nickel and copper sulfides. The authors studied the distribution of the precious metals within the button using a microanalyzer; it has been found that the distribution is unsuitable for XRF, because of segregation of nickel metal and the magnetite phase, $FeFe_2O_4$, in the lead. It is thus recommended to dissolve the button and employ wet analysis.

D. Collection of Common Metals into a Lead Button

Application of classical fire assay with a lead collector has been recommended by Artem'ev et al.[610-613] for separation and NAA determination of selenium, tellurium, bismuth, and molybdenum. It is known that selenium and tellurium, together with sulfur, arsenic, and antimony, exert unfavorable influence on fire assay of gold. In addition to matte, various intermetallic alloys with gold may be formed and cause losses of gold. On the other hand, collection of selenium and tellurium has been utilized for quantitative separation of these elements.[610] In fire assay of minimal amounts of ores, the authors found using radioisotopes that 50 to 70% selenium and 60 to 80% tellurium, depending on the fusion agent composition, passes into lead. The best results have been obtained in fusion of 1-g samples with 5 g of litharge, 0.5 g of soda, 0.1 g of borax, and 0.5 g of sugar as the reductant. After mixing, the substances in the pot are wetted with 2 mℓ of a solution of 0.2 g of lead(II) acetate and dried. The mixture is fused for 15 min at 1000°C; more than 90% of the two chalcogenides thus passes into the lead. Relatively large losses occur during scorification of the regulus with borax. In practical analyses of geological objects, the material is first irradiated by neutrons with a sufficiently intense flux, and fire assay is carried out after the hardest radiators have decayed. The button obtained is dissolved in 7 *M* nitric acid, and silver and lead are separated by radiochemical procedures. Collectors are added, and the isotopes of the two chalcogenides are obtained by selective reduction. The element separated is collected on a filter and its activity is measured at suitable analytical lines of the γ-spectra. The method has been verified on sedimentary sandstone ores mineralized with chalcocite and bornite.

In another variant,[611] the chalcogenides are first collected in the lead, and the button obtained is irradiated by neutrons. After 10 days, the ^{131}I activity, produced by the reaction in Equation 1,

$$^{130}Te(n,\gamma) \rightarrow {}^{131}Te \xrightarrow[(1.2\ days)]{-\beta} {}^{131}I \qquad (T/2 = 8.1\ days) \qquad (1)$$

is measured with a suitable detector. The activity of the selenium isotope is only measured after ^{203}Pb has decayed, using the intensity of the photopeak with the energy, 265 keV. The decomposition process can be accelerated by decreasing the sample and fusion agent weights two to three times. The sensitivity of the detection of tellurium was increased by radiochemical separation of ^{131}I that is sublimed from the solution, suitably trapped, and its activity measured. The minimum determinable contents of selenium and tellurium are $1 \times 10^{-8}\%$ and $1 \times 10^{-6}\%$, respectively.

Bismuth readily passes into lead buttons, its collection is quantitative, and no loss occurs during fusion and scorification (even with a tenfold decrease in the button weight). In determining traces of this element, the button is activated by alpha particles in a cyclotron, with formation of ^{209}At and ^{210}At astatine. By activation of lead, ^{207}polonium is simultaneously formed. As the two sources coincide in the measurement, they must be radiochemically separated. The irradiated button is fused in a quartz tube at 800°C; the astatine volatilizing from the melt is trapped in a lump of silver-coated copper wire whose activity is measured. The separation coefficient of astatine and polonium is *circa* 10^3. Without separation, $10^{-2}\%$ Bi can be directly determined in the regulus. By sublimation from the molten lead, 95 $\pm$ 5% of astatine is separated, and the sensitivity of the determination increases[612] to $10^{-6}\%$ Bi.

To determine traces of molybdenum in geological materials, the activity of ^{99}Mo is utilized. The radioisotope is produced by bombarding natural molybdenum with alpha particles. However, isotopes of rhenium, selenium, iron, and others interfere in direct measurement. Therefore, the nuclear reaction, $^{94}Mo(\alpha,n) \rightarrow {}^{97}Ru$ is used to advantage. The interfering activities are removed during fire assay, using the procedure described in Reference 610, and the ruthenium isotopes are extracted into the lead button. The button is cast into a special mold, permitting the obtaining of buttons with a standard geometry in the standard fusion procedure. The method of the sample activation and the activity measurement permits molybdenum to be determined down to a content of $1 \times 10^{-5}\%$. Using an analogous method of NAA combined with fire assay, traces of uranium can be determined utilizing the decay products ^{132}Tc and ^{103}Ru.[613]

E. Extraction of Precious Metals with Tin

It is well known that many PGM form intermetallic compounds with tin. A large number of these phases are formed in the tin-platinum binary system, involving phases from Pt_3Sn

to $PtSn_4$. With other PGM, the compounds $RuSn_3$, $RhSn_3$, and $IrSn_2$ have been described. Palladium also forms compounds with tin; these, however, have not been isolated. No intermetallic compound of osmium with tin is known. Most of the above phases are insoluble in hydrochloric acid. It is assumed that solid solutions of platinum in tin, up to a content of 30% Pt, are readily dissolved in dilute hydrochloric acid, but the decomposition is slower when the platinum content increases. Aqua regia is required for decomposition with high platinum contents above 80%.

Faye[614] has utilized easy extractability of the PGM into tin metal and has developed a new fire assay procedure. Further advantages are easy reducibility of tin(II) compounds, the low melting point of the metal formed, easy separation of the button from the slag, and favorable behavior of the metal during parting. Faye and Moloughney[615] have elaborated the determination procedure and have provided a comprehensive analytical scheme for determination of the PGM, gold, and silver after collection of these metals in a tin button.

Their procedure may be summarized as follows. Prior to the fusion, sulfur, arsenic, and antimony must be removed, depending on the sample character, to prevent the formation of matte and speiss. This is done by roasting the original material (possibly with an addition of silica) at 750 to 800°C. However, osmium volatilizes during this procedure. Copper and nickel, which are also readily reduced and form various alloys with tin, must be removed by preliminary leaching of the sample with concentrated hydrochloric acid in the presence of ammonium chloride. Metallic meteorites and other metallic materials should be dissolved in this acid prior to the fusion, the solution evaporated to dryness, and the dried chlorides used for the further operations.

The sample weight depends on the absolute contents of the precious metals and on the analytical method (usually up to 1 assay ton). The sample is mixed with the fusion agent of 35 to 40 g SnO, 50 g of soda, 10 to 20 g of silica, 10 g of borax, and 40 g of flour or 6 to 8 g of coke as the reductant. After thorough mixing, the substances are transferred into a pot, placed in a preheated furnace, and fused 1.5 hr at 1250°C. The contents are stirred occasionally. After the collection, the melt is cast into a steel mold and allowed to cool, and the button is separated and mechanically cleaned from the slag. Then it is refused in a nitrogen atmosphere, and the melt is poured into circa 3 ℓ of cold water. The metal granulates on rapid cooling, or becomes brittle and can be pulverized mechanically; leaching is easier in the granular form. The dissolution and the following chemical procedure for separation of the PGM can be seen on a flow-sheet given in Reference 615. Notes to fire assay are

1. The total amount of the sum of copper and nickel in the sample should not exceed 4 g. As both the metals increase the melting point of the button, it is suitable to add 10 g of tin metal to the fusion mixture. Iron quantitatively remains in the silicate slag.
2. The blank must be carried out for each new batch of tin(II) oxide and tin metal, as both the substances contain silver and gold.
3. It is unnecessary to modify the fusion agent for materials with various matrices from acidic to basic. Modifications are only required with certain separated mineral phases. Magnetite and ilmenite are readily decomposed in amounts of up to 15 g; the amounts of the reductants given suffice not only for the button formation, but also for reduction of larger amounts of Fe(III) and titanium(IV) to lower valence states. Chromites are not decomposed by the fusion agent and must be predecomposed by sintering with sodium peroxide at 700°C, on an ignition plate placed on 10 g of silica.
4. Osmiridium and iridosmine, natural solid solutions of osmium and iridium that belong among chemically highly stable minerals, are readily dissolved in molten tin.

The above procedure has been applied to various types of African platinum ores, Cu-Ni matte, flotation sulfidic concentrates, black sand concentrate (mostly ilmenite), and the products of precious metal refining. The results have been compared with a control determination by independent laboratories, with a satisfactory agreement. The proposed method removes the difficulties of the classical fire assay connected both with the collection and with the cupellation of lead; the collection in tin is much simpler.

The collection of silver with tin has been verified in analyses of ores and metallurgical products.[616] Fire assay is carried out under the above conditions with the same fusion mixture. If the copper and nickel contents in the sample are greater than 3 g, 15 to 25 g of tin lumps are added to the charge to decrease the melting point of the Cu-Ni-Sn alloys. The button is dissolved in hydrochloric acid, the silver chloride separated is decomposed by the same acid with hydrogen peroxide, and the silver is determined by AAS. The method is rapid and suffers from no interelemental interference. The results of control analyses are in good agreement with those obtained by classical fire assay.

The collection of platinum and palladium by molten tin was later improved by adding elemental tellurium to the fusion mixture.[617] Tellurium reacts with the above metals with formation of intermetallic compounds $PtTe_2$ and $PdTe_2$ and functions as the collector in the fusion and parting. The sample pretreatment by leaching and roasting in the air, as well as the fusion and granulation of the button, are identical with the procedure in Reference 615. In dissolution of the button in HCl, the precious metal tellurides remain in the insoluble residue which is filtered off and dissolved in a mixture of hydrochloric acid and hydrogen peroxide, followed by aqua regia; the metals are then determined by AAS.

The efficiency of the whole procedure of the collection of the PGM and gold according to Reference 614 was studied by Palmer and Watterson[570] using radioisotopes. The authors have found that a considerable part of the precious metal is lost during the fusion by passage into the slag; at the sample weight of 2 assay tons, the losses of Au, Ir, Ru, and Os amount to 6.8, 4.8, 7.3, and 4.9% of the original content, respectively. The losses double if the sample weight decrease to one half (1 assay ton). An improvement was attained on reassaying the slag with borax; the losses decreased to about 3% of each metal. Reduction of precious metals in button solutions by formic acid has been found to have a poor efficiency. Especially high losses were encountered with Ir, Ru, and Os — 44.5, 73.1, and 79%, respectively. Direct separation of osmium by distillation from the solution of the button in hydrochloric acid with hydrogen peroxide also yielded low results.[618] Therefore, collection of the PGM by nickel(II) sulfide is preferred.[570,618]

Ryabinina et al.[619] used tin as the PGM collector in assays of nickel-copper ores. As losses of some PGM may occur during preignition and leaching of the material, they have omitted these operations and modified the fusion by adding potassium nitrate and iron filings to the mixture. Copper and nickel metals formed by the reduction yield intermetallic compounds with some PGM, and these are rapidly separated into the button. The collection completeness also depends on the slag viscosity. With strongly acidic slags, nickel and copper are reoxidized on prolonged fusion, which decreases the collection efficiency. The procedure has been verified by analyses for platinum, palladium, rhodium, and gold on standard samples. At low contents (0.1 to 5 ppm) the losses of the PGM and gold do not exceed 10% of the original contents.

F. Collection by Copper and a Cu-Ni-Fe Alloy

It was early found that greatest difficulties in classical fire assay with the lead collector are caused by higher contents of copper and nickel in the original material, as these elements are readily reduced and alloyed with lead. Therefore, other procedures have been sought, using phases present in the material as the collectors, especially copper, nickel, and iron. Their alloys readily react with many PGM with formation of solid solutions, whose existence has been confirmed for platinum, palladium, rhodium, and osmium. If the weight of their sum is less than 10 mg, all the alloys, including the PGM, are easily dissolved in hydrochloric acid. At higher contents, above 20 mg, an insoluble residue remains, containing mainly iridium and ruthenium, which suggests mechanical collection of the metals. A Fe-Ni-Cu alloy dissolves even chemically resistant minerals well, such as Tasmanian osmiridium and iridosmine. The alloys formed are, again, readily soluble in mineral acids. Fusion cannot be directly applied to sulfidic ores, as matte would be formed that would prevent separation of the metallic and silicate phase. These materials should first be decomposed by roasting in an oxygen-free, e.g., hydrogen, atmosphere.

Mixtures of soda, borax, and amorphous graphite are used as fusion agents. If the ore does not contain a sufficient amount of oxides of the button-forming elements, then these oxides must be added in a suitable amount. The empirical ratio of the oxygens from the acidic and basic components of the mixture should be within the range, 0.56 to 0.8. Copper is probably the active component of the alloy and quantitatively collects iridium, rhodium, palladium, and platinum. Analogously, nickel is an effective collector of the PGM, and its presence or absence in the slag affects the losses of the precious metals. As the oxides of the two metals are easily reduced, they decrease the PGM content in the slag so that the slag usually need not be refused.[525,526]

This procedure is performed: 100-g sample (3.42 assay tons) of a Cu-Ni ore or concentrate is roasted in H_2 stream in a porcelain dish for 2 hr at 980°C, with occasional stirring. The cooled product is mixed with 42.4 g of soda, 27 g of borax, and 9.5 g of fine amorphous carbon. The mixture is ground to pass through a 45-mesh sieve and is thoroughly mixed on a plastic sheet. It is then transferred into a pot and placed in an oven heated to 1200°C. The temperature is slowly decreased, and the reaction is allowed to proceed until the batch volume decreases to one third. Then the temperature is increased, using a gas-air flame, to 1200°C within 45 min and to 1450°C within 1.5 hr. The pot is then removed, cooled, broken, and the button of *circa* 35 g is separated. Its surface is mechanically cleaned and parting is carried out, usually with warm hydrochloric acid.[525]

Beamish[525] studied in detail the mechanism of the PGM collection by these alloys. Special attention has been paid to the collection of osmium, rhodium, and ruthenium. Van Loon and Beamish[620] followed the extraction of osmium from a flotation concentrate with the above fusion agent, to which sufficient amounts of Fe_2O_3 and CuO were added. Danilova et al.[621] described the behavior of osmium during roasting of sulfidic Cu-Ni ores. Their results have shown that in ignition in the air at temperatures from 500 to 850°C the losses of osmium increase from 18 to 45%, with the ignition time increasing from 3 to 10 hr. These losses can be substantially decreased with addition of 5% portions of ammonium chloride and carbonate and charcoal to poor ores with 2% of sulfur. A 10% addition of these reagents effectively retains osmium from massive ores, rich in sulfur (up to 26%). The ignition is always carried out in the air, at temperatures of up to 850°C, for 2 to 3 hr, when the loss of osmium is minimized to 0.5%.

Collection of rhodium by copper metal has been described.[622] For the copper-rhodium binary system, solid solutions have been demonstrated in ranges of 0 to 20 and 90 to 100% Rh, as well as the existence of the phases Rh_3Cu, RhCu, and $RhCu_3$. The behavior of rhodium during the collection by copper was followed using radioisotopes.[623] It has been shown that the collection of the metal amounts to 99.9%, the loss of 0.1% being uniformly divided between the slag and the pot walls. The efficiency of the ruthenium collection by lead and a Cu-Ni alloy was compared.[624] The low volatility of ruthenium permits its separation by volatilization of the metallic matrix by ignition at a temperature immediately below the collector boiling point (1550°C for copper). From a roller-flattened Cu button, a part of 6 to 15 mg is separated, evaporated in a suitable cuvette as a matrix vapor tension of $p_M \lesssim$ 50 Pa, and ruthenium is determined by laser photoionization spectroscopy. An OES determination of the PGM in Cu-Ni ores was studied by Danilova et al.[625] The ore is roasted at 850 to 900°C without additives, and the Cu/Ni ratio in the product is adjusted to at least 2 by adding CuO. Various mixtures of soda, borax, glass, and starch as the reductant are used for variously basic initial materials. The completeness of the collection depends on the absolute amount of the collector, which must amount to 6 to 7% of the sample weight in analyses of products with 8 to 20% of Cu + Ni, or to 0.5 to 2% with poor ores containing 0.3 to 3% of Cu + Ni. After dissolving the button in various mixtures of acids, the PGM are separated from ballast ions by selective sorption on a synthetic sorbent. The ashes from the burning of the sorbent are mixed with carbon and analyzed by OES.

The distribution of many metals of silicate rocks and meteorites in a Fe-Ni alloy after reductive fusion in a nitrogen atmosphere was followed by the NAA method.[626] The two phases were separated and dissolved in acids, and the following distribution was found. The metallic phase contains W, As, Cu, Ni, Ir, Au, and Re, the silicate phase contains Ca, Th, Ta, and REE; whereas Fe and Cr are distributed between the two phases.

G. Collection by Nickel(II) Sulfide

It has been known from the metallurgical practice that matte strongly retains the PGM during fusion, and its presence is undesirable for the obtaining of the noble metals. However, it took a long time before matte was intentionally used as the collector. The first mention in this respect can be found in the method for determination of osmium in ores from Witwatersrand which contain, in addition to gold, also the PGM bound mainly to osmiridium. The residue after the cyanide leaching of gold is flotated, and the concentrate obtained is ignited at 950 to 1000°C in a reducing atmosphere. The roasted product is leached with hydrochloric acid to separate iron. The PGM are then concentrated by fusion to a NiS matte. The matte is again leached with hydrochloric acid, and the insoluble residue containing the noble metals is fused with sodium peroxide.[627] A systematic study of NiS as the collector of noble metals was carried out by the workers of the N.I.M.[628,629] Their experience has led to the following procedure for the collection of the PGM and gold. The sample, whose weight depends on the total amount of the PGM (>1000 ppm/0.5 g; <50 ppm/50.0 g), is placed on a sheet of a glossy paper and 60 g of fused borax, 30 g of anhydrous soda, 32 g of nickel(II) carbonate, and 12.5 g of sulfur powder are added. After thorough mixing, the substances are transferred to a pot. The paper is wiped with a piece of silk paper, which is then placed over the surface of the mixture. The pot is placed in a furnace heated to 1000°C and fused for 75 min. The pot contents are then cast into an iron mold and allowed to cool. The button formed should weigh *circa* 25 to 30 g. It is crushed in a hydraulic press, and the lumps are carefully pulverized. The NiS powder is then decomposed by heating with 400 mℓ of concentrated hydrochloric acid for *circa* 16 hr. The noble metals remain undissolved in the form of the sulfides.

It has been found that fusion agent composition has no substantial effect on the course of the fusion and the collection. The sample should contain 10 g of silica, which must be added when not present in the original material. The fusion agent decomposes most rock-forming minerals, except for chromite. The button quality is adversely affected by the sulfur present, as its excess causes brittleness of the button. The amount of sulfur in the batch must thus be added to the recommended amount of 12.5 g. The slag-forming additives must be highly pure chemically and must not contain the PGM. An advantage of fire assay with a NiS button is a low fusion temperature and the possibility of the obtaining of all the six PGM. It can be directly applied to samples with high contents of sulfur and nickel. Drawbacks involve a long time of the parting and the fact that the recovery of gold from ores is 10 to 20% lower than with the collection by lead.

Dixon et al.[630] studied the efficiency of NiS as the collector for platinum, iridium, gold, and silver, using radioactive tracers, with the fusion and parting according to Reference 628. The slags obtained after separation of the NiS button were refused with another portion of nickel(II) carbonate and sulfur. The distribution of the noble metals has shown that their recovery in the first fusion is on average 88% and amounts to 97 to 99% after reassaying. The losses in the slag are 1 to 3%, except for gold, whose loss is 3.5 to 5.5%. Silver can also be recovered by this procedure (96.6%), but more than 90% of it is dissolved during the parting of the button in hydrochloric acid, so that it cannot be determined together with the other noble metals.

Robért and Van Wyk[631] studied the matrix effect on the collection efficiency under the above conditions. It has been found that the building elements of rock-forming minerals have no effect on the PGM recovery and remain in the slag. Tin causes large negative errors with all the noble metals. Chalcophilic elements, such as Pb, Bi, Co, Sb, Cr, and Cu, form the appropriate sulfides during the fusion, which pass into the button. They do not affect the collection, but most antimony sulfides do not dissolve on the parting. Copper sulfides are dissolved, provided that the absolute amount of copper in the sample does not exceed 3.2 g. At higher contents they remain undissolved and pass into solution only with the noble metals, interfering in their determination (e.g., by AAS).

Analogously, As, Se, and Te (2.5 g of each) do not affect the fusion and collection. However, 40 to 70% of each element enters the button in the form of the sulfide, and these sulfides remain undissolved. Attempts to separate larger amounts of these elements by further heating of the insoluble residues have been unsuccessful, and thus, the noble metals cannot be determined in their presence.

In fire assay of chromites, a maximum of 25-g sample can be used with the fusion agent amount given in Reference 628. For slag reassaying, 30 g of borax, 10 g of soda, 7 g $NiCO_3$, and 3 g of sulfur are added. For parting of the joint buttons, hydrochloric acid is used at the temperature of boiling water bath for 3 to 4 hr. An amount of 8 to 12 mg of chromium remains in the insoluble residue. At a PGM sum of less than 100 ppm, several 25-g portions of the mineral must be treated.[596]

The losses of ruthenium, iridium, osmium, and gold were also followed in the NiS collection, using radioisotopes.[570] The efficiency of the tin button has also been evaluated. It has been found that the losses in the slag and during parting (in the parentheses) were 1.1 to 8.3 (1.4), 4.3 to 43 (1.2), 0.6 to 84.4 (2.3), and 1.9 to 87 (4.5)% for gold, iridium, ruthenium, and osmium, respectively. The magnitude of the losses depends on the sample size; the lower values hold for 2 assay tons, and the larger ones for 8 assay tons. In general, NiS is a more efficient collector than tin metal, including the parting procedure. An analogous study of determination of gold and the PGM after collection of a Pb button, NiS and chemical leaching was carried out by Bowditch.[632] He applied the decompositions to three Australian Ni-sulfidic ores and a basic silicate of the Merensky Reef horizon, Bushveld complex, South Africa. He found the method with the NiS collection as the most efficient for a good recovery of the PGM (and gold), with an advantage of the melt readily decomposing refractory minerals, such as osmiridium. The procedure is especially convenient for Australian concentrates, in which nickel and sulfur need not be removed by preliminary leaching and roasting. The fact that the sum of the noble metals cannot be directly determined is considered as a drawback by the author.

Determination of osmium in NiS buttons was studied.[618,633] Using the ^{185}Os tracer, the collection of the element by the NiS button and its behavior during parting followed by fusion of the insoluble residue with sodium peroxide were again followed. To decrease the loss of osmium in oxidative fusion, the isolated sulfides of the PGM were burned in a stream of elemental hydrogen in a Rose crucible. In fusion of a 0.75-g residue with 1 g Na_2O_2 the loss of osmium is less than 3.7%. A similar procedure has been applied to determination of osmium in a platinum-rich material.

The favorable properties of NiS as the noble metal collector was later utilized[634,635] for the noble metal determinations in poor ores and rocks. A negative effect of the Na-Fe matte has been pointed out,[634] which is readily dissolved in the slag and causes losses of gold and silver. Collection on NiS was also used in analyses of sweeps with high noble metal contents.[593] In contrast to Reference 631, no adverse effect of tin on the behavior of the PGM during dissolution in HCl was observed. Nickel(II) carbonate was further replaced by nickel metal, which reacts easier with sulfur. The incomplete collection of gold, which partially remains in the slag, is solved by reassay to a lead button.

To determine the noble metals in the button or in the insoluble residue after parting, NAA was used,[636-638] or the OES-ICP method after complete dissolution of the residue.[639] For deposition of the PGM from solution, a selective sorbent has been proposed.[640]

H. Reductive Fusion of Metals Other Than the Platinoids

This type of fusion has not found wide application in analytical chemistry. An example is reductive fusion with potassium cyanide used for decomposition of tin dioxide — cassiterite. A similar effect is attained in fusion with sodium carbonate with finely dispersed carbon. If the melt is protected from the air, the reduced tin settles at the pot bottom. The tin can be converted into an alloy with copper that is added to the fusion agent in the form of an oxide.

Other reducing agents involve elemental metals, both in heterogeneous mixtures in sintering and in alkaline melts. Sintering of cassiterite with zinc powder and ammonium chloride leads to selective reduction of Sn(IV) to Sn(II), which is readily soluble in acids. By heating coal with magnesium powder, all the forms of sulfur present can be reduced to sulfidic S(II) and the nitrogen compounds to the ammonium ion. The reagent also reacts with silicates, with formation of alkali metal compounds soluble in dilute ethanol. Resistant minerals can be decomposed by fusion with a mixture of aluminum powder and an alkali hydroxide.[641] To reduce metal oxides, fusion with sodium hydroxide containing a few percent of sodium hydride can be used.[642]

The reducing properties of elemental carbon have recently been utilized to decompose barium and strontium sulfates and their natural analogues. By 1-hr heating of the same amounts of the material and graphite at 1000°C and repeating this procedure under the same conditions, more than 95% of a sulfide soluble in hydrochloric acid can be obtained from the sulfates.[643] A mixture of sodium carbonate and carbon (5 + 1) was used to decompose chloridation roasted concentrates containing silver. From the melt, in which silver is reduced to the metal, chloride ions can be quantitatively leached with hot water and determined by ISE.[644]

Fusion of ferroalloys with iron or nickel metals can also be considered as reductive fusion. The procedure is used to prepare buttons for XRF. As follows from the phase diagrams for the iron-alloying element systems, ferroalloys are mostly eutectics with highly heterogeneous textures. On fusion of a ferroalloy with excess iron, the product composition shifts to the region of existence of a homogeneous solid solution of iron with the alloying element.[645] The fusion is carried out in corundum crucibles with an iron inset in a hf oven that can be evacuated or purged with an inert gas.[646] The fusion temperature is around 1600°C, and the fusion time is several tens of seconds.[647] The ferroalloy-to-iron ratio depends on the kind of the alloy; it is 1 + 1 to 3 for ferrochromium, ferrotungsten, ferromanganese, ferroniobium, and ferrovanadium. Ferrosilicium and ferrotitanium are difficult to fuse, and a 1 + 10 dilution is required.[648]

Reductive fusion is also employed to determine oxygen in various metallurgical materials. The fusion takes place in a high vacuum in a graphite crucible placed in a hf oven with programmed heating. The oxygen present reacts with graphite with formation of carbon monoxide that is determined by direct spectrometry or gas chromatography (GC). To analyze tungsten carbide, WC, various metals and alloys are employed as fusion agents, e.g., the alloy 75% Ni + 25% Sn, tin alone, or platinum. The fusion is carried out at 2150 to 2250°C in a helium stream as the CO carrier.[649]

I. Alkaline Fusion in the Presence of Sulfur

This fusion procedure, also called the *Freiberg decomposition,* is a rapid and effective method for decomposition of natural sulfides, polysulfides, and sulfosalts that form polymetallic ore formations. The fusion agent also reacts, to a lesser extent, with natural oxides such as cassiterite. The active component of the fusion agent consists of alkali polysulfides that are formed during the first stage of fusion by mild heating of a mixture of sodium or potassium carbonate with sulfur at a ratio of 4:3 (sodium carbonate) or 5:3 (potassium carbonate). This stage must take place at a low temperature to prevent burning of the sulfur and lasts *circa* 15 to 20 min. The temperature is then increased, so that the mixture is liquefied. In this stage most metals are converted into crystalline, insoluble sulfides, whereas arsenic, antimony, tin, germanium, molybdenum, and partially also vanadium and tungsten are converted into soluble sulfosalts. The decomposition itself takes 10 to 15 min and is carried out almost exclusively in porcelain crucibles, as platinum is strongly corroded by the melt. After the decomposition, the crucible is cooled cautiously, to avoid mechanical damage. The cooled contents of the crucible are leached with hot water, thus dissolving the sulfosalts of the above elements. The dissolution must be performed in a small volume of the liquid to prevent hydrolysis of the sulfides of iron, nickel, cobalt, and other metals. If hydrolysis did occur, the solution with the precipitate is digested on a boiling water bath, with a few grams of KCl or NH_4Cl added, until the solution becomes clear. The solid phase is filtered off and washed with a 1% solution of sodium sulfide containing a small amount of potassium chloride. Fusion with a sulfur-potash mixture is primarily used to separate higher contents of copper and mercury, whose sulfides tend to form soluble sulfosalts in the presence of excess Na^+ ions. With ores and minerals rich in mercury, part of the mercury may be lost owing to a high volatility at an elevated temperature. The ratio between the sample to be decomposed and the fusion agent is selected depending on the reactivity of the phases, mostly within a range of 1:6 to 8. An up to 15-fold excess of the fusion agent is recommended for nickel and cobalt arsenides.

The Freiberg decomposition is a historical procedure and at present is used only exceptionally. The classical procedures of separation and determination of nonferrous metals have been largely replaced mainly by AAS, and crystallochemical study of sulfidic phases employs primarily the electron microprobe. The decomposition has been applied to lead sulfoantimonites, bournonite, boulangerite, jamesonite, and other minerals in this group. In the same way, noble ores of silver, consisting of miargyrite, pyrargyrite, proustite, polybazite, etc., can be decomposed. Sulfostannates of copper, iron, and zinc — stannite, koesterite, and hocartite, as well as germanium minerals — are readily decomposed. The procedure is especially advantageous for decomposition and simple separation of antimony from polymetallic ores and for analyses of bismuth arsenides.[650] The procedure has been successfully used to decompose nickel and cobalt arsenides of the safflorite-skutterudite group and iron arsenide, loellingite and arsenopyrite.

Tetraedrites, natural sulfoantimonites, and sulfoarsenites of copper, silver, and mercury are economically important sources of these metals. The Freiberg decomposition is especially well suited for total chemical analysis of these phases, although the determination of mercury (schwazites contain up to 20% Hg) must be carried out using another analytical method.

Small amounts of cassiterite can also be decomposed by this method; the decomposability of the mineral probably depends on the amount of heterogeneous inclusions. The alkaline fusion in the presence of sulfur has also been proposed for a determination of tin in rocks and poor ores. However, determination of this element in galena and other sulfides is not quantitative and is subject to an error of up to 50%.[641]

REFERENCES

1. **Mitchell, J. W.,** *Int. Lab.,* January-February, 1982, 12.
2. British Standards Institution, B.S. 6070, Part 0-7, 1981; *Anal. Abstr.,* 41, 3B36, 1981.
3. **Baechmann, K., Spachidis, C., and Weitz, A.,** *Fresenius Z. Anal. Chem.,* 301, 3, 1980.
4. **Bock, R. and Thier, W.,** *Fresenius Z. Anal. Chem.,* 253, 123, 1971.
5. **Bock, R. and Jacob, D.,** *Fresenius Z. Anal. Chem.,* 200, 81, 1964.
6. **Trofimov, I. V. and Busev, A. I.,** *Zavod. Lab.,* 49(3), 5, 1983.
7. **Shevchuk, I. A., Simonova, T. N., Kovalenko, L. I., Smirnova, L. M., and Sagirova, L. S.,** *Zh. Anal. Khim.,* 31, 1289, 1976.
8. **Flaschka, H. and Myers, G.,** *Fresenius Z. Anal. Chem.,* 274, 279, 1975.
9. **Norwitz, G. and Gordon, H.,** *Talanta,* 24, 159, 1977.
10. **Huka, M.,** personal communication.
11. **Uchida, H., Iwasaki, K., Tanaka, K., and Iida, Ch.,** *Anal. Chim. Acta,* 134, 375, 1982.
12. **Schinkel, H.,** *Fresenius Z. Anal. Chem.,* 317, 10, 1984.
13. **Yoshida, K. and Haraguchi, H.,** *Anal. Chem.,* 56, 2580, 1984.
14. **Iwasaki, K., Fuwa, K., and Haraguchi, H.,** *Anal. Chim. Acta,* 183, 239, 1986.
15. **Donaldson, E. M.,** Methods for the analysis of ores, rocks, and related material, in *Mines Branch Monogr. 881,* Energy Mines Branch Resources Canada, Ottawa, 1974, 41.
16. **Yakovleva, A. F. and Chupakhin, M. S.,** *Zavod. Lab.,* 41, 185, 1975.
17. **Young, R. S.,** *Analyst (London),* 107, 721, 1982.
18. **Ishizuka, T., Uwamino, Y., and Tsuge, A.,** *Bunseki Kagaku,* 33, 576, 1984; *Anal. Abstr.,* 47, 11B110, 1985.
19. **Thévenot, F. and Goeuriot, P.,** *Analusis,* 6, 359, 1978.
20. **Hoede, D. and Das, H. A.,** *J. Radioanal. Chem.,* 35, 167, 1977.
21. **Donaldson, E. M.,** *Talanta,* 30, 497, 1983.
22. **Yoshimura, K., Kaji, H., Yamaguchi, E., and Tarutani, T.,** *Anal. Chim. Acta,* 130, 345, 1981.
23. **Koljonen, T.,** *Suom. Kemistil.* B, 46, 133, 1973; *Chem. Abstr.,* 79, 114393, 1973.
24. **Das, A. K. and Das, J.,** *J. Indian Chem. Soc.,* 60, 67, 1983; *Anal. Abstr.,* 45, 5B128, 1983.
25. **Sprenz, E. C. and Prager, M. J.,** *Analyst (London),* 106, 1210, 1981.
26. **Sixta, V. and Šulcek, Z.,** *Sklar a Keram.,* 28, 364, 1978.
27. **Fuge, R.,** *Chem. Geol.,* 17, 37, 1976.
28. **Shkrobot, E. P. and Tolmacheva, N. S.,** *Zh. Anal. Khim.,* 31, 1491, 1976.
29. **Shiraishi, N., Hisayuki, T., and Kodama, K.,** *Jpn. Analyst,* 23, 453, 1974; *Anal. Abstr.,* 28, 6B178, 1975.
30. **Troll, G., Farzaneh, A., and Cammann, K.,** *Chem. Geol.,* 20, 295, 1977.
31. **Khalizova, V. A., Polupanova, L. I., Bebeshko, G. I., and Alexeva, A. Ya.,** *Zh. Anal. Khim.,* 30, 2201, 1975.
32. **Ackermann, H.,** *Interceram,* 27, 404, 1978; *Anal. Abstr.,* 38, 6B126, 1980.
33. **Bebeshko, G. I., Roze, V. P., and Khalizova, V. A.,** *Zh. Anal. Khim.,* 34, 507, 1979.
34. **Avsec, H. and Kosta, L.,** *Mikrochim. Acta,* 229, 3(3-4), 1984.
35. **Thomas, J., Jr. and Gluskoter, H. J.,** *Anal. Chem.,* 46, 1321, 1974.
36. **Wilson, S. A. and Gent, C. A.,** *Anal. Lett.,* 15, 851, 1982.
37. **Wilson, S. A. and Gent, C. A.,** *Anal. Chim. Acta,* 148, 299, 1983.
38. **Yasinskene, E. I. and Umbrazheyunaite, P.,** *Zh. Anal. Khim.,* 30, 962, 1975.
39. **Akaiwa, H., Kawamoto, H., and Hagesawa, K.,** *Talanta,* 26, 1027, 1979.
40. **Akaiwa, H., Kawamoto, H., and Hagesawa, K.,** *Talanta,* 27, 909, 1980.
41. **Owens, J. W., Gladney, E. S., and Knab, D.,** *Anal. Chim. Acta,* 135, 169, 1982.
42. **Walsh, J. N.,** *Analyst (London),* 110, 959, 1985.
43. **van Raaphorst, J. G. and Lingerak, W. A.,** *Fresenius Z. Anal. Chem.,* 267, 26, 1973.
44. **Cook, C. J., Dubiel, S. V., and Hareland, W. A.,** *Anal. Chem.,* 57, 337, 1985.
45. **Malhotra, P. D., Prasada, G. H. S., and Rao, V.,** *Rec. Geol. Surv. India,* 93, 215, 1963.
46. **Reichen, L. E.,** *U.S. Geol. Surv. Prof. Pap.,* 750B, B163, 1971.
47. **Bock, R. and Tschoepel, P.,** *Fresenius Z. Anal. Chem.,* 246, 81, 1969.
48. **Qureshi, M., Rathore, H. S., and Thakur, J. S.,** *Talanta,* 25, 232, 1978.
49. **Kustova, L. V., Larkina, A. N., and Smirnova, N. V.,** *Zavod. Lab.,* 52(4), 15, 1986.
50. **Kuteinikov, A. F., Mashkovich, L. A., Kirevina, T. P., Pekaln, L. A., and Gryukan, V. S.,** *Zavod. Lab.,* 44, 666, 1978.
51. **Wahlberg, J. S.,** *Chem. Geol.,* 33, 155, 1981.
52. **Kožlicka, M. and Wojtowicz, M.,** *Chem. Anal. (Warsaw),* 16, 739, 1971.
53. **Mishchenko, V. T., Tselikov, E. I., Shilova, L. P., Mukomel, V. L., and Poluektov, N. S.,** *Zavod. Lab.,* 51(7), 4, 1985.

54. **Roy, N. K., Das, A. K., and Ganguli, C. K.,** *At. Spectrosc.*, 6, 166, 1985.
55. **Russell, B. G., Spangenberg, J. D., and Steele, T. W.,** *Natl. Inst. Met. Repub. S. Afr.* Rep., No. 193, 1967.
56. British Standards Institution, B.S. 4140, Part 4, 1986; *Anal. Abstr.*, 48, 7B69, 1986.
57. **Slavcheva, Y., Popova, E., and Dimitrova, V.,** *Khim. Ind. (Sofia)*, 52, 166, 1980; *Anal. Abstr.*, 39, 6B155, 1980.
58. **Gomez, C. A. and Valle, F. J.,** *Appl. Spectrosc.*, 39, 24, 1985.
59. **Kolihová, D., Dudová, N., Janoušková, J., and Sychra, V.,** *Chem. Listy*, 69, 613, 1975.
60. **Ilsemann, K. and Bock, R.,** *Fresenius Z. Anal. Chem.*, 274, 185, 1975.
61. **Sizonenko, N. T., Antipova, G. Ya., Gudzenko, L. V., and Obodynskaya, N. V.,** *Zavod. Lab.*, 49(12), 16, 1983.
62. **Nadezhda, A. A., Ivanova, K. P., and Lukyanenko, T. A.,** *Khim. Promyst. Ser. Metody Anal. Kontrolya*, 1980, 33; *Anal. Abstr.*, 42, 3B261, 1982.
63. **Tserkovnitskaya, I. A. and Diabi, L.,** *Vestn. Leningr. Univ. Fiz. Khim.*, No. 10, 96, 1980; *Anal. Abstr.*, 40, 1B61, 1981.
64. **Boix, A. and Debras-Guédon, J.,** *Chim. Anal. (Paris)*, 53, 459, 1971.
65. **Sugawara, K. F., Sin, S.-Y., and Strzegowski, W. R.,** *Talanta*, 25, 669, 1978.
66. **Jawaid, M. and Ingman, F.,** *Talanta*, 22, 1037, 1975.
67. **Mizuno, K., Suzuki, T., Itakura, M., and Kodama, K.,** *Jpn. Analyst*, 25, 128, 1976; *Anal. Abstr.*, 31, 3B188, 1976.
68. **Grechko, L. I. and Fartushnaya, I. A.,** *Zavod. Lab.*, 46, 565, 1980.
69. **Novoselova, I. M.,** *Zavod. Lab.*, 50(10), 5, 1984.
70. **Komárková, E.,** *Natl. Inst. Met. Repub. S. Afr.* Rep., No. 1770, 1975.
71. **Vasileva, L. N., Yustus, Z. P., Rozhkova, L. S., and Zasadych, S. G.,** *Zh. Anal. Khim.*, 33, 1567, 1978.
72. **Vasileva, L. N., Yustus, Z. P., and Zasadych, S. G.,** *Zh. Anal. Khim.*, 32, 273, 1977.
73. **Piryutko, M. M. and Kostyreva, T. G.,** *Zh. Anal. Khim.*, 37, 1644, 1982.
74. **Weiss, D. and Korečková, J.,** unpublished data.
75. **Burdo, R. A. and Wise, W. M.,** *Anal. Chem.*, 47, 2360, 1975.
76. **Ishikawa, K. and Fukuda, S.,** *Bunseki Kagaku*, 28, 323, 1979; *Anal. Abstr.*, 38, 1B154, 1980.
77. **Piryutko, M. M. and Mironovich, V. Ya.,** *Zavod. Lab.*, 41, 395, 1975.
78. **Bakhmatova, T. K., Dedkov, Yu. M., and Ershova, V. A.,** *Zh. Anal. Khim.*, 31, 292, 1976.
79. **Panteleeva, E. Yu. and Polikarpova, N. V.,** *Zavod. Lab.*, 46, 1008, 1980.
80. **Boniface, H. J. and Jenkins, R. H.,** *Analyst (London)*, 102, 739, 1977.
81. **Nadkarni, R. A.,** *Anal. Chem.*, 52, 929, 1980.
82. **Adelantado, J. V. G., Martinez, V. P., Moreno, A. C., and Reig, F. B.,** *Talanta*, 32, 224, 1985.
83. **Minin, A. A., Filippova, L. P., and Plyusnina, V. N.,** *Uchen. Zap. Perm. Univ.*, No. 289, 148, 1973; *Anal. Abstr.*, 29, 2B50, 1975.
84. **Povondra, P.,** unpublished data.
85. **Smith, A. Y. and Lynch, J. J.,** *Geol. Surv. Can. Pap.*, 69-40, 1969.
86. **Veselsky, J. C. and Woelfl, A.,** *Anal. Chim. Acta*, 85, 135, 1976.
87. **Kaschani, D. T. and Brauns, A.,** *GIT Fachz. Lab. Chromatogr.*, Suppl., 29, 1981.
88. **Wilde, H. E.,** *Anal. Chem.*, 45, 1526, 1973.
89. **Vokhrysheva, L. E. and Gladysheva, K. F.,** *Zavod. Lab.*, 50(10), 16, 1984.
90. **Bock, R. and Herrmann, A.,** *Fresenius Z. Anal. Chem.*, 248, 180, 1969.
91. **Floyd, M. A., Fassel, V. A., and D'Silva, A. P.,** *Anal. Chem.*, 52, 2168, 1980.
92. **Fabbri, B. and Donati, F.,** *Analyst (London)*, 106, 1338, 1981.
93. **Desborough, G. A., Pitman, J. K., and Huffman, C., Jr.,** *Chem. Geol.*, 17, 13, 1976.
94. **Russell, D. S., McPherson, H. B., and Clancy, V. P.,** *Talanta*, 27, 403, 1980.
95. **Josephson, M., Cook, E. B. T., and Dixon, K.,** *Natl. Inst. Met. Rep. S. Afr.*, No. 1886, 1977.
96. **Unni, C. K. and Schilling, J.-G.,** *Anal. Chim. Acta*, 96, 107, 1978.
97. **Zhang, G.-Y. and Pan, Z.-H.,** *Ti Ch'in Hua Hsueh*, p. 353, 1979; *Anal. Abstr.*, 40, 6B184, 1981.
98. **Rubeška, I. and Hlavinková, V.,** *At. Absorpt. Newsl.*, 18, 5, 1979.
99. **Arbab-Zavar, M. H. and Howard, A. G.,** *Analyst (London)*, 105, 744, 1980.
100. **Oliveira, E., McLaren, J. W., and Bergman, S. S.,** *ICP Inf. Newsl.*, 11(11), 703, 1984.
101. **Brown, F. W., Simon, F. O., and Greenland, L. P.,** *J. Res. U.S. Geol. Surv.*, 3, 187, 1975.
102. **Smith, R. G., van Loon, J. C., Knechtel, J. R., Fraser, J. L., Pitts, A. E., and Hodges, A. E.,** *Anal. Chim. Acta*, 93, 61, 1977.
103. **Smith, B. F. L. and Bain, D. C.,** *Commun. Soil Sci. Plant. Anal.*, 13, 185, 1982; *Anal. Abstr.*, 43, 3G19, 1982.
104. **Malysheva, V. I., Glukhova, N. V., Chupaeva, M. L., Kozhevnikova, V. P., and Fokina, N. F.,** *Khim. Promst. Ser. Metody Anal. Kontrolya*, p. 62, 1981; *Anal. Abstr.*, 42, 1B164, 1982.

105. **Vinarova, L. I., Antonovich, V. P., Stoyanova, I. V., and Malyutina, T. M.,** *Zh. Anal. Khim.*, 40, 1645, 1985.
106. **Leinz, R. W. and Grimes, D. J.,** *J. Res. U.S. Geol. Surv.*, 6, 259, 1978; *Anal. Abstr.*, 36, 3B131, 1979.
107. **Belova, T. Ya. and Volkova, L. P.,** *Zavod. Lab.*, 50(9), 5, 1984.
108. **Hashitani, H., Takeo, A., and Yoshida, H.,** *Anal. Chim. Acta,* 76, 85, 1975.
109. **Smith, J. D.,** *Anal. Chim. Acta,* 57, 371, 1971.
110. **Martinez, V. P., Adelantado, J. V. G., and Reig, F. B.,** *Fresenius Z. Anal. Chem.*, 314, 665, 1983.
111. **Adelantado, J. V. G., Martinez, V. P., and Reig, F. B.,** *Int. Ceram.*, 33(4), 40, 1984; *Chem. Abstr.*, 102, 66246, 1985.
112. **Tikhonov, V. N. and Grigorovich, L. F.,** *Zavod. Lab.*, 43, 1450, 1977.
113. **Goguel, R.,** *Anal. Chim. Acta,* 169, 179, 1985.
114. **Ginzburg, S. I., Ezerskaya, N. A., Prokofeva, I. V., Fedorenko, N. V., Shlenskaya, V. I., and Belskii, N. K., Eds.,** *Analiticheskaya Khimiya Platinovykh Metallov,* Nauka, Moskow, 1972, 25.
115. **Alekseeva, I. I., Gromova, A. D., Golysheva, L. T., and Khvovostukhina, N. A.,** *Zavod. Lab.*, 42, 650, 1976.
116. **Tan, K.,** *Fenxi Hauxue,* 9, 498, 1981; *Anal. Abstr.*, 42, 3B209, 1982.
117. **Tagaki, H., Kimura, T., Kobayashi, H., Iwashima, K., and Yamagata, N.,** *Bunseki Kagaku,* 33, 582, 1984.
118. **Rybakov, A. A. and Ostroumov, E. A.,** *Zh. Anal. Khim.*, 39, 2168, 1984.
119. Laboratory crucibles, Prospect B-J Scientific Products, Inc. Manufacturer information leaflet, Universal Scientific Inc., Albany, Ore., 1979.
120. **Mashkovich, L. A., Kuteinikov, A. F., Pekaln, L. A., Kirevina, T. P., Tashilova, L. P., and Litvinov, B. F.,** *Zh. Anal. Khim.*, 37, 1528, 1982.
121. **Dymova, M. S., Kozina, G. V., and Titova, T. V.,** *Zavod. Lab.*, 50(6), 18, 1984.
122. **Kuteinikov, A. F., Kirevina, T. P., Mashkovich, L. A., and Stepanova, A. N.,** *Zavod. Lab.*, 50(6), 16, 1984.
123. Sample preparation bomb, Bulletin 1100, Parr Instrument Co., Moline, IL.
124. **Jefferey, P. G. and Hutchison, D.,** *Chemical Methods of Rock Analysis,* 3rd ed., Pergamon Press, Oxford, 1981, 134.
125. **Huffman, C., Jr., van Shaw, E., and Thomas, J. A.,** *U.S. Geol. Surv. Prof. Pap.*, 750B, B185, 1971.
126. **Russell, G. M. and Watson, A. E.,** *Natl. Inst. Met. Repub. S. Afr. Rep.,* No. 1934, 1977.
127. **Donaldson, E. M.,** *Talanta,* 27, 779, 1980.
128. **Ginsburg, S. I.,** *Khimcheskii Analiz Platinovykh Metallov,* Nauka, Moscow, 1972, 526.
129. **Novák, J.,** unpublished data.
130. **Kallmann, S. and Maul, C.,** *Talanta,* 30, 21, 1983.
131. **Shifris, B. S. and Kolpakova, N. A.,** *Zh. Anal. Khim.*, 41, 502, 1986.
132. **Strilchenko, T. G., Kabanova, O. L., and Danilova, F. N.,** *Zavod. Lab.*, 52(8), 11, 1986.
133. **Igoshchina, E. V. and Talalaev, B. M.,** *Zh. Anal. Khim.*, 38, 1648, 1983.
134. **Lazareva, V. I. and Lazarev, A. I.,** *Zavod. Lab.*, 51(12), 1, 1985.
135. **Khvostova, V. P. and Golovnya, S. V.,** *Zavod. Lab.*, 48(7), 3, 1982.
136. **Ryspekova, Z. A. and Azimova, Z. Kh.,** *Zavod. Lab.*, 52(6), 32, 1986.
137. **Dixon, K., Krueger, M. M., and Radford, A. J.,** *Natl. Inst. Met. Repub. S. Afr. Rep.,* No. 1654, 1975.
138. **Apt, K. E. and Gladney, E. S.,** *Anal. Chem.*, 47, 1484, 1975.
139. **Rysev, A. P., Zhitenko, L. P., and Nadezhdina, V. A.,** *Zavod. Lab.*, 47(6), 20, 1981.
140. **Kolpakova, N. A. and Shvets, L. A.,** *Zh. Anal. Khim.*, 38, 1470, 1983.
141. **Haskell, R. J. and Wright, J. C.,** *Anal. Chem.*, 59, 427, 1987.
142. **Romanovskaya, L. E., Samulenkova, I. E., Khomutova, E. G., Khvorostukhina, N. A., and Alekseeva, I. I.,** *Zavod. Lab.*, 51(8), 8, 1985.
143. **Pohlandt, C. and Steele, T. W.,** *Talanta,* 21, 919, 1974.
144. **Lilipenko, A. T., Lukovskaya, N. M., Terlitskaya, A. V., Bogoslovskaya, T. A., and Kushevskaya, N. F.,** unpublished data.
145. **Robért, R. V. D., van Wyk, E., and Dixon, K.,** *Natl. Inst. Met. Repub. S. Afr. Rep.,* No. 1580, 1973.
146. **Ehmann, W. D. and Gillum, D. E.,** *Chem. Geol.*, 9, 1, 1972.
147. **Shkrobot, E. P. and Sherbarshina, N. I.,** *Zh. Anal. Khim.*, 36, 1986, 1981.
148. **Lyle, S. J. and Zatar, N. A.,** *Fresenius Z. Anal. Chem.*, 313, 313, 1982.
149. **Qing-Lie, H., Hughes, T. C., Haukka, M., and Hannaker, P.,** *Talanta,* 32, 495, 1985.
150. **Chen, D. and Liu, X.,** *Fenxi Huaxue,* 11, 617, 1983; *Anal. Abstr.*, 46, 7B63, 1984.
151. **Li, S. and Hu, W.,** *Huaxue Xuebao,* 41, 1073, 1983; *Anal. Abstr.*, 46, 8B70, 1984.
152. **Sun, P. and Chen, Y.,** *Guangpuxue Yu Guangpu Fenxi,* 5, 47, 1985; *Anal. Abstr.*, 48, 6B57, 1986.
153. **Brenner, I. B., Steele, T. W., Watson, A. E., Jones, E. A., and Goncalves, M.,** *Spectrochim. Acta Part B,* 36 B, 785, 1981.

154. **Guest, R. J. and MacPherson, D. R.**, *Anal. Chim. Acta,* 71, 233, 1974.
155. **Donaldson, E. M.**, *Talanta,* 31, 997, 1984.
156. **Qian, A. and Liu, J.**, *Fenxi Huaxue,* 11, 372, 1983; *Anal. Abstr.*, 46, 5B106, 1984.
157. **Harley, M. L.**, *At. Spectrosc.*, 3, 76, 1982.
158. **Vilkova, O. M. and Ivanov, V. M.**, *Zh. Anal. Khim.*, 33, 1785, 1978.
159. **Yi, L.-Y. and Guo, T.-Z.**, *Ti Chih K'o Hsueh,* 359, 1979; *Anal. Abstr.*, 40, 2B96, 1981.
160. **Wang, Q., Li, S., and Liu, Z.**, *Fenxi Huaxue,* 13, 195, 1985; *Anal. Abstr.*, 47, 12B132, 1985.
161. **Donaldson, E. M.**, *Talanta,* 26, 999, 1979.
162. **Donaldson, E. M.**, *Talanta,* 26, 1119, 1979.
163. **Halicz, L. and Russell, G. M.**, *Analyst (London),* 111, 15, 1986.
164. **Watson, A. E., Russell, G. M., and Balaes, G.**, *Natl. Inst. Met. Repub. S. Afr. Rep.*, No. 1815, 1976.
165. **Kinnunen, J., Mericanto, B., and Wennerstrand, B.**, *Kem. Kemi,* 6, 457, 1979; *Anal. Abstr.*, 38, 4B1, 1980.
166. **Chen, M. and Fu, B.**, *Fenxi Huaxue,* 12, 384, 1984; *Anal. Abstr.*, 47, 2B136, 1985.
167. **Shiryaeva, M. B., Lyubimova, L. N., Salmin, Yu. P., Ryumina, K. N., and Tatarkin, M. A.**, *Zavod. Lab.*, 50(9), 3, 1984.
168. **Bhargava, O. P., Gmitro, M., and Hines, W. G.**, *Talanta,* 27, 263, 1980.
169. **Young, R. S.**, *Talanta,* 33, 561, 1986.
170. **Bhargava, O. P., Alexiou, A., and Hines, W. G.**, *Talanta,* 25, 357, 1978.
171. **Bhargava, O. P. and Hines, W. G.**, *Anal. Chem.*, 48, 1701, 1976.
172. **Chakraborti, D. and Adams, F.**, *Anal. Chim. Acta,* 109, 307, 1979.
173. **Kallmann, S. and Komárková, E.**, *Talanta,* 29, 700, 1982.
174. **Bhargava, O. P.**, *Talanta,* 26, 146, 1979.
175. **Bhargava, O. P.**, *Analyst (London),* 101, 125, 1976.
176. **Bozhkov, O. D., Jordanov, N., Borissova, L. V., and Fabelinskii, Yu. I.**, *Fresenius Z. Anal. Chem.*, 321, 453, 1985.
177. Standards Association of Australia, *Australian Stand.*, A.S. 2489, 10, 1984; *Anal. Abstr.*, 47, 2B169, 1985.
178. **Dixon, K., Royal, S. J., Komárková, E., Austen, C. E., and Watson, A. E.**, *Natl. Inst. Met. Repub. S. Afr. Rep.*, No. 1830, 1977.
179. **Donaldson, E. M.**, *Talanta,* 27, 499, 1980.
180. **Budesinsky, B. W.**, *Analyst (London),* 105, 278, 1980.
181. **Basitova, S. M., Yurina, R. D., and Vakhobova, R. U.**, *Zh. Anal. Khim.*, 34, 935, 1979.
182. **Wei, X. and Wang, H.**, *Fenxi Huaxue,* 13, 450, 1985; *Anal. Abstr.*, 48, 3B130, 1986.
183. **Gallorini, M., Orvini, E., Rolla, A., and Burdisso, M.**, *Analyst (London),* 106, 328, 1981.
184. **Westland, A. D. and Kantipuly, Ch. J.**, *Anal. Chim. Acta,* 154, 355, 1983.
185. **Westland, A. D. and Kantipuly, Ch. J.**, *Talanta,* 30, 751, 1983.
186. **Rigin, V. I.**, *Zh. Anal. Khim.*, 40, 253, 1985.
187. **Marranzino, A. P. and Wood, W. H.**, *Anal. Chem.*, 28, 273, 1956.
188. **Barbaro, M., Passariello, B., Sbrilli, R., and Milozzi, P.**, *At. Spectrosc.*, 4, 155, 1983.
189. **Gowda, H. S. and Shakunthala, R.**, *Analyst (London),* 103, 1215, 1978.
190. **Norris, J. D.**, *Analyst (London),* 109, 1475, 1984.
191. **Mazzucotelli, A., Vannucci, R., Vannucci, S., and Passaglia, E.**, *Talanta,* 31, 185, 1984.
192. **Chow, C.**, *Analyst (London),* 104, 154, 1979.
193. **Iyer, S. and Pillai, C. K.**, *Indian J. Technol.*, 21, 444, 1985; *Anal. Abstr.*, 47, 3B107, 1985.
194. **De Benzo, Z. A., De Bierman, M. H., Ceccarelli, C. M., and La Brecque, J. J.**, *At. Spectrosc.*, 5, 83, 1984.
195. **Pillai, C. K., Natarajan, S., and Venkateswarlu, Ch.**, *At. Spectrosc.*, 6, 53, 1985.
196. **Barbina, T. M. and Polezhayev, Yu. M.**, *Zavod. Lab.*, 50(6), 12, 1984.
197. **van der Walt, T. N. and Strelow, F. W. E.**, *Anal. Chem.*, 57, 2889, 1985.
198. **Viets, J. G., O'Leary, R. M., and Clark, J. R.**, *Analyst (London),* 109, 1589, 1984.
199. **Donaldson, E. M.**, *Talanta,* 28, 461, 1981.
200. **Grossmann, O.**, *Fresenius Z. Anal. Chem.*, 321, 442, 1985.
201. **Balakshina, A. V. and Sokolova, M. A.**, *Zavod. Lab.*, 42, 152, 1976.
202. **Kustova, L. V., Larkina, A. N., and Smirnova, N. V.**, *Zavod. Lab.*, 52(4), 15, 1986.
203. **Pietri, C. E. and Wenzel, A. W.**, *Annual Progress Report 1967—1968,* New Brunswick Laboratory, NBL-247, 1969, 18.
204. **Angeletti, L. M. and Bartscher, W. J.**, *Anal. Chim. Acta,* 60, 238, 1972.
205. **Binghan, C. D., Scarbororough, J. M., and Pietri, C. E.**, Paper IAEA-SM-201/22, presented at Symp. Safeguarding of Nuclear Materials, Vienna, October, 1975.
206. **Sill, C. W. and Willis, C. P.**, *Anal. Chem.*, 49, 302, 1977.

207. **Sill, C. W.,** *Health Phys.*, 33, 393, 1977.
208. **Matsumoto, K., Misaki, Y., Hayashi, K., and Terada, K.,** *Fresenius Z. Anal. Chem.*, 312, 542, 1982.
209. **Matsumoto, K., Hayashi, K., and Terada, K.,** *Fresenius Z. Anal. Chem.*, 313, 562, 1982.
210. **Milner, G. W. C., Wood, A. J., Weldrick, G., and Phillips, G.,** *Analyst (London)*, 92, 239, 1967.
211. **Milner, G. W. C., Phillips, G., and Fudge, A. J.,** *Talanta*, 15, 1241, 1968.
212. **Yamamoto, M.,** *J. Radioanal. Nucl. Chem.*, 90, 401, 1985.
213. **Hashiba, M., Miura, E., Nurishi, Y., and Hibino, T.,** *Bunseki Kagaku*, 27, 362, 1978; *Anal. Abstr.*, 36, 3B97, 1979.
214. **Hashiba, M., Miura, E., Nurishi, Y., and Hibino, T.,** *Bunseki Kagaku*, 29, 323, 1980.
215. **Headridge, J. B.,** *CRC Crit. Rev. Anal. Chem.*, p. 461, January 1972.
216. **Sill, C. W.,** Problems in sample treatment in trace analysis, *Natl. Bur. Stand. Spec. Publ. 422*, p. 463, 1976.
217. **Arkhangelskaya, A. S. and Molot, L. A.,** *Zavod. Lab.*, 46, 883, 1980.
218. **Gladilovich, D. B. and Stolyarov, K. P.,** *Zh. Anal. Khim.*, 40, 653, 1985.
219. **Gladilovich, D. B., Grigorev, N. N., and Stolyarov, K. P.,** *Zh. Anal. Khim.*, 33, 2113, 1978.
220. **Everest, D. A.,** *The Chemistry of Beryllium*, Elsevier, New York, 1964, 104.
221. **Ramados, K. and Arora, H. C.,** *Indian J. Technol.*, 21, 211, 1983; *Anal. Abstr.*, 46, 7B79, 1983.
222. **Huka, M.,** unpublished data, 1980.
223. **Steele, T. W., Vollenweider, M., King, R. H., Wall, G. H., and Stoch, H.,** *Natl. Inst. Met. Repub. S. Afr. Rep.*, No. 998, 1970.
224. **Jones, E. A. and Watson, A. E.,** *Natl. Inst. Met. Repub. S. Afr. Rep.*, No. 1428, 1972.
225. **Stoch, H. and Dixon, K.,** *Natl. Inst. Met. Repub. S. Afr. MINTEK Spec. Publ.*, No. 4, 1983.
226. **Brenner, I. B., Jones, E. A., Watson, A. E., and Steele, T. W.,** *Chem. Geol.*, 45, 135, 1984.
227. **Walsh, J. N., Buckley, F., and Barker, J.,** *Chem. Geol.*, 33, 141, 1981.
228. **Sill, C. W., Hindman, F. D., and Anderson, J. I.,** *Anal. Chem.*, 51, 1307, 1979.
229. **Byster, S. E.,** *Report 1980*, NBL-296, U.S. Department of Energy, Argonne, Ill.; *Anal. Abstr.*, 42, 2B98, 1982.
230. **Hahn, P. B., Bretthauer, E. W., Altringer, P. B., and Matthews, N. F.,** EPA-600/7-77-078, Office of Research and Development, U.S. Environmental and Protection Agency, 1977.
231. **Sill, C. W., Puphal, K. W., and Hindman, F. D.,** *Anal. Chem.*, 46, 1725, 1974.
232. **Sill, C. W.,** *Anal. Chem.*, 50, 1559, 1978.
233. **Sill, C. W.,** *Anal. Chem.*, 49, 618, 1977.
234. **Sill, C. W.,** *Anal. Chem.*, 52, 1452, 1980.
235. **Sill, C. W. and Williams, R. L.,** *Anal. Chem.*, 53, 412, 1981.
236. **Martin, D. B.,** *Anal. Chem.*, 51, 1968, 1979.
237. **Hindman, F. D.,** *Anal. Chem.*, 58, 1238, 1986.
238. **Krtil, J.,** *Jad. Energ.*, 29, 440, 1983.
239. **Biskupski, V. S.,** *Anal. Chim. Acta*, 33, 333, 1965.
240. **Gorsuch, T. T.,** *The Destruction of the Organic Matter*, Pergamon Press, Oxford, 1970, 41.
241. **McQuaker, N. R. and Fung, T.,** *Anal. Chem.*, 47, 1462, 1975.
242. Borate fusion for xRF analysis, Corp. Scient., Claise, Sainte-Foy, Qué Canada, 1984.
243. **Banerjee, S. and Olsen, B. G.,** *Appl. Spectrosc.*, 32, 576, 1978.
244. **Halas, S. and Wolacewiz, W. P.,** *Anal. Chem.*, 53, 686, 1981.
245. **Din, V. K.,** *Anal. Chim. Acta*, 159, 387, 1984.
246. **Mulder, B. J.,** *Anal. Chim. Acta*, 72, 220, 1974.
247. **Mironov, I. A., Khalimon, G. M., Saposhnikov, Yu. P., and Pevtsova, N. I.,** *Zh. Anal. Khim.*, 40, 473, 1985.
248. **Zief, M. and Mitchell, J. W.,** *Contamination Control in Trace Element Analysis*, John Wiley & Sons, New York, 1976, 161.
249. **Winterkorn, M., Schulze, K., and Toelg, G.,** *Mikrochim. Acta*, Suppl. 7, 27, 1977.
250. **Davy, H.,** *Phil. Trans.*, 231, 1805.
251. **Jannasch, P.,** *Ber. Dtsch. Chem. Ges.*, 28, 2822, 1895.
252. **Jannasch, P. and Heidenreich, O.,** *Z. Anorg. Chem.*, 12, 208, 1896.
253. British Standards Institution, Methods of test boric acid, boric oxide, disodium tetraborates, sodium perborates and sodium borates for industrial use, in *BS 5688*, Parts 1 to 21, British Standards Institution, London, 1979.
254. **Hoffman, J. J. and Lundell, G. E. F.,** *J. Res. Natl. Bur. Stand.*, 3, 581, 1929.
255. **Bennett, H.,** *Analyst (London)*, 102, 153, 1977.
256. **Brown, R. B., Bruce R., and Doremus, R. H.,** *J. Am. Ceram. Soc.*, 59, 510, 1976.
257. **Maessen, F. J. M. S. and Boumans, P. W. J. M.,** *Spectrochim. Acta Part B*, 23B, 739, 1968.
258. **Maxwell, J. A.,** *Mineral and Rock Analysis*, Interscience, New York, 1968, 98.

259. **Galanova, A. P. and Gurin, P. A.,** *Nauchn. Tr. Irkutsk. Nauchno Issled. Inst. Redk. Met.*, No. 11, 52, 1963, *Anal. Abstr.*, 12, 1759, 1965.
260. **Tsubaki, I.,** *Jpn. Analyst,* 16, 610, 1967.
261. **Archibald, A. K. and McLeod, M. E.,** *Anal. Chem.*, 24, 222, 1952.
262. **Itakura, M. and Kodama, K.,** *Bunseki Kagaku,* 27, 531, 1978.
263. **Cook, E. B. T. and Gereghty, A.,** *Natl. Inst. Met. Repub. S. Afr. Rep.*, No. 1145, 1971; *Chem. Abstr.*, 75, 14648, 1971.
264. **Chuchmarev, S. K. and Kamyshov, V. M.,** *Zavod. Lab.*, 30, 1068, 1964.
265. **Levin, E. N., Robbins, C. R., and McMurdie, H. F.,** *Phase Diagrams for Ceremists,* American Ceramic Society, Columbus, Ohio, 1966, 188.
266. **Bennett, H. and Oliver, G. J.,** *Analyst (London),* 101, 803, 1976.
267. **Oliver, G. J.,** *Proc. Anal. Div. Chem. Soc.*, 16, 151, 1979.
268. **Bock, R., Jacob, D., Fariwar, M., and Frankenfeld, K.,** *Fresenius Z. Anal. Chem.*, 200, 81, 1964.
269. **Neuberger, A., Schoeffmann, E., and Herkenhoff, K.,** *Arch. Eisenhuettenwes.*, 31, 91, 1960.
270. **Botto, R. I.,** *Jarrell Ash Plasma Newsl.*, 2, 4, 1979.
271. **Donaldson, E. M.,** *Mines Br. Monogr. 881,* Energy Mines Branch Resources Canada, Ottawa, 99.
272. **Donaldson, E. M.,** *Mines Br. Monogr. 881,* Energy Mines Branch Resources Canada, Ottawa, 95.
273. **Broekaert, J. A. C., Leis, F., and Laqua, K.,** *Spectrochim. Acta Part B,* 34, 73, 1979.
274. **Davoine, P., Briand, B., and Germanique, J. C.,** *Methods Phys. Anal.*, 7, 349, 1971.
275. **Jeffery, P. G.,** *Analyst (London),* 82, 67, 1957.
276. **Tsuge, S., Leary, J. J., and Isenhour, T. L.,** *Anal. Chem.*, 45, 198, 1973.
277. **Tsuge, S., Leary, J. J., and Isenhour, T. L.,** *Anal. Chem.*, 46, 106, 1974.
278. **Lung, W.-H. and Liu, T. I.,** *Acta Geol. Sin.*, No. 1, 92, 1975.
279. **Borneman-Starynkevich, I. D.,** *Dokl. Akad. Nauk SSSR,* 24(4), 335, 1939.
280. **Weiss, D.,** *personal communication, 1980.*
281. **Silńichenko, V. G. and Gritsenko, M. M.,** *Zavod. Lab.*, 31, 657, 1965.
282. **Sizonenko, N. T., Gudzenko, L. V., Antipova, G. Ya., and Obodynskaya, N. V.,** *Zavod. Lab.*, 49, 16, 1983.
283. **Sizonenko, N. T., Khukhryanskii, A. K., Egorova, L. A., and Gaiduk, O. V.,** *Zh. Anal. Khim.*, 40, 466, 1985.
284. **Watarai, H. and Suzuki, N.,** *Anal. Chim. Acta,* 159, 283, 1984.
285. **Grosskreutz, W.,** *Fresenius Z. Anal. Chem.*, 258, 208, 1972.
286. **Richards, C. S. and Boyman, E. C.,** *Anal. Chem.*, 36, 1790, 1964.
287. **British Standards Institution,** *Methods of Testing Refractory Materials. Part 2C,* Chemical analysis of chrome-bearing materials, BS 1902, Part 26, 1974.
288. **Sidiropoulos, N.,** *Analyst (London),* 94, 389, 1969.
289. **Elinson, S. V. and Petrov, K. I.,** *Tsirkonyi, Khimicheskie i Fizicheskie Metody Analiza,* Atomizdat, Moscow, 1960, 36.
290. **Romantseva, T. I., Shmanenkova, G. I., Kakhaeva, T. V., Kolenkova, M. A., and Sazhina, V. A.,** *Zh. Anal. Khim.*, 36, 1529, 1981.
291. **Jeczalik, A. and Morawska, T.,** *Chem. Anal. (Warsaw),* 14, 363, 1969.
292. **Knyazeva, D. N., Nikitina, I. B., Korsakova, N. V., and Volchenkova, V. A.,** *Zavod. Lab.*, 48, 16, 1982.
293. **Liteanu, C. and Paniti, M.,** *Analusis,* 1, 492, 1972.
294. **Shawky, S. M.,** *Anal. Chem.*, 49, 451, 1977.
295. **Jeffery, P. G.,** *Chemical Methods of Rock Analysis,* Pergamon Press, Oxford, 1970, 30.
296. **Cobb, W. D. and Harrison, T. S.,** *Analyst (London),* 96, 764, 1971.
297. **O'Gorman, J. V. and Suhr, N. H.,** *Analyst (London),* 96, 335, 1971.
298. **Hazel, W. M.,** Silicate analysis, in rep. of round-table discussion held by Division of Analytical Chemistry at 119th Meeting, American Chemical Society, Boston, Mass., April 1951.
299. **Bennett, H. and Reed, R. A.,** *Analyst (London),* 92, 466, 1967.
300. **Terashima, S.,** *Bull. Geol. Soc. Jpn.*, 21, 693, 1970.
301. **Terashima, S.,** *Bull. Geol. Soc. Jpn.*, 23, 287, 1972.
302. **Ishikawa, K. and Fukuda, S.,** *Bunseki Kagaku,* 28, 323, 1979.
303. **Terashima, S.,** *Jpn. Analyst,* 22, 1317, 1973.
304. **Jullietti, R. J. and Williams, D.R.,** *Analyst (London),* 106, 794, 1981.
305. **Boix, A. and Debras-Guédon, J.,** *Chim. Anal. (Paris),* 53, 459, 1971.
306. **Kastela-Macan, M. and Cerjan-Stefanovic, S.,** *Chromatographia,* 14, 415, 1981.
307. **Piryutko, M. M. and Kostyreva, T. G.,** *Zh. Anal. Khim.*, 37, 1644, 1982.
308. **Komárková, E.,** *Natl. Inst. Met. Repub. S. Afr. Rep.*, No. 1770, 1975; *Anal. Abstr.*, 31, 5B157, 1976.
309. **Ilsemann, K. and Bock, R.,** *Fresenius Z. Anal. Chem.*, 274, 185, 1975.
310. **Keith, M. L.,** Ph.D. thesis, Pennsylvania State University, University Park, 1963.

311. **Ingamells, C. O.,** *Talanta,* 11, 665, 1964.
312. **Ingamells, C. O.,** *Anal. Chem.,* 38, 1228, 1966.
313. **Ingamells, C. O.,** *Anal. Chim. Acta,* 52, 323, 1970.
314. **Suhr, N. H. and Ingamells, C. O.,** *Anal. Chem.,* 38, 730, 1966.
315. **Ingamells, C. O., Suhr, N. H. Tan, F. C., and Anderson, D. H.,** *Anal. Chim. Acta,* 53, 345, 1971.
316. **Saavedra, J., Garcia Sanchez, A., and Rodriguez Perez, S.,** *Chem. Geol.,* 13, 135, 1974.
317. **Cremer, M. and Schlocker, J.,** *Am. Mineral.,* 61, 318, 1976.
318. **Feldman, C.,** *Anal. Chem.,* 55, 2451, 1983.
319. **Šulcek, Z.,** personal communication, 1984.
320. **Shapiro, L.,** *U.S. Geol. Surv. Prof. Pap.,* No. 575B, B187, 1967.
321. **Abbey, S.,** *Geol. Surv. Can. Pap.,* 70, 23, 1970.
322. **Abbey, S., Lee, N. J., and Bouvier, J. L.,** *Geol. Surv. Can. Pap.,* 74, 19, 1974.
323. **Abbey, S., Aslin, G. E. M., and Lachance, G. R.,** *Rev. Anal. Chem.,* 3, 181, 1977.
324. **Gill, R. C. O. and Kronberg, B. I.,** *At. Absorpt. Newsl.,* 14, 157, 1975.
325. **Burdo, R. A. and Wise, W. M.,** *Anal. Chem.,* 47, 2360, 1975.
326. **Barredo, F. B. and Diez, L. P.,** *Talanta,* 23, 859, 1976.
327. **Barredo, F. B. and Diez, L. P.,** *Talanta,* 27, 69, 1980.
328. **Kiss, E.,** *Anal. Chim. Acta,* 140, 197, 1982.
329. **Bastius, H.,** *Fresenius Z. Anal. Chem.,* 288, 344, 1977.
330. **Bodkin, J. B.,** *Analyst (London),* 101, 44, 1976.
331. **Moyano, M. A., Guerrero, M. A., and Tobias, M.,** *Bol. Geol. Min.,* 91, 490, 1980; *Anal. Abstr.,* 1B89, 1982.
332. **Storms, L., Bonne, A., and Viaene, W.,** *J. Geochem. Explor.,* 13, 51, 1980.
333. **Van Loon, J. C. and Parissis, C. M.,** *Anal. Lett.,* 1, 519, 1968.
334. **Van Loon, J. C. and Parissis, C. M.,** *Analyst (London),* 94, 1057, 1969.
335. **Yule, J. W. and Swanson, G. A.,** *At. Absorpt. Newsl.,* 8, 30, 1969.
336. **Medlin, J. H., Suhr, N. H., and Bodkin, J. B.,** *At. Absorpt. Newsl.,* 8, 25, 1969.
337. **Medlin, J. H., Suhr, N. H., and Bodkin, J. B.,** *Chem. Geol.,* 6, 143, 1970.
338. **Foscolos, A. E. and Barefoot, R. R.,** *Geol. Surv. Can. Pap.,* 70, 16, 1970.
339. **Boar, P. L. and Ingram, L. K.,** *Analyst (London),* 95, 124, 1970.
340. **Voronkova, M. A., Butkina, T. A., Pyatova, V. N., Stepanova, N. A., Vorobev, V. S., and Kostyukova, L. M.,** *Metody Khimicheskogo Analiza,* VSEGEI, Leningrad, 1982, 88.
341. **Tardon, S.,** *Chem. Prum.,* 30, 195, 1980.
342. **Verbeek, A. A., Mitchell, M. C., and Ure, A. M.,** *Anal. Chim. Acta,* 135, 215, 1982.
343. **Owens, J. W. and Gladney, E. S.,** *At. Absorpt. Newsl.,* 15, 95, 1976.
344. **Salles, L. C. and Curtius, A. J.,** *Mikrochim. Acta,* No. 2, 125, 1983.
345. **Bailey, N. T. and Wood, S. J.,** *Anal. Chim. Acta,* 69, 19, 1974.
346. **Ronda, R. A. and Montero, R. P.,** Preparation y utilisation de lateritas, bauxitas, y susminerales constituyentes, *Forum de Analisis de Rocas y Minerales,* Ciudad de La Habana, Cuba, 1981.
347. **Ronda, A. R.,** Interferncias en la determinacion de Cr, Mn, Fe, Co y Ni en lateritas cubanas y suis principales minerales constituentes mediante espectrometria de absorcion atomica, Ph.D. thesis, Charles University, Prague, Czechoslovakia, 1982.
348. **Korte, N., Hollenbach, M., and Donivan, S.,** *At. Spectrosc.,* 3, 79, 1982.
349. **Rao, P. D.,** *At. Absorpt. Newsl.,* 11, 49, 1972.
350. **Davey, J. and Nicholson, N. M.,** Determination of trace elements in fluorspar mining survey samples, Report GS/EX/14/72/C, British Steel Corporation, Sheffield, 1973, 1.
351. **Poluektov, N. S., Meshkova, S. B., and Nikonova, M. P.,** *Zavod. Lab.,* 35, 166, 1969.
352. **Donnay, G., Ingamells, C. O., and Mason, B.,** *Am. Mineral.,* 51, 198, 1966.
353. **Zolotovitskaya, E. S., Potapova, V. G., and Ekel, V. A.,** *Zavod. Lab.,* 47, 18, 1981.
354. **Rice, T. D.,** *Talanta,* 23, 359, 1976.
355. **Rice, T. D.,** *Anal. Chim. Acta,* 91, 221, 1977.
356. **Zolotovitskaya, E. S. and Potapova, V. G.,** *Zh. Anal. Khim.,* 37, 415, 1982.
357. **Hall, A.,** *Chem. Geol.,* 30, 135, 1980.
358. **Subramanian, K. S. and Sastri, V. S.,** *Talanta,* 27, 469, 1980.
359. **Subramanian, K. S.,** *Int. Lab.,* p. 32, October 1981.
360. **Chan, Ch. Y. and Baig, M. W. A.,** *Anal. Chim. Acta,* 136, 413, 1982.
361. **Zhou, L., Chao, T. T., and Meier, A. L.,** *Talanta,* 31, 73, 1983.
362. **Weiss, D.,** personal communication, 1985.
363. **Zhou, L., Chao, T. T., and Meier, A. L.,** *Anal. Chim. Acta,* 161, 369, 1984.
364. **Bodkin, J. B.,** *Analyst (London),* 102, 409, 1977.
365. **Keller, E. and Parsons, M. L.,** *At. Absorpt. Newsl.,* 9, 92, 1970.
366. **Nuhfer, E. B. and Romanosky, R. R.,** *At. Absorpt. Newsl.,* 18, 8, 1979.

367. **Roubault, M., de la Roche, H., and Govindaraju, K.,** *Sci. Terre,* 9, 339, 1964.
368. **Govindaraju, K.,** *Appl. Spectrosc.,* 20, 302, 1966.
369. **Govindaraju, K.,** *Appl. Spectrosc.,* 24, 81, 1970.
370. **Govindaraju, K.,** *Anal. Chem.,* 40, 24, 1968.
371. **Govindaraju, K. and L'homel, N.,** *At. Absorpt. Newsl.,* 11, 115, 1972.
372. **Govindaraju, K.,** *Analusis,* 2, 367, 1973.
373. **Govindaraju, K., Mevelle, G., and Chouard, Ch.,** *Anal. Chem.,* 48, 1325, 1976.
374. **Govindaraju, K. and Montanari, R.,** *X-Ray Spectrom.,* 7, 148, 1978.
375. **Govindaraju, K.,** *Analusis,* 3, 116, 1975.
376. **Omang, S. H.,** *Anal. Chim. Acta,* 46, 225, 1969.
377. **Das, C. R. and Chauhan, R. S.,** *Trans. Indian Ceram. Soc.,* 32, 57, 1973.
378. **Lu, Y.-Z., Liang, F.-X., and Bai, Y.-L.,** *Ti Ch'iu Hua Hseuch,* No. 3, 282, 1980; *Anal. Abstr.,* 42, 1B99, 1982.
379. **Degre, J. P.,** *Silic. Ind.,* 47, 17, 1982.
380. **Bartuška, M., Hlaváč, J., and Procházka, S.,** *Silikáty,* 1, 87, 1957.
381. **Braicovich, L. and Landi, M. F.,** *Spectrochim. Acta,* 17, 51, 1957.
382. **Young, P. N. W.,** *Analyst (London),* 99, 588, 1974.
383. **Mochizuki, T. and Kuroda, R.,** *Analyst (London),* 107, 1255, 1982.
384. **Oguma, K., Ishino, S., and Kuroda, R.,** *Bunseki Kagaku,* 33, 280, 1984.
385. **Kuroda, R. and Ida, I.,** *Fresenius Z. Anal. Chem.,* 316, 53, 1983.
386. **Kuroda, R., Ida, I., and Oguma, K.,** *Mikrochim. Acta,* No. 1, 377, 1984.
387. **Kuroda, R., Ida, I., and Kimura, H.,** *Talanta,* 32, 353, 1985.
388. **Manoliu, C., Tomi, B., Panovici, I., and Popescu, O.,** *Chim. Anal. (Paris),* 52, 1270, 1970.
389. **Ohlweiler, O. A., Meditsch, J. O., and Piatnicki, C. M. S.,** *Anal. Chim. Acta,* 84, 431, 1976.
390. **Ohlweiler, O. A., Meditsch, J. O., da Silveira, C. L. P., and Silva, S.,** *Anal. Chim. Acta,* 61, 57, 1972.
391. **Ohlweiler, O. A., Meditsch, J. O., Santos, S., and Oderich, J. A.,** *Anal. Chim. Acta,* 69, 224, 1974.
392. **Ohlweiler, O. A., Meditsch, J. O., and Piatnicki, C. M. S.,** *Anal. Chim. Acta,* 67, 283, 1973.
393. **Hamilton, E. I., Minski, M. J., and Cleary, J. J.,** *Analyst (London),* 92, 257, 1967.
394. **Bennett, H. and Oliver, G. J.,** *Analyst (London),* 96, 427, 1971.
395. **Bettinelli, M.,** *Anal. Chim. Acta,* 148, 193, 1983.
396. **Bettinelli, M.,** *At. Spectrosc.,* 4, 5, 1983.
397. **Luecke, W.,** *Neues Jahrb. Mineral, Monatsh.,* p. 263, 1971.
398. **Zolotukhina, N. M. and Erenpreis, T. M.,** *Zavod. Lab.,* 45, 297, 1979.
399. **Brown, D. F. G., Mac Kay, A. M., and Turek, A.,** *Anal. Chem.,* 41, 2091, 1969.
400. **Rechenberg, W.,** *Zem. Kalk Gips,* 25, 410, 1972.
401. **Rechenberg, W.,** *Fresenius Z. Anal. Chem.,* 263, 333, 1973.
402. **Rechenberg, W.,** *Zem. Kalk Gips,* 25, 496, 1972.
403. **Nadkarni, R. A.,** *Anal. Chem.,* 52, 929, 1980.
404. **Nadkarni, R. A.,** Characterization of oil shales, Paper ACS 230, No. 27, in Symp. Geochemistry and Chemistry of Oil Shales, American Chemical Society, Seattle, Wash., 1983, 477.
405. **Nadkarni, R. A., Botto, R. I., and Smith, S. E.,** *At. Spectrosc.,* 3, 180, 1982.
406. **Bennett, H. and Reed, R. A.,** *Chemical Methods of Silicate Analysis,* Academic Press, New York, 1975.
407. **Rigin, V. I. and Simkin, N. M.,** *Zavod. Lab.,* 45, 713, 1979.
408. **Malyutina, T. M., Namvrina, E. G., and Shiryaeva, O. A.,** *Zavod. Lab.,* 47, 8, 1981.
409. **Rigin, V. I., Simkin, N. M., Tolkachnikov, Yu. B., and Kolosova, M. M.,** *Zavod. Lab.,* 45, 291, 1979.
410. **Welsch, E. P.,** *Talanta,* 32, 996, 1985.
411. **Shapiro, L.,** *U.S. Geol. Surv. Bull.,* No. 1401, 1975.
412. **Shapiro, L.,** *J. Res. U.S. Geol. Surv.,* 2, 357, 1974.
413. **Watkins, P. J.,** *Analyst (London),* 104, 1124, 1979.
414. **Wise, W. M. and Solsky, S. D.,** *Anal. Lett.,* 10, 273, 1977.
415. **Wise, W. M. and Solsky, S. D.,** *Anal. Lett.,* 9, 1047, 1976.
416. **Koenig, K.-H. and Neumann, P.,** *Anal. Chim. Acta,* 65, 210, 1973.
417. **Whitehead, D. and Malik, S. A.,** *Analyst (London),* 101, 485, 1976.
418. **Kruidhof, H.,** *Anal. Chim. Acta,* 99, 193, 1978.
419. **Rigin, V. I.,** *U.S.S.R. Patent* 983, 497, 1982.
420. **Rigin, V. I.,** *Zh. Anal. Khim.,* 40, 630, 1985.
421. **Rigin, V. I.,** *Zh. Anal. Khim.,* 38, 462, 1983.
422. **Yoshida, Y. and Kanji, I.,** *Bunseki Kagaku,* 28, 733, 1979.
423. **Jeanroy, E.,** *Chim. Anal. (Paris),* 54, 159, 1972.
424. **Fung, D. K., Dubois, J. P., and Kubler, B.,** *Analusis,* 11, 291, 1983.

425. **Jeanroy, E.,** *Analusis,* 2, 703, 1973.
426. **Voinovitch, I. A.,** *Bull. Liason Lab. Ponts Chaussées,* 79, 81, 1975.
427. **Riandey, C., Alphonse, P., Gavinelli, R., and Pinta, M.,** *Analusis,* 10, 323, 1982.
428. **Campbell, D. E. and Passmore, W. O.,** *Anal. Chim. Acta,* 76, 355, 1975.
429. **Foner, H. A.,** *Isr. J. Chem.,* 8, 541, 1970.
430. **Foner, H. A.,** *Analyst (London),* 109, 1469, 1984.
431. **Plug, C. and Niekerk, J. N.,** *J. S. Afr. Chem. Inst.,* 18, 71, 1965.
432. **Tingle, W. H. and Matocha, C. K.,** *Anal. Chem.,* 30, 494, 1958.
433. **Danielsson, A., Lundgren, F., and Sundkvist, G.,** *Spectrochim. Acta,* No. 2, 122, 1959.
434. **Bowditch, D. C., Cowan, E., and Lamman, R. F.,** Rock analysis by direct-reacing emission spectrography, *Proc. Australas. Inst. Min. Met.,* No. 229, 121, 1969.
435. **Welday, E. E., Baird, A. K., McIntyre, D. B., and Madlem, K. W.,** *Am. Mineral.,* 49, 889, 1964.
436. **Reckziegel, M. and Staats, G.,** *Arch. Eisenhuettenwes.,* 35, 633, 1964.
437. **Peterson, M. J. and Zink, J. B.,** Report 6132, U.S. Bureau of Mines, Washington, D.C., 1962.
438. **Jiang, G. and Li, S.,** *Fenxi Huaxue,* 12, 794, 1984; *Anal. Abstr.,* 47, 5B162, 1985.
439. **Besnus, Y. and Rouault, R.,** *Analusis,* 2, 111, 1973.
440. **Wittmann, A. A. and Willay, G.,** *ICP Inf. Newsl.,* 9, 33, 1984.
441. **Burman, J. O., Pontér, C., and Boström, K.,** *Anal. Chem.,* 50, 679, 1978.
442. **Bankston, D. C., Humphris, S. E., and Thompson, G.,** *Anal. Chem.,* 51, 1218, 1979.
443. **Walsh, J. N.,** *Spectrochim. Acta Part B,* 35, 107, 1980.
444. **Brenner, I. B., Watson, A. E., Russell, G. M., and Goncalves, M.,** *Chem. Geol.,* 28, 321, 1980.
445. **Hoffer, D., Brenner, I. B., and Halicz, L.,** *ICP Inf. Newsl.,* 9, 494, 1984.
446. **Sinex, S. A., Cantillo, A. Y., and Helz, G. R.,** *Anal. Chem.,* 52, 2342, 1980.
447. **Cantillo, A. Y., Sinex, S. A., and Helz, G. R.,** *Anal. Chem.,* 56, 33, 1984.
448. **Bowker, P. C. and Manheim, F. T.,** *Appl. Spectrosc.,* 36, 378, 1982.
449. **Bankston, D. C. and Fisher, N. S.,** *Anal. Chem.,* 49, 1017, 1977.
450. **Crock, J. G., Lichte, F. E., and Wildeman, T. R.,** *Chem. Geol.,* 45, 149, 1984.
451. **Crock, J. G. and Lichte, F. E.,** *Anal. Chem.,* 54, 1329, 1982.
452. **Brenner, I. B., Watson, A. E., Steele, T. W., Jones, E. A., and Goncalves, M.,** *Spectrochim. Acta Part B,* 36, 785, 1981.
453. **Wünsch, G. and Czech, N.,** *Fresenius Z. Anal. Chem.,* 317, 5, 1984.
454. **Claisse, F.,** Preliminary Report No. 327, Department of Natural Resources, Quebec, Canada, 1956.
455. **Claisse, F.,** Tetraborate fusion for X-ray sample preparation, *Spex Speaker,* 18, 7, 1973.
456. **Claisse, F.,** Borate Fusion for xRF analysis, Corp.scient.Claisse leaflet, Sainte-Foy, Qué., Canada, 1984.
457. **Turmel, S., LeHouillier, R., and Claisse, F.,** *Can. J. Spectrosc.,* 23, 125, 1978.
458. **Ohls, K. and Becker, G.,** *Fresenius Z. Anal. Chem.,* 279, 183, 1976.
459. **Willgallis, A. and Schneider, G.,** *Fresenius Z. Anal. Chem.,* 246, 115, 1969.
460. **Norrish, K. and Hutton, J. T.,** *Geochim. Cosmochim. Acta,* 33, 431, 1969.
461. **Emmermann, R. and Obi, D. V. C.,** *Fresenius Z. Anal. Chem.,* 254, 1, 1971.
462. **Petin, J., Bentz, F., Wagner, A., and Arbed, E.-B.,** PERL-xPhilips Analytical Group, Science and Industry Division, p. 141, Philips, Eindhoven Netherlands.
463. **LeHouillier, R. and Turmel, S.,** *Anal. Chem.,* 46, 734, 1974.
464. **Pella, P. A.,** *Anal. Chem.,* 50, 1380, 1978.
465. **VanWilligen, J. H. H. G., Kruidhof, H., and Dahmen, E. A. M. F.,** *Talanta,* 18, 450, 1971.
466. **DeJongh, W. K. and Willy, K.,** *Int. Lab.,* 11, 56, 1980.
467. **Exnar, P.,** *Chem. Listy,* 78, 920, 1984.
468. **Palme, Ch. and Jagoutz, E.,** *Anal. Chem.,* 49, 717, 1977.
469. **Sastry, B. S. R. and Hummel, F. A.,** *J. Am. Ceram. Soc.,* 43, 23, 1960.
470. **Jurczyk, J., Smolec, W., and Stankiewicz, G.,** *Chem. Anal. (Warsaw),* 24, 1005, 1979.
471. **Fabbi, B. P.,** *Am. Mineral.,* 57, 237, 1972.
472. **Kinson, K., Knott, A. C., and Belcher, C. B.,** *Talanta,* 23, 815, 1978.
473. **Dixon, K.,** *Mintek Report 1985,* M184, Council Min. Technol. Repub. S. Afr, MINTEK, Randburg, South Africa.
474. **Padfield, T. and Gray, A.,** Major element rock analysis by X-ray fluorescence — a simple fusion method, *Bulletin Philips FS 35,* Philips, Eindhoven, Netherlands, August 1971.
475. **Ambrose, A. D., Rutherford, R., and Muir, Ṡ.,** *Metallurgia,* 82, 119, 1970.
476. **Pietrosz, J. and Kuboň, Z.,** *Hutn. Listy,* 34, 660, 1970.
477. **Feret, F.,** *X-Ray Spectrom.,* 11, 128, 1982.
478. **Nieuwenhuizen, C.,** The analysis of ferro-alloys, Philips Export B.V., Application laboratories for X-ray and Nuclear Analysis, Philips, Eindhoven, Netherlands.
479. **Boix, A., Dragnaud, M., and Debras-Guédon, J.,** *Bull. Soc. Fr. Ceram.,* 106, 69, 1975.

480. **Vaeth, E. and Griessmayr, E.,** *Fresenius Z. Anal. Chem.*, 303, 268, 1980.
481. **Haukka, M. T. and Thomas, I. L.,** *X-Ray Spectrom.*, 6, 204, 1977.
482. **Thomas, I. L. and Haukka, M. T.,** *Chem. Geol.*, 21, 39, 1978.
483. **Thomas, I. L. and Haukka, M. T.,** *Proc. Australas. Inst. Min. Met.*, 267, 55, 1978.
484. **Smith, T. K.,** *Trans. Inst. Min. Met. B*, 81, 156, 1972.
485. **Simakov, V. A., Sorokin, I. V., and Zemtsova, L. I.,** *Zh. Anal. Khim.*, 36, 1847, 1981.
486. **Smales, A. A., Mapper, D., Webb, M. S. W., Webster, R. K., and Wilson, J. D.,** *Science*, 167, 509, 1970.
487. **Lüschow, M.-H. and Schäfer, H. P.,** *Fresenius Z. Anal. Chem.*, 250, 317, 1970.
488. **Malyutina, T. M., Sharova, N. A., Makarova, R. F., Shestakov, V. A., and Shvartsman, S. P.,** *Zh. Anal. Khim.*, 38, 2137, 1983.
489. **DeJongh, W. K.,** XRF analysis of major constituents in silicate rocks, in *Philips Application Report*, Philips, Eindhoven, Netherlands.
490. **Raschka, H., Lodziak, J., Li, G., Liang, D., and Zhang, B.,** *Fenxi Huaxue*, 10, 609, 1982; *Anal. Abstr.*, 45, 2B92, 1983.
491. **Ashley, D. G. and Andrews, K. W.,** *Analyst (London)*, 97, 841, 1972.
492. **Augustin, F. and Borsier, M.,** *BRGM Rapport* 81 SGN 305 M6A, Bureau Recherches, Géologiques et Minieres, Orléans, France, 1981.
493. **Webber, G. R. and Newbury, M. L.,** *Can. Spectrosc.*, 16, 90, 1971.
494. **Jagoutz, E. and Palme, Ch.,** *Anal. Chem.*, 50, 1555, 1978.
495. **Pelikánová, M.,** *Fresenius Z. Anal. Chem.*, 320, 338, 1985.
496. **Mahan, K. I. and Leyden, D. E.,** *Anal. Chim. Acta*, 147, 123, 1983.
497. **Diaz-Guerra, P. J., Bayon, A., and Roca, M. A.,** *An. Quim.*, 73, 688, 1977.
498. **Pella, P. A., Lorber, K. E., and Heinrich, K. F. J.,** *Anal. Chem.*, 50, 1268, 1978.
499. **Musikas, N. and Vantighem, G.,** *Cim. Betons. Platres. Chaux*, 706, 135, 1977.
500. **Farne, G., Randi, G., and Grimaldi, R.,** *ICP Newsl.*, 12, 31, 1984.
501. **Grimaldi, R., Ametrano, E., Battini, A., and Grassi, B.,** Generalised methods of analysis for ferroalloys, in *Report Italsider*, Italsider, Genova, Italy, 1980.
502. **Giles, H. L. and Holmes, G. M.,** *X-Ray Spectrom.*, 7, 2, 1978.
503. **Jurczyk, J., Stankiewicz, G., and Kojder, J.,** *Chem. Anal. (Warsaw)*, 26, 1027, 1981.
504. **Rowledge, H. P.,** *J. R. Soc. West Aust.*, 20, 165, 1934.
505. **Kolthoff, I. M., Sandell, E. B., Meehan, E. J., and Bruckenstein, S.,** *Quantitative Chemical Analysis*, Macmillan, London, 1969, 524.
506. **Biskupsky, V. S.,** *Anal. Chim. Acta*, 33, 333, 1965.
507. **Russel, B. G., Spangenberg, J. D., and Steele, T. W.,** *Talanta*, 16, 487, 1969.
508. **Bond, A. M., O'Donell, T. A., Waugh, A. B., and McLaughlin, R. J. W.,** *Anal. Chem.*, 42, 1168, 1970.
509. **Weiss, D.,** *Geochem. Geochem. Methods Data*, 1, 137, 1971.
510. **Peták, P.,** personal communication, 1985.
511. **Owen, L. B. and Faure, G.,** *Anal. Chem.*, 46, 1323, 1974.
512. **Biskupski, V. S.,** *Chemist Analyst*, 56, 49, 1967.
513. **Rubeška, I.,** *At. Absorpt. Newsl.*, 12, 724, 1973.
514. **Cook, E. B. T. and Shelton, B. J.,** *Natl. Inst. Met. Repub. S. Afr. Rep.*, No. 1840, 1976; *Anal. Abstr.*, 33, 6B81, 1978.
515. **Hitchen, A. and Zechanowitsch, G.,** *Talanta*, 27, 383, 1980.
516. **Florence, T. M. and Farrar, Y. J.,** *Anal. Chem.*, 42, 271, 1970.
517. **Luginin, V. A. and Tserkovnitskaya, I. A.,** *Zh. Anal. Khim.*, 26, 1593, 1971.
518. **Mikhailova, Z. M., Mirskii, P. B., and Yarushkina, A. A.,** *Zavod. Lab.*, 30, 407, 1964.
519. **Novikova, Ju. N.,** *Zh. Anal. Khim.*, 23, 1057, 1968.
520. **Hey, M. H.,** *Mineral. Mag.*, 26, 117, 1941.
521. **Meyrowitz, R.,** *Anal. Chem.*, 42, 1110, 1970.
522. **Cook, E. B. T. and Steele, T. W.,** *Natl. Inst. Met. Repub. S. Afr. Lab. Methods*, No. 26/27, 1970.
523. **Dixon, K., Cook, E. B. T., and Silverthorne, D. F.,** *Natl. Inst. Met. Repub. S. Afr. Rep.*, No. 1394, 1972; *Anal. Abstr.*, 24, 2792, 1973.
524. **Snetsinger, K. G.,** *Am. Mineral.*, 51, 1623, 1966.
525. **Beamish, F. E.,** *The Analytical Chemistry of the Noble Metals*, Pergamon Press, Oxford, 1966.
526. **Beamish, F. E. and Van Loon, J. C.,** *Recent Advances in the Analytical Chemistry of the Noble Metals*, Pergamon Press, Oxford, 1972.
527. **Beamish, F. E. and Van Loon, J. C.,** *Analysis of Noble Metals. Overview and Selected Methods*, Academic Press, New York, 1977.
528. **Ginzburg, C. I., Ezerskaya, N. A., Prokofeva, I. V., Fedorenko, N. V., Shlenskaya, V. I., and Belskii, N. K.,** *Analiticheskaya Khimiya Platinovykh Metallov*, Izdat. Nauka., Moskow, 1972.

529. **Haffty, J., Riley, L. B., and Goss, W. D.,** *U.S. Geol. Surv. Bull.*, p. 1445, 1977.
530. **Hosking, J. W.,** Review of Analytical Methods for Determining Trace Amounts of Gold in Ores and Process Streams, in 32nd Symp. Ser. Carbon Pulp Technol. Extr. Gold, Australasian Institute of Mining Methods, Bentley, West Australia, 1982, 351.
531. **Khvostova, V. P. and Golovnaya, S. V.,** *Zavod. Lab.*, 48, 3, 1982.
532. **Kalinin, S. K. and Vinnitskaya, E. G.,** *Zh. Anal. Khim.*, 35, 2226, 1980.
533. **Sen Gupta, J. G.,** *Min. Sci. Eng.*, 5, 207, 1973.
534. **Williams, C. J. and Seidemann, H. J.,** *Am. Lab.*, No. 8, 63, 1975.
535. **Heady, H. H. and Broadhead, K. G.,** *U.S. Bur. Min. Inform. Circ.*, revised ed. No. 7695, 1976; *Anal. Abstr.*, 33, 1B41, 1977.
536. **Van Loon, J. C.,** *Pure Appl. Chem.*, 49, 1495, 1977; *Anal. Abstr.*, 34, 5B15, 1978.
537. **Young, R. S.,** *Gold Bull.*, 13, 9, 1980; *Anal. Abstr.*, 40, 5B48, 1981.
538. **Williams, B. J. and Munro, H. C., Eds.,** *Assaying and Analytic Techniques for the Determination of the Noble Metals,* Inst. Assayers and Analysts, Johannesburg, S. Afr., 1980.
539. **Van Loon, J. C.,** *TrAC Trends Anal. Chem.*, 3, 272, 1984; *Anal. Abstr.*, 47, 7B165, 1985.
540. **Beamish, F. E.,** *Talanta,* 5, 1, 1960.
541. **Kallmann, S.,** *Anal. Chem.*, 56, 1020A, 1984.
542. **Johnson, W. M. and Maxwell, J. A.,** *Rock and Mineral Analysis,* John Wiley & Sons, New York, 1981, 390.
543. **van Wyk, E. and Dixon, K.,** *Counc. Min. Technol. Repub. S. Afr. MINTEK Rep.*, M88, Randburg, South Africa, 1983.
544. **Koleva, E. G. and Arpadjian, S. H.,** *Talanta,* 17, 1018, 1970.
545. **Lupan, S., Protopopescu, M., and Ponta, T.,** *Rev. Chim. (Bucharest),* 25, 833, 1974; *Chem. Abstr.*, 83, 37098, 1975.
546. **Zdorova, E. P., Popova, N. N., and Kondulinskaya, M. A.,** *Zavod. Lab.*, 43, 926, 1977.
547. **Berkovich, T. B. and Chechulina, L. I.,** *Zavod. Lab.*, 44, 140, 1978.
548. **Kallmann, S. and Hobart, E. W.,** *Talanta,* 17, 845, 1970.
549. **Harris, A. M., Lengton, J. B., and Farrell, F.,** *Talanta,* 25, 257, 1978.
550. **Fishkova, N. L., Zdorova, E. P., and Popova, N. N.,** *Zh. Anal. Khim.*, 30, 806, 1975.
551. **Hoehn, R. and Jackwerth, E.,** *Erzmetall,* 29, 279, 1976; *Anal. Abstr.*, 33, 5B31, 1977.
552. **Prudnikov, E. D., Kolosova, L. P., Kalachev, V. K., Shapkina, Yu. S., Novatskaya, N. V., and Bychkov, Yu. A.,** *Zh. Anal. Khim.*, 33, 468, 1978.
553. **Moloughney, P. E.,** *Talanta,* 24, 135, 1977.
554. **Moloughney, P. E.,** *Talanta,* 27, 365, 1980.
555. **Diamantatos, A.,** *Anal. Chim. Acta,* 148, 293, 1983.
556. **Diamantatos, A.,** *Anal. Chim. Acta,* 165, 263, 1984.
557. **Rowe, J. J. and Simon, F. O.,** *U.S. Geol. Surv. Circ.* No. 599, 1968.
558. **Finkelshtein, Y. B., Leipunskaya, D. I., Savosin, S. I., Drynkin, V. I., Aliev, A. I., and Popova, N. N.,** *Zavod. Lab.*, 37, 1471,1971.
559. **Simon, F. O. and Millard, H. T., Jr.,** *Anal. Chem.*, 40, 1150, 1968.
560. **Coxon, C. H., Verwey, C. J., and Lock, D. N.,** *J. S. Afr. Min. Met. Inst.*, 62, 546, 1962; *Anal. Abstr.*, 9, 5200, 1962.
561. **Sinclair, W. A.,** *J. S. Afr. Inst. Min. Met.*, 64, 333, 1964.
562. **Trokowicz, J.,** *Chem. Anal. (Warsaw),* 15, 1147, 1970.
563. **Yaguchi, K. and Kaneko, J.,** *Jpn. Analyst,* 21, 601, 1972.
564. **Wall, S. G. and Chow, A.,** *Anal. Chim. Acta,* 69, 439, 1974.
565. **Wall, S. G. and Chow, A.,** *Anal. Chim. Acta,* 70, 425, 1974.
566. **Zolotov, Yu. A.,** *Zavod. Lab.*, 50(1), 3, 1984.
567. **Guest, R. N., Debbo, P., Bushell, L. A., and Levin, J.,** *Natl. Inst. Met. Repub. S. Afr. Rep.*, No. 1942, 1977; *Anal. Abstr.*, 36, 4B146, 1979.
568. **Anisimov, S. M., Petrenko, V. I., and Savalskii, S. I.,** *Analiz Blagorodnykh Metallov,* Izd. Akad. Nauk U.S.S.R., Moscow, 1965, 142.
569. **Semenova, N. Ya. and San'ko, A. Z.,** *Analiz blagorodnykh metallov,* Izd. Cvetmetniiinformaciya, Moskow, 1965, 169.
570. **Palmer, R. and Watterson, J. I. W.,** *Natl. Inst. Met. Repub. S. Afr. Rep.*, No. 1185, 1971; *Chem. Abstr.*, 75, 58205, 1971.
571. **Watterson, J. I. W., Robért, R. V. D., and van Vyk, E.,** *Natl. Inst. Met. Repub. S. Afr. Rep.*, No. 1048, 1970; *Chem. Abstr.*, 74, 106768, 1971.
572. **Georgiev, G. T. and Apostolov, D.,** *Zh. Anal. Khim.*, 27, 506, 1972.
573. **Daiev, Ch., Georgiev, G. T., Jovtschev, M., Stefanov, G., and Apostolov, D.,** *Isotopenpraxis,* 7, 138, 1971.
574. **Artenmev, O. I. and Stepanov, V. M.,** *Zavod. Lab.*, 48(8), 1, 1982.

575. **Robért, R. V. D. and van Wyk, E.,** *Natl. Inst. Met. Repub. S. Afr. Rep.*, No. 977, 1970.
576. **Coombes, R. J., Chow, A., and Wageman, R.,** *Talanta*, 24, 421, 1977.
577. **Tello, A. and Sepulveda, N.,** *At. Absorpt. Newsl.*, 16, 67, 1977.
578. **Donnelly, T. H.,** *Proc. Australas. Inst. Min. Met.*, No. 236, 61, 1970; *Anal. Abstr.*, 21, 1144, 1971.
579. **Schnepfe, M. M. and Grimaldi, F. S.,** *Talanta*, 16, 591, 1969.
580. **Greenland, L. P., Rowe, J. J., and Dinnin, J. I.,** *U.S. Geol. Surv. Prof. Pap.*, 750-B, B175, 1971.
581. **Broadhead, K. G., Piper, B. C., and Heady, H. H.,** *Appl. Spectrosc.*, 26, 461, 1972.
582. **Haffty, J., Haubert, A. W., and Page, N. J.,** *U.S. Geol. Surv. Prof. Pap.*, 1129-A-1, G1-G4, 1980.
583. **Anisimov, S. M., Semenova, A. Ya., San'ko, A. Z., and Gordanov, V. I.,** *Metody Analiza Platinovykh Metallov, Zolota i Serebra*, Metallurgizdat, Moskow, 1960, 171.
584. **Whitehead, A. B. and Heady, H. H.,** *Appl. Spectrosc.*, 24, 225, 1970.
585. **Cooley, E. F., Curry, K. J., and Carlson, R. R.,** *Appl. Spectrosc.*, 30, 52, 1976.
586. **Alexander, P. W., Hoh, R., and Smythe, L. E.,** *Talanta*, 24, 549, 1977.
587. **Coombes, R. J., Chow, A., and Flint, R. W.,** *Anal. Chim. Acta*, 91, 273, 1977.
588. **Coombes, R. J. and Chow, A.,** *Talanta*, 26, 991, 1979.
589. **Rowe, J. J. and Simon, F. O.,** *Talanta*, 18, 121, 1971.
590. **van Wyk, E.,** *Natl. Inst. Met. Repub. S. Afr. Rep.*, No. 2068, 1980; *Chem. Abstr.*, 94, 24402, 1981.
591. **Olkhovich, P. F. and Latysh, I. K.,** *Zavod. Lab.*, 46, 299, 1980.
592. **Samchuk, A. I. and Latysh, I. K.,** *Ukr. Khim. Zh.*, 48, 638, 1982.
593. **Kallmann, S. and Maul, C.,** *Talanta*, 30, 21, 1983.
594. **Haffty, J. and Riley, L. B.,** *Talanta*, 15, 111, 1968.
595. **Grimaldi, F. S. and Schnepfe, M. M.,** *Talanta*, 17, 617, 1970.
596. **Robért, R. V. D., van Wyk, E., and Ellis, P. J.,** *Natl. Inst. Met. Repub. S. Afr. Rep.*, No. 1905, 1977; *Anal. Abstr.*, 35, 4B17, 1978.
597. **Turkstra, J. and de Wet, W. J.,** *Talanta*, 16, 1137, 1969.
598. **Turkstra, J., Pretorius, P. J., and de Wet, W. J.,** *Anal. Chem.*, 42, 835, 1970.
599. **Kolosova, L. P., Novatskaya, N. V., and Vinnitskaya, E. G.,** *Zavod. Lab.*, 42, 508, 1976.
600. **Kolosova, L. P.,** *Zavod. Lab.*, 48(7), 8, 1982.
601. **Kolosova, L. P., Novatskaya, N. V., Ryshova, R. I., and Aladyshkina, A. E.,** *Zh. Anal. Khim.*, 39, 1475, 1984.
602. **Rakovskii, E. E., Zdorova, E. P., Kuligin, V. I., Popova, N. N., Fishkova, L. P., and Shvedova, N. B.,** *Zavod. Lab.*, 48(8), 11, 1982.
603. **Georgiev, G. T.,** *Zh. Anal. Khim.*, 33, 740, 745, 1978.
604. **Diamantatos, A.,** *Anal. Chim. Acta*, 90, 179, 1977.
605. **Diamantatos, A.,** *Anal. Chim. Acta*, 91, 281, 1977.
606. **Diamantatos, A.,** *Anal. Chim. Acta*, 92, 171, 1977.
607. **Diamantatos, A.,** *Anal. Chim. Acta*, 94, 49, 1977.
608. **Diamantatos, A.,** *Anal. Chim. Acta*, 98, 315, 1978.
609. **Grinzaid, E. L., Kolosova, L. P., Nadezhina, L. S., and Novatskaya, N. V.,** *Zavod. Lab.*, 44, 683, 1978.
610. **Artem'ev, O, I. and Stepanov, V. M.,** *Zh. Anal. Khim.*, 31, 724, 1976.
611. **Artem'ev, O. I., Stepanov, V. M., and Kovel'skaya, G. B.,** *Zh. Anal. Khim.*, 33, 493, 1978.
612. **Artem'ev, O. I., Kiselev, B. G., and Stepanov, V. M.,** *Zh. Anal. Khim.*, 33, 2163, 1978.
613. **Artem'ev, O. I., Kiselev, B. G., and Pozdnyakov, S. V.,** *Zh. Anal. Khim*, 34, 2227, 1979.
614. **Faye, G. H.,** *Mines Branch Can. Rep.*, No. R-154, The Queen's Printer, Ottowa, Canada, 1965.
615. **Faye, G. H. and Moloughney, P. E.,** *Talanta*, 19, 269, 1972.
616. **Moloughney, P. E. and Graham, J. A.,** *Talanta*, 18, 475, 1971.
617. **Moloughney, P. E. and Faye, G. H.,** *Talanta*, 23, 377, 1976.
618. **Jones, E. A., Kruger, M. M., and Wilson, A.,** *Natl. Inst. Met. Repub. S. Afr. Rep.*, No. 1232, 1971; *Chem. Abstr.*, 75, 71029, 1971.
619. **Ryabinina, G. N., Strat'ev, A. I., and Kirillov, V. P.,** *Zavod. Lab.*, 48(7), 12, 1982.
620. **Van Loon, J. C. and Beamish, F. E.,** *Anal. Chem.*, 37, 113, 1965.
621. **Danilova, F. I., Orobinskaya, V. A., Khudolei, G. N., and Dmitrieva, G. A.,** *Zh. Anal. Khim.*, 29, 2276, 1974.
622. **Banbury, L. M. and Beamish, F. E.,** *Fresenius Z. Anal. Chem.*, 218, 263, 1966.
623. **Francois, J. P., Gijbels, R., and Hoste, J.,** *Talanta*, 21, 780, 1074.
624. **Bekov, G. I., Kurskii, A. N., Letokhov, V. S., and Radayev, V. N.,** *Zh. Anal. Khim.*, 40, 2208, 1985.
625. **Danilova, F. I., Fedotova, I. A., and Nazarenko, R. M.,** *Zavod. Lab.*, 48(8), 9, 1982.
626. **Rammensee, W. and Palme, H.,** *J. Radioanal. Chem.*, 71, 401, 1982.
627. **Williamson, J. E. and Savage, J. A.,** *J. S. Afr. Inst. Min. Met.*, 65, 343, 1965; *Anal. Abstr.*, 13, 2376, 1966.

628. **Robért, R. V. D., van Wyk, E., and Palmer, R.,** *Natl. Inst. Met. Repub. S. Afr. Rep.*, No. 1371, 1971; *Anal. Abstr.*, 23, 1474, 1972.
629. **Robert, R. V. D., van Wyk, E., Palmer, R., and Steele, T. W.,** *J. S. Afr. Chem. Inst.*, 25, 179, 1972; *Chem. Abstr.*, 78, 92104, 1973.
630. **Dixon, K., Jones, E. A., Rasmussen, S., and Robért, R. V. D.,** *Natl. Inst. Met. Repub. S. Afr. Rep.*, No. 1714, 1975; *Anal. Abstr.*, 31, 2B245, 1976.
631. **Robért, R. V. D. and van Wyk, E.,** *Natl. Inst. Met. Repub. S. Afr. Rep.*, No. 1705, 1975; *Anal. Abstr.*, 29, 4B227, 1975.
632. **Bowditch, D. C.,** *Natl. Inst. Met. Technol. Memo.*, National Institute of Metallurgy, Johanesburg, South Africa, June 27, 1972.
633. **Dixon, K., Krueger, M. M., and Radford, A. J.,** *Natl. Inst. Met. Repub. S. Afr. Rep.*, No. 1654, 1975; *Anal. Abstr.*, 31, 3B180, 1976.
634. **Kuznetsov, A. P., Kukushkin, Yu. N., and Makarov, D. F.,** *Zh. Anal. Khim.*, 29, 2155, 1974.
635. **Davidova, I. Yu., Kuznetsov, A. P., Antokol'skaya, I. I., Nikol'skaya, N. N., and Ezhkova, Z. A.,** *Zh. Anal. Khim.*, 34, 1145, 1979.
636. **Hoffman, E. L., Naldrett, A. J., Van Loon, J. C., Hancock, R. V. G., and Manson, A.,** *Anal. Chim. Acta,* 102, 157, 1978.
637. **Bortwick, A. A. and Naldrett, A. J.,** *Anal. Lett.*, 17, 265, 1984.
638. **Annegarn, H. J., Erasmus, C. S., Sellschop, J. P. F., and Tredoux, M.,** *Nucl. Instrum. Methods Phys. Res.*, 218, 33, 1983; *Anal. Abstr.*, 46, 12B157, 1984.
639. **Wemyss, R. B. and Scott, R. H.,** *Anal. Chem.*, 50, 1694, 1978.
640. **Myasoedova, G. V., Antokol'skaya, I. I., and Saavin, S. B.,** *Talanta,* 32, 1105, 1985.
641. **Šulcek, Z., Povondra, P., and Doležal, J.,** *CRC Crit. Rev. Anal. Chem.*, 6, 255, 1977.
642. **Plesek, J. and Hermaněk, S.,** *Sodium Hydride,* CRC Press, Cleveland, 1968, 27.
643. **Janoušek, K.,** personal communication, 1984.
644. **Peták, P.,** personal communication, 1977.
645. **Nieuwenhuizen, C.,** *The Analysis of Ferro-Alloys,* Philips Export B.V., Application laboratories for X-ray analysis, Philips, Eindhoven, Netherlands.
646. **Johansson, G.,** The Remelting of Ferro-Alloys, Report 1974, No. 4511, SKF Steel Hellefors, Sweden.
647. **Schwarz, W., Kandlbauer, W., and Schwarz, K.,** Probenvorbereitung und Roentgenfluoreszenzanalyse von Ferrolegierungen, *13th XRF-Symp.*, de Gruyter, New York, 1981.
648. ASPECT, *Report 1975,* No. 10/24/01, Department Spectrographe.
649. **Colombo, A. and Vivian, R.,** *Talanta,* 27, 881, 1980.
650. **Korostelev, P. P.,** *Fotometricheskii i Kompleksometricheskii Analiz v Metallurgii,* Izd. Metallurgiya, Moscow, 1984, 135.

Chapter 9

DECOMPOSITION BY SINTERING

I. INTRODUCTION

Fusion decomposition, even if it can be used for most substances, suffers as universal procedures from two limitations: (1) a great excess of the fusion agent introduces large amounts of neutral salts that may exhibit various salt effects during subsequent analytical procedures; and (2) fusion almost always leads to corrosion of the decomposition vessel materials and often to irreversible exchange of impurities between the vessel wall and the melt. These effects that adversely influence the analytical results are considerably minimized when sintering procedures are used in which the reagent amount is limited, and melts are not formed. The mixture is only sintered, with minimal interaction with the crucible material. The completeness of the decomposition depends on the reagent amount, the temperature, and the heating time. As the reactions occurring at the interface are rather complicated and have not been quantitatively formulated, the sintering decompositions are empirical. Any particular conditions hold usually only for a certain type of material and cannot be mechanically transferred to other materials.

Sintering and reactions of substances in the solid state can be explained in terms of thermal motion of atoms or ions. As each atom in the surface of a solid has a higher energy than that of the atoms inside the crystal structure, powders with large surface areas (in powder metallurgy the surface areas of the materials are as large as 50 m^2/g) are in a nonequilibrium state and exhibit a tendency toward decreasing the surface area, in order to attain the arrangement with the lowest surface energy. This process is imperceptible at normal temperatures, due to a low rate of diffusion and/or a high viscosity. However, if a sufficient thermal energy is supplied for the oscillating atoms or ions to overcome the particular crystal energy, the atoms or ions are exchanged at the contact areas among the grains, such that the final distribution corresponds to the statistical normal distribution. The whole process is relatively slow, but even small inhomogeneities in the structure (e.g., dislocations and vacancies) accelerate it considerably. In addition to this internal bulk diffusion, diffusion also occurs along the grain boundaries and in the surfaces, the ratio of their rates being dependent on the temperature. At lower and medium temperatures, diffusion along the boundaries predominates and is several orders of magnitude faster than the bulk diffusion. On the other hand, the latter predominates at high temperatures. The ratio of the bulk (volume) diffusion coefficients, D_v, and the grain boundary diffusion coefficients, D_b, depends on the temperature according to the equation,

$$D_v/D_b = k \cdot \exp(-\Delta U_v - \Delta U_b) \cdot RT \quad (1)$$

where ΔU are the activation energies of diffusion, R is the gas constant, and T is the absolute temperature (K). Simultaneously with changes in the ionic positions at the interfaces, chemical changes also occur that perpetrate diffusion into the bulk of the crystal. The reaction is often connected with the formation of new crystal structures, with recrystallization. The lowest necessary temperature for excitation of the crystal lattice is termed the internal diffusion temperature and is expressed in terms of a fraction of the absolute temperature of melting of the substance, which amounts to 50 to 70% for most compounds. The internal diffusion removes the reaction products from the contact zone, so that the reactants continue to react. The diffusion acts as stirring of a mixture of solids.

The diffusion rate is increased in the presence of a gas or a liquid. Even if these phases do not participate in the reaction, they penetrate into the pores and dislocations, increase the surfaces of the solids, and thus enhance their reactivity. Last but not least, internal diffusion is a function of the activities of the reactants, especially those that are formed "in statu nascendi" during the heating. For example, the oxides formed during sintering by thermal decomposition of carbonates, hydroxides, and peroxides are much more reactive than the oxides alone. This difference is probably due to disordering and numerous defects in the newly formed crystal structure of the oxide formed. Moreover, the gas evolved also plays a role. Significant effects are also exerted by those ions present in the decomposition agent that cause heterovalent substitutions in the resultant phases, with formation of vacancies.

Sintering is accompanied by changes in the shapes and number of the reacting grains and, thus, also of pores. Sintering leads to formation of connecting necks among the grains, the grains approach one another so that the pores among them begin to close, and the volume of the whole system decreases. The neck material is transported by the above diffusion mechanisms, including evaporation and material condensation. A sinter results and is used for further chemical treatment.

The great variety in the combinations of physical and chemical processes accompanying reactions among solids considerably complicates the possibilities of discovery of general laws governing the kinetics of these reactions. Therefore, only qualitative effects of some of the above factors on the course and rate of sintering can be evaluated. For this reason, the use of these decomposition procedures in analytical practice has been rather limited, although the application of these principles in powder metallurgy has brought about a revolution in the technology of preparation of materials. Analytical practice has mainly employed sintering with alkali carbonates and peroxides, either alone or mixed with some oxides, mostly those of divalent cations.

II. SINTERING WITH CARBONATES AND THEIR MIXTURES WITH OTHER AGENTS

Sintering with sodium carbonate, alone or mixed with other alkali carbonates, is mainly important in decomposition of silicates, for several obvious reasons: (1) a decrease in the excess of sodium ions that prevent complete separation of silicic acid in solutions of carbonate melts; and (2) a decrease in their effect on the subsequent determinations, e.g., chelatometric or spectrometric. Provided that the test material is very finely pulverized and intimately mixed with the reagent, the decomposition reactions between the reagent and quartz or feldspars are very rapid at temperatures below the melting point of the reagent. To decompose rocks containing the above minerals, about a 20-min sintering is sufficient. Basic rocks with dark minerals predominating (pyroxenes, amphiboles, biotites) require about 3 hr of sintering at a suitable temperature. Using this procedure, aluminum-rich minerals can also be decomposed, e.g., cordierites, garnets, tourmalines, and light micas. It has been shown by X-ray diffraction and thermal studies that alkali monosilicates, aluminates, and some oxides are formed in the reactions of the above minerals.[1] Only certain accessoric minerals of rocks remain undecomposed, e.g., sillimanite, kyanite, and zircon. The following universal procedure has been recommended for decomposition of rocks: a sample (usually 500 mg) is mixed with a twofold excess of sodium carbonate (containing 5% of sodium nitrate) in a platinum crucible and covered with *circa* 200 mg of the mixture. The crucible is covered and placed in a cold furnace, whose temperature is increased to 750 to 780°C and then maintained for 180 min. The sinter is readily loosened from the crucible after the decomposition, transferred to a dish, wetted with a few milliliters of water, and dissolved in 15 mℓ of concentrated hydrochloric acid with a few drops of hydrogen peroxide.

The high reactivity of cesium ions in destruction of silicate structures has been utilized in sintering decomposition with cesium carbonate. The sintering proceeds at *circa* 700 to 800°C, with a 1.8- to 2.2-fold excess of the agent over the sample, in a platinum crucible, using the flame of a Meker burner. Higher excesses are used in decompositions of basic rocks and minerals; lower ones suffice for rocks rich in silica. The sintering occurs at a lower temperature for *circa* 5 min, and then the reaction is completed by a 15-min heating with full flame. The corrosion of the platinum is then minimal. The sinter or melt formed is easily leached with water and acidified, and silicic acid is precipitated in the usual way. The filtrate is appropriately diluted and serves as the master solution for spectrophotometric, titration, and AAS determinations. The cesium ions present act as an ionization buffer in flame spectrometry.[2]

A 3 + 1 mixture of sodium carbonate and peroxide has been used to decompose rare earth elements (REE) raw materials containing bastnaesite, parisite, barites, and a silicate matrix. After a preliminary decomposition with nitric and hydrofluoric acid, the dried residue is mixed with an eightfold excess of the mixture and sintered for 1 hr at 480°C. The sinter is extracted with water containing a few drops of hydrogen peroxide, the REE are quantitatively precipitated, as the hydroxides which are then dissolved in acids and further treated.[3,4]

Sintering with soda has often been used to decompose some types of chromites and aluminum, manganese, and iron oxides. Oxidic iron ores can be rapidly decomposed by a 1-min sintering at 1000 to 1100°C. A 5 + 1 mixture of sodium carbonate with boric acid has been tested[5] for the same purpose, sintering the material for 20 min at 900°C. However, completeness of the decomposition has not been verified,[6] even when the temperature was increased, the reaction time prolonged, or the reagent amount increased.

As the sinter formed is relatively porous and mildly alkaline, the sulfidic sulfur is rapidly oxidized by atmospheric oxygen to sulfate ions; tellurides and selenides are similarly oxidized to compounds with the elements in valence states IV to VI. The oxidation with atmospheric oxygen in the presence of soda has been employed for combustion of organic matter in coal (400°C) and coke (600°C), in determination of arsenic in these materials.[7] The oxidizing effect in determination of sulfidic sulfur has been enhanced by sintering with a five- to tenfold excess of a 1 + 1 mixture of sodium carbonate and potassium permanganate at 800°C for 30 to 40 min.[8]

Mineral raw materials are more often decomposed using mixtures of sodium carbonate with divalent cation oxides, especially those of magnesium, zinc, and calcium. Decompositions with sodium carbonate and cadmium oxide have also been tested,[9] as the latter is more reactive than the above oxides. However, platinum vessels cannot be used, because platinum is readily alloyed with cadmium, and the great toxicity of the agent also prevents general use of this mixture. The composition of the sintering mixtures is usually empirical and is adapted according to the character of the materials to be decomposed. In general, mixed reagents exhibit two effects: (1) alkali carbonates are the fusion agents proper and make the mixture weakly alkaline, thus preventing volatilization of acid-forming components, such as halides, sulfur oxides, selenium, arsenic, boron, etc.; (2) the divalent cations then mainly react with silicic acid and alkali aluminates, forming sparingly soluble salts that can be readily separated from the other components by leaching the sinter with water. Moreover, their presence causes the sinter to be porous, with a large surface area on which the oxidation by atmospheric oxygen is faster and more complete. This has been utilized in the oxidation of sulfur, selenium, and tellurium, and in combustion of organocarbon in solid caustobiolites. The solubility of divalent cations is limited by the formation of insoluble carbonates, and thus, their contents in aqueous extracts are minimal. However, the presence of these carbonates affects the behavior of phosphate ions in solution; they are trapped on the solid through adsorption (and partly by ion exchange), so that up to 50% losses in the phosphates occur in the extract.[10]

The sintering decomposition has most often been applied to determinations of halides in silicate products, rocks, coal, meteorites, soils, and similar materials. For liberation of fluoride, 2 to 6 + 1 mixtures of sodium carbonate and zinc oxide have given good results when used in a fivefold excess. The sample is thoroughly mixed with the reagent, usually in a platinum crucible, and the mixture is covered with a small amount of the reagent and sintered for *circa* 30 min at 900°C in an electric furnace. The sinter is leached with water, the ions of manganic and chromic acid, if present, are reduced with ethanol, and an aliquot part of the solution is used for a determination, mostly using ion selective electrode (ISE) or photometry. The procedure has been applied to determination of fluorine in rocks,[11-14] sphalerite concentrates,[15] and cement works raw materials and products.[16] Chlorine can be determined in rocks under the same decomposition conditions.[11]

A mixture of sodium and calcium carbonates has been recommended[13] for decomposition of silicates and silicon alloys in determination of silica. A sample of 0.25 to 0.5 g is mixed with 0.35 g of soda and 1.5 g of calcium carbonate. The mixture is transferred to a platinum crucible containing 1.5 g of the latter salt, the crucible is placed in a furnace preheated to 400°C, and the temperature is increased to 900°C within 30 min. After a 5-min heating to 950°C, the mixture is ignited for 50 min at 1150°C. The sinter is dissolved in concentrated hydrochloric acid in a beaker, perchloric acid is added, and silica is precipitated in the classical manner. An advantage of the procedure is the fact that low concentrations of alkali metal salts are used, thus minimizing the amount of dissolved silicic acid. Oxidants are not used in the decomposition and, thus, contamination of the system by the crucible material is prevented.

The Eschka mixture (Na_2CO_3-MgO = 1 + 2) has been employed for similar decompositions. It has originally been proposed for combustion of coal, but is also applicable to sulfidic materials. In combustion of organic carbon, the sample is mixed with a twofold excess of the reagent in a porcelain crucible and slowly heated, first at 200°C and then at a temperature increasing up to 800°C at a rate of 7°C·min^{-1}. The latter temperature is maintained for 2 hr. Chlorides and sulfates are determined by ion chromatography in a water extract.[17] Other applications of reagents of this type involve determination of iodine and bromine in soils, rocks, geochemical samples, and meteorites.[18-21] The final determination is usually carried out electrochemically (ISE) or by MS. Up to 70% of the iodide is oxidized to iodate during the sintering.[21]

Sintering with the Eschka mixture and similar mixtures is a classical procedure for the oxidation of sulfides, selenides, and tellurides that pass into an aqueous extract of the sinter and are thus separated from almost all elements that interfere in their gravimetric or titration determination. Manganates, if they are formed, must be reduced by, e.g., sodium peroxide in an alkaline medium. These procedures have more recently been applied to determination of total sulfur in coal,[17,22] oil shales,[22] and coke.[23]

To determine arsenic in silicates with carbonaceous admixtures, sintering of the sample with an eightfold excess of a 3 + 1 mixture of potassium carbonate and magnesium oxide has been recommended, at 650°C for 30 min and then at 900°C, until the organic phase is completely destroyed.[24] Novák[25] has decomposed coal by sintering with a fourfold excess of the Eschka mixture for 3 hr at 815°C. An aqueous extract of the sinter is dissolved in acids and diluted, using the solution for an AAS determination of As, Sb, and Se with the hydride generation.

As an aqueous extract of the sinter contains all of the boron in the form of an alkali borate, this decomposition procedure with a 3 to 4 + 1 mixture of Na_2CO_3 and ZnO can be employed for destruction of silicates[26] and boron carbide[27] in determination of total boron. A 4 + 1 mixture readily decomposes natural borosilicates, such as tourmalines of the schorl-dravite series or axinite.[28] With lithium-aluminum tourmalines (elbaites), the sample must first be mixed with silica powder to attain an Al/Si ratio of less than unity; otherwise the mineral is undecomposed.

subversive activities. His appointment was therefore to be expected.[3]

Owing to geographical reasons, the province of Heilungkiang was the last to be submitted to the Japanese military forces. Only after the fall of Tsitsihar into the Japanese hands was General Chang Ching-huei asked to be the civil governor. It is interesting to know that in the ceremony of his assumption of office on January 6th, 1932, the official language used on the occasion was Japanese. Later the Japanese were anxious to win over General Ma Chan-shan who had made himself famous at the Nonni River Bridge, and the governorship was offered to him. In the provincial organizations as in the local or district governments, the Chinese appointees were, in practically every case, compelled to assume office under severe penalties. They were merely puppets, and the actual power of the government resides either in the Japanese advisers or in the Special Affairs Departments which also exist in the provincial governments and are presided over by Japanese appointed by the Japanese military headquarters.

15. **(3) The Period of Inter-Provincial Conferences.** (3) The Period of Inter-Provincial Conferences. The time was now ripe for something like a unified government for all of the conquered territory. Japanese influence had been extended to every district where the Japanese troops had appeared. The control had become well-nigh complete, and the time had come for the establishment of a unified government. Accordingly, on February 16th, a conference was called at Mukden where Chinese representing the Three Provinces were compelled to take part. The whole thing from beginning to

[3] For details, see *ibid.*, Part III, p. 3.

end was arranged beforehand. The time and place for the meeting were decided by the headquarters of the Kwantung Army. And even the agenda of the conference with their resolutions were all previously decided upon. All that the Chinese needed to do, and were supposed to do, was to attend the conference and endorse whatever was passed. A greater mockery of "independence" and "self-determination" has never been known. Both Chang Ching-huei and Ma Chan-shan were present. They could not do otherwise. As usual, the Special Affairs Committee saw to it that they were safely conducted to Mukden under adequate Japanese military protection. They flew in the Japanese army planes, and, on their arrival at Mukden, they were sedulously guarded in everything that they said or did.

The result of that extraordinary conference was the "decision to organize a North-Eastern Administrative Council which immediately came into being and which was to be dissolved at the birth of the new regime. On February 18th, at the conclusion of the conference, the council issued a statement which, among other things, made it perfectly clear that a new "State" would be formed carrying out to the full the "Principle of the Open Door and of equal opportunity for all nations.

16. (4) The Formation of the bogus government.

(4) The Formation of the bogus government. When all the machinery for the new "government" was sufficiently well prepared, the "Manchukuo" was ready to be proclaimed. The great event came off on March 1st of the present year, and the ceremony of installing the executive Henry Puyi took place on March 9th. He had been previously kidnapped from Tientsin. On November 2nd, Dohihara saw Puyi and asked him to proceed to Mukden. Puyi refused, whereupon Dohihara reminded him that

there might be possible danger if his advice was not accepted. On the 7th, a basket of fresh fruit was brought to Puyi's residence in which there was discovered a bomb. On the night of the 8th, Dohihara fomented the Tientsin troubles, and in the midst of the disorder, two days afterwards, Puyi had no alternative but to proceed in a Japanese steamer to Dairen whence he was brought to Tankantze, an interior resort on the S.M.R. where he was kept for nearly three months when, after all the arrangements were made, he was installed as the Chief Executive of the "Manchukuo."

It was originally planned to have him assume office on the 1st of the month. The delay came as a result of reverses suffered by the Japanese troops at Shanghai and of controversy as to whether the capital should be at Mukden or Changchun. To follow the traditional way of the early emperors of China, the Chief Executive must be approached three times before he was supposed to agree to assume the post. He was first approached on February 29th by a group of eleven people elected by the Japanese. Puyi refused, whereupon he was asked for the second time on March 4th. Again he refused and it was on March 7th, when he was approached for the third time, that he consented to become the Chief Executive.

17. **Puyi assuming office.** The ceremony took place at 3 p.m. on March 9th when Puyi was conducted to his office by General Honjo and Count Uchida, then the President of the South Manchuria Railway. The Japanese comprised 70% of the people present. No one was admitted without the express consent of the Kwantung Army Headquarters, as all the invitations were issued by that military organization. The eighth item on the

program of the ceremony was the reception of foreign guests. Puyi alighted from the platform and bowed once to Honjo, who it was arranged did not reply. On the same evening, on his return to Mukden, Honjo requested that Puyi should see him off. It was through the present "prime minister's" good office that Puyi was spared the trouble. Puyi on the same day gave out two statements which had been prepared in advance by the Kwantung Army Headquarters. These Japanese prepared statements were usually written in Chinese, and to the Chinese it is clear that they could only be drafted by the Japanese because of certain linguistic peculiarities never found in Chinese statements drafted by the Chinese themselves.

18. **Japanese rejoicing.** The only people who rejoiced in the formation of this state were, of course, the Japanese as it was by them that it was created. The Chinese people evinced no interest. They had been present under compulsion, and were not consulted as to their own wishes. But the joy of the Japanese was genuine and spontaneous. Telegrams congratulating the formation of the new "state" came in by the thousands from Japanese everywhere.

Two days after the ceremony, on March 11th, the Japanese in Mukden offered prayers at the Japanese temple, followed by one of the most elaborate parades that ever took place. The nature of that celebration may best be seen from the following program which was printed and distributed to all the Japanese residents:

> "1. At 11 a.m. on March 11th local residents must appear before the Mukden shrine to offer prayers.
>
> "2. After the prayer, they must repair to the Memorial (erected in honour of the Japanese who

died on Manchurian soil) and yell three times "Long live the Japanese Empire" and "Long live the 'Manchukuo.' "

"3. As regards decoration, the following details should be observed:

a. There shall not be less than 6,000 lanterns on the main thoroughfares of the Japanese section of the city.

b. Everybody on that day must hoist the flag of the new State.

c. There shall be not less than 25,000 flags of the new State which shall be distributed by the Young Men's Club and the Youth League to pedestrians on the main thoroughfares, the time for such distribution being from 8 a.m. to 10 a.m. The flags should likewise be distributed to carriages and other vehicles, to be hoisted on them.

d. At the junctions of the main streets, there shall be a Japanese flag hoisted together with the five-coloured flag of the new State."

19. **The "Self-Government Guidance Department": its nature and activities.** Before concluding this section on the various stages leading to the formation of the puppet government, there is one organization that needs to be emphasized and described, and that is the so-called "Self-Government Guidance Department." That organ is the embodiment of the spirit of General Honjo and of the Kwantung Army Headquarters. The successive stages of the "independence" movement were planned out by that organ. All the slogans and the wording of the handbills and statements in the name of Puyi and the Chinese officials were all decided upon by that organ. The systematic manufacture of "public opinion," so as to give the impression that it was a real "independence" movement was the work of that extraordinary organ.

This organ originally constituted the fourth division

of the Kwantung Army Headquarters, and began its activities from the beginning of the September incident, although it was not formally organized as an independent organ till November 10th with Yu Chung-han as its Chairman. He is the only Chinese in the organ and a figurehead, all the rest being Japanese. It has branches in every local district which compel the local residents to select representatives to clamour for "independence." In the larger cities, they compel the various organizations to express themselves on the same subject. Refusal to obey means immediate arrest and the severest form of punishment.

The flags used on different public occasions were ordered by this "Self-Government Guidance Department". They were either made in Japan or in the Japanese quarter of Mukden. The shops were compelled to buy these flags, some of which cost as much as Yen 6.50 each, and to display them. The slogans on posters were not only worded by this organ but also posted in different localities of the Three Eastern Provinces under its direction. People who were caught tearing these slogans down were subjected to the most cruel forms of punishment.

Many cases occurred where the people were compelled by agents of this organ to say that they were citizens of the "Manchukuo." On March 20th, an employee of the S.M.R., evidently connected with the organ, while on duty, distributed printed matter to the passengers extolling the virtues of the new "government" and compelling them to say that they were "Manchukuo" citizens. Japanese police likewise, on various occasions, would ask Chinese what nationality they belonged to. Upon their refusal to say that they were "Manchukuo" citizens they were kicked about and severely manhandled until they said that they were. Chinese passengers on the S.M.R. trains would also be asked what nationality they belonged to. On one occasion a Chinese passenger in reply said he was Chinese, whereupon a railway guard heavily struck him on his head and caused severe bleeding. Another case as illustration might be mentioned. Two young students on their way to the Fourth Liaoning Provincial Primary School on February 28th, passed by the Japanese Consulate from which three soldiers came out and asked them what country they belonged to. The boys replied that they were Chinese, whereupon the soldiers asked again. The boys then shouted that they were citizens of the great Republic of China, whereupon the soldiers kicked and knocked them about that they almost died.

20. **Examples of parades staged by the department.** "The Self-Government Guidance Department" must be given the credit for successfully handling two major celebrations, one towards the end of February when the formation of the new "state" was going to be announced and the other after Puyi assumed office.

From February 20th, all the major cities of the Three Eastern Provinces were ordered by the Department to hold new "National Establishment" demonstrations. Those in Mukden were held under the direct control of the department and took place on three successive days. On the 27th of February there was the Mukden People's Assembly. On the 28th there was the assembly for the representatives from the various localities of Mukden or Fengtien Province. On the 29th was held the assembly for representatives covering the whole of the Three Eastern Provinces. The first of these meetings took place in the North-eastern Theatre and only a handful of poor people participated. There was a generous distribution of biscuits which was what actually attracted the poor. There was a large number of slogans prepared by the department. The second day saw the presentation of a prepared statement on the Mukden provincial constitution, and dances and dramatic performances were given to attract the crowds. In the third assembly of "delegates" representing the whole of the Three Eastern Provinces, two prepared resolutions were read and were at once passed. The first was that the Kwantung Army Headquarters should be asked to establish a new "state". The second was that twelve delegates should be asked to request Puyi to become the Chief Executive. In the afternoon there was an elaborate parade where large numbers of Japanese appeared in Chinese clothing holding flags on which were written that they were delegates from such and such a locality. At the same time, Japanese photographers were busy at work later to show to the world that the parade was an enthusiastic Chinese affair. A few delegates from the provinces did appear and were paid by the Japanese according to the following scale: Yen 500 for each person from Heilungkiang, Yen 300 from Kirin, and Yen 100 from Liaoning, and even local riffraffs were paid Yen 10 or 20 a person for taking part in the parade.

Similar events were arranged after Puyi assumed office, when there was a three-day celebration. The most striking part came off at Harbin where the Mayor, Pao Kwan-chen, under the influence of the Japanese, staged a dramatic show in the open air to which a large crowd were attracted by the announcement that it was a big athletic meet. No sooner had they arrived than they were immediately roped in. The camermen then got busy, but the crowd turned their backs to them whereupon the Mayor yelled "Hats off" and only three people went through the performance. The Mayor further yelled "Three bows to the new national flag," and only two others accompanied him in the act. The last part of the performance was the usual yelling of "Long live the 'Manchukuo'," and the same three shouted. The Mayor went away in disgust.

When the Commissioners were in Mukden, there was an athletic meet organized by the same "Self-Government Guidance Department" and middle school boys who should be in schools, but could not because such schools had been closed down, were made to take part in the games along with the Japanese. It was intended to impress the Commission with the enforced enthusiasm of the Chinese youths by having them sing the "Manchukuo" song, but they did not. The lack of interest on the part of the Chinese students was only too evident.

III. The "Government" under Japanese Domination.

21. **Composition and structure of the "government."** It is interesting to enquire briefly into the composition and structure of this unique "government" of the Japanese, by the Japanese, and for the Japanese which was set up after the puppet sovereign has gone through his manipulated movements. The "government" is supposed to be based on a constitution of 39 articles divided into six chapters which on paper appears not unlike one of the normal documents of its kind. But the question is in whom is the actual power of the "government" vested. In every branch of the "government", although the chief is nominally a Chinese, there is what is known as the Director of the Department of General Affairs who is the substantial source of authority for that organization. That Director is always a Japanese. It is extraordinary to call the "government" a "Manchukuo" government when the personnel consists of Japanese who all occupy the key positions. The "government" is best represented in the following chart:

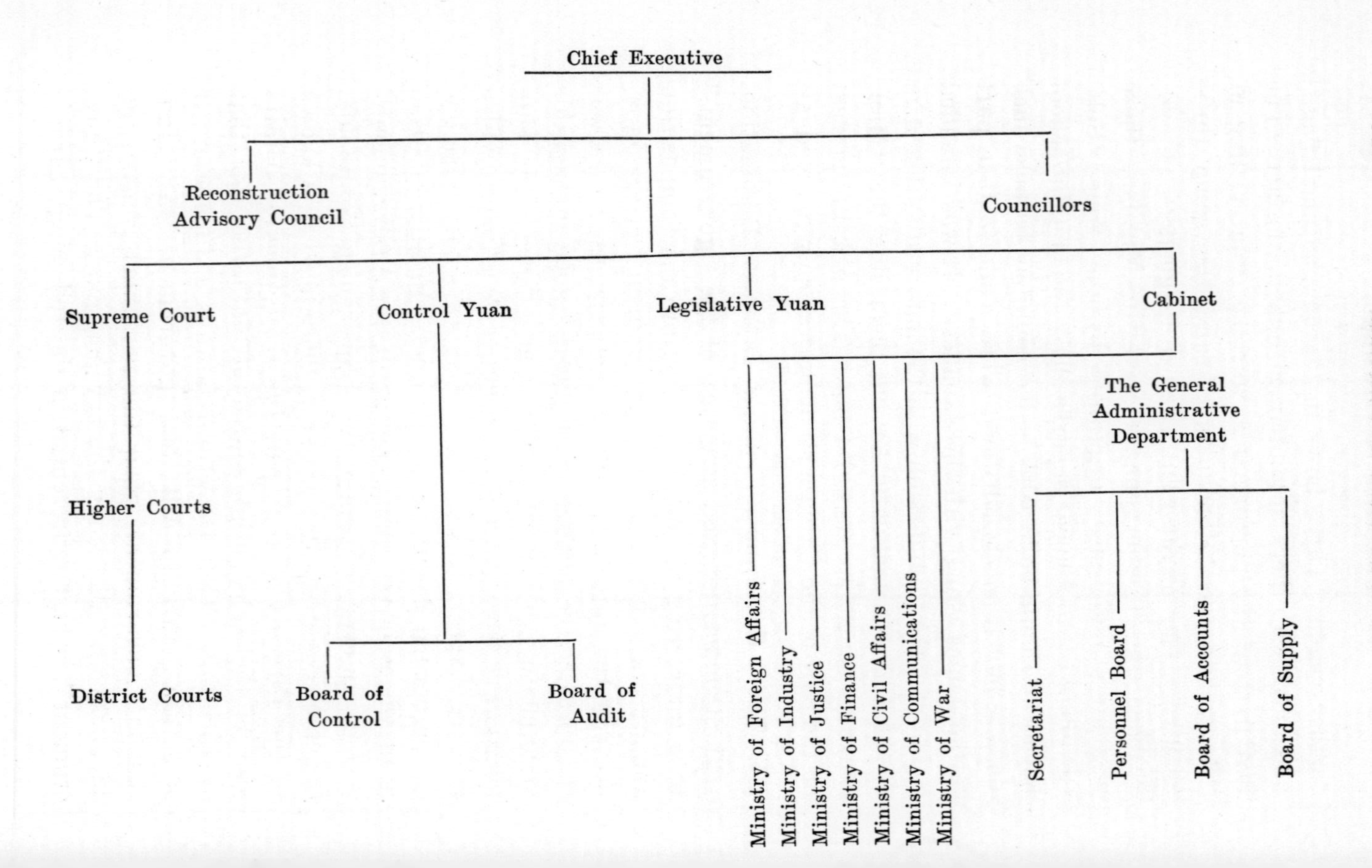

Chief Executive
Reconstruction Advisory Council
Councillors
Supreme Court
Control Yuan
Legislative Yuan
Cabinet
The General Administrative Department
Higher Courts
District Courts
Board of Control
Board of Audit
Ministry of Foreign Affairs
Ministry of Industry
Ministry of Justice
Ministry of Finance
Ministry of Civil Affairs
Ministry of Communications
Ministry of War
Secretariat
Personnel Board
Board of Accounts
Board of Supply

22. **Japanese occupying key positions in the "government."** The names of the Japanese officials in the "government" will be found in the appendix (A.) For the present, it is necessary to mention what positions in the different branches of the "government" are held by the Japanese.

In the office of the chief executive, there is a Japanese adviser. The executive is a titular head and has no administrative power anyhow. The personnel in his office is therefore not large and there is no need for any large number of Japanese.

In the Councillors' office the chief secretary is a Japanese who attends to such affairs as there are in an office which likewise has no executive functions.

The Cabinet is a different thing. It is the final organ of control for the "government", and one would naturally expect it to be entirely dominated by the Japanese. There is a General Administrative Department which in fact controls all important matters in the cabinet and it is presided over by a Japanese. The work of that department is distributed among four bureaus or boards and everyone of them is under a Japanese.

There are seven ministries which are directly responsible to the Cabinet. The ministers are all Chinese, but like the premier, they are no more than figure-heads. In every one of the ministries, there is a Department of General Affairs similar to what is found in the Cabinet's office, and the Directors of these departments are all Japanese. It is they who actually control the different ministries. Without their approval nothing can be done. Without their counter-signature no document can be sent out.

In the Ministry of Foreign Affairs, the Vice-Minister is a Japanese. So also is the Director of the Department of Publicity. In the Ministry of Finance, a Japanese is also the chief of the department of the Customs in addition to having a Japanese high adviser. In the Ministry of Communications the Directors of the Departments of Railway and the Post are Japanese in addition to the usual Japanese Director of General Affairs. The Director of the Mining Department of the Ministry of Industry is a Japanese. In the Ministry of Justice, the head of the Department of Judicial Affairs is a Japanese, while the usual Director of General Affairs is concurrently high adviser to the Supreme Court.

The Control Yuan, like the Ministries, has a Japanese Director of General Affairs, while the two bureaus, those of audit and control, are placed under two Japanese officials. The President of the Law Codification office is a Japanese. The Director of the Department of Police Affairs is a Japanese as also the Director of the Police Training Headquarters. The Central Bank has a Japanese Vice-President and three Japanese members on the Board of Trustees.

From the list so far given which includes only the most important positions in the Central "government" at Changchun, is it a "Manchukuo" "independent" government or is it a Japanese Government that has come into being?

Likewise in the provincial and district governments, there is such a plethora of Japanese advisers, high advisers, Directors of General Affairs, Directors of Special Affairs, and even Japanese chiefs of sections that it is difficult to see where the "independent" "Manchukuo" government comes in.

23. **Completion in Japanese subjection of the "government."** Up to the present, although the "government" at Changchun is entirely nominated and run by

the Japanese, it remains a separate entity. It will not be long before it will be amalgamated with the other Japanese administrative units in the Three Eastern Provinces. What more will then be needed for the annexation of a territory of 382,000 square miles and a population of 30 millions? The proposed appointment of an "ambassador plenipotentiary" by the Japanese Government who shall be the administrative representative of the Mikado for the Kwantung Leased Territory and the whole of the recently conquered territory has already been discussed on more than one occasion, and the Changchun "government" will simply become part of a bigger whole. The so-called chief executive with all the paraphernalia of his "government" will then take his place according to the following chart:

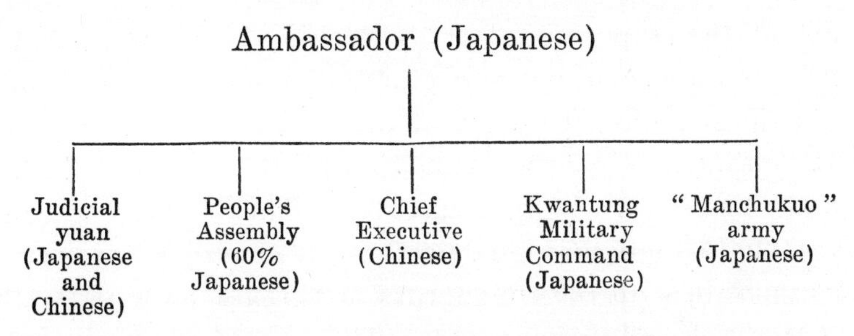

The enlargement of the Japanese Empire and the conquest of the Continental mainland will then be more or less completed!

In order to show still more adequately that the Japanese never meant to establish an "independent" Manchukuo Government but have every desire to transform the Three Eastern Provinces into a colonial possession, we give in another appendix (B.) a set of eighteen articles embodying general principles for the foundation and establishment of the new "government". The document is from a very reliable source and is frank and outspoken on the whole Japanese outlook towards China's north-eastern question.

IV. Japanese Control over the New "State."

24. **Japanese confiscating private property and compelling co-operation.** It is clear, from what we have observed, that the control of the Japanese over the new "state" is well-nigh complete. There is no reason to believe that that control will be relaxed by the Japanese themselves. How thoroughly Japanese control has been enforced in the Three Eastern Provinces may be shown by a series of drastic measures which were adopted from the very beginning when military conquest was becoming a grim reality. The number of such measures has become indeed many, and we shall content ourselves with only some of the more important ones. The first act of the Japanese after capturing the Chinese cities was to arrest and detain Chinese of wealth and position. All the personal property of people like Marshal Chang Hsueh-liang, General Wan Fu-lin, General Po Yu-lin and others have been confiscated. Those who were unfortunate enough to be present were either forced to co-operate with the conquerors or threatened with death if they refused to co-operate. They would at least be imprisoned and submitted to all kinds of indignities. Mr. Yung Ho, formerly chairman of the Commission of Finance in Kirin, Mr. Kan Shao-shih, formerly Governor of Jehol and a number of others were thus arrested. On the night of April 1st, thirteen people, including the police chief of a section in the Kirin and Tunhua Railway, the presidents of the Commercial and Agricultural Associations of Tunhua and Chiao-ho and others were actually sent from the Japanese Consulate at Kirin to a place known as Chu Men Ko (the mouth of the Nine

Gates) and brutally put to death while the Japanese soldiers talked and laughed among themselves. The chief of police of Huaiteh district fared just as badly, as he was sawed to death.

25. **Control over army and police.** At present, when the "Manchukuo" government has been established, every aspect of its activity is under effective Japanese control. The first thing to be organized under Japanese influence is the army and with that the police and espionage system. The Ministry of War in the "Government" is now dominated by the Japanese. But that is of comparatively less importance. It is the actual military forces that count, and these with their various divisions have all Japanese advisers appointed by the Kwantung Military Headquarters from which they receive orders. The army in the "Manchukuo" and the Japanese forces there are in fact one and indivisible. The present number of Japanese troops in the Three Eastern Provinces is much greater than the number of railway guards whom the Japanese claim the right to station by virtue of the treaties.

The police likewise is in Japanese hands. The so-called "Manchukuo" police is not relied upon by the Japanese who provide them with no equipment dangerous enough to create trouble, while the whole of the Three Eastern Provinces is now under a network of espionage.

26. **Over publicity.** Closely connected with espionage is the work of the Japanese intelligence and publicity officials. All Chinese independent papers in the Three Eastern Provinces have been suppressed, and the present Chinese or, if you please, "Manchukuo" papers must only publish news approved by the Japanese. The Chinese are therefore not allowed to know anything that

goes on in China Proper. All Chinese papers published in Shanghai, Tientsin, Peiping, etc., are now not admitted into the occupied territory, and it is said that only six foreign papers published in China Proper are permitted to reach the foreign consulates. Direct communication with China proper through letters and telegrams are similarly censored. No people have experienced stricter censorship and enjoyed less freedom of speech than the unfortunate 30 millions of Chinese in the Three Eastern Provinces to-day. Yet it was they, we have been told, who started an "independence" movement!

27. **Over education.** The most thorough-going censorship exists however, in the realm of education. The Japanese indeed look far into the future. The younger generation of the population must be taught that they form a part of the Japanese Empire and have nothing to do with China. All Chinese educational institutions have been closed down or reorganized under Japanese supervision. The major emphasis is on the elementary and middle schools. The first subject in the scheme of "reorganization" is the enforcement of the teaching of the Japanese language. The *Isaka Shimbun* for July 6th, 1932 carries the following news item, "all the elementary schools and colleges for preceptors in Liaoning Province have already taken up the study of the Japanese language. In a conference of the district magistrates of Heilungkiang Province, it was resolved that the study of Japanese was to be made compulsory in all the elementary schools and that the budget for the teaching of the language was decided upon. The various police stations of that province also issued instructions to police officers to commence the study of the language." A Rengo despatch of July 15th, from Mukden says: "a

decision has been reached by the Bureau of Education of Fengtien Province to teach Japanese language to students of middle schools, commencing in September when a fresh school term opens''. The text-books in Chinese used in the different schools have been changed and even historical facts have been twisted to suit Japanese purposes. Everything that is taught must, in a word, be favourable to the Japanese and subversive of any attachment or loyalty to China. Patriotism for Japan of 30 millions of Chinese and their descendants must be made to grow at any cost.

28. **Over the railways.** The railways too and other means of communication are now in Japanese hands. All the Chinese built railways, including the so-called ''parallel'' railways to which the Japanese have taken such strong exception, are now placed in one way or another under the control of the South Manchuria Railway. More distant railways are being forced into Japanese control by fair means or foul. General Ma Chanshan in his telegram to the Commission before their departure for their tour in the Three Eastern Provinces gave out the authentic information that the Hulan-Hailun Railway was forced to be mortgaged to the Japanese for 50 years with $3,000,000.

The Mukden-Shanhaikwan part of the Peining Railway is now completely under the control of the Japanese and the budget is being prepared upon the basis of a decreasing revenue, as the trade is being purposely diverted to the South Manchuria Railway. The Kirin-Hailung Railway has now been combined with the Kirin-Changchun Railway Administration, or in other words, under the control of the South Manchuria Railway. It was formerly a provincial railway. The Kirin-Changchun and the Kirin-Tunhua Railways which were previously under separate administrations have now been combined, and all employees above the rank of a chief of section are Japanese. There is a representative from the South Manchuria Railway who controls everything. The Director simply signs the documents. Trains now run all the way from Dairen to Kirin. The loan contract of $2,400,-000 for the Kirin-Tunhua Railway which has been dealt with in another memorandum,[4] has been combined with the loan contract of the Kirin-Changchun Railway to form one contract for $36,300,000 for a period of fifty years.

29. **Over the banks.** All the Chinese banks have been prevented from carrying on normal business. The North-eastern Provincial Bank, the Frontier Bank at Mukden, and the provincial banks of Kirin and Heilungkiang were all under Japanese military occupation, and their capital has been seized to form the Central Bank with its note issue of Yen 20 million. That and the Bank of Chosen and the Yokohama Specie Bank which circulate silver notes and yen notes have now the entire control of the financial life of the Three Eastern Provinces. They have agents in the remotest villages, and loans are granted to farmers against the delivery of grain so that even the crops of the following season are placed under Japanese control.

V. Concluding Remarks.

30. **The new order against real public opinion.** It is clear from the above description that with the exception of a handful of disgruntled politicians and political adventurers who are mostly of alien origin, the bulk of the population in the Three Eastern Provinces are against the new order. They are willing to admit that

[4] *Vide Memorandum on the Kirin-Hueining Railway.*

previous Chinese administrations could be healthier and stronger, but even these they are not willing to barter away for a condition of virtual slavery. In the meanwhile, the doors of the Three Eastern Provinces are being closed, and all the gibberish about the "principle of the Open Door and of equal opportunity" is becoming no more than "sound and fury signifying nothing", since foreign investments in the so-called "Manchukuo" must be made with previous Japanese sanction. Thus an imposing edifice is being erected in defiance of all international agreements and of all conceptions of civilized society which mankind through centuries of painful struggle has been trying to establish. The Chinese Government has, it must be admitted, shown the greatest forbearance and patience as it believes implicitly that a peaceful solution can be found based on the principles upon which many of the recent international treaties and the League Covenant itself were constructed. But it is clear that it cannot entirely, nor should, under the present circumstances, suppress the activities spontaneously and voluntarily sponsored by the people against the oppression which the Japanese choose to impose upon them.

31. **Forces of resistance.** General Ma Chan-shan in the northern part of the Three Eastern Provinces was among the earliest to unfurl the banner of resistance. And his call has been answered with telling effect by other generals, including General Li Tu and General Ting Chao, while large numbers of volunteers organized and recruited from the population of the invaded territory have been exerting their utmost to overthrow the Japanese yoke in the southern part of the Provinces.

32. **General Ma Chan-shan.** General Ma Chan-shan from the very beginning prevented the Japanese from

extending their influence to Heilungkiang, the province of which he is Chairman. "On the pretext of repairing the Nonni River Bridge," he said in his telegram of April 12th, 1932 to the Commissioners, "the Japanese began to invade the province of Heilungkiang and engaged in active hostilities against our troops. I immediately put up a stubborn resistance." That resistance has been sustained for the last nine months. Although, in a purely military sense, the resistance has not entirely achieved the object in mind because of the lack of sufficient military equipment and other conditions necessary for such achievement, the courage and determination and, above all, the intense spirit of patriotism which he has evinced, have given strength to, and moved the hearts of, millions of his fellow-countrymen in the common cause to resist the alien invader. A lonely figure in the wilds of this northernmost province of China, General Ma has an epic grandeur which inspired the heroes who fell before fearful odds in the battle-fields of Shanghai and its vicinity. Some day a history of this noble soul, representing the best and finest essence of the Chinese nation, will be written.

33. **Generals Ting Chao and Li Tu.** What is being done in Heilungkiang is also being done in Kirin Province under the leadership of Generals Ting Chao and Li Tu. An account of their activities was submitted in their own telegram to the Commission. "It is", they said, our duty to resist in self-defence. We have therefore formed the Kirin Self-Defence Army . . . and the Chinese Eastern Railway Guards." Their forces include "the 24th Brigade of Li Tu, the 28th Brigade of Ting Chao, the 22nd Brigade of Chao Ni, the 25th Brigade of Ma Hsi-chang, the 26th Brigade of Sun Wen-tsung, the 29th Brigade of Wang Zung-hwa, the Provisional First

Brigade of Feng Chu-hai, and the brigades under Kung Chang-hai and Yao Tien-chen." From early January of the present year, these soldiers, "determined to take Manchuria back by force under any circumstances irrespective of the length of time," have been fighting with the Japanese forces of occupation, and the latest information shows that their strength has been increasing, and because the movement is inspired by purely patriotic motives, it has given a stimulus to the volunteer armies which have been spreading all over the occupied territory.

34. **The Volunteers.** These volunteers have now appeared in the very midst of the regions where the Japanese are supposed to have the greatest influence. The South Manchuria Railway has been threatened and cut and the Nantai power station which supplies electricity to the Railway has been blown up. Yingkow, Koupangtse, the Tahusan-Tungliao Railway are all on the verge of being re-occupied by these volunteers. According to a recent Reuter message, their number within a radius of 20 miles of Mukden is between 5000 and 6000. "Whenever possible," the message in its earlier parts said, "the volunteers carry out their attacks at night. These tactics, it is said, are rendering the Japanese countermeasures tedious and difficult. . . . Hundreds of villages along the railway lines have been destroyed by Japanese planes on suspicion of harbouring volunteers."

35. **Resistance growing.** There is no question that these forces of resistance are growing. As opponents to a supreme wrong committed by one nation against another, they are fighting as champions of justice and will not suspend their activities unless they are assured that international treaties and the League of Nations are maintained and respected.

Peiping, August 5th, 1932.

APPENDIX.

A. LIST OF JAPANESE OFFICIALS IN THE "MANCHUKUO" GOVERNMENT.

I. Central Government.

Name of Organ	*Name of Official Position*	*Name of Japanese Incumbent*
Premier's Office (cabinet)	Director of the General Affairs Department	Tokuzo Komayi
"	Chief Adviser	Otsuhiko Yichiku
"	Director of the Interior Affairs Department	Sohiji Ohira
"	Counsellor	Hitakichi Nakajima
Department of General Affairs (cabinet)	Assistant Director of Department	Kiyichi Sakatani
"	Chief Secretary of the Department	Toyoji Minakawa
"	Chief of the Board of Personnel	Heiji Seko
"	Chief of the Board of Statistics	Toshiro Mukai
"	Chief of the Board of Accounts	Kokuye Murasumi
"	Chief of the Board of Supply	Noboru Kumamoto
"	Chief Accountant	Kokuye Murasumi
Councillor's Office	Chief Secretary	Shizao Arayi
Bureau of Law Codification	Director	Isami Matsumoto
Bureau of Political Training (formerly Bureau for Guidance in Self-Government)	Director	Koitsu Nakano
"	Chief of Section	Yaginuma
Bureau of Peace Preservation	Assistant Director	Jitsuryu Kikudake
Control *Yuan*	Director of the Department of General Affairs	Seitaro Yuigi

Name of Organ	*Name of Official Position*	*Name of Japanese Incumbent*
Ministry of Civil Affairs	Director of the General Affairs Department	Koitsu Nakano
"	Director of the Department of the Police	Masahiko Amakasu
"	Superintendent of the Police Training Service	Nishimura
"	Commander-in-Chief of the Frontier Police Inspection Corps	Nakamura
Ministry of Foreign Affairs	Vice-Minister	Tordayichi Ohashi
"	Director of the General Affairs Department	Shotoku Tajiro
"	Director of the Information and Intelligence Department	Torao Kawa
Ministry of Finance	Director of the Customs Revenue Department	Matsuzo Motoda
"	Adviser of Superior Rank	Rentaro Mizuno
Ministry of Communications	Director of the Railway Department	Seigen Morita
"	Director of the General Affairs Department	Chikaro Osachi
"	Director of the Postal Department	Yasuaki Fujiwara
"	Expert	Yoshizo Toyoda
Ministry of Industry	Director of the General Affairs Department	Kazuo Fujiyama
"	Director of the Department of Agriculture and Mines	Kan Matsushima
Ministry of Justice	Director of the General Affairs Department	Kenji Abiru
"	Director of the Judicial Department	Shigeji Kuriyama
Supreme Court	Adviser	Kenji Abiru
Central Bank	Vice-Governor	Kyoroku Yamanari
"	Trustee	Fukuo Takeyasu
Central Bank	"	Isoichi Washio
"	"	Yasuji Iarashi
Office of the Chief Executive	Chief Aide-de-camp	Tadashi Kudoh

II. Liaoning Provincial Government.

Name of Organ	Name of Official Position	Name of Japanese Incumbent
Office of the Provincial Government	Director of the General Affairs Department	Shoji Kanayi
"	Chief of the Bureau of Personnel	Nagao
"	Chief of the Police Department	Kiyoshi Mitani
"	Section Chief in the Police Department	Kozaka
"	Counsellor	Issho Kuroyanagi
"	"	Kotaro Yamazaki
Department of Industry	Chief of the Section on General Affairs	Tsunenori Takai
"	Chief of the Section on Mines	Shigeono Arayi
Department of Education	Chief of the Section on General Affairs	Yokichi Hirakawa
"	Chief of the Section on Schools	Ando
"	Chief of the Section on Social Affairs	Takemura
Department of Education	Adviser	Mitsugi Yirobe
"	"	Nobuhiko Oya

III. Other Major Organs in Liaoning Province.

Name of Organ	Name of Official Position	Name of Japanese Incumbent
Municipal Government (Shenyang)	Adviser	Toshio Goto
Bureau of Public Safety (Shenyang)	Chief of the Section of Special Affairs	Teiyichiro Tanaka
"	Attaché	Kameji Hashikami
Office of the Commissioner of Salt Transportation for the Three Eastern Provinces at Yinkow	Adviser	Kujiro Nagata
"	"	Kohji Kimura
Office of the Superintendent of Customs at Shanhai-	"	Shigeichi Ozawa
The Light and Power House (Shenyang)	Superintendent of the Light and Power House	Yoshio Oyiso

Name of Organ	*Name of Official Position*	*Name of Japanese Incumbent*
Peace Preservation Corps	Commander	Tsuyoshi Wata
"	Chief of Staff	Shin Miyamoto
Government Bank of the Three Eastern Provinces	Adviser	Masatoshi Sutoh
"	"	Tokusaburo Takeuchi
"	"	Tesuba Sakayi
"	Counsellor	Ishimatsu Kawakami
"	"	Yano
"	"	Sado Kurasaki
The Frontier Bank	"	Noriji Fukuta
"	"	Kenzo Shibata
"	"	Sakayo
Commission on Communications in the North-Eastern Provinces	"	Murakami
"	"	Kanayi
"	"	Yamaguchi
"	"	Satoh
"	"	Kojima
"	"	Yamamoto
Commission on Communications in the North-Eastern Provinces	"	Ogasawara
"	"	Ozawa
"	"	Hamamoto
The Shenyang-Hailun Railway Company	Chairman of the Board of Directors	Taisaku Kawamoto
"	Vice-Chairman of the Board of Directors	Nariyuki Morita
"	Adviser to the General Affairs Department	Tadashi Tanaka
"	"	Masatoshi Imazawa
"	Adviser to the Traffic Department	Masaji Ohashi
"	"	Yoshikore Ikebara
"	Adviser to the Accounts Department	Shokichi Tsugawa
"	"	Hatsutaro Kazama
"	Adviser to the Works Department	Churyo Inouye
"	"	Genichi Horie

Name of Organ	*Name of Official Position*	*Name of Japanese Incumbent*
The Shenyang-Hailun Rail-Way Company	Adviser to the Bureau of Railway Guards	Tsugimori Wata
"	"	Jiro Watase
"	"	Shoto Yoshikawa
"	Secretary	Miss Inouye
"	"	Miss Nakajima
The bogus "Fengtien-Shanhaikwan" Railway	Adviser	Yamaguchi
"	"	Furuyama
"	Member of the Administration	Yetchu
"	"	Kojima
"	"	Yamamoto
"	"	Ozawa
"	"	Kawano
"	"	Kato

IV. District Administrations in Liaoning Province.

Name of Place	*Name of Official Position*	*Name of Japanese Incumbent*
Yinkow	Directing Official	Kenkai Tsuki
"	"	Nobujiro Takatsuna
"	"	Yoji Itoh
Fu-Hsien	"	Umitaro Arakawa
"	"	Yasahi Fukui
"	"	Kunizo Samejima
Pan-chi	"	Sadao Nakajima
"	"	Masaichi Kohno
"	"	Tetsuo Sasaji
Chuang-Ho	"	Kansaburo Ohya
"	"	Hidetoshi Matsuzaki
Kai-Ping	"	Morinosuke Kageyama
"	"	Usaburo Sasayama
Antung	"	Nao Motojima
"	"	Saji Kanayi
Hai-Cheng	"	Shomei Kamata
"	"	Tashiji Kobayashi
"	"	Tsutome Kobayashi

Name of Place	*Name of Official Position*	*Name of Japanese Incumbent*
Kai-Yuan	Directing Official	Motoyoshi Takoyi
"	"	Tamio Fujiyi
"	"	Tetsuba Sawai
Fushuh	"	Hajime Tawahisa
"	"	Kitsuzo Yamashita
"	"	Yasushi Nakamura
Shenyang	"	Tatsuzo Nagao
"	"	Isaburo Yamagata
Teh-ling	"	Ryoryu Ishigaki
"	"	Seiji Kahi
"	"	Yeizo Suebiro
Feng-Cheng	"	Toshio Nakagawa
"	"	Kiyoshi Senpa
Chueh-Yuen	"	Kohachiro Matsuoka
"	"	Masashi Nakao
"	"	Ishiro Okamura
Liao-Yang	"	Shizuo Kojima
"	"	Morita Ohkushi
"	"	Teizo Sekiya
"	"	Kichijiro Nagata
Tao-Nan	"	Torao Satoh
"	"	Tomota
"	"	Teruo Takahashi
Huei-Teh	"	Yeijiro Takazuke
"	"	Hanyei Kiso
"	"	Kojima
Ly-Shu	"	Minoru Inouye
"	"	Masashi Nakagawa
"	"	Terafumi Murakami
Hsin-Min	"	Shigetoshi Takaoka
"	"	Takayuki Yamane
Chang-Tu	"	Noronobu Tara
"	"	Teiichiro Kohritsuya
"	"	Jiro Kawara
Chin-Hsien	"	Tatsuo Niwakawa
"	"	Tasuo Nishikisaki
Chin-Hsien	"	Ichiro Tanaka
Hei-Shan	"	Kensuke Zeidokoro
"	"	Heijiro Kondo
Liao-Chung	"	Masuro Sakakibara
"	"	Koichi Shioda
"	"	Masashi Kawamura
Fa-Ku	"	Hitojiro Nishioka

Name of Place	Name of Official Position	Name of Japanese Incumbent
Chin-Si	"	Masuki Uemura
Liao-Yuan	"	Tatsuki Nakazawa
"	"	Murataro Niigata
"	"	Nobu Yamashita
Yih-Hsien	"	Genjiro Murata
"	"	Takeo Tokira
Hsin-Wu	"	Toshio Nishizaki
Pan-Shan	"	Kanaji Ansai
"	"	Tatsuo Shikayoshi
Sui-Chung	"	Tetsuba Sawai
"	"	Ishiro Tanaka
Shing-Cheng	"	Matasuke Shigeoka
"	"	Naotaka
Pei-Chên	"	Naotaka Yotsumoto
"	"	Keizo Masuta
Si-An	"	Yusuke Yoshita
"	"	Kengen Nagaai

V. Kirin Provincial Government.

Name of Organ	Name of Official Position	Name of Japanese Incumbent
Office of Provincial Government	Director of the General Affairs Department	Haratake
"	Adviser	Mihashi
Department of Police	Chief of the Section of Special Affairs	Tsunoyasu Ito
"	Chief of the Section of Police	Ko Yasuta
Changchun Municipal Government	Adviser	Okuda

VI. Heilungkiang Provincial Government.

Name of Organ	Name of Official Position	Name of Japanese Incumbent
Office of Provincial Government	Director of the General Affairs Department	Murata
"	Adviser	Iwazaki
"	Adviser to the Department of Industry	Mizubuka
"	Adviser to the Finance Department	Haruta
"	Director of the Police Department	Tatsuo Nawakawa

N. B.—The above list only includes the names of the Japanese officials of first importance occupying positions in the Cabinet, the various *Yuan,* the different Ministries, the Provincial Governments of the Three Eastern Provinces, the Municipal Governments, the District Administrations and other organizations under the bogus Government of the "Manchukuo." Those of secondary importance and those whose names and positions have not yet been reported are not included. Many changes have been recently made in the personnel of the organizations, but if anything, it has been a change for a still greater number of Japanese officials than are found in the present list.

B. *PRINCIPLES FOR THE ORGANIZATION OF THE "MANCHUKUO" GOVERNMENT.*

1. Japan, with a view to maintaining and developing her rights in Manchuria, shall establish an organ of control, which shall unify all the existing organs, to direct the affairs of the "Manchukuo" The organ shall, in addition, be given sufficient power so that the new state may rely upon the Japanese for guidance and follow the lines of development as laid down by Japan who, on her side, shall exercise her control with sincerity and goodwill in order that the new state may not have any feelings of distrust. Our attitude should be just and fair.

2. The form of the government shall not be republican. Established on the basis of the "kingly principle," the new state shall adopt a monarchical form of government.

3. The new state shall exercise absolute authority over its internal affairs, and under the direction of Japan, it shall carry out policies based on monarchical principles.

4. With a view to realizing the above object, Japan shall provide the royal family ruling over the new state with sufficient military power to control Manchuria and Mongolia.

5. The political system of the new state and its various organizations shall be modelled after those of Japan with modifications according to local conditions in the new state. This principle holds good not only for organizations of a political nature, but also for social institutions as, for instance, customs and usages, which should be gradually assimilated with those of Japan. In the execution of this principle, however, simplicity and practica-

bility should be emphasized so as to avoid the complexity of Japanese laws.

5 The Chinese people do not have the same sense of loyalty to the Imperial House as the Japanese. The authorities should therefore be severe in their rule of the people so that they may have reverence and awe towards their rulers.

7. If the necessity of drafting the Constitution arises, the Constitution shall be modelled after that of Japan. The representative assembly shall be nothing more than an advisory council and an organ to publish statements respecting the finance of the state. (Toward the last decade of the Tsing Dynasty, the evils of the Advisory Council which was established to prepare for a constitutional government were too obvious).

8. Legislative and executive powers shall reside with the supreme authorities, but the judicial power shall be respected in accordance with laws so that its independent spirit may be maintained.

9. With regard to the laws, Japanese laws, and particularly Japanese procedural laws, shall be made use of. As to the laws of the family, the authorities shall take into consideration the special customs and usages of the population. In cases where foreign nationals are involved, it seems better that Japan should assume the responsibility of supervising the final decision.

10. All executive acts shall be performed in accordance with the existing customs and usages, and simplicity and practicability shall be emphasized. The complexity of Japanese local organizations shall be avoided and the corruption of officials shall be eliminated.

11. The police force of the new state shall be under the control of a strong organ which shall also be given the power of bandit suppression. The force shall be organized on a semi-military basis and well distributed in the state so that under strict supervision and able direction bandit activities may be eliminated.

12. The national defence of the new state shall be left entirely to Japan. In order to complete its nationhood, the new state is not yet in a position to defend itself from China and Russia under its present conditions. Moreover, the national defence of the state coincides with that of Japan; therefore, under no circumstance, shall the new state have the right of defending itself.

13. The foreign relations of the new state shall be entrusted to Japan so that serious diplomatic blunders may be avoided and national safety assured.

14. The military forces shall be maintained to a degree sufficient to suppress internal disorders. The royal family ruling over the new state shall directly command the forces so that the symbol of authority may be deeply impressed upon the imagination of the people. But the ultimate strength of the new state resides in the military power of Japan, so that there is no necessity of maintaining a large force. Whenever the need of suppressing internal disorders arises, the new state can always apply to Japan for aid, because she has special rights and duties of maintaining order in Manchuria.

15. The South Manchuria Railway shall have the exclusive right of railway enterprises in Manchuria. The new projected railways shall be always in the form of joint enterprises so that the spirit of the idea of "mutual dependence to promote mutual glory" may be realized.

16. As regards the right of existence in Manchuria, the Japanese people shall enjoy the same rights and privileges as the nationals of the new state. Besides concluding treaties with the new state to that effect, Japan shall devise means to facilitate Japanese immigration.

17. Because of many complications, Japan will consider carefully the desirability of appointment of Japanese people as the officials of "Manchukuo." But in any case, Japan shall exercise actual control and disciplinary supervision in the execution of the basic policies of the government in order to prevent corruption and achieve political success. The present organization should be submitted for revision, but in the selection of the personnel, if sufficient care is not taken, there will be grave consequences. Japan shall therefore take very serious consideration of the matter.

18. With regard to military organs, there shall be careful selection of Japanese people to enforce discipline and supervise the training of the soldiers. As regards matters relating to the distribution of Japanese officials in the various executive organs, to their appointment and dismissal, their control and supervision, the new unified Japanese organ in Manchuria shall assume full responsibility of deciding upon them.

MEMORANDUM

ON

The Japanese Seizure of the Chinese Postal Administration in the Three Eastern Provinces

Document No. 27 Peiping, August 1932

MEMORANDUM ON THE JAPANESE SEIZURE OF THE CHINESE POSTAL ADMINISTRATION IN THE THREE EASTERN PROVINCES.

---

1. Japanese troops seizing Liaoning office, September 19th, 1931.
2. Seizure of branch offices at Shenyang same day.
3. First class office at Yingkow seized same day.
4. Hopei branch office at Yingkow seized same day.
5. Liaoning office telegraphic codes taken, September 20th.
6. Interference with offices at Yingkow, Yungyoh, Ssupingkai, etc. same day.
7. Liaoning postal inspectors arrested, September 23rd.
8. Censorship of mails in Liaoning and branch offices, October 2nd.
9. Attempt to detain funds of Liaoning office, October 9th.
10. Changchun office Commissioner imprisoned, October 24th.
11. Establishment of Japanese military post office since November 13th.
12. Japanese establishing air-mail service in December.
13. Arrest of Lishan postmaster, December 8th.
14. Fakumen postmaster in difficulty, December 29th.
15. Report of Mukden branch office on Japanese plans, January 29th, 1932.
16. Disloyalty of Japanese employees of Chinese Post Office reported on February 24th.

17. Japanese bribing Chinese postal employees, February 28th.
18. *Mukden Daily News* carries a report, March 2nd.
19. Exclusive use of Japanese postage stamps on Chinese mails.
20. Cruelties on Chinese postal employee, March 7th.
21. Post offices compelled to hoist "Manchukuo" flags.
22. Tokyo representatives investigating postal service, March 20th.
23. Antung office searched and codes taken, March 28th.
24. Seizure of Kirin-Heilungkiang office, March 29th. Mr. Poletti's protest, April 2nd.
25. Schemes to assassinate Mr. Poletti.
26. Forcing loyalty to "Manchukuo" upon Mr. Poletti.
27. Another postal employee imprisoned, April 7th.
28. "Instructions" from the bogus Communications Ministry.
29. Another postal employee imprisoned, April 24th.
30. Murder of Ngemo postmaster, May 27th.
31. Japanese military post offices established.
32. Japanese gendarmes arrested Chinese employees, June 15th.
33. Interference with the office at Suihwa, June 21st.
34. Ching Chen postmaster captured, July 13th.
35. Japanese appointing postal inspectors, July 13th.
36. Japanese threat to Commissioner Poletti. Pressure upon Deputy Commissioner Liu Yao-ting.
37. Bogus administration issued stamps. Great difficulties in maintaining the service.
38. Japanese gendarmes searched administrative office and residence of Deputy Commissioner.
39. Employee of service deprived of freedom.

MEMORANDUM ON THE JAPANESE SEIZURE OF THE CHINESE POSTAL ADMINISTRATION IN THE THREE EASTERN PROVINCES

The following are the facts concerning the seizure of the Postal Administration of the Three Eastern Provinces by Japan:

1. **Japanese troops seizing Liaoning office, September 19th, 1931.** On September 19th, 1931, at 2.45 a. m., over 20 Japanese troops broke open the gate of the Liaoning office at Shenyang and then entered the office. As soon as the Commissioner Mr. Poletti and the Deputy Commissioner Mr. Liu Yao-ting heard the news, they went straight to the office but were both forced to surrender. One postal motor truck and nine postal bicycles were taken away. The gate-keeper, Kin Kuo-chun, was arrested and badly beaten. He recovered only after receiving medical treatment for several days. Protests were later lodged with the Japanese troops, and the motor truck was returned on October 4th, 1931, two of the bicycles were found from among the Japanese residents, and the other seven bicycles have not yet been returned up to the present.

2. **Seizure of branch offices at Shenyang same day.** On the morning of September 19th, 1931, the Japanese troops entered the branch post office at the Great West Gate of Shenyang, and sealed its safe with the label of the Japanese military headquarters. Its key was taken away. Not until the 24th of the same month did the chief of the Japanese gendarmerie at that place

agree to have that safe opened, but the post office was forced to fly the Japanese flag at the gate. It was not taken down until June, 1932.

3. **First class office at Yingkow seized same day.** On September 19th, 1931, the gate of the first class post office at Yingkow was also guarded by several Japanese soldiers under instructions from their officer, and automatic rifles were placed on the counter for demonstration. According to the military officer's statement, he was instructed by his higher authorities to close the post office, and without special permission, no mails could be delivered. At 2.30 p.m. on the same day, the Japanese troops began to withdraw. The secretary of the Japanese Consulate declared that only mails would be accepted, but even these could not be delivered without having first the permission of the Japanese military authority. 36 bags of mails were received from other places, and all of them were detained by the Japanese post office. Through repeated negotiations with the Japanese Consulate, the Japanese post office agreed to release the mails on the following day after they had been either censored or opened.

4. **Hopei branch office at Yingkow seized same day.** On September 19th, 1931, the Hopei branch post office at Yingkow was occupied by the Japanese troops. All mail service stopped and was not resumed until March 1st.

5. **Liaoning office telegraphic codes taken, September 20th.** On September 20th, 1931, at 10 a.m., Captain Murasakishiba of the Kwantung Garrison Headquarters went to the Liaoning post office with two armed guards for a search and demanded the Commissioner to hand over the office's telegraphic codes to him. He took away the secret codes for two days.

6. **Interference with offices at Yingkow, Yungyoh, Ssupingkai, etc. same day.** On September 20th, 1931, a Japanese named Yiwasa calling himself a member of the staff of the Japanese Consulate at Yingkow went to the first class post office there to investigate the stamps, the sources of the public funds and the ways of their disposal. At the same time he ordered that public funds were not to be remitted to Liaoning post office or to any other place. The chief of the Japanese post offices at Yungyoh and Ssupingkai went to the Chinese post offices in those places to study mail transactions and their financial conditions. The Japanese garrison troops also investigated a number of mails in the post offices at Kungchulin, Taonan, Fencheng and Changpeh.

7. **Liaoning postal inspectors arrested, September 23rd.** On September 23rd, 1931, the Liaoning postal inspectors, Chang Yung-fuh and King Shao-tien, went on duty to the first class post office at Yingkow with official passes identifying them as the postal inspectors and also with a letter of introduction in Japanese from a Japanese employee, Tanaka Kango, of the Chinese post office to the Japanese police bureau. Upon arriving there they were arrested together with the chief of that post office Wu Shu and taken to the Japanese police station for trial. Through the negotiation of Tanaka Kango on the telephone, they were released after being detained for half a day; but Chang Yung-fuh fainted after a brutal trial. The serious injuries which he sustained did not recover for many days.

8. **Censorship of mails in Liaoning and branch offices, October 2nd.** On October 2nd, 1931, the Japanese gendarmes arrived at the Liaoning post office and censored all the mails. Thence they went to the first class, sec-

ond class and third class post offices for the same measure. All the mails either opened or detained were stamped with the name of the Japanese gendarmerie. Since April 18th, 1932 the Japanese gendarmes for censoring the mails have changed their uniforms and are now in plain clothes. All the mails opened and detained were then stamped with the name of the Bureau of Safety of the "Manchukuo." The Japanese Government does not consider its invasion of the Three Eastern Provinces as an act of formal international warfare nor does the Japanese army consider Shenyang as a military area under martial law. Yet, in spite of protests lodged with the Japanese Consul-General and the military headquarters by the Commissioners of the Kirin and Heilungkiang post offices regarding the illegal censorship of mails, there has been no result.

9. **Attempt to detain funds of Liaoning office, October 9th.** On October 9th, 1931, Captain Kawamato, the chief of the Japanese sub-police station in the west part of Shenyang city, went to the Liaoning post office with three Japanese gendarmes demanding the Commissioner, Mr. Poletti, that no public funds or cash shall be deposited with the local bank or transmitted to places outside of the Three Eastern Provinces. After three hours' protracted explanation on the need of such postal remittances, the gendarmes waived their demand.

10. **Changchun office Commissioner imprisoned, October 24th.** On October 24th, 1931, the Commissioner of the Changchun post office was condemned by the Japanese gendarmerie as a reactionary because he lodged protests with the headquarters against the illegal censorship of Chinese mails. He was then detained in a cold cell for thirty-six hours. All the orders despatched

by the Kirin post office were taken away by the gendarmes.

11. **Establishment of Japanese military post office since November 13th.** On November 18th, 1931, stamps bearing the word "military" were seen at Liaoyuan. On the 26th, a military post office was established in a Japanese firm for the sale of that kind of military stamps. The Liaoning post office immediately ordered a Japanese member of the staff to make an investigation into that matter. It was reported that since November 13th, the Japanese troops had already established a military post office on the following basis: (1) All the ordinary mails not over 20 grams each which are despatched by the Japanese soldiers to Japan are free of charge; (2) The same privilege is not granted to the mails despatched from Japan to the soldiers in active service; (3) All mails or parcels which are of an official nature and despatched by the soldiers are also free of charge.

12. **Japanese establishing air-mail service in December.** During the first part of December, 1931, the Japanese established a Sino-Japanese air service at Mukden with a view to installing a postal air service in Mukden, Pinkiang, Lunkiang, Taonan, Ssupingkai, etc.

13. **Arrest of Lishan postmaster, December 8th.** On December 8th, 1931, the Japanese gendarmes who were then stationed at Lishan arrested the local postmaster and sent him to Anshan for unlawful trial, because he had delivered the *Ta Kung Pao* of Tientsin. The post office of Liaoning sent Komatsu Koichi a Japanese official to explain the matter, and the postmaster was then released.

14. **Fakumen postmaster in difficulty, December 29th.**

On December 29th, 1931, the Postal Commissioner of Liaoning secretly reported that Japanese troops, during their invasion of Fakumen, forced the local postmaster together with the gentry to sign a declaration to the effect that the conduct of the Japanese troops was beyond criticism and that he and the gentry should report to the Japanese troops the military movements of Chinese soldiers.

15. **Report of Mukden branch office on Japanese plans, January 29th, 1932.** On January 29th, 1932, the Postal Commissioner of Liaoning received a report from Ma Tuh-ying, postmaster of the branch office of the Mukden station of the South Manchuria Railway to the following effect: "A former postal employee of Tientsin, named Tan Kuyan went to Mukden upon the request of Dohihara to investigate the possibility of seizing the postal service in the Three Eastern Provinces. This Tan has formulated plans to be presented to the headquarters of the Kwantung Army at Mukden."

16. **Disloyalty of Japanese employees of Chinese Post Office reported on February 24th.** On February 24th, 1932, Tanaka Kango, a Japanese employee of the Liaoning post office, reported to Mr. Poletti, Postal Commissioner of Liaoning, that Sakuray, Japanese postmaster at Dairen, told Tanaka in person that some postal employees of Liaoning were thinking of giving up their duties and leaving the district and that the frontier authorities (meaning Japan and the bogus Government) would take necessary measures to prevent them from going away. Tanaka further asked the opinion of the Postal Commissioner what, if the bogus government would pay him in cash the money he should receive as retiring allowance after his many years of service in

the Chinese Postal Administration, would be his attitude with regard to the plan of seizure of the postal service by the bogus government. (Tanaka became a renegade on April 1st and joined the postal service of the bogus government to seize our Postal Administration in the Three Eastern Provinces.)

17. **Japanese bribing Chinese postal employees, February 28th.** On February 28th, 1932, Ma Tuh-ying, Postmaster of the branch office of the Mukden station of the South Manchuria Railway secretly reported to Mr. Poletti, Postal Commissioner of Liaoning, that a certain Japanese sub-postmaster bribed him to report to No. 5000 of the Japanese local telephone service as regards conditions of the postal service. Later it was found out that the number of this telephone belonged to the local Japanese postmaster, Okibe.

18. ***Mukden Daily News* carries a report, March 2nd.** On March 2nd, 1932, the Japanese-owned *Mukden Daily News* published a news item that the Japanese postmaster at Mukden would take over the Postal Administration of the Three Eastern Provinces. The Japanese postmaster did not deny the report.

19. **Exclusive use of Japanese postage stamps on Chinese mails.** On March 2nd, 1932, the Chinese Directorate of Posts at Shanghai received from the Japanese post office at Dairen registered and ordinary mails delivered by the Japanese post offices at Changchun, Antung, Mukden, Yingkow, and Liaoyang with only Japanese postage stamps. According to the Sino-Japanese Postal Convention of 1910, these mails should be also affixed with Chinese postage stamps. What was done was therefore a violation of the said Convention.

20. **Cruelties on Chinese postal employee, March 7th.**

On March 7th, 1932, Chao Chang-chen an employee of the Liaoning post office was arrested and detained by the Japanese gendarmes because he wrote to his friends at Peiping slightly touching upon the Japanese aggression in the Three Eastern Provinces. He was released after a few hours' detention upon signing a pledge not to commit any further indiscretion. Later, the Ta Kung Pao published the news, and on March 24th, he was arrested and tried through torture to tell the names of those who were engaged in the anti-Japanese movement in the Postal Administration. He was further threatened that if he revealed the trial, he would be put to death. He was released after being detained for 24 hours. The Postal Commissioner of Liaoning, Mr. Poletti, reported the case to the Consuls of the United States, Great Britain, France, and Japan. The Japanese Consul explained that he could render some assistance to the case by hushing it up, because Chao would be in greater danger if the case were to be reported to the high Japanese military authorities. If the gendarmes knew that Chao had revealed his trial they would take appropriate steps to punish him most severely.

21. **Post offices compelled to hoist "Manchukuo" flags.** On March 9th, 1932, when the so-called chief executive of the bogus government took office, the post offices of Liaoning were obliged to hoist the flags of the bogus government and to affix special seals on the mails as a mark of respect.

22. **Tokyo representatives investigating postal service, March 20th.** On March 20th, 1932, four high officials of the Japanese Ministry of Posts and of the Japanese post office at Dairen, Fuji, Midzukami, Yoshiwora, and Nakao, came to the Three Eastern Provinces to in-

vestigate the conditions of the Chinese and Japanese postal services. The Japanese employee of the Chinese postal service Tanaka told Mr. Poletti, Postal Commissioner of Liaoning, that they did not come for a friendly visit but with more serious ideas. Their purpose was to investigate the local postal condition and report to Tokyo. They had come to investigate for the second time as to how the dual system of postal service might be eliminated.

23. **Antung office searched and codes taken, March 28th.** On March 28th, 1932, the Antung first class post office was searched by the Japanese gendarmes, and its postmaster was arrested and released after a few hours' detention. During his detention he was repeatedly questioned as to the reason why Chien Chu-fan, postal employee of the Antung office, was transferred elsewhere. On May 6th, telegraphic codes in the English language belonging to the Antung first class post office were forcibly taken by the Japanese gendarmes and returned after six hours. It was reported from reliable sources that the codes were photographed.

24. **Seizure of Kirin-Heilungkiang office, March 29th. Mr. Poletti's protest, April 2nd.** On March 29th, 1932, Shimizu, Junji and Fuziwara, high officials of the Ministry of Posts at Tokyo, came to Mukden. It was said that they came to advise the Japanese Headquarters of the Kwantung Army to seriously consider the seizure of the Postal Administration of the Three Eastern Provinces. But on March 31st, Tanaka, the Japanese employee, left the Chinese postal service to receive the appointment of the bogus government. On April 1st, he led about 20 policemen to the post office of Kirin and Heilungkiang at Pinkiang to force Mr. Smith, Postal

Commissioner, to hand over the control of his office. On the same day, the bogus government sent a number of representatives to take over control of the post office. Among them was a Japanese named Komatsu Koichi, who was also an employee of the Chinese postal service, but he refused to act as the bogus government ordered. This shows that Japanese with some sense of justice could not be persuaded to perform a mean task. The representatives sent to take over the control were headed by a puppet Chinese Yui Chen-tu, who was followed by four Japanese, Miki Taku, Kusano Jiro, Koyizumi Kosuke and Nara Hidaji.

On April 2nd, 1932, when Yui Chen-tu carried on negotiations with Mr. Poletti, the latter handed out a protest written in English. Yui Chen-tu could not understand English, but his followers conferred in Japanese, saying that people in the local Japanese post office could understand English. As Mr. Poletti understood Japanese, he knew what the remark meant. Moreover, these representatives were always found going in and out of the local Japanese post office, which was undoubtedly where they had their headquarters.

25. **Schemes to assassinate Mr. Poletti.** From April 1st, 1932, the representatives of the bogus government every day threatened the Postal Commissioner to hand over the postal service of the Three Eastern Provinces and ordered him to stop making remittances to the Directorate. But on the 27th, the Postal Commissioner offered to stop the postal money order service entirely as a means of protest, and the Japanese gave up that threatening attitude for a while. The postal money order service, which was actually suspended from April 1st, was resumed from April 28th. The officials returned to

Tokyo to report to the Ministry upon the failure of the seizure. At the time when Mr. Poletti refused to hand over the Postal Administration and when the situation was very serious, some Japanese lawless elements plotted against his life. On April 25th, a Chinese of the Mukden Public Safety Bureau secretly informed Mr. Poletti that the director of the police at Mukden, a Japanese named Mitani Kiyoshi, would send Japanese lawless elements to assassinate him because he had obstinately refused to hand over the control. Thereupon, Mr. Poletti personally protested to the puppet Governor of Fengtien and at the same sime reported the matter to the local consular body and the Italian Commissioner of the League of Nations Inquiry Commission.

26. **Forcing loyalty to "Manchukuo" upon Mr. Poletti.** On April 15th, 1932, the representatives from the bogus government brought with them a pledge or a declaration of oath and forced the Postal Commissioner to sign in order to show his loyalty and obedience to the "Manchukuo." Mr. Poletti refused.

27. **Another postal employee imprisoned, April 7th.** On April 7th, 1932, a Chinese employee of the Postal Administration, Chang Ssu-wen, had a slight quarrel with the Japanese gendarmes who had been sent to his office as mail censors. He was arrested the next day by the Japanese gendarmes in front of his own residence. He was put in a motor car and disappeared. Mr. Poletti asked the Japanese Consul-General and the Japanese military and police authorities to investigate the matter, but they all replied that they had no knowledge of the matter. Chang was released on May 1st after a detention of 23 days.

28. **"Instructions" from the bogus Communications**

Ministry. On April 20th, 1932, the authorities of the bogus government ordered the Postal Commissioner, Mr. Poletti, to change the date in the postal seal into the "first year of Tatung." On May 5th, the Ministry of Communications of the bogus government formally notified the Liaoning and Kirin-Heilungkiang post offices to carry out the order.

29. **Another postal employee imprisoned, April 24th.** On April 24th, 1932, Chen Chuan-chi, a postal official, came to the Yamato Hotel situated in the railway area of the South Manchuria Railway to visit his friend Koo Chi-chun (secretary in the Chinese Assessor's office). After a brief conversation of a few minutes with his friend, he came out of the hotel but was immediately arrested by Japanese plain clothes police. He was detained in the Japanese police station for four days, and it was only after Mr. Poletti negotiated with the Japanese Consul-General and the police department that he was released on the 28th.

30. **Murder of Ngemo postmaster, May 27th.** On May 27th, 1932, the Postal Commissioner reported that the postmaster of Ngemo, Yang Chi-chen, was kidnapped by the local Japanese soldiers and murdered at Tunhua. The post office at Fanchen was sealed by the Japanese troops.

31. **Japanese military post offices established.** On June 10th, 1932, the Postal Commissioner reported that after the Japanese occupation of the Three Eastern Provinces by force, a number of Japanese military post offices have been established. Owing to the lack of exact information at the moment, report could not be made with regard to this matter.

32. **Japanese gendarmes arrested Chinese employees, June 15th.** On June 15th, 1932, the Postal Commissioner of Liaoning, Mr. Poletti, reported that on the same day there came some Japanese gendarmes with two interpreters stating that some employees of the postal service had been discourteous in talking to them over the telephone and they had to arrest them. Effort was then made to take the matter up with their superior officer. But during the moment of confusion, three men were taken away by them. The Commissioner, Mr. Poletti, again reported on June 17th that the three men arrested by the Japanese were employees of the service, namely Chao Tien-kuo, Yung Cheng-chin and King Yun-yuan. They were all subjected to torture and one of them, Chao Tien-kuo, was almost choked to death by a rope tightly put around his throat. They were later released.

33. **Interference with the office at Suihua, June 21st.** On June 21st, 1932, Captain Terata of the Japanese general staff sent a Korean named Tang Fu-li to the post office at Suihua, alleging that the bandits were active and it was necessary to borrow some postmen to do the spying work. Although the request was courteously declined, the Japanese compelled the office to hire a Chinese for him and took away by force a complete uniform of the postman.

34. **Ching Chen postmaster captured, July 13th.** On July 13th, 1932, the Kirin-Heilungkiang Postal Commissoiner reported that the postmaster at Ching Chen named Tsui Siu-ting was captured by the Japanese after the city fell into their hands, but was later released.

35. **Japanese appointing postal inspectors, July 13th.** On July 13th, 1932, inspectors appointed by the bogus

government, namely, Nakamura Kuhira and Togura Katsujiu, arrived at Shenyang. A Chinese named Tan Ko-yuan who had been dismissed from the postal service at Tientsin on a charge of smuggling morphine and smoking opium was also appointed by the Japanese. The three inspectors appointed for the territory of Kirin and Heilungkiang, namely, Tsung Yu-ching (Chinese), Yito Tasuke, and Nakajim Ugemi also arrived at the Pinkiang office on July 15th.

36. **Japanese threat to Commissioner Poletti. Pressure upon Deputy Commissioner Liu Yao-ting.** On July 13th, 1932, the Japanese gendarmerie at Shenyang summoned an employee of the postal service named Tuan Ching-jen to the sub-station of the gendarmerie and forced him to furnish inside information of the service. He was also told by the Japanese that they would drive by force the Commissioner Mr. Poletti out of the territory and compel the Deputy Commissioner, Mr. Liu Yao-ting, to swear his allegiance to the bogus government so that he could succeed Mr. Poletti as Postal Commissioner. In the afternoon of July 15th, the bogus inspector Togura Katsujiu secretly induced Liu Yao-ting to surrender himself to the bogus "Manchukuo" in order to take Mr. Poletti's place and promised him various privileges after the change. Finding himself under the heavy pressure of the Japanese, Mr. Liu then got permission from the Commissioner secretly to flee across Shanhaikwan. The Commissioner, Mr. Poletti, then lodged a strong protest with the local Japanese Consul regarding this matter.

37. **Bogus administration issued stamps. Great difficulties in maintaining the service.** On July 16th, 1932, the bogus inspector began to send new stamps to the

Liaoning and Kirin-Heilungkiang post offices and distributed them to the post offices of various grades. They set August 1st as the date for the new stamps to be used. On July 18th, the Japanese seized the savings department of the postal service. Repeated protests from the Chinese postal authorities were all in vain. Since the incident of September 18th, 1931, the authorities of the Postal Administration of the Three Eastern Provinces have, for the interest of both the Chinese and foreigners, gone through all kinds of difficulties and sufferings for a period of more than ten months in order to maintain the postal service. But finally, under instructions from the Directorate, they were compelled to declare on July 24th and 25th a complete suspension of the service. The whole personnel of the service was moved inside Shanhaikwan. Strong protest was duly sent to the local Japanese Consul, and the British and American Consuls were informed of the development of the real situation. The new stamps issued by the bogus government were then put in use by the inspectors on the 26th.

38. **Japanese gendarmes search administrative office and residence of Deputy Commissioner.** On July 25th, 1932, the Japanese gendarmes at Shenyang entered, without permission, the premises of the Liaoning post office and the residence of the Deputy Commissioner Liu Yaoting. They made a careful search and tore off the notice by the postal commissioner regarding the suspension of the service. This may serve as a positive proof as to how the Japanese military participated in the plan of the bogus government in seizing the Postal Administration of the Three Eastern Provinces.

39. **Employee of service deprived of freedom.** On

July 27th, 1932, according to the report of Mr. Poletti, the chief of the section of registration of his office, Mr. Hsu Kwai-lun, was arrested by the Japanese gendarmes and forced to furnish a list of names and addresses of the employees of the postal service so that their movement might be watched over and prevented from leaving the service. He was released only after a strong protest with the local Japanese Consul by Mr. Poletti. Up to the end of July, most of the employees of the postal service of the Three Eastern Provinces were deprived of their freedom and few of them could escape.[1]

Peiping, August 17th, 1932.

---

[1] The bogus postal commissioner Fujiwara had repeatedly told Mr. Poletti that those employees who were unwilling to work under the Japanese might be allowed to leave. No force would be used to prevent them from doing so. But when the employees made known their intention of leaving the service, the Japanese police and gendarmes in the Three Eastern Provinces interfered with their movements and forced them to remain in the service or compelled them to swear allegiance to the bogus "Manchukuo."

MEMORANDUM

ON

JAPANESE SEIZURE OF SALT LOAN FUNDS IN THE THREE EASTERN PROVINCES

Document No. 28 Peiping, August 1932

MEMORANDUM ON JAPANESE SEIZURE OF SALT LOAN FUNDS IN THE THREE EASTERN PROVINCES.

1. Extent of Japan's seizure of the salt administration.
2. Failure to regain loan quota and special surtax.
3. Manchurian authorities reported to set aside portion of salt revenue.
4. Japan making incorrect representations.
5. Japan disclaiming responsibility.
6. Testimony for Japanese military interference everywhere.
7. Japan misrepresenting loans secured on salt.
8. Chinese Government's record in maintaining obligations.

MEMORANDUM ON JAPANESE SEIZURE OF SALT LOAN FUNDS IN THE THREE EASTERN PROVINCES [1]

1. **Extent of Japan's seizure of the salt administration.** The Chief Inspectors of Salt Revenue, who are charged with the responsibility for maintaining the loan service, advise that since the forcible seizure of the Salt Administration in the Three Eastern Provinces around April 1st, no funds whatever for loan service have been remitted from the Three Eastern Provinces. Even the foreign loan quota and surtax have been seized, along with the rest of the revenue, in spite of statements that Manchuria's share of loan payments would be forthcoming. Japan's action materially prejudices the interests of holders of foreign loans secured upon the salt revenue, at the same time impairing the ability of the National Government to meet its other commitments.

The foreign loan quota was paid from September 1931 to March 1932, inclusive, at the rate of $217,800 monthly, plus the proceeds of the special loan surtax of $0.30 per picul. The total under these heads amounts to over $300,000 monthly. The arrears of loan quota and special surtax for the five months from April to August inclusive are estimated to be over $1,500,000.

2. **Failure to regain loan quota and special sur-tax.** Immediately following the dispossession of the Salt Inspectorate in these Provinces, the Bank of China at the

[1] *Vide* also *Memorandum on Japan's Seizure of the Salt Gabelle in the Three Eastern Provinces.*

instance of the Chief Inspectors endeavoured to arrange for the continued remittance of at least the loan quota and special surtax, but the Bank's efforts were fruitless. Subsequently, a representative of the Chief Inspectors was sent to Mukden and Changchun with instructions to endeavour to arrange for these remittances. His mission was equally fruitless.

3. **Manchurian authorities reported to set aside portion of salt revenue.** While the Manchurian authorities claim to have been setting aside a certain sum out of the salt revenue for the purpose of eventually contributing to the service of foreign loans, available information indicates that the amounts claimed to be set aside fall considerably short of Manchuria's fair share. However, even if funds are being set aside such action avails nothing if in practice such funds are not forthcoming to assist in meeting the loan payments as they fall due.

4. **Japan making incorrect representations.** In this connection, it is necesary to refute certain incorrect representations made by the Japanese Government concerning (1) Japan's responsibility, and (2) the position of loans secured upon the salt revenue.

5. **Japan disclaiming responsibility.** (1) Under date of April 14th, 1932, the Japanese Minister addressed a letter to the Chief Inspectors of Salt Revenue, replying to their letter of April 1st, with respect to forcible measures affecting the position of the Salt Inspectorate in the Three Eastern Provinces. The Japanese Minister stated that he had no information concerning interference with the functions of the Salt Inspectorate offices at Newchwang, and that his Government had no concern with it. The Minister stated further:

"No Japanese authorities have either motivated

or given sanction to the forcible occupation of a Salt Inspectorate Office or to any activities of the Manchukuo authorities, and the Japanese Government are not in a position either to hold themselves responsible for, or to put any restraint upon, what the Manchukuo authorities do.''

6. **Testimony for Japanese military interference everywhere.** Officials of the Inspectorate are prepared to support by personal testimony that the initial acts of armed visitation, inspection, intimidation and interference which occurred at the Newchwang Inspectorate on and subsequent to September 19th, 1931 were taken or dictated by Japanese military officers; that the interference which later occurred at Changchun on, and subsequent to, November 6th, 1931 was also taken or dictated by Japanese military officers; that the restraint upon the funds of District Inspectors lodged at the local banks was imposed by Japanese military officers; that the subsequent withdrawal of these funds for local use was demanded and forcibly effected by a Japanese subject with the direct support of Japanese military officers; that the receipt for the sum of $672,709.56 initially expropriated at Newchwang on October 30th was signed by that Japanese subject alone; that the subsequent restraint and other irregular acts leading to the expropriation, to the present date, of salt revenue funds whose amount may be estimated at about $14,000,000 have all been carried out or dictated by Japanese subjects; and that the forcible occupation, closure, and sealing of the Newchwang District Inspectorate which occurred on April 15th, ostensibly at the demand of the Salt Commissioner (who fled to Darien), were similarly conducted. It is clear, moreover, that there is a direct connection between all these acts, whether taken in the name of the Japanese military authorities, the ''Peace Preservation Committees'' insti-

tuted and directed by Japanese advisers, or persons purporting to represent the so-called Manchukuo authorities.

It may further be pointed out that the Chief Inspectors' efforts to withdraw salt revenue from Newchwang to Shanghai for the express purpose of meeting payments urgently required for foreign loan service, at the time when the Newchwang revenues were solely under Japanese military restraint (that is to say, long before the creation of Manchukuo), were directly frustrated by the same military, who persistently refused to consider the proposals, accompanied by safeguards, which were advanced by the Chief Inspectors' representatives in the interests of the bondholders. There is little doubt that if these and subsequent funds had been released in accordance with the reiterated requests made to the Japanese authorities then in direct and open control of the situation, the subsequently constituted "Manchukuo authorities" would not have ventured to embark upon new measures disadvantageous to the security of the loans.

In the circumstances, it is quite clear that the Japanese Government is responsible for the acts stated.

7. **Japan misrepresenting loans secured on salt.** (2) The Japanese Minister's letter contains the following further statement:

> "In November 1928, the Chinese Government, despite the Japanese Government's protest, and in contravention of the stipulation of the Reorganization Loan Agreement, unilaterally introduced a change in the established practice of allocating the whole of the salt revenue to the service of the loans secured on that revenue, and since then only a certain fixed amount forming a small part of the revenue has annually been devoted to loan repayment, while the greater part thereof has been appropriated by the central and local governments, with the result that

> the repayment of the loans secured on the revenue has often been in arrear and that the loans secured on the surplus of the revenue has since been entirely in default."

This statement involves a misapprehension of the situation, concerning which it is necessary to comment.

There are now three foreign loans secured upon and being paid from the salt revenue. They are: (1) the Anglo-French Loan of 1908; (2) the Hukuang Loan of 1911; and (3) the Crisp Loan of 1912. The agreements under which all of these loans were contracted were concluded before the reorganization of salt revenue collection pursuant to the Reorganization Loan Agreement of 1913.

From 1913 to 1917 the Reorganization Loan was paid out of the salt revenue, but since 1917 it has been paid out of the Customs revenue pursuant to the provisions of Article IV of the agreement (except that prior to 1923 payments required for the certain adjustments of the loan service account were made out of the salt revenue). Payments of service of this loan have been regularly effected at all times. Because of the position which this loan enjoys by reason of being charged on the Customs revenue, and the ample security thereby afforded, which has proved sufficient during fifteen years, it scarcely seems conceivable that there should be occasion to revert to salt revenue security for service of this loan.

It is alleged in the Minister's note of April 14th that the changes made in November 1928, led to "the result that the repayment of the loans secured on the revenue has often been in arrear and that of the loans secured on the surplus of the revenue has since been entirely in default."

As just stated this statement is incorrect so far as concerns the service of the Reorganization Loan. The arrears of payments due from salt revenue in respect of the three other loans mentioned above, which loans antedate the agreement of 1913 for organization of the Inspectorate originated in the period prior to November 1928, during which, because of detention of revenue by local authorities, the Inspectorate was unable to collect sufficient funds to effect the loan service.

In the fall of 1928 the National Government, after full consideration of the situation, concluded that the only practical way to bring about resumption of the loan service was through the plan then adopted for the restoration of the collection functions of the Inspectorate. Accordingly, the Inspectorate's functions were restored and arrangements made whereby funds are drawn from all parts of China, and the loan service thereupon resumed. So well did these arrangements work that in September 1929, the Minister of Finance was in a position to announce a plan which not only made provision for current service of the three loans but also for payment of arrears. Beginning from September 1929, the amount of arrears has been steadily reduced. All arrears of Interest of the Anglo-French and Crisp Loans were wiped out by reason of extra payments in 1929 and 1930, and a beginning has been made toward payment of arrears of principal. Furthermore, since 1928 the arrears of the salt revenue contribution of Hk. Tls. 950,000 yearly for service of the Hukuang Loan have been entirely repaid, and also additional payments in respect of this loan have been made over and above the amounts due from salt revenue under the loan contract in order that one semi-annual interest coupon might be paid each year.

8. **Chinese Government's record in maintaining obligations.** These things have been accomplished despite the unprecedented slump in the price of silver, the flood dev-

astation in central China, and internal disturbances. The record of the National Government since 1928 in maintaining the service of salt-secured obligations in the face of economic and financial difficulties will bear favourable comparison with that of other governments in Europe and Latin America during the present difficult period.

This accomplishment would have been impossible without the efforts of the Chief Inspectorate of Salt Revenue, the integrity of which as an agency operating throughout China the National Government has endeavoured to maintain in the interest alike of bondholders and of the National Government. Persistence in the present attempt to dispossess the Inspectorate would be a serious blow to its position.

The Japanese Minister refers also to arrears in the payment of loans secured upon the "surplus" of salt revenue. The Chief Inspectorate of Salt Revenue, it is understood, has not been charged with responsibility for the service of these loans, the arrears of which in any event date from the period prior to 1928, and in no way result from the measures taken in connection with the salt revenue service in the fall of 1928. The obligations form a part of the general problem of obligations in arrears, a matter of great complexity and difficulty, with which the National Government intends to deal at the earliest opportunity. Maintenance of the integrity of the Salt Inspectorate will contribute toward dealing with this problem in due course.

> The Japanese Minister further stated that "so far as the Manchukuo authorities pay due respect to international relations and take sufficient care not to interrupt the existing practice of the repayment of the foreign loans of China, the Powers interested will not be placed in a less favourable position than hitherto."

The seizure and retention of loan quota funds since March, as above set forth, constitute sufficient comment upon the implication that the "Manchukuo" authorities will take care not to interrupt the existing practice of the repayment of the foreign loans of China.

Peiping, August 25th 1932.

APPENDIX.

COPY OF A LETTER FROM THE JAPANESE MINISTER TO CHIEF INSPECTORS OF SALT REVENUE.

Japanese Legation in China.
Shanghai, April 14, 1932.

Dear Sirs,

I have the honour to acknowledge the receipt of your letter of the 1st instant, in which you state that, according to a message received from the District Inspectors at Newchwang on the 29th March, a person served on them on the preceding day "an order purporting to come from the Minister of Finance for Manchukuo" and "demanded the transfer of collecting functions and took forcible possession of the Inspectorate Office," and that "there is every reason to believe that the forcible occupation of the Inspectorate Office on the 28th March was motivated and given sanction by the same i. e. Japanese) military force." It is further requested in your letter that I should "take early measure to make restitution for and the restoration of the offices and functions of the Inspectorate of Salt Revenue in Manchuria" and "make such representations to the Japanese Government as are necessary to prevent further interference with Inspectorate functions in the Three North Eastern Provinces."

In reply I have the honour to state that I have no information regarding the alleged interference in the functions of the Salt Inspectorate Office at Newchwang, but that if such action has been taken by the Manchukuo authorities, the Japanese Government have no concern with it. No Japanese authorities have either motivated or given sanction to the forcible occupation of a Salt Inspectorate Office or to any activities of the Manchukuo authorities, and the Japanese Government are not in a position either to hold themselves responsible for, or to put any restraint upon, what the Manchukuo authorities do. I regret, therefore, that I am unable to take such steps as you request in connection with the functioning of the Inspectorate of Salt Revenue in Manchuria.

I may add that in November, 1928, the Chinese Government despite the Japanese Government's protest, and in contravention of the stipulation of the Reorganization Loan Agreement, unilaterally introduced a change in the established practice of allocating the whole of the salt revenue to the service of the loans secured on that revenue, and since then only a certain fixed amount forming a small part of the revenue has annually, been devoted to loan repayment, while the greater part thereof has been appropriated by the central and local Governments, with the result that the repayment of the loans secured on the revenue has often been in arrear and that of the loans secured on the surplus of the revenue has since been entirely in default. Under the circumstances, it is considered by the Japanese Government, which have as large interest in the protection of the loans secured on the salt revenue as other Powers concerned, that so far as the Manchukuo authorities pay due respect to international relations and take sufficient care not to interrupt the existing practice of the repayment of the foreign loans of China, the Powers interested will not be placed in a less favourable position than hitherto.

I have the houour to be,

Sirs,

Your obedient servant,

(*Signed*) M. SHIGEMITSU

Japanese Minister to China.

T. C. Chu, Esq.,
Dr. F. A. Cleveland,
Chief Inspectors of Salt Revenue,
Shanghai.

MEMORANDUM

ON

THE SALE AND SMUGGLING OF NARCOTIC DRUGS IN CHINA BY JAPANESE SUBJECTS AND FIRMS

Document No. 29 Peiping, August 1932

MEMORANDUM ON THE SALE AND SMUGGLING OF NARCOTIC DRUGS IN CHINA BY JAPANESE SUBJECTS AND FIRMS.

1. Narcotic drugs in China and Japanese policy.
2. The Opium Suppression Commission on facts and figures.
3. Chinese Customs giving facts and figures.
4. Japanese subjects smuggling and selling narcotic drugs caught at Peiping.
5. Japanese caught in making heroin and "white powder" at Peiping.
6. Japanese arrested at Tientsin.
7. Japanese arrested at Paoting.
8. Japanese arrested at Tsingtao.
9. Japanese arrested in other parts of Shantung.
10. Japanese arrested at Foochow and Amoy.
11. Shansi Province.
12. Japanese control of drug traffic in the Three Eastern Provinces.

APPENDIX:

A. Table showing manufacturing and selling of narcotics by Japanese in China.[1]

B1. Table showing kind and amount of narcotics smuggled by Japanese boats and captured from July 1st, 1929, to June 30th, 1932, inclusive.[1]

B2. Table showing kind and amount of narcotics smuggled by Japanese and Koreans and captured from July 1st, 1929, to June 30th, 1932.

C. Table showing cases of smuggling and selling of morphine,

[1] Information furnished by the National Opium Suppression Commission.

heroin, etc. by Japanese and Koreans in Peiping as discovered by the Bureau of Public Safety (Peiping).[2]

D. Cases regarding Japanese subjects caught manufacturing narcotic drugs at Peiping by the Public Safety Bureau.[1]

E. Table showing Japanese subjects caught smuggling and selling narcotics in Hopei Province.[1]

F. Table showing Japanese subjects caught smuggling and selling narcotics at Tsingtao.[1]

G. Table showing Japanese subjects caught manufacturing and selling narcotic drugs in Shantung Province.[1]

---

[2] Information furnished by the National Opium Supression Commission.

MEMORANDUM ON THE SALE AND SMUGGLING OF NARCOTIC DRUGS IN CHINA BY JAPANESE SUBJECTS AND FIRMS

1. **Narcotic drugs in China and Japanese policy.** In view of the large number of Japanese nationals (including Koreans) and firms involved in the smuggling and selling of narcotics in China, it is difficult to believe that the Japanese Government is not behind the policy of selling poisonous drugs to undermine the health of the Chinese race. The fact that smuggling on such a large scale is being carried on by the Japanese and that so many Japanese nationals are devoting their energy and time to invent new drugs for consumption by the Chinese people gives additional cause for suspecting that official Japanese encouragement and facilities must be playing an important part.

The invention of "white powder" (baimien) by the Japanese is ingenious as it is dangerous. The powder is not only more effective than any other kind of narcotics and more poisonous than opium, but it can also be carried about and smoked with the cigarette. It does not require all the paraphernalia of opium smoking, is much less expensive and can therefore be more generally used.

Some facts together with many cases regarding the activities of Japanese subjects in smuggling and selling narcotic drugs, such as morphine, heroin, cocaine, etc. in China through their abuse of extraterritorial privileges, the connivance of Japanese consular and diplomatic officials and their deliberate failure in enforcing effective measures in checking such smuggling have already been mentioned in another memorandum.[1] The following pages deal with additional cases according to information received from various parts of China and especially from Peiping and other districts of Hopei Province, Shansi Province Tsingtao and other districts of Shantung Province, Fukien Province and the Three Eastern Provinces.

2. **The Opium Suppression Commission on facts and figures.** During a period of about two years (April 1929, May 1931), the National Opium Suppression Commission [2] dealt with 16 important cases concerning the illegal trade of drugs by Japanese in China, regarding most of which the Waichiaopu has lodged strong protests with the Japanese consular and diplomatic representatives. To few of these protests, however, have any satisfactory replies been received.

The serious nature of these cases may be readily understood if we notice the large number of Japanese engaged in certain of these cases, and the large amount of money involved. One case shows that there were more than 200 Japanese firms and shops connected with smuggling narcotics at Tsinan, while another shows there were more than 20,000 Japanese engaged in selling the poisonous drugs along the Kiaochow-Tsinan Railway.[3] Still another case reveals that the amount of poisonous drugs smuggled by a Japanese firm to be shipped from Dairen to Tientsin and other places was valued at Yen 4,600,000.[4]

3. **Chinese Customs giving facts and figures.** According to information furnished by the Chinese Customs

[1] See *Memorandum on Japan's Violations of Treaties and Infringements of Chinese Sovereignty.*

[2] See Appendix (A.).

[3] *Ibid.*

[4] *Ibid.*

House, during a period of about three years (July 1st, 1929—June 30th, 1932) there were 104 captures of Japanese boats engaged in smuggling narcotic drugs into the various ports of China.[5] The total quantity was 73,249.92 *liang* (about 1.33 ozs.) in addition to 12 boxes and 500 tubes. The total value was 328,899.69 taels, (including 4,300 gold units). During the same period there were 68 cases in which Japanese subjects were caught smuggling narcotic drugs into China.[6] The total quantity was 18,144.10 *liang* in addition to 34 boxes and 36 bottles while the total value was 125,781.69 taels (including 6,202.50 gold units).

The narcotic drugs smuggled by the Japanese boats and nationals into China include morphine, heroin, cocaine, opium, etc. Opium and heroin usually constitute by far the largest portion.

4. **Japanese subjects smuggling and selling narcotic drugs caught at Peiping.** Within a period of about two years (February 18th, 1929—November 12th, 1931), there were over 19 cases of smuggling and selling of narcotic drugs by Japanese subjects (including Koreans) in Peiping. All the offenders were turned over to the Japanese Consulate at Tientsin.[7]

In addition there were 336 cases (November 20th, 1929—April 29th, 1932) in which the Japanese subjects were involved in administering narcotic drugs to Chinese addicts.

5. **Japanese caught in making heroin and "white powder" at Peiping.** On March 1st, 1932, five Japanese subjects including one woman were captured by the police manufacturing heroin at 39 Hu Shen Miao and were sent to the Japanese Legation. All the raw materials and the

[5] Appendix (B.). [6] *Ibid.* [7] See Appendix (C.).

apparatus were seized together with some boxes of the finished products of "white powder."[8]

On May 30th, 1932, the Peiping Public Safety Bureau caught one Japanese manufacturing heroin in 3 Tsuihua Street and handed him over to the Japanese Legation, 44 *liang* of finished "white powder" and 160 *liang* of finished "yellow powder" were seized together with the tools and raw materials. A large quantity of finished heroin had already been smuggled to and sold at other places.[9]

On June 2nd, 1932, the Public Safety Bureau raided 26 Hsi-chi-men-ta-chia and arrested three Japanese manufacturing heroin. They were duly handed over to the Japanese Legation. The tools and raw materials were confiscated. A large amount of finished heroin had already been sent to other places.[10]

On July 8th, 1932, four Japanese and on July 21st, one Japanese, were caught manufacturing poisonous drugs and "white powder" at 10 North Tsungpu Hutung and 14 Hsin-kai-lou respectively. They were all handed over to the Japanese Legation. Their arrest was carried out by the Peiping Public Safety Bureau together with Japanese police from the Japanese Legation, and the tools, raw materials and finished products were taken away by them.[11]

6. **Japanese arrested at Tientsin.** According to the investigation made in 1931, there were 60 Japanese firms at Tientsin selling heroin and morphine.

During a period of about six months (January 1st—June 24th, 1932) there were 59 cases of people caught smuggling and selling "white powder" (*baimien*) and other narcotic drugs at Tientsin. A large number of these were Japanese.

[8] See Appendix (D.).

[9] *Ibid.*

[10] *Ibid.*

[11] *Ibid.*

7. **Japanese arrested at Paoting.** During a period of about three years (February 27th, 1928—September 11th, 1931), there were 9 cases of Japanese caught smuggling and selling morphine, heroin, "white powder" etc. at Paoting (Capital of Hopei Province). Many of them were disguised as Chinese.[12]

8. **Japanese arrested at Tsingtao.** During a period of about two years (April 24th, 1930—June 29th, 1932), there were 50 cases of Japanese subjects and firms caught smuggling and selling narcotic drugs including morphine, heroin, "white pills," opium pills, etc. Regarding most of these cases, strong protests were duly lodged wth the Japanese Consular officials at Tsingtao by the local Chinese authorities, but few satisfactory replies were received.[13]

9. **Japanese arrested in other parts of Shantung.** During a period from August 19th, 1929 to July 29th, 1931, there were 18 arrests of Japanese engaged in smuggling and selling narcotic drugs at Tsinan. Most of the arrests were effected by the Provincial Public Safety Department together wth police or members of the Japanese Consulate. Almost all of the offenders were immediately turned over to the Japanese Consulate.[14]

10. **Japanese arrested at Foochow and Amoy.** According to recent information, there are at Foochow, 183 firms and shops and, at Amoy, 81 firms and shops, owned by Japanese subjects coming from Formosa who are engaged in smuggling and selling narcotic drugs. A complete list of the names and addresses of those owners and firms has been kept on file by the Public Safety Bureaus of Foochow and Amoy.

[12] See Appendix (E.). [13] See Appendix (F.). [14] See Appendix (G.).

11. **Shansi Province.** Owing to rigid suppression of opium smoking in Shansi Province, the Japanese find it convenient to smuggle various kinds of narcotic drugs there as substitutes for opium. The chief sources of supply of these drugs are Shanghai, Tientsin and Shichia-chuang (Hopei Province).

The principal routes for smuggling the drugs are along the Chengting-Taiyuan Railway and the Peiping-Suiyuan Railway and *via* the road through the mountains in the South-east of the Province. The Japanese either guarantee to deliver the finished drugs to the Chinese dealers, or they themselves would bring in the products. Sometimes the Japanese would come to the capital and induce Chinese to manufacture the drugs together with them. As the Japanese are under protection of extraterritoriality, their illegal activities can not be easily checked. When they are arrested and handed over to the Japanese consular officials, the latter usually take no action, and often release these lawbreakers without any punishment.

The total amount and value of the drugs smuggled into Shansi may be given in the following table:

Kinds	Quantity (liang)	Value	Percentage
Powder (*Liao-mien*).....	500,000	$25,000,000	50%
Golden pills (*King-tan*)..	200,000	20,000,000	40%
Opium pills (*Che-chi-pao*)	1,000,000 pa [15]	5,000,000	10%

12. **Japanese control of drug traffic in the Three Eastern Provinces.** Japanese settlements in the various parts of the Three Eastern Provinces have served as centres for traffic in narcotic drugs chiefly in opium, morphine and heroin.

In the Kwantung Leased Territory there is a government opium monopoly. The so-called "opium retail

[15] 1 *pa* is equal to 100 pills.

shops" are in no way different from ordinary opium-smoking dens. There are 105 government opium dens. They have over 30,000 daily customers and the profit is about $18,000 every day. The Japanese government in Kwantung has been deriving from this opium traffic a revenue of between Yen 1,500,000 and 3,000,000 every year. Besides, there are many restaurants and brothels which also serve as places for opium-smoking.

The Kwantung Leased Territory is also a great centre for morphine traffic. While a large portion of the morphine manufactured there is smuggled into other parts of China, the Kwantung Leased Territory itself consumes a good portion of it. The morphine addicts are mostly Chinese, but the morphine dens are usually owned by the Japanese.

Japanese settlements or communities in various cities of the Three Eastern Provinces and along the stations of the South Manchuria Railway also serve as important places for the traffic of narcotic drugs. According to information available, the Japanese settlement at Yingkow has more than 10 morphine dens; that at Mukden, over 200 opium dens and more than 10 morphine dens; that at Changchun, between 500 and 1,000 opium dens; and that at Antung, from 2,000 to 5,000 opium dens and 21 morphine dens.

Peiping, August 27th 1932.

APPENDIX.

A. TABLE SHOWING MANUFACTURING AND SELLING OF NARCOTICS BY JAPANESE IN CHINA.[16]

Date	Offender	Place of discovery	Facts and settlement of the case
April, 1929	Akiyoshi and Tabiro (Japanese)	In the lane adjacent to Republican Theatre in French Settlement, Hankow	A Japanese named Akiyoshi was caught manufacturing morphine by the Garrison Headquarters and turned over to the Japanese Consulate. Later he was again caught manufacturing morphine at Chang Tsin Li No. 96, by the French police. He then removed to Kwan Hou Li, No. 69, in S.A.D., No. 1, and through raid two Japanese named Tabiro and Akiyoshi were arrested and deported to Japan by the Consulate.
October, 1929	Chofu Maru (Japanese Steamer)	Woo Soong Port	Woo Soong Garrison Headquarters searched the Japanese steamer Chofu Maru at Woo Soong port and seized 21 pieces of opium on the ship. This case was taken up by the Waichiaopu with the Japanese Consul-General in Shanghai.
December, 1929	Iiji Shoten (Japanese)	En route from Hamburg to Liaoning	The Liaoning Post Office discovered over 120 pieces of heroin smuggled and sent by parcel post to Liaoning (Mukden from Hamburg, Germany, by the Japanese. They were seized and turned over to the Liaoning Provincial Government for destruction.
December, 1929	Over 20,000 Japanese involved	Along the stations of Kiaso-Tsi Railway	Over 20,000 Japanese engaged in selling narcotic drugs at different stations along the Kiao-Tsi Rail-

[16] Information furnished by the National Opium Suppression Commission.

			way, such as Chow Chan, Wei Hsien, Pu San, Chang Tien and other small stations. The drug was smuggled in from the Japanese stores in Tsingtao. This case was taken up by the Waichiaopu with the Japanese authorities.
December, 1929	Over 200 Japanese stores involved	In the city of Tsinan, Shantung	Over 200 Japanese stores havings signs of pharmacy, shop and hotel engaged in selling narcotics as their regular business. The Opium Suppression Commission have investigated the facts and requested the Waichiaopu to lodge strong protest with the Japanese authorities.
December, 1929	Over 100 Japanese stores involved	Kwantung peninsular, especially Port Arthur and Dairen	Over 100 Japanese stores under the protection of Kwantung Japanese authorities engaged in selling opium, especially in Port Arthur and Dairen. Investigation was made by the Opium Suppression Commission and the case was referred to the Liaoning Provincial Government for taking the matter up with the Japanese authorities.
January, 1930	Mitaka (Sanryu) Shokai	66 Ta-shan Road, Tsingtao	64 kilos of heroin and cocaine in 101 packages at the value of $50,000 were smuggled to Tsingtao from Switzerland and Germany through parcel post and seized by the Tsingtao Postal Office when the receiver, a Japanese, went to the post office in person for the parcels. After the seizure of the parcels, the Opium Suppression Commission requested the Waichiaopu to take the matter up. The Kuomintang Headquarters at Tsingtao and official representatives then undertook to burn the drugs.

Date	Offender	Place of discovery	Facts and settlement of the case
February, 1930	A Japanese named Harata of Dairen Stock Exchange	From Dairen to Tientsin, Shenyang, Shihchiachuang, Kirin, etc.	Narcotics amounting to over Yen 4,600,000 was smuggled to Tientsin, Shenyang, Shihchiachuang, Kirin, etc. The case was referred to the Waichiaopu which duly lodged strong protest with the Japanese Legation.
April, 1930	9 Japanese firms involved	Changli District (Hopei)	Some Japanese in Changli posted as doctors were actually engaged in selling morphine, cocaine and heroin. The Opium Suppression Commission requested the Waichiaopu to take up the matter as Changli is not a trade-port and the Japanese are not allowed to trade over there. The Japanese Minister was accordingly notified of the fact.
April, 1930	Japanese or Japan's naturalized subjects from Formosa	Amoy	There were over 300 opium dens owned by Japanese subjects at Amoy. List of their names and addresses is kept on file by the Opium Suppression Commission. The case was referred to the Waichiaopu which lodged strong protests with the Japanese Minister and duly reported it to the League of Nations.
May, 1930	Timine Company	Changchun	147 packages of heroin were smuggled to Liaoning from Hamburg, Germany, through parcel post, and they were seized by the Liaoning Postal Office. These narcotics were burned in the presence of the representatives of the Government upon their seizure. Pictures of the destruction of the narcotics were taken and the parcel covers were handed over to the Opium Suppression Commission by the Liaoning Provincial Government.

Date	Offender	Place of discovery	Facts and settlement of the case
November, 1930	Chima Gennosuke	Tsinan	A Japanese named Shima Gennosuke, was engaged in selling narcotics and opium. This case was taken up by the Waichiaopu with the Japanese Chargé d'Affaires for action.
November, 1930	Japanese	On board a German ship discovered by Shanghai Customs House	Shanghai Customs House discovered a secret letter mentioning the smuggling of narcotics by the Japanese and consequently seized 17,400 pounds of Persian opium at the value of $1,000,000.00 on a German ship Cruser. The Japanese authority claimed that the opium seized was to be shipped to Dairen for manufacturing narcotics and demanded of its return. The Opium Suppression Commission requested the Ministry of Finance to order the Customs House to burn the opium.
November, 1930	Tanaka Yoko	On board an Italian ship discovered by Shanghai Customs House	518 pounds of heroin, 334 pounds of acid heroin and 623 pounds of morphine were smuggled from Shanghai to Tsingtao by the Japanese on board an Italian ship, and they were labelled as raisins and strawberry jam. The Opium Suppression Commission took picture of these narcotics seized and requested the Waichiaopu to instruct the Chinese Delegation at Geneva to report the case to the International Bureau of Opium Suppression of the League of Nations.
February, 1931	Fu Ta Hong owned by a Japanese	Nai Tai Hsia Hong Street	A Japanese firm, named Fu Ta Hong was engaged in wholesale of opium at Foochow. Over 2,950 *Liang* of raw opium and 45 *Liang* of opium juice were seized by the Bureau of Public Safety in co-operation with the Japanese Consulate upon the instruction of the

Date	Offender	Place of discovery	Facts and settlement of the case
			Opium Suppression Commission. This case was referred to the Waichiaopu to take the matter up with the Japanese authorities.
May, 1931	Japanese and Korean	Antashan	Koreans and Japanese in Antashan, west of Harbin, had over 60 firms engaged in narcotics business. These narcotics were sent through the Japanese post office in the Railway Zone of South Manchuria Railway. Special messengers were employed in delivering the narcotics to the various stations along the Chinese Eastern Railway. The Opium Suppression Commission requested the Heilungkiang Provincial Government to make a thorough investigation and found the true facts as reported.

B1. TABLE SHOWING KIND AND AMOUNT OF NARCOTICS SMUGGLED BY JAPANESE BOATS AND CAPTURED FROM JULY 1, 1929, TO JUNE 30, 1932, INCLUSIVE.[17]

From July 1st to September 30th, 1929.

Customs House	Kind	Quantity (Liang)	Value (HK. Tls.)	Name of boat
Lungkow	Morphine	10	400.00	Takamatsu Maru
	Heroin	10	233.30	" "
	Opium	72	360.00	" "
Kiaohai	"	216	432.00	Sakaki Maru

From October 1st to December 31st, 1929.

Customs House	Kind	Quantity (Liang)	Value (HK. Tls.)	Name of boat
Tsinhai	Heroin	514	4,774.00	Nagashiwa Maru
	Cocaine	36	480.00	No. 84 Fishing boat
Lungkow	Opium	57	114.00	Takamatsu Maru
Kiaohai	Native opium	20	30.00	Kwazan Maru
	" "	50	75.00	" "
Kianghai	" "	5,661	11,322.00	Gakuyo Maru
	" "	3,243	4,324.00	" "
	" "	18	24.00	Chofu Maru
	" "	52	69.33	Taifu Maru
	" "	2,084	4,168.00	Hoyo Maru
	" "	13,070	17,426.66	" "

From January 1st to March 31st, 1930.

Customs House	Kind	Quantity (Liang)	Value (HK. Tls.)	Name of boat
Tsinhai	Opium juice	4	10.00	Basho Maru
	" residue	28	10.00	" "
	Heroin	2,605.5	24,300.00	Tenshin Maru
Tunghai	Opium	24	48.00	Kitami Maru
	"	144	288.00	Kyodo No. 36
	Cocaine	.77	2.31	" " 26
Kiaohai	Opium	48.	96.00	Harata Maru
	Native opium	300	450.00	Dairen Maru
	" "	257	358.50	Hoten Maru
Chinkiang	" "	100	156.25	Hoyo Maru
	" "	272	425.00	Suiyo Maru
Kianghai	" "	48	96.00	Zuiyo Maru
	" "	52	69.33	" "
	" "	874	1,156.33	Namyo Maru
	" "	10,400	13,866.67	Rakuyo Maru

[17] Information furnished by the Bureau of Customs, Ministry of Finance.

Customs House	Kind	Quantity (Liang)	Value (HK. Tls.)	Name of boat
	Native Opium	3,296	6,592.00	Rakuyo Maru
Amoy	Medicine containing morphine	77.4 grams	79.00 (Gold unit)	Kanton Maru

From April 1st to June 30th, 1930.

Customs House	Kind	Quantity (Liang)	Value (HK. Tls.)	Name of boat
Tsinhai	Heroin	1,412	3,530.00	Tozan Maru
	"	180	3,444.00 (Gold unit)	Muryo Maru
	"	18	360.00	Hokuryo Maru
Lungkow	Opium	24	48.00	Takamotsu "
	Native opium Juice	14	18.00	Kaiju Maru
	Opium	36	72.00	Ryuku Maru
Kiaohai	Native opium	250	375.00	Chohu Maru
Changsha	" "	64	64.00	Shoko Maru
Chinkiang	" "	560	875.00	Nauyo Maru
	" "	34	28.00	Taitie Maru
	" "	48	75.00	Nanyo Maru
Kianghai	" "	388	517.33	Zuiyo Maru
Kianghai	Native opium	2,563	3,417.33	Rakuyo Maru
	" "	246	492.00	" "
	Native opium powder	196	392.00	" "
Yuhai	" "	36	49.50	Rozan Maru

From July 1st to September 30th, 1930.

Customs House	Kind	Quantity (Liang)	Value (HK. Tls.)	Name of boat
Tunghai	Opium	48	144.00	Fukuju Maru
	"	12	36.00	Kyodo Maru No. 26
	Native opium	7	21.00	Fukuju Maru
	Opium	516	1,548.00	Kyodo Maru No. 26
Kiaohai	Native opium	250	375.00	Choku Maru
	" "	216	324.00	Tozan Maru
Chinkiang	Opium	141	235.00	Daikichi Maru
	Native opium	106	141.34	Zuiyo Maru

From October 1st to December 31st, 1930.

Customs House	Kind	Quantity (Liang)	Value (HK. Tls.)	Name of boat
Antung	Native opium	560	525.00	Chosan Sapan
Kianghai	" "	2	2.67	Chohu Maru
	" "	14	18.67	Hoyo Maru

Customs House	Kind	Quantity (Liang)	Value (HK. Tls.)	Name of boat
	Native Opium	108	144.00	Jakuyo Maru
	Opium	464	644.77	Chosa Maru
	Native opium	46	61.33	Zuiyo Maru
	" "	40	80.00	" "
	Opium pills	8	4.00	Hyosen Maru

From January 1st to March 31st, 1931.

Customs House	Kind	Quantity (Liang)	Value (HK. Tls.)	Name of boat
Tsinhai	Native opium	234	587.50	Tensho Maru
Minghai	" "	5.5	51.00	Sukyo Maru

From April 1st to June 30th, 1931.

Customs House	Kind	Quantity (Liang)	Value (HK. Tls.)	Name of boat
Kianghai	Heroin	4,497	147,333.00	Tohon Maru

From July 1st to September 30th, 1931.

Customs House	Kind	Quantity (Liang)	Value (HK. Tls.)	Name of boat
Kiukiang	Native opium	15	15.00	Manyo Maru
	" "	10	10.00	Rakuyo Maru
	Pills for curing opium disease	12 boxes	8.00	Gakuyo Maru
Kianghai	Native opium	3,896	7,532.26	Nanyo Maru
	" "	20	38.67	Rakuyo Maru
	" "	592	947.20	Hoyo Maru
	" "	112	179.20	Zuiyo Maru
	Native opium	52	83.20	Nanyo Maru
	" "	20	32.00	Hoyo Maru
	Opium juice	17	45.33	Rakuyo Maru
	" residue	1	.40	" "
	Morphine	990	15,840.00	" "
Tsinhai	Heroin	24	192.00 (G. Units)	Tensho Maru

From October 1st to December 31st, 1931.

Customs House	Kind	Quantity (Liang)	Value (HK. Tls.)	Name of boat
Amoy	Morphine	24	585.00 (G. Units)	Canton Maru
	Morphine injections	500 (tube)	75.00	" "
Tunghai	Heroin	10.75	112.00	Kyodo Maru
Chinkiang	Native opium	256.00	400.00	Shoyo Maru
	" "	12.00	19.00	Daikichi Maru
	" "	168.00	262.50	Kichian Maru
Dairen	Opium	180	350.00	Sukyo Maru
Kianghai	"	280	989.33	Dairen Maru
	"	1,248	6,656.00	" "

Customs House	Kind	Quantity (Liang)	Value (HK. Tls.)	Name of boat
	Opium	360	1,008.00	Dairen Maru
	"	24	84.80	Choshun Maru
	"	23	81.27	" "
Kianghai	Heroin	596	19,866.67	Heiryu Maru
	Opium seeds	12	1.50	Choshun Maru
	Opium	8	1.00	Tsukusbigi "

From January 1st to March 31st, 1932.

Customs House	Kind	Quantity (Liang)	Value (HK. Tls.)	Name of boat
Kiukiang	Native opium	21	21.00	Shoyo Maru
Kianghai	Morphine pills	74	49.33	" "
Tsinhai	Native opium	2,336	4,380.00	Tozan Maru
	" "	1,696	3,180.00	Saitsu Maru
	Ground morphine	216	2,719.53	Tenshin Maru
	Native opium	1,744	3,270.00	Saitsu Maru

From April 1st to June 30th, 1932.

Customs House	Kind	Quantity (Liang)	Value (HK. Tls.)	Name of boat
Tsinhai	Native opium	592	1,110.00	Tensho Maru
	" "	86	154.80	Tenshin Maru
Kianghai	Dye stuff containing morphine	38	330.58	Ama Maru

B2. TABLE SHOWING KIND AND AMOUNT OF NARCOTICS SMUGGLED BY JAPANESE AND KOREANS AND CAPTURED FROM JULY 1ST, 1929, TO JUNE 30TH, 1932.

From July 1st to September 30th, 1929.

(Haikwan Taels.)

Customs House	Kind	Quantity (Liang)	Value	Smuggler
Yenki	Opium	100	200.00	Korean
	"	13	86.00	"
Dairen	Heroin	75	1,500.00	Japanese
	Morphine	18.8	300.00	"
Shanghai	Cocaine medicine	56	61.95	"
	" "	.09	2.55	"
Tsinhai	Salt acid morphine	4 (box)	2.40	"
Kiaohai	Heroin	78	1,209.00	"

From October 1st to December 31st, 1929.

Customs House	Kind	Quantity (Liang)	Value	Smuggler
Pingkiang	Morphine	131	968.99	Japanese
	"	25	62.16	"
	"	368	2,713.18	"
	"	79	532.95	"
	"	192	1,453.49	"
Yenki	Opium	688	1,500.00	Korean
Kiaohai	Heroin	123	1,906.50	Japanese
Kianghai	Native opium	832	1,109.33	"
	Morphine	3,933.6	5,463.00	"

From January 1st to March 31st, 1930.

Dairen	Heroin	93.75	1,406.25	Japanese

From April 1st to June 31st, 1930.

Shenyang	Morphine	474	6,763.00	Japanese
Kiaohai	"	19	294.50	"
	"	19	294.50	"
Kiaohai	Heroin	60	930.00	"
	"	119	5,505.37	"
	Cocaine	196	3,778.19	"
Yohai	Opium medicine	30 (box)	16.67	Japanese
	Opium pills	36 (bottle)	12.00	"

From July 1st to September 30th, 1930.

Yenki	Opium	64	192.00	Korean
Dairen	Heroin	18.8	300.00	Japanese
	"	169.2	2,707.20	"
Kiaohai	"	119	5,505.37	"
	Cocaine	196	3,778.19	"
	Heroin	114	5,412.00	"

From October 1st to December 31st, 1930.

Pingkiang	Mixed opium	960	150.00	Korean
Kiaohai	Heroin	44	1,760.00	Japanese

From January 1st to March 31st, 1931.

Customs House	Kind	Quantity (Liang)	Value	Smuggler
Pingkiang	Morphine	4	92.00	Japanese
	Heroin	25	362.40	"
Yenki	Opium	96	216.00	Korean
	"	56	129.50	"
Tsinhai	Morphine	2.5	12.50 (G. Unit)	Japanese
	Opium (extract)	925.96	2,448.12	"
Kiaohai	Heroin	115	6,366.00	"
	"	500	27,680.00	"
	"	19	821.00	"
Minhai	Native opium	329	329.00	"
	Opium	823	1,646.00	"
Minhai	Native opium	148	148.00	"
	Opium	311	622.00	"

From April 1st to June 30th, 1932.

Customs House	Kind	Quantity (Liang)	Value	Smuggler
Kiaohai	Heroin	100	4,800.00 (G. Unit)	Japanese
Yenki	Opium	800	3,200.00	Korean
Kianghai	" Aiconi "	96	3,413.33	Japanese

From July 1st to September 30th, 1931.

Customs House	Kind	Quantity (Liang)	Value	Smuggler
Pingkiang	Native opium	21	48.25	Japanese
Wunchun	Opium	136	480.00	Korean
	Morphine	18	600.00	Japanese
Yenki	Opium	800	3,743.00	Korean

From October 1st to December 31st, 1931.

Customs House	Kind	Quantity (Liang)	Value	Smuggler
Kiaohai	Heroin	40	2,000.00	Japanese
Yenki	Opium	697	1,390.00	Korean

From January 1st to March 31st, 1932.

Customs House	Kind	Quantity (Liang)	Value	Smuggler
Antung	Opium	556	1,390.00	Korean
			(G. Unit)	
	"	32.50	68.75	"
	"	162	405.00	"
	"	126	315.00	"
Wunchun	"	228	570.00	"
			(Kwanping Tael)	
	"	100	250.00	"
	"	87	174.00	"
Yenki	Opium	41	49.20	"
	"	182	240.00	"

From April 1st to June 30th, 1932.

Customs House	Kind	Quantity (Liang)	Value	Smuggler
Kiaohai	Heroin	78	3,744.00	Japanese
Yenki	Opium	16	16.00	Korean

C. TABLE SHOWING CASES OF SMUGGLING AND SELLING OF MORPHINE, HEROIN, ETC., BY JAPANESE AND KOREANS IN PEIPING AS DISCOVERED BY THE BUREAU OF PUBLIC SAFETY (PEIPING).[18]

Date	Location	Offender	Name, Quantity and Place of Contrabands
Feb. 18, 1929.	Outer Police Station, No. 2	Hiyaguchi (Japanese)	A small package of morphine was discovered in a Japanese's residence in Chang Sho Hutung, as he confessed.
Sept. 3, 1929.	Inner Police Station, No. 3	Takaki Zendo (Japanese) Chujo Fujukichi (Japanese)	18.7 *liang* of heroin were brought to Peiping for sale by two Japanese, as they confessed.
Mar. 11, 1929.	Inner Police Station, No. 1	Kin Ying-tsun (Korean)	8 *chien* 8/10 *liang* of "white powder" were discovered at the residence of a Korean named Kin Ying-tsun, as he confessed.
July 1, 1930.	Inner Police Station, No. 1	Kin Chi-hsien (Korean)	In Kin Chi-hsien's residence, Soochow Hutung, No. 96, three small packages of "white powder" and another paper bag were discovered for sale. The search and capture were carried out in co-operation with a member of the Japanese Legation named Okano.
July 11, 1930.	Inner Police Station, No. 1	Kin Yi-ching (Korean)	9 *liang* of heroin were discovered in Kin's residence, Poh Pao Hutung, No. 7, in co-operation with a member of the Japanese Legation named Okano.

[18] Information furnished by the Peiping Public Safety Bureau.

July 17, 1930.	Inner Police Station, No. 1	Kin Tao-hsien (Korean)	22 packages of heroin were discovered for sale in a Korean's residence, Hou Wei Hutung, in co-operation with a member of the Japanese Legation named Okano.
July 19, 1930.	Inner Police Station, No. 1	Temiyasu Maruzo (Japanese)	9 *chien* of heroin was discovered in a Japanese's residence, Yang Yi Hutung, in co-operation with a member of the Japanese Legation named Okano.
Aug. 17, 1930.	Inner Police Station, No. 1	Yamazaki (Japanese)	3.7 *liang* of "white powder" and 8 morphine injectors were discovered in Yamazaki's residence, West Piao Pai Hutung, No. 44, for sale as he confessed.
Aug. 27, 1930.	Inner Police Station, No. 1	Tso Yin-pi (Korean)	7.2 *chien* of "white powder" and 6 *liang* of caffeine were discovered in Tso's residence, Yao Chi Kuo Hutung, No. 10.
Dec. 2, 1930.	Inner Police Station, No. 1	Po Tsun-sin (Korean)	11 small packages of "white powder" were discovered for sale in a Korean's residence, San Yuan Ang, as he confessed.
Dec. 2, 1930.	Inner Police Station, No. 1	Kiang Tsun-sun (Korean)	Half a bag of "white powder" was discovered for sale in a Korean's residence, Kuan Mao Hutung, as he confessed.
April 3, 1931.	Inner Police Station, No. 1	Chang Fo-tien (Korean)	A burned opium pill and 4 packages of "white powder" weighing 2.25 *liang* were discovered in a Korean's residence, Tang P'u Hutung, No. 6.
April 5, 1931.	Inner Police Station, No. 1	Tsai Fong-yuh (Korean)	According to Chen Pao-chien's confession "white powder" was bought from a Korean named Tsai Fo-tien, in Yao Chu Kuo Hutung, No. 10, and 3 small packages of "white powder" were discovered. The raid was carried out in co-operation with Okano of the Japanese Legation.

Date	Location	Offender	Name, Quantity and Place of Contrabands
April 17, 1931.	Inner Police Station, No. 1	Kin Li-sui (Korean)	With the co-operation of a Japanese Legation policeman, 127 *liang* of "white powder" and one narcotic balance were discovered in Kin Li-sui's residence.
April 15, 1931.	Inner Police Station, No. 1	Ko Yuen-chi (Korean)	Sale of narcotics was discovered at 13 Ma Sien Hutung and .4 *liang* of "white powder" and 2 *liang* of morphined sugar and a narcotics balance were discovered in Ho's residence in co-operation with a Japanese policeman, Ishibashi, of the Japanese Legation.
April 20, 1931.	Inner Police Station, No. 1	Chang Tung-kun (Korean)	Sale of narcotics was discovered through searching Chang's residence in co-operation with the Japanese Legation policeman, Ishibashi, and consequently 1 *liang* of "white powder" and a narcotics balance were found.
May 11, 1931.	Inner Police Station, No. 1	Kin Yin-suo (Korean)	1.5 *liang* of "white powder" and a narcotics balance were seized in Kin's residence in co-operation with the Japanese Legation policemen.
May 19, 1931.	Inner Police Station, No. 1	Sun Tsiang-ti (Korean)	Opium smoking apparatus and opium residue were seized in Sun's residence, Fan Tse Ping Hutung, No. 26, in co-operation with the Japanese Legation policeman, Ishibashi, and a half of the seized contrabands were taken to the Legation.
Nov. 12, 1931.	Inner Police Station, No. 1	Chang Chi-ta (Korean)	Chang sold "white powder" at Fan Tse-ping's residence to Mrs. Wang and quarrelled with her regarding accounts.

The 19 cases listed above were forwarded to the Japanese Consulate in Tientsin for action by the Japanese Legation.

D. CASES REGARDING JAPANESE SUBJECTS CAUGHT MANUFACTURING NARCOTIC DRUGS AT PEIPING BY THE PUBLIC SAFETY BUREAU.[19]

Case I.

Name of Offender	Nationality	Age	Organ to which offenders were handed over
Lu Pu-sze (female)	Korean	37	Japanese Legation
Ishiyi Takeichi (male)	Japanese	41	" "
Chen Chi-sia (male)	Korean	38	" "
Tsui Chia-ku (male)	"	26	" "
Li Ming-nieu (male)	"	48	" "

Remarks: The seizure was made by the Inner Police Station No. 1, at 39 Hushenmiao on March 1, 1932. A complete outfit of apparatus for the manufacture of narcotics were seized. The raw materials consisted of 160 *liang* of opium, 2 boxes of opium powder and a bag of pig's hair. The finished products consisted of 1 box of "yellow powder" and two and half bags of "white powder" and about 4 boxes of other powder narcotics.

Case II.

Name of Offender	Nationality	Date
Kino (male)	Japanese	May 30th, 1932.

Remarks: The seizure was carried out by the Public Safety Bureau on May 30th, 1932, at 3 Tsui Hua Street. There was a complete set of apparatus for the manufacture of narcotics. The raw materials consisted of large quantity of various kinds of opium and the finished products consisted of 44 *liang* of "white powder" and 160 *liang* of "yellow powder." It was found out that a large quantity of finished heroin had already been sent to other places. The offender escaped.

[19] Information furnished by the Peiping Bureau of Public Safety.

Case III.

Name of Offender	Nationality	Age	Organ to which offenders were handed over
Sakurayi Tetsuo	Japanese	40	Japanese Legation
Nori Gaikichi	Japanese	40	" "
Wang Tse-ming	Korean	39	" "

Remarks: The seizure was made by the Inner Police Station, No. 4 at 26 Hsi-chi-men-ta-chieh, on June 2, 1932. There was a complete set of apparatus for the manufacture of narcotics and a large quantity of raw opium. The finished products had already been smuggled to other places.

Case IV.

Name of Offender	Nationality	Organ to which offenders were handed over	Date
Kobayashi Shigenori (male)	Japanese	Japanese Legation	July 8, 1932
Shimodera Sadame (male)	"	" "	"
Kaneko Goro (male)	"	" "	"
Hai Tsai (male)	Korean	" "	"

Remarks: The seizure was carried out by the Inner Police Station No. 1 together with the Japanese Legation Police at 10 North Tsungpu Hutung on July 8, 1932. There was a complete set of apparatus for the manufacture of narcotic drugs and a large quantity of raw opium. The finished products were taken away by the Japanese police.

E. TABLE SHOWING JAPANESE SUBJECTS CAUGHT SMUGGLING AND SELLING NARCOTICS IN HOPEI PROVINCE.[20]

Name	Nationality	Name and quantity of drugs	Place of seizure	Date	Disposal of the case
Ogawa Fujisamburo Ogawa Shushi Matsushita Suekichi	Japanese	"gold" pills	Shihmen	February 27th, 1928	Case was handed over to the Japanese Consul at Tientsin.
King Hong-zu Pei Shou-loh	Korean	narcotics	Shihmen	November 1st, 1928	Case was handed over to the district government of Wuloh.
Lu Zoong-hong Li Wen-shu Wen Yung-long King Lui-chen Tso Tsu-sui Li Zoong-san	Korean	narcotics	8 Tun An Hutung, Shihmen	January 8th, 1929	Case was handed over to the Japanese Consul at Tientsin.
Tomu Sato Koto	Japanese	"gold" pills	8 Tu Tse street Yingtai District	March 21st, 1929	Case was handed over to the Japanese Consul at Tientsin.
Itagaki Tokutaro	Japanese	10.46 *Liang* of morphine and 2.65 *Liang* of narcotics.	Changli District	March 21st, 1929	Case was handed over to the district government.

[20] Information furnished by the Hopei Provincial Government.

Name	Nationality	Name and quantity of drugs	Place of seizure	Date	Disposal of the case
Chao Ming-tuh Con Yung-kwen	Korean	30 *Liang* of heroin	Tientsin Station of Peiping-Liaoning Railway	October 21st, 1930	Case was handed over to the Japanese Consul at Tientsin.
Higuchi Iohiro	Japanese	8.5 *Liang* of heroin	Tanta	November 9th, 1930	Case was handed over to the Japanese Consul at Tientsin.
Takayama Isami	Korean	4 *Liang* of heroin	Shunchia Shih-men	September 4th, 1931	Case was handed over to the Japanese Consul at Tientsin.
Okta (woman)	Japanese	76 pieces of morphines and 8 small packets of narcotics.	Lulung District	September 11th, 1931	Case was handed over to the Japanese Consul at Tientsin.

F. TABLE SHOWING JAPANESE SUBJECTS CAUGHT SMUGGLING AND SELLING NARCOTICS AT TSINGTAO.[21]

32

Date	Name of offender	Nationality	Quantity of narcotics seized	Disposal of the case
April 24th, 1930	King Sui-wan	Korean	14 packages of morphine	The Japanese Consul replied that offender had been severely punished.
Feb. 26th, 1930	Shoryu Yakubo Drug Store	Japanese	1 package of morphine and 26 morphine-powder pills	The Japanese Consul replied that license of the store had been cancelled.
July 1st, 1930	Chang Chang-ming	Korean	One half bottle of morphine, 4 packages of phorphine, and 1 opium ball	The Japanese Consul replied that offender's store had been closed.
July 1st, 1930	Li Tuh-ming	Korean	95 packages of heroin	
June 30th, 1930	Tsai Kwan-ming	Korean	25 packets and 6 *liang* of heroin	Case turned over to the Japanese Consulate.
July 14th, 1930	Ni Hong-chong	Korean	40 packages of morphine and heroin	The Japanese Consul did not reply.
August 21st, 1930	The Tah Loh Store	Japanese	1 package of heroin	
October 25th, 1930	Kuo Cha-sun	Korean	A small quantity of heroin, morphine, and opium	The Japanese Consul replied that the guilty had been punished.

[21] Information furnished by the Tsingtao Municipal Government.

Date	Name of offender	Nationality	Quantity of narcotics seized	Disposal of the case
Nov. 18th, 1930	Hai Ming-lun	Korean	5 small packages of heroin	The Japanese Consul replied that there was no evidence, and no fine could be imposed.
Nov. 4th, 1930	Toyogawa Drug Store	Korean	1 small package of heroin	The Japanese Consul did not reply.
Nov. 4th, 1930	Ito Yoko	Japanese	1 small package of heroin	The Japanese Consul replied that after careful search, there was no evidence.
Dec. 3rd, 1930	Akigama Dentaro	Japanese	10 packages of heroin 1 pakage of opium 1 package of opium residue	The Japanese Consul replied that the guilty had been punished.
Aug. 21st, 1930	Shine-Sui Store	Japanese	3 packages of heroin and 7 packets of morphine	
Aug. 21st, 1930	Li Chin-chen	Korean	3 packages of heroin	These cases were referred to the Japanese Consul, who has not yet replied.
Aug. 21st, 1930	Hong Ni-chen	Korean	1 package of heroin	
Sept. 13th, 1930	Kiang Bing-sui	Korean	Several packages of morphine	According to the Japanese Consul, the guilty had been punished.

Date	Name of offender	Nationality	Quantity of narcotics seized	Disposal of the case
Aug. 24th, 1930	Ideki Shoten	Japanese	A small quantity of heroin	The case was referred to the Japanese Consul, who did not reply.
Oct. 8th, 1930	Ta Chang Store	Japanese	160 small packages of heroin	The Japanese Consul did not reply.
Oct. 16th, 1930	Watanchi Uge	Japanese	16 packets of morphine powder	According to the Japanese Consul, the guilty had been punished.
Oct. 7th, 1930	Tsai Chou-sin	Korean	16 packages of morphine powder	According to the Japanese Consul, the guilty had been punished.
Oct. 19th, 1930	Li Hua-tung	Korean	A small quantity of heroin	The case was referred to the Japanese Consul, who replied that evidence was not sufficient to impose a fine and that the store was under strict supervision.
Oct. 22nd, 1930	Chao Ping-kuo	Korean	6 packages of morphine and morphine injections.	The Japanese Consul replied that the guilty had been punished.
Oct. 6th, 1930	Okuta Unosuke	Japanese	1 package of heroin about 15 *liang*	The Japanese Consul replied that the guilty had been punished.

Date	Name of offender	Nationality	Quantity of narcotics seized	Disposal of the case
Dec. 24th, 1930	King Tien-pin	Korean	1 injection instrument and 70 small packets of heroin	The Japanese Consul replied that the guilty was deported.
Jan. 26th, 1931	King Yung-kwai	Korean	About 120 packets of heroin	The Japanese Consul replied that the guilty was punished.
Feb. 10th, 1931	King Yi and others	Korean	About 37 packages of heroin, morphine and white morpine pills	The Japanese Consul replied that the guilty were all punished.
March 30th, 1931	Chan Nan-er and others	Korean	77 packages of morphine	The Japanese Consul replied that the guilty were punished.
March 24th, 1931	Li Wen-chi	Korean	13 packages of morphine and heroin	Li was a seller of these narcotics. The case was referred to the Japanese Consul, who did not reply.
March 30th, 1931	Chen Chan-chu	Korean	73 packages of heroin	The Japanese Consul did not reply.
April 23rd, 1931	Chan Li-chen	Korean	216 small packages of heroin	Chan was a seller. The case was referred to the Japanese Consul, who did not reply.
April 17th, 1931	Kuboike Susumi	Japanese	Instruments for opium smoking	Kuboike was a proprietor of an opium smoking establishment. He was fined.

Date	Name of offender	Nationality	Quantity of narcotics seized	Disposal of the case
May 13th, 1931	Pu Ni-ta	Korean	21 small packages of white heroin and 2 big packages of yellow heroin	Pu was a seller. The case was referred to the Japanese Consul, who replied that Pu was punished.
June 27th, 1931	Hung Chen-chin	Korean	74 small packages of heroin	Hung was a seller. The case was referred to the Japanese Consul, who replied that the guilty was fined.
July 6th, 1931	Li Chen-hai	Korean	36 packages of heroin	The Japanese Consul did not reply.
July 13th, 1931	Hai Chi-fu	Korean	88 small packages of morphine	Hai was a seller. The Japanese Consul replied that he was fined.
July 22nd, 1931	Chang Chi-yen	Korean	104 small packages of heroin and 2 morphine injection instruments	Chang was a seller of morphine. The case was referred to the Japanese Consul, who has not yet replied.
July 22nd, 1931	Chao Chu-kwan and others	Korean	89 packages of heroin	Chao was a seller. The Japanese Consul did not reply.
July 24th, 1931	King Hen-kun	Korean	1 small package of heroin	The Japanese Consul did not reply.

Date	Name of offender	Nationality	Quantity of narcotics seized	Disposal of the case
July 29th, 1931	Tsin Kwan-ming	Korean	66 small packages of heroin	The Japanese Consul did not reply.
July 29th, 1931	Tsai Chan-tsi	Korean	24 packages of heroin	The Japanese Consul did not reply.
Aug. 2nd, 1931	Cho Yi-san	Korean	7 packages of morphine and 1 injection instrument	The Japanese Consul did not reply.
Aug. 11th, 1931	King Yoh-ku	Korean	122 packages of heroin and 1 packet of narcotics	The Japanese Consul did not reply.
Aug. 10th, 1931	Pu Mo-yi	Korean	5 packages of heroin	The Japanese Consul did not reply.
Aug. 13th, 1931	King Yoh-chen	Korean	13 packages of heroin	The Japanese Consul did not reply.
Aug. 18th, 1931	Hung Yi-chen	Korean	60 packages of heroin	The Japanese Consul did not reply.
Dec. 16th, 1931	Sato Yasutoshi	Japanese	4 packages of heroin	The Japanese Consul did not reply.
Dec. 27th, 1931	King Chung-shin	Korean	5 opium balls and instruments for opium smoking	The Japanese Consul did not reply.

Date	Name of offender	Nationality	Quantity of narcotics seized	Disposal of the case
Feb. 23rd, 1932	Ishiyi Banjiro	Japanese	Opium smoking instruments	Ishiyi was a proprietor of an opium smoking establishment. The case was referred to the Japanese Consul, who replied that the guilty was fined.
March 6th, 1932	Suto Minoru	Japanese	Instruments for opium smoking	The Japanese Consul did not reply.
March 14th, 1932	Hai Ding-chu	Korean	4 small packages and 2 big packages of heroin	The Japanese Consul did not reply.
March 25th, 1932	Yamaguchi Koki	Japanese	Small quantity of opium residue and instruments for opium smoking	The Japanese Consul did not reply.
April 15th, 1932	Hirota Kakujiro	Japanese	Opium smoking instruments and opium liquid	The Japanese Consul did not reply.
April 27th, 1932	Wu Chia-kun	Korean	A small quantity of opium juice, 1 package of heroin, 1 piece of opium and instruments for smoking	The Japanese Consul did not reply.
June 29th, 1932	Chung Chi-yen	Korean	2 packages of heroin and opium smoking instruments	The Japanese Consul did not reply.

G. TABLE SHOWING JAPANESE SUBJECTS CAUGHT MANUFACTURING AND SELLING NARCOTIC DRUGS IN SHANTUNG PROVINCE.[22]

Place	Organ and Date of Seizure	Kind and Quantity of Narcotics	Offender	Disposal of the Case
Commercial District, No. 2	Seizure on Aug. 19th, 1929, by the Bureau of Public Safety in co-operation with the police of Kiaotsi Railway.	8 catties (Chinese) of morphine.	Kobayoshi Shingo.	Turned over to the Bureau of Foreign Affairs.
Third District in city. No. 1	Seizure on Sept. 3rd, 1929, by No. 3 Police Station in co-operation with the Japanese Consulate.	17 packages of heroin, 10 packages of morphine pills and 4 *liang* of morphine.	Wataichi Mikazu.	Turned over to the Japanese Consulate.
Commercial District, No. 1	Seizure on Sept. 21st, 1929, by the Bureau of Public Safety in co-operation with Bureau of Foreign Affairs and Jananese Consulate.	Half a package of morphine pills and 14 pieces of tool for manufacturing narcotics.	Ikari Toragi, manager of Kwahoku Co.	Turned over to the Japanese Consulate.
Commercial District, No. 1	Seizure on Sept. 22nd, 1929, by the Bureau of Public Safety in co-operation with Bureau of Foreign Affairs and Japanese Consulate.	About 10,000 morphine pills, 2 packages of morphine powder and a tool for manufacturing morphine pills.	Kaneyori, manager of Choryu Co.	Turned over to the Japanese Consulate.

[22] Information furnished by the Shantung Provincial Government.

Place	Organ and Date of Seizure	Kind and Quantity of Narcotics	Offender	Disposal of the Case
Commercial District, No. 1	Seizure on Sept. 22nd, 1929, by the Bureau of Public Safety in co-operation with Bureau of Foreign Affairs and Japanese Consulate.	A package of morphine pills and a tool for manufacturing the same.	Ikoma Kyoashi, manager of Tairyu Co.	Turned over to the Japanese Consulate.
Commercial District, No. 1	Seizure on Sept. 22nd, 1929, by the Bureau of Public Safety in co-operation with Bureau of Foreign Affairs and Japanese Consulate.	A big package of raw morphine, 7 small packages of milk sugar and an outfit of machine for manufacturing morphine pills.	Yamoura Torao, manager of Yamoura Co.	Turned over to the Japanese Consulate.
Commercial District, No. 1	Seizure on Sept. 22nd, 1929, by the Bureau of Public Safety in co-operation with Bureau of Foreign Affairs and Japanese Consulate.	1 tube of morphine powder (for making pills), heroin papers and 31 pieces of tools for manufacturing narcotics.	Tajima, manager of Tenchi Yoko.	Turned over to the Japanese Consulate.
Commercial District, No. 1	Seizure on Sept. 22nd, 1929, by the Bureau of Public Safety in co-operation with Bureau of Foreign Affairs and Japanese Consulate.	10 packages of morphine pills of 5,000 each, iron case of morphine powder, a half tube of milk sugar mixed with raw powder, a package of acid, a half tube of sugar extract, ¾ tube of purple "gold" pill material, 276 packages of heroin, and 35 pieces of narcotics tools.	Kyotami Tomoshichi, manager of Gisii Yoko.	Turned over to the Japanese Consulate.

Place	Organ and Date of Seizure	Kind and Quantity of Narcotics	Offender	Disposal of the Case
Commercial District, No. 1	Seizure on Sept. 22nd, 1929, by the Bureau of Public Safety in co-operation with Bureau of Foreign Affairs and Japanese Consulate.	Over 10 packages of narcotics and 22 pieces of tools for manufacturing morphine pills.	Nonaka, manager of Tsinan Mineral Water Co.	Turned over to the Japanese Consulate.
Commercial District, No. 1	Seizure on Sept. 22nd, 1929, by the Bureau of Public Safety in co-operation with Bureau of Foreign Affairs and Japanese Consulate.	A half catty of pill perfume, 2 tubes of coffee extract, plus 27 packages of candy, a bag of powder, a case of acid, a tube of sugar powder, a tube of quinine, 2 tubes of malted sugar, over 30 dozens of milk sugar, 5 pieces of morphine apparatus, and 1 stove.	Omori, manager of Kwaishun Co.	Turned over to the Japanese Consulate.
Commercial District, No. 4	Seizure on Sept. 1st, 1929, by Bureau of Public Safety.	5 bags of morphine pills.	A Korean, Chang Li-chung	Turned over to the Bureau of Foreign Affairs.
Commercial District, No. 4	Seizure on Sept. 20th, 1929, by Sub-Police Office in No. 4 Commercial District, in co-operation with the Japanese Consulate.	60.035 catties of purple "gold" pill material, 3 catties of milk sugar, 25 catties of morphine pill material, 8 *liang* of heroin, 10 catties of morphine powder, 27 bags of	Yoshimori Masayomi	Turned over to the Japanese Consulate.

Place	Organ and Date of Seizure	Kind and Quantity of Narcotics	Offender	Disposal of the Case
		purple "gold" pills, 5 catties of ground coffee, a piece of machine for manufacturing purple "gold" pills, narcotics sorter and other apparatus consisting of altogether over 60 pieces.		
Commercial District, No. 4	Seizure on Oct. 4th, 1929, by Sub-Police Office in No. 4 Commercial District in cooperation with the Japanese Consulate.	265 *liang* of heroin pills, 11 *liang* of purple "gold" pills, 3 *liang* of opium, 24 *liang* of raw powder, etc.	2 Koreans, Kin Yung-yi and Chang Min-yuen.	Turned over to the Japanese Consulate.
Commercial District, No. 4	Seizure on Oct. 10th, 1929, by Sub-Police Office in No. 4 Commercial District in cooperation with the Japanese Consulate.	24 *liang* of morphine pills.	A Korean, Kin Dai-lien	Turned over to the Japanese Consulate.
Commercial District, No. 3	Seizure on Nov. 19th, 1930, by Sub-Police Office in No. 3 Commercial District in cooperation with the Japanese Consulate.	2,583 *liang* of narcotic material, 38 bags of purple "gold" pills, a whole set of apparatus, consisting of 69 pieces for manufacturing purple "gold" pills.	A Japanese, Tanamatsu	Turned over to the Japanese Consulate.

Place	Organ and Date of Seizure	Kind and Quantity of Narcotics	Offender	Disposal of the Case
Commercial District, No. 4	Seizure on Feb. 24th, 1931, by the Sub-Police Office, in No. 4 Commercial District.	7 small packages of heroin and tools for manufacturing narcotics.	A Japanese, Konouchi	Turned over to the Provincial Government.
South-west District,	Seizure on June 9th, 1931, by the Special Corps of the Bureau of Public Safety.	4.5 *liang* of heroin and 4.5 *liang* of narcotics.	A Japanese, Yoshida.	Turned over to the Provincial Government.
Commercial District, No. 1	Seizure on July 29th, 1931, by Special Corps of the Bureau of Public Safety.	18 *liang* of heroin.	2 Japanese, Mizuto Yoshimi Iitaka Sakae.	Turned over to the Provincial Government.